COLONIZATION, SUCCESSION AND STABILITY

COLONIZATION, SUCCESSION AND STABILITY

THE 26TH SYMPOSIUM OF
THE BRITISH ECOLOGICAL SOCIETY
HELD JOINTLY WITH
THE LINNEAN SOCIETY OF LONDON

EDITED BY

A.J. GRAY
Institute of Terrestrial Ecology
Furzebrook Research Station, Dorset

M.J. CRAWLEY
Imperial College at Silwood Park
Ascot, Berkshire

P.J. EDWARDS
Department of Biology
University of Southampton

BLACKWELL SCIENTIFIC PUBLICATIONS
OXFORD LONDON EDINBURGH
BOSTON PALO ALTO MELBOURNE

and published for them by
Blackwell Scientific Publications
Editorial offices:
Osney Mead, Oxford, OX2 0EL
8 John Street, London, WC1N 2ES
23 Ainslie Place, Edinburgh, EH3 6AJ
52 Beacon Street, Boston
Massachusetts 02108, USA
667 Lytton Avenue, Palo Alto
California 94301, USA
107 Barry Street, Carlton
Victoria 3053, Australia

First published 1987

Set in Linotron 202 by
Setrite Typesetters Ltd. Hong Kong
and printed and bound in Great Britain

DISTRIBUTORS

USA and Canada
Blackwell Scientific Publications Inc
PO Box 50009, Palo Alto
California 94303

Australia
Blackwell Scientific Publications
(Australia) Pty Ltd
107 Barry Street,
Carlton, Victoria 3053

British Library
Cataloguing in Publication Data

British Ecological Society *Symposium (26th)*
Colonization, succession and stability: the 26th symposium of the British Ecological Society held jointly with the Linnean Society of London.—(British Ecological Society special publication,
ISSN 0262-7027)
1. Biotic communities
I. Title II. Linnean Society of London
III. Gray, A.J. IV. Crawley, Michael J.
V. Edwards, P.J. VI. Series
574.5′247 QH541

ISBN 0-632-01631-0

Library of Congress
Cataloging-in-Publication Data

British Ecological Society. Symposium
(26th: 1984: Southampton, Hampshire)
Colonization, succession, and stability.

Bibliography: p.
Includes index.
1. Ecological succession—Congresses.
2. Colonies (Biology)—Congresses.
I. Gray, A.J. (Alan J.) II. Crawley,
Michael J. III. Edwards, Peter J.
IV. Linnean Society of London. V. Title.
QH540.B75 1984 574.5′248 86-26873
ISBN 0-632-01631-0

CONTENTS

PREFACE

When planning this symposium the central importance of succession in ecology created something of a problem. Most ecologists encounter it, in theory or practice, and the entire meeting could easily have been given over to case studies, anecdotes and species lists. It was therefore decided to exclude papers dealing solely with field studies, the classic examples of which, e.g. on islands such as Surtsey and the Krakataus, are well documented elsewhere. Instead we asked our contributors to set their studies in the context of specific questions, our aim being to see what new light could be shed on a very old subject by recent advances in disciplines ranging widely from modelling, population dynamics and community theory to physiology, genetics and quaternary studies.

The first three chapters serve as an introduction. John Miles traces the history and development of the concept of succession, noting in particular the search for an elusive unifying theory, before presenting the wide range of current views of successional processes and showing how an empirical approach can aid vegetation management. Michael Usher reviews various models of succession, exploring in detail one type, Markov chain models, using the classic Breckland data of A.S. Watt. In Chapter 3 Martin Mortimer focuses on recent studies of the dynamics of population renewal after disturbance and their importance in understanding the early stages of succession.

The next six chapters deal mainly with colonization, covering the attributes of colonizing and early successional plants and animals (Chapters 4–7) and the special problems of ephemeral habitats (Chapter 8) and colonization and speciation (Chapter 9). Peter Grubb demonstrates that the great diversity among colonizing green plants is paralleled by that among colonizing fungi and emphasizes that useful generalizations cannot be made about colonizers unless particular types of substratum are specified. Michael Fenner reviews the seed characteristics of early, mid and late-successional plants, and Tony Brown and Jeremy Burdon consider the diversity of mating systems and their effect on the genetic structure of populations of successful colonizing plant species. In three chapters dealing mainly with animals Peter Parsons discusses the genetic and phenotypic features of colonizing species, Ilkka Hanski considers the population dynamics of insects colonizing ephemeral resources such as dung, carrion and fungal fruiting bodies and, drawing on extensive studies of the

Hawaiian islands, Hampton Carson asks why some colonizers remain as single species whilst others proliferate to form many.

Chapters 10–16 consider aspects of succession at the population, community, and ecosystem level. Peter Vitousek and Lawrence Walker contrast the patterns of change in nutrient availability in primary versus secondary succession and discuss how plant growth, colonization, and species' replacement patterns can be keyed to these changes. John Lawton's treatment of the structure of successional communities examines whether assembly rules can be detected, arising principally from species interactions, and looks at the effects of chance and determinism on community assembly. Drawing on a very large body of field and experimental data Fahkri Bazzaz re-examines the concept of niche, particularly in relation to successional plant populations, and demonstrates how useful insights can be gained and many predictions tested from such an approach. His discussion of the dimension of genetic variation is taken up in the next chapter where Alan Gray examines the evidence that successional change is a factor maintaining genetic diversity, and therefore a potential microevolutionary force, in plant species populations. In Chapter 14 Peter Edwards and Michael Gillman emphasize that herbivores are not simply adjuncts of succession which may deflect or interrupt it but, via their effects at all stages of the regeneration cycle, have a significant role in the organization and structure of plant communities. In a complementary contribution Valerie Brown and Richard Southwood examine the trends in community patterns and the attributes of associated organisms to be found along a successional gradient, their observations being based on a comprehensive field study of a secondary succession on sandy soils in southern England. In contrast, in his study of the sessile marine organisms colonizing hard substrata off the coast of California, Joseph Connell finds little evidence of orderly successional change and is prompted to re-examine the notion of fugitive species.

The chapters in the final group (17–21) share a broader perspective, dealing with some of the longer-term and larger-scale implications of inexorable successional change. In addressing the question 'Are communities ever stable?' Mark Williamson emphasizes the importance of time-scales in assessing stability and considers what observable reactions to perturbation can tell us about the stability of individual communities. Margaret Bryan Davis discusses the rapid postglacial range expansion of shade-tolerant, long-lived and generally slow-growing late-successional tree species and their different rates of invasion of forest communities in N. America. In a detailed analysis of current tree replacement patterns in a tropical forest, Stephen Hubbell and Robin Foster focus on the link between spatial and local-temporal variation in generating a mosaic in a species-rich community. The maintenance of species richness is a theme taken up in Philip Grime's contribution, in which he suggests that the way in which different types of dominant species determine the fate of subordinates is a major cause of variation in the numbers of species maintained within plant communities. Lastly, the popular notion that stable, equilibrium communities are more resistant to invasion is one of several ideas called into question by Michael Crawley in his discussion of both the invasibility of communities and the attributes of successful invaders.

It would be invidious to attempt to summarize this rich diet of ideas and information. Nor did the meeting seek a unifying theory or a grand synthesis. At the same time we have noted the way in which problems and concepts commonly recur and run as themes throughout the volume. For example, the important distinctions between primary and secondary succession emerge in several contributions. Perhaps these processes are so different that it is misleading to give them similar names! It may well be that the study of succession has been hindered in the past by overgeneralization and the unquestioning acceptance of particular ideas, and in many chapters we see a reappraisal of much of the received wisdom. Several chapters make the link between invasibility and disturbance by the relatively rare event, be it windthrow or fire in a forest, or the overturning of a submarine boulder. Predicting such events is difficult. Predicting structural patterns of communities from the population dynamics of the constituent species clearly remains a vital unsolved problem. This inability to make predictions is bemoaned by more than one author and we are delighted to see in many chapters the increasing emphasis on an experimental approach. Another feature of this volume is the recognition of the genetical dimension to successional change. In general we believe that several genuinely new ways of looking at the subject have emerged.

We are indebted to members of the organizing committee, 'Sam' Berry, Tony Bradshaw, Alistair Fitter and Mike Hassell, for helping to plan the meeting and choose the speakers, and to a small army of referees who ensured that our choice appears a wise one. We are grateful to the Linnean Society of London and the University of Southampton for their help and hospitality. Finally we are pleased to thank the contributors for easing our task by the high quality of their productions.

ALAN J. GRAY
MICHAEL J. CRAWLEY
PETER J. EDWARDS

1. VEGETATION SUCCESSION: PAST AND PRESENT PERCEPTIONS

JOHN MILES
Institute of Terrestrial Ecology, Hill of Brathens, Banchory, Kincardineshire, AB3 4BY

INTRODUCTION

The aim of this paper is to introduce the notion of succession, summarizing its history, present state and intellectual trappings. I shall confine my remarks to vegetation, where the concept has its origin, and ignore planktonic, sea-bed and littoral successions, and also successions of heterotrophic organisms. These latter mostly have little in common with vegetation succession, and seem to have generated fewer intellectual lathers in their savants.

Although the concept of succession is both simple and useful, succession as a class of phenomena seen in the field is, like Medawar's (1969) description of biology, 'complex, messy and richly various'. It is thus perhaps not surprising that the literature about the idea reflects the confusion of the subject matter. I cannot better summarize the disorder enmeshing the concept than by quoting the opening remarks from the most comprehensive and thoughtful review of succession (McIntosh 1980) that I know: 'Succession is one of the oldest, most basic, yet still in some ways, most confounded of ecological concepts. Since its formalization . . . in the early 1900s, thousands of descriptions of, commentaries about and interpretations of succession have been published and extended inconclusive controversy has been generated. Withal, no effective synthesis of the divergent observations from many different ecosystems, terrestrial and aquatic, has produced a body of laws and theories which ecologists, generally, have embraced. Repeated symposia on succession and the corollary problems of the succession concept, . . . have not produced notable convergence of thought. Problems of conceptualization and terminology are still evident after three quarters of a century.'

The confusion in the literature, and the absence of an acceptable unifying theory, seem to spring largely from an unachievable search for an ecological Shangri-La. Even passing acquaintance with the subject matter shows that vegetation change in time stems from many different causes. Yet all these changes have at some time or other been termed succession. It should surprise no-one that changes from disparate causes cannot be explained in a common way. A further reason for the unattainability of a unifying theory of succession derives from the individualistic nature of vegetation, first argued so clearly by Gleason (1917, 1926) over 60 years ago. These matters have been common knowledge for so long that I can only conclude that the compulsive search by so many workers this century for a universal generalization of succession, a search that appears still to be in progress (Finegan 1984), arises from emotion rather than logic.

The phenomena of succession are too numerous to review here, while the phenomenon of the study of succession was recently reviewed in detail by McIntosh (1980, 1981). Therefore, in addition to discussing the development of the concept of succession, I shall develop the argument that it is pointless to pursue the quest for the ecological grail of a single universal generalization of succession that will be both useful and widely accepted. Finally, I shall outline my particular approach to using knowledge of succession in land management.

HISTORICAL AWARENESS OF SUCCESSION

'Succession' implies a sequence of something in time or place. In the ecological literature it has mostly been used to denote sequences of change in time. The first to use the term in this sense may have been De Luc (1806, cited by Clements 1916) in his description of hydroseres in northern Germany. However, an awareness that vegetation tends to change with time, and particularly that change follows disturbance, must be far older. The effects of forest and prairie fires caused by lightning, and of other gross disturbances, will have been seen by the earliest specimens of *Homo sapiens*, and seen more frequently by those still remote, hunter-gatherer ancestors who first chanced upon the use of fire for driving game or clearing forest. The early practitioners of slash-and-burn cultivation will have seen the successions they caused. They were active in Britain in the postglacial Atlantic period for certain (Dimbleby 1962; Romans & Robertson 1975), and possibly in the Boreal (Smith 1970), and active in Mediterranean regions even earlier (Naveh & Dan 1973).

In classical times, Theophrastus (*c.* 300 BC) wrote about vegetation changing with time (Hort 1916), and several early Roman writers were aware of succession (Spurr 1952). Clements (1916) reviewed the post-renaissance literature, and traced observations on successions back to 1685. The tendency for individuals of particular tree species to be replaced after death by those of different species was remarked on early. Cochon (1846) claimed the phenomenon was first recognized by Telles d'Acosta in the eighteenth century. However, the most remarkable early study comprised the felling experiments of Dureau de la Malle (1825), an inquisitive and observant Normandy landowner, who concluded: '. . . the alternative succession in the reproduction of plant species, especially when they are forced to live socially, is a general law of nature, a condition essential to their conservation, to their development. This rule applies equally to long-lived, high forest trees, to shrubs and undershrubs, controls the vegetation of social plants, of artificial and natural grasslands, of perennial, biennial and annual species, living socially or even isolated'. Awareness of change, and of one cause, competition, is shown by De Candolle's remark in 1820 (cited by Warming & Vahl 1909) that: 'All the plants of a country, all those of a given place, are in a state of war, the ones relative to the others.'

DEVELOPMENT OF THE CONCEPT

From the mid-nineteenth century, successions were noted increasingly in the literature. While foresters in North America at least seemed to assume that forest succession was a general phenomenon (Dawson 1847; Thoreau 1860; Douglas 1875, 1889), the published views of botanists studying vegetation and plant geography developed more slowly. Braun-Blanquet (1932) claimed that Kerner (1863) was 'the real founder of the doctrine of the development of communities', from his studies in the Danube Basin. Clements (1916) ignored Kerner in his historical review, but cited Hult's 1885 publication on the vegetation of Blekinge in Finland as evidence that Hult was 'the first to fully recognize the fundamental importance of development in vegetation'. Hult argued that the distribution of different kinds of vegetation could only be understood with knowledge of their development. However, Warming (1895) seems to have been the first to stress that vegetation change was universal and unceasing.

Warming did not elaborate his ideas further, and the concept of succession was only developed in detail at the turn of the century in North America by Cowles and Clements, at first apparently independently. Cowles, a physical geographer, studied the dynamics of sand dunes at the head of Lake Michigan (Cowles 1899, 1901), and found that although the dunes were formed and destroyed by wind, their development between formation and destruction was largely controlled by their vegetation. From these and other studies, Gleason argued that vegetation in general will change as the physiography changes, though he also noted that vegetation can often change faster than geomorphology. Cowles' (1911)

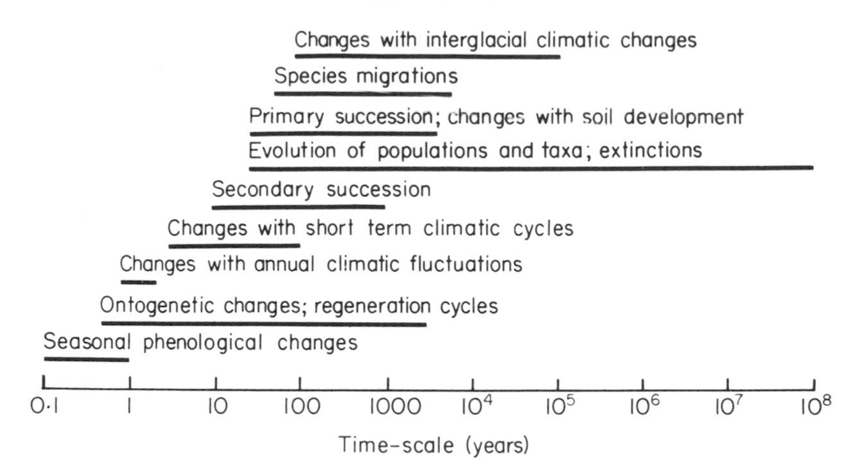

FIG. 1.1. Time-scales of vegetation change from different causes.

conclusions were soundly based on observation, and bear restating. On succession, Cowles wrote, 'the original plant formations in any habitat give way in a somewhat definite fashion to those that come after.' Cowles also recognized that changes in vegetation caused by climatic, topographic and biotic changes occurred at different rates and concurrently, 'this complex of cycle within cycle, each moving independently of the others and at times in different directions', a point that is still too infrequently stressed. Fig. 1.1 shows the relative time-scales of various kinds of vegetation change.

Concurrently with Cowles, Clements began to study the Nebraska prairies. In 1898, Pound and Clements wrote, 'formations and subordinate groups are in stable equilibrium only rarely and usually for a comparatively short time'. Shortly after, Clements (1904) published an excellent empirical and analytical account of the causes of succession, in which he clearly distinguished between primary succession on newly exposed surfaces which had never borne vegetation before, and secondary succession where vegetation on soils already developed had been damaged or destroyed. In 1916, however, Clements published the monumental work on succession by which he is most commonly remembered. This is a mix of scholarship and sound observation curiously blended with a purely deductive and hypothetical classification of succession based on the concept of 'climax' vegetation, an attempt to create a unifying theoretical framework of succession. His speculative theory was fatally flawed from the outset, and has been criticized often and in detail (Gleason 1917, 1926, 1927, 1937; Colinvaux 1973; Drury & Nisbet 1973;

Miles 1979). Yet his influence on the development of plant ecology in America and Britain was profound. Gleason (1975), Clements' contemporary, explained his motives thus: 'He was a classicist, and Latin and Greek are precise languages, with an elaborate system of conjugations and declensions. Perhaps this led him to attempt an equally precise and elaborate classification of the whole content of ecology'. Egler (1951) explained the phenomenon in similar vein: 'We have Clements the uncompromising idealist, the speculative philosopher, driven by some demon to set up a meticulously orderly system of nature, as neatly organized and arranged as the components of Dante's Inferno.'

Why did Clements' theory command such support? Doubtless one reason is the human yearning for order, to discern patterns in apparent chaos, to transform collections of facts into predictive theory. Parts of Clements' schema fitted with prevailing sentiment at the time. The 'organismic concept' was a widely accepted tradition in natural history years before Clements took it in hand (McIntosh 1980). His views on the persistence of climax vegetation, and on the convergence of successions to the regional climax, fitted with the then current geomorphological theory of Davis (1909), and with the lack of present knowledge on climatic variability in time. Perhaps above all it was his prestige as a practical man in the field that won him support. Egler (1951) wrote: 'We have Clements the practical realist, the Clements who impressed the soil conservationists, agronomists and range managers with his down-to-earth understanding of land management, the Clements who could discuss and write convincingly about such problems with almost complete absence of technical terminology and intellectual gyrations. . . . It is easy to see how Clements gained the respect and enthusiasm of the practical field man, who would then "accept" the complicated philosophy, not as something which he really comprehended, but as something which must be "right" since it came from Clements the realist.'

The ideas in Clements' (1916) work proved seminal, though perhaps less for their intrinsic merit than for the opposing ideas they created, which might not otherwise have been developed and presented so clearly! However, as Watt (1964) noted, Clements was largely responsible for injecting a dynamic principle into a largely static ecology.

VARYING INTERPRETATIONS OF SUCCESSIONAL PHENOMENA

What is succession? Recent definitions have tended to be broad, to exclude only seasonal, fluctuational and cyclic changes from the gamut of recognized vegetation changes, and to be mutually consistent: 'a progressive alteration in the structure and species composition of the vegetation' (Grime 1979); 'directional change away from an initial state. ... changes which markedly alter the appearance of a patch [of vegetation] such that it can be considered to have changed into a different type' (Miles 1979); 'the directional change with time of the species composition and vegetation physiognomy of a single site' (Finegan 1984). Succession thus hinges on the occurrence of gross, directional changes in species composition. Such changes may occur when species die and are replaced by others, but, in secondary succession and the later stages of primary succession, may just reflect changing patterns of dominance by species present through most or all of the succession period as minor components, or as propagules. This latter can be represented schematically as:

$$A_{BCD} \rightarrow {}_{A}B_{CD} \rightarrow {}_{AB}C_{D} \rightarrow {}_{ABC}D$$

This definition is broader than that adopted by Clements, but narrower than Gleason's. Clements (1916) defined succession as 'a sequence of plant communities marked by the change from lower to higher life-forms.' He also held that reverse successions, from higher to lower life-forms, were impossible. 'Succession is inherently and inevitably progressive. ... Regression, an actual development backwards, is just as impossible for a sere as it is for a plant.' Many instances have now been published in which environmental variability has caused just that; for example, fluctuating ecotones between forest and prairie, fluctuating montane treelines, northern forests turning into bogs (Drury 1956; Ugolini & Mann 1979). Gleason (1953, 1975) recorded finding one reverse succession which seems to have been a milestone in the development of his ideas: 'I found a pond developing on top of a sand dune, not a stone's throw from the side of the dune and 30 feet or so above the adjoining land. It started in a blow-out, waterproofed its own bottom with blue-green algae and the remains of Polytrichum, and then these zones of hydrophytes marched steadily up the hill. Hydrophytes moving centrifugally! Impossible. Unheard of.' (Gleason 1975).

Gleason's (1927, definition of succession was very broad: 'The successional phenomena of vegetation include all types of change in time, whether they are merely fluctuating or produce a fundamental change in the association.' Because succession has traditionally been used to describe vegetation change at a site level, Gleason's definition seems unnecessarily broad. Although the distinctions are sometimes arbitrary, fluctuations and regeneration (which is cyclic) can commonly and usefully be distinguished from directional change, i.e. succession (Miles 1979).

That being so, it is sensible to use the definition of succession with the greater information content.

Does succession have any fixed attributes, other than the defined floristic change? Drury & Nisbet (1973) argued not, and I agree (Miles 1979). However, attempts have been made to define succession in terms of functional attributes of ecosystems, notably by Margalef (1958, 1963, 1968) and Odum (1969). I find these interesting, but not giving useful insights in terms of plants or vegetation. Thus, Margalef (1958) wrote: '... succession may be defined as a gradual, irreversible change in the structure of a mixed population in the direction of a replacement of systems slightly structured and having a rapid dynamics, made up of relatively small organisms, having a high productivity/biomass relationship, adapted to the rapid utilization of the resources of the medium, by other, more stable, communities made up of larger organisms with a greater thermodynamic output, adapted to an efficient utilization of the resources and having a lower productivity/biomass relationship.' This is just what Clements said, that succession is a sequence from lower to higher life-forms! The rest follows automatically from the differing biology of varying life-forms. Later in the same paper, Margalef gave a terser definition: 'In thermodynamics, succession can be described as the acquisition of greater efficiency in exploiting the medium and reducing to the minimum the dissipation of energy.' Then in 1968 he wrote: '... succession ... appears to be a process of self-organization occurring in every cybernetic system with the properties of an ecosystem. ... The process of succession is equivalent to a process of accumulating information.' The briefest 'systems' definition I have encountered is Noble's (1981) statement, 'Succession is the time-dependent integration of all ecological processes at a site.' Such definitions may give insights to a systems expert, though Margalef's stress on the irreversibility of succession will certainly mislead, but I see no relevance to a vegetation manager.

However, vegetation or ecosystem properties are of great interest to most vegetation managers, particularly yields of timber and other crops. Fig. 1.2 shows the general pattern of net ecosystem production (NEP) during secondary succession. Initially after disturbance, organic matter decomposition exceeds production, so NEP has negative values. Closely correlated with NEP are element output rates. Most elements are lost in surface run-off and drainage waters. Any sudden increase in output rates after disturbance to vegetation (because uptake by the vegetation is greatly reduced) are of potential concern to water and fisheries boards.

There are other interesting ways of looking at vegetation during succes-

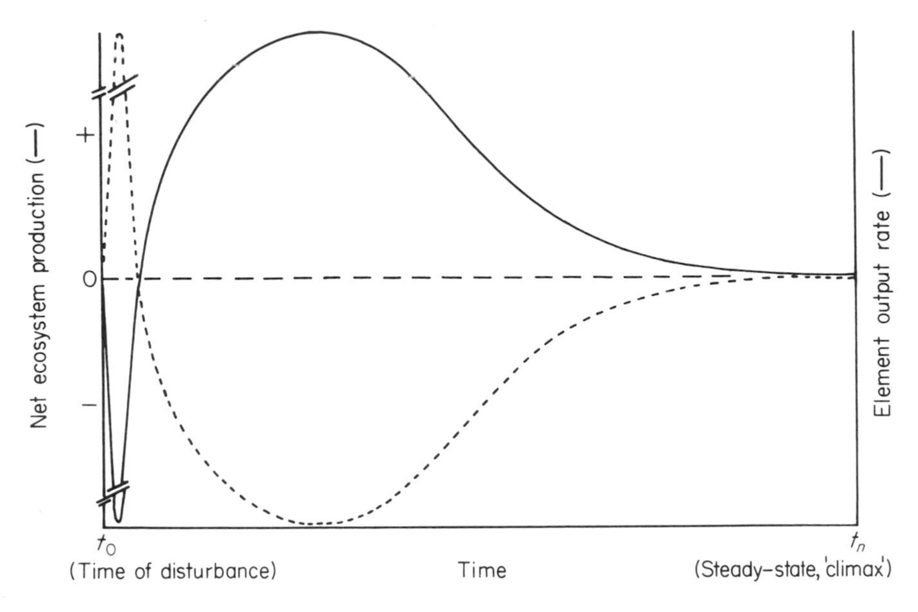

FIG. 1.2. Patterns of change in net ecosystem production and element output rate (assuming constant input rates) during secondary succession. (After Gorham, Vitousek & Reiners, 1979.)

sion. Grime (1977, 1979 and this volume) has argued that evolution in plants has produced three main 'strategies' in response to competition, stress and disturbance, and that knowing the potential productivity of a site makes it possible to recognize the probable sequence of life-forms there during succession.

Regardless of plant strategies, an important determinant of the course of any succession, and the prime determinant of secondary succession, is the range of species present at the outset and migrating to the site early in succession. The proximity of seed sources is thus important (Kellman 1970; Miles 1979). Lepart & Escarre (1983) have shown how the relative importance of different plant dispersal mechanisms varies with time during succession of abandoned fields around Montpellier in southern France (Fig. 1.3). As expected, most early colonizers were wind-dispersed, but after about 10 years there was an increase in the proportion of species dispersed via faeces, and of species with large seeds or fruits without regular dispersal mechanisms other than gravity or chance dispersal by animals (e.g. *Quercus coccifera* acorns).

Clements (1916) recognized the importance of species accessibility, especially in secondary succession, where a productive soil had already

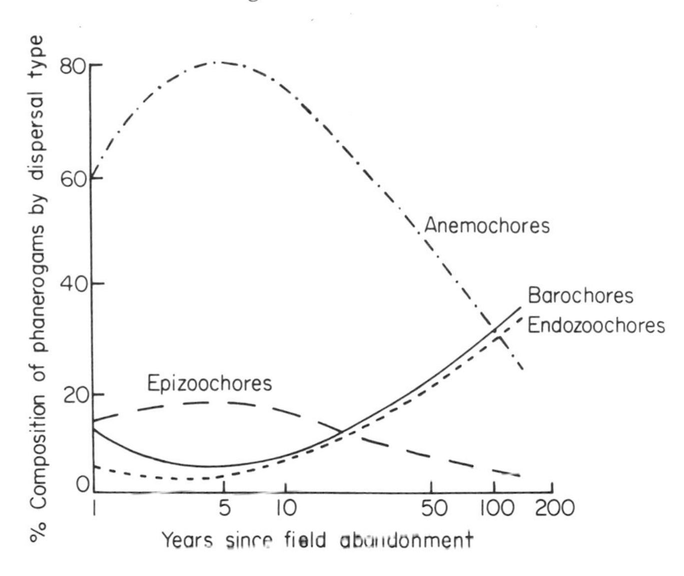

FIG. 1.3. Relative frequency of four methods of propagule dispersal in the phanerogam flora of abandoned fields around Montpellier, southern France, at different times after abandonment. (After Lepart & Escarre 1983.)

developed, and did not necessarily expect site modification to be important in secondary succession, nor progressive colonization by species: 'Secondary bare areas ... normally possess viable germules of more than one stage, often in large number, retain more or less of the proceeding reactions, and consequently give rise to relatively short and simple seres. ... In some cases it seems that the seeds and fruits for the dominants of all stages, including the climax, are present at the time of initiation.' Clements thus stressed the importance in secondary succession of the species composition at the time of disturbance, an idea usually attributed to Egler (1954).

One particularly interesting approach is to analyse which properties of species determine their successful reproduction and survival following disturbance. Gill (1975) and Naveh (1975) listed attributes that help species to survive recurrent fires. Noble & Slatyer (1977, 1979, 1980) termed such properties 'vital attributes', showed how these could be used to predict a qualitative replacement sequence during regeneration after fire, and point out the major changes (again qualitatively) in dominance and composition. They showed that the scheme worked for several different vegetation types, while Hobbs, Mallik & Gimingham (1984) recently showed that the scheme also depicted the sequence of relative abundance in *Calluna vulgaris* dominated heathland in north-east Scotland. Noble and Slatyer's classification is a clever formalization of species properties that ecologists have long used in an unsystematic way in predicting succession on the basis of experience. Its two major disadvantages are that it does not allow quantitative predictions, and is effectively dependent on a vegetation type in which all species occurring in the succession are present at the outset in some life-stage or other, establish together if not already established, and subsequently co-exist. Unlike Grime's (1977, 1979) model, there is no way of indicating if a particular species, arriving at the site some years after the disturbance, can compete effectively and establish. Nor does the scheme allow for the implications of site modification. It has been tested on the special case of vegetation adapted to recurrent fire. The process of canopy redevelopment of such vegetation after fire is probably best thought of as 'regeneration' or 'cyclic change' (Miles 1979), rather than succession. It is thus somewhat misleading to suggest the scheme at its present state of development can be used to predict successional change (see Noble & Slatyer's (1980) and Noble's (1981) titles). The scheme can however be used to examine multiple pathways of post-fire regeneration (Cattelino *et al.* 1979).

Clements' concept of succession included three major points apart from directionality and irreversibility: succession proceeded (i) deterministically (because it always went in the same direction to the same climax), (ii) to climax vegetation (i.e. the end-point of succession), and (iii) was driven by reaction (i.e. the effects that the plants had on their environment, now commonly termed site modification). The determinism arose from Clements' (1916) extreme views on the nature of vegetation: 'The developmental study of vegetation necessarily rests upon the assumption that the unit or climax formation is an organic entity. As an organism the formation arises, grows, matures, and dies. ... each climax formation is able to reproduce itself, repeating with essential fidelity the stages of its development.' Braun-Blanquet (1932) called this view 'a flight of imagination', while Gleason (1917) published an immediate and cogent rebuttal, later repeated and amplified (Gleason 1926, 1937). However, the 'organismic concept' survives, barely disguised, in the 'systems approach' of the computer age (Margalef 1963, 1968; Odum 1969; Whittaker & Woodwell 1972, Patten 1975), an age when 'the things that count are those that can be counted' (Egler 1970).

What is climax vegetation? Clements (1916) defined it as follows: 'Every complete sere ends in a climax. This point is reached when the occupation and reaction of a dominant are such as to exclude the invasion of another dominant.' However, he did not regard climax as unchanging:

'No climax area lacks frequent evidence of succession, and the greater number present it in bewildering abundance. ... the most stable association is never in complete equilibrium, nor is it free from disturbed areas in which secondary succession is evident.' Later, in the context of his assumption that a climax formation was an organism, he wrote, 'succession is reproduction'. Cowles (1901) regarded climax as something rather less tangible: 'The condition of equilibrium is never reached, and when we say there is an approach to the mesophytic forest, we speak only roughly and approximately. As a matter of fact we have a variable approaching a variable rather than a constant.' Cooper (1926) stated forthrightly that the concept of the climax was, 'more or less subjective and artificially imposed upon the facts of nature', although he added that it was exceedingly useful.

Yet like succession, the climax concept, which is simple enough on a site basis, has caused a deal of pother and heart searching (Whittaker 1951, 1953). Over 70 years ago Harper (1914) noted that every different soil type seemed to have its own characteristic successional and climax vegetation. Further, acceptance of Gleason's ideas about the nature of vegetation must lead to rejection of any notion of a widespread climatic climax. Fifty years ago Braun-Blanquet (1932) suggested that Clements had overemphasized the concept of climax, because 'the areas today occupied by climax stages have become greatly reduced, and sometimes have become almost completely obliterated.' With this even truer today, and with increasing knowledge that long-term stability in vegetation is the exception rather than the rule, the concept approaches irrelevance. Certainly the term seems out of favour, except where, as in range management (Meeker & Merkel 1984), it still has practical utility.

How important is reaction as a driving force in succession? We should remember that Clements (1916) mostly referred to primary succession, i.e. succession on new, previously unvegetated bare areas such as lakes, rocks, fresh sand dunes, lava flows, moraines or mine and quarry waste, all initially sterile, and all with undifferentiated and usually unweathered soils. In all such cases, a degree of site modification is essential for succession to proceed. At the least, lakes must fill with sediment, and soils must accumulate the level of nitrogen capital needed to sustain non nitrogen-fixing plants (Bradshaw *et al.* 1982); plant growth on site is usually important in bringing about these changes. It is in secondary succession that the extent to which reaction can act as a driving force is uncertain. Many soil properties inevitably change during secondary succession (Miles 1981a, 1985b), perhaps pH and rates of nutrient cycling in particular, but there is little substantive evidence that they materially influence the sequence of species changes. The best documented example may be when *Betula pendula* and *B. pubescens* colonize *Calluna*-dominated moorland (Miles & Young 1980; Miles 1985a). Gross changes can then occur in soil chemical and structural properties which permit many plant species to colonize the field layer that cannot grow in the moorland soil.

SUCCESSIONAL PATHWAYS

Multiple pathways

Patterns of vegetation change in time show a bewildering variety. Cooper (1926), a student of Cowles, implied this in his analogy: '... the vegetation of the earth presents itself as a flowing stream, undergoing constant change. It is not a simple stream but a "braided" one, of enormous complexity, with its origin in the far distant past. Its more or less separate and definite elements branch, interweave, anastomose, disappear, reappear. ... Vegetation as we see it today is thus a mere cross section of this complex stream.' A year later, Gleason (1927) wrote, more definitely: '... succession is an extraordinarily mobile phenomenon, whose processes are not to be stated as fixed laws, but only as general principles of exceedingly broad nature, and whose results need not and frequently do not ensue in any definitely predictable way.'

For all these remarks, it seems to have taken Walker's (1970) now classic synthesis of palynological studies of post-glacial hydroseres, which showed that 'variety is the keynote of hydroseral succession', for the notion of multiple pathways of succession to gain even limited acceptance. Now people are analysing successional data in ways that can detect multiple directions of change, and are finding them. Londo (1974) found a complex network of successional transitions from mapping 10 ha of dune slack vegetation successively in 1956, 1963 and 1968. Fig. 1.4 gives a simplified graphic account. Most of the vegetation units showed retrogressive trends to some extent. The most important unit that did not was *Hippophaë* scrub, but it is probable that insufficient time had passed for this either to degenerate with stand senescence, or to be succeeded by a different woody species. Dorp *et al.* (1985) found similar complex successional networks on a larger area of dunes, and Hogeweg (cited by Maarel & Werger, 1978) found them on saltmarsh. All three instances, and Walker's (1970) study, were of primary successions, but secondary successions behave similarly. Abrams, Sprugal & Dickmann (1985) found several directions of change during natural regeneration of burnt and felled

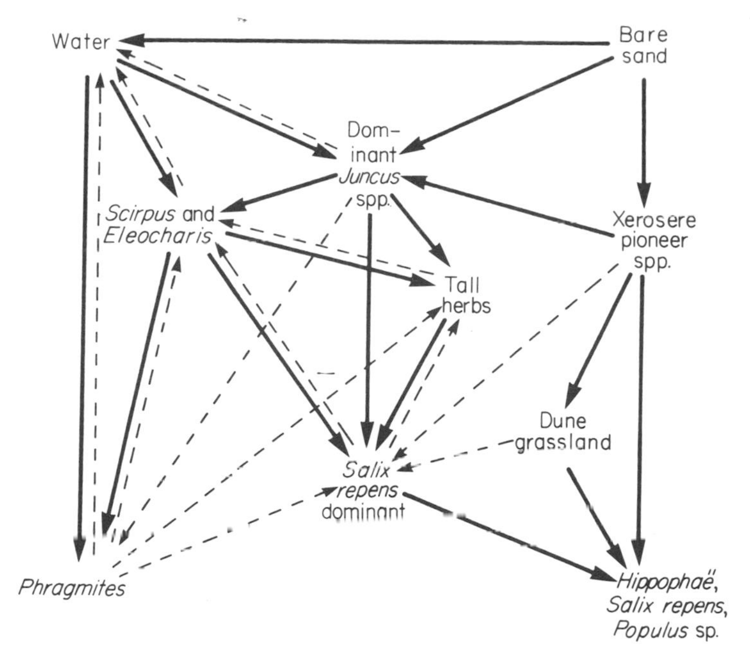

FIG. 1.4. Simplified successional trends during 1956–68 in dune slacks near Haarlem, the Netherlands. Solid lines represent major trends, dashed lines, minor trends. (From Londo 1974, courtesy of Dr W. Junk Publishers.)

stands of *Pinus banksiana*. Bakker *et al*. (1983) found multiple transitions when they subjected 11 ha of heathland, woodland and pasture to sheep grazing, at a mean stocking rate of 3 sheep ha^{-1} $year^{-1}$, during 1972–77. Table 1.1 gives the transition probabilities among four aggregate vegetation units from heathland only. Only the grassland dominated by *Molinia caerulea* did not show appreciable change to at least one other type.

Successional sequences and mechanisms

There has been considerable recent interest in mechanisms of succession. Egler (1954) stimulated the interest, Horn (1976) and Whittaker & Levin (1977) suggested examples, but Connel & Slatyer's (1977) paper, devoted largely to suggested mechanisms, caught the ecological community's imagination, and its terminology is now widely used. Fig. 1.5 indicates the replacement sequences suggested for succession by these authors. Fig. 1.5a is the 'model' implied by Clements (1916) and explicitly stated by other workers for primary succession. To my knowledge, no-one has ever

TABLE 1.1. Transition probabilities among four aggregated vegetation units* in 3 ha of grassy heathland during 1972–77 (from Bakker *et al.* 1983)

		Vegetation in 1972			
		1	2	3	4
	1	0·36	0·04	0	0·01
Vegetation in 1977	2	0·26	0·39	0·04	0·01
	3	0·16	0·57	0·68	0
	4	0·22	0	0·28	0·98

* Type 1, *Calluna vulgaris* and *Erica tetralix* predominant, with >50% cover; type 2, *Calluna* and *E. tetralix* predominant, with 20–50% cover; type 3, *Molinia caerulea* predominant, with >50% cover; type 4, grassland with <50% cover of *Molinia*.

shown that this pattern of successive waves of colonizing species ever occurs in secondary succession. 'Direct' succession (Fig. 1.5b) occurs in tundra and desert vegetation, where succession can be just the direct recolonization by the earlier occupants because no 'successional' species grow there (Shreve 1942; Muller 1952; Babb & Bliss 1974).

The model in Fig. 1.5c is that described by Egler (1954) for succession in abandoned fields, a sequence that was also implied by Clements (1916) for secondary succession. Egler noted that there was a tendency for most species seen during succession either to be present at the outset, as buried seed, rhizomes or other perennating organs, or to colonize within a few years. Many studies have now shown that the initial floristic composition is an important factor. In areas regenerating after fire, it seems usually to be the predominant factor, which is why Noble & Slatyer's (1977, 1979, 1980) system of 'vital attributes' can predict the regeneration sequence. Fig. 1.6 gives an example, the generalized sequence, averaged from a large number of sites, of vegetation change after burning *Calluna*- dominated moorland in part of north-east Scotland. Of the major components, only the pleurocarpous mosses were not recorded in the first year after burning. This model does not clearly differ from that in Fig. 1.5d, though Egler (1954) and others have suggested that distinct phases of dominance occur because of differences in the growth rates of different groups of plants, typically in the sequence: annual herbs, perennial herbs, short-lived woody plants, long-lived woody plants.

The model in Fig. 1.5d seems to occur commonly during old-field succession when species characteristic of later phases of the sequence are not present at the outset, but invade during the succession. Because they

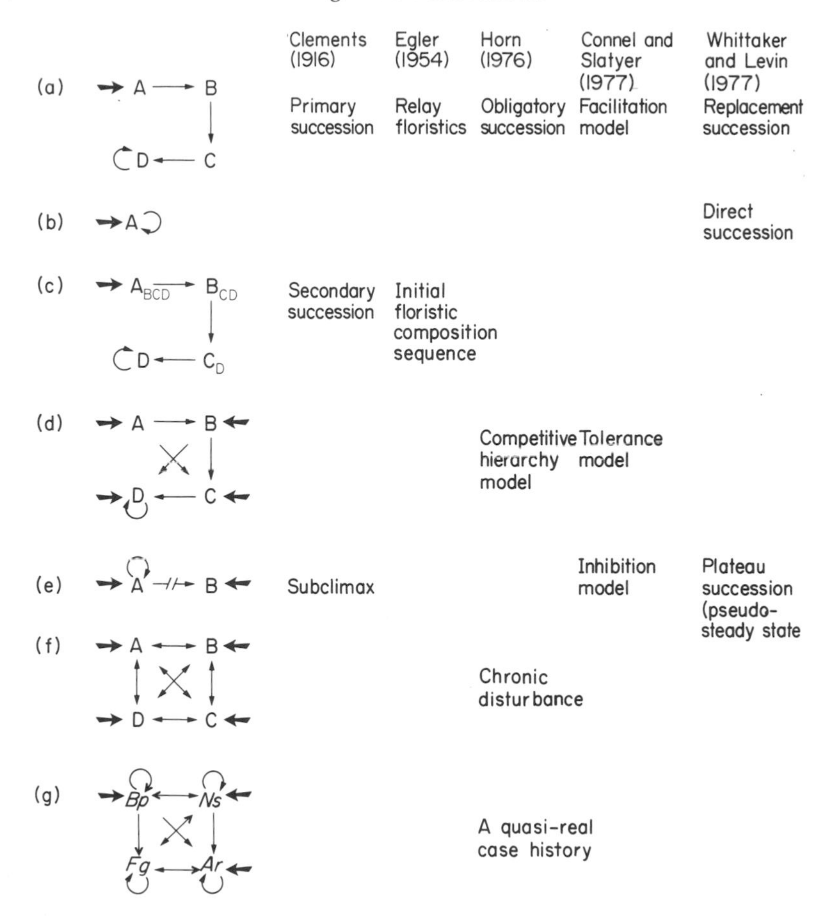

FIG. 1.5. Replacement sequences in succession proposed by different authors. The letters A–D in (a)–(f) represent hypothetical vegetation types or dominant species; subscript letters in (c) indicate that species are present as minor components or as propagules—for simplicity, they have been omitted from (d)–(g); thin arrows represent species or vegetation sequences in time; bold arrows represent alternative starting points for succession after disturbance; in (g), *Bp*, *Ns*, *Ar* and *Fg* represent *Betula populifera*, *Nyssa sylvatica*, *Acer rubrum* and *Fagus grandifolia* respectively, and the three open-headed arrows represent less frequent transitions. (After Noble 1981.)

can invade, Connel & Slatyer (1977) called this the 'tolerance' model. The 'tolerance' model certainly operates to some extent during primary succession, but as the soil at the beginning of secondary succession is never without propagules of some species, it will never occur then without an element of the initial floristic composition model.

Fig. 1.5e depicts what happens when a stand of vegetation somehow

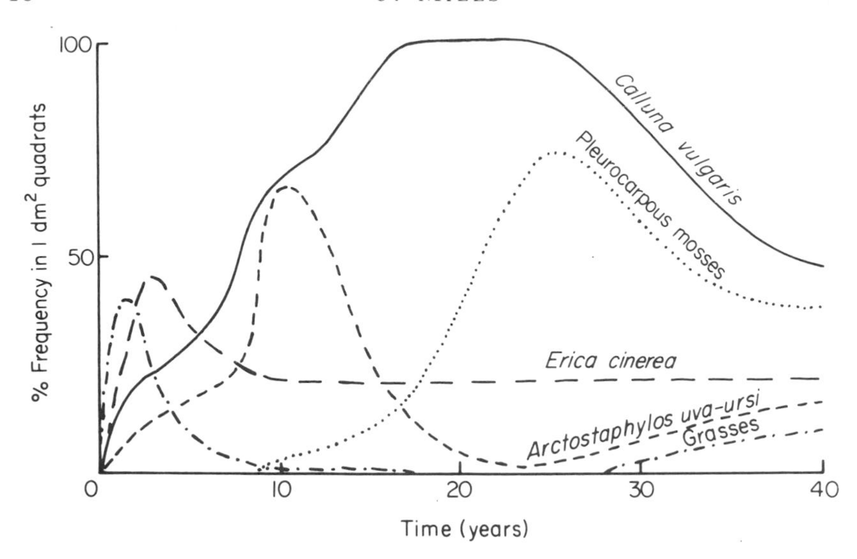

FIG. 1.6. Generalized sequence of vegetation change after fire on Dinnet Moor, north-east Scotland. (After Hobbs, Mallik & Gimingham 1984.)

'resists' invasion by plants of higher life-form. Connel & Slatyer called this 'inhibition'; Whittaker & Levin (1977) termed it 'plateau succession' in reference to the persistence of a non-climax stage. Clements (1916) called any vegetation prevented, temporarily or permanently, from developing into climax vegetation, 'subclimax', and noted that subclimax could be caused by competition.

Horn (1976) suggested that chronic, patchy disturbance could give a situation where any species could, if propagules were present, invade an opening resulting from the death of any other species (Fig. 1.5f). Clearly such a mechanism could operate at the level of individual plants or small patches. It is likely that such purely probabilistic succession never occurs as the sole mechanism in any stand, but Horn suggests that some of the celebrated local diversity of tropical forests could be due to this mechanism.

It seems clear that no succession, primary or secondary, ever occurs in which any single one of these models operates alone. Fig. 1.7 gives the generalized sequence of vegetation change when *Betula pendula* or *B. pubescens* colonizes *Calluna*-dominated moorland in the Scottish Highlands. The early changes follow the 'initial floristic composition' sequence. *Calluna* cannot regenerate vegetatively under a developing *Betula* canopy, and as it dies of old age it is replaced by species that can spread

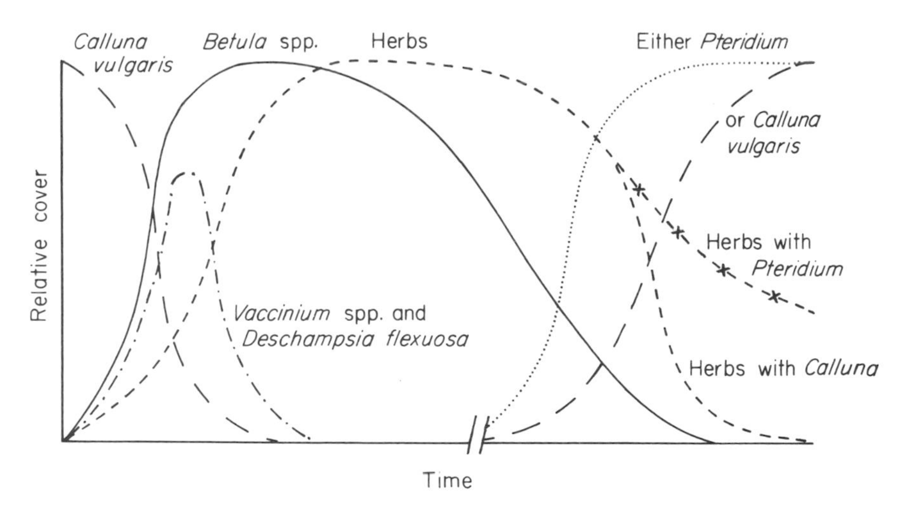

FIG. 1.7. Generalized sequence of vegetation change during the life-cycle of a Scottish Highland birchwood. (After Miles 1981b.)

by vegetative means; variously *Deschampsia flexuosa*, *Vaccinium myrtillus* and *V. vitis-idaea*. A dense substratum of pleurocarpous mosses also forms. Experimental sowings have shown that the field layer then prevents establishment of woodland herbs that can otherwise grow in that soil; this is 'inhibition'. Eventually, however, herbs characteristic of woodland and grassland do invade, spread, and except at sites with very poor soils, largely suppress the earlier dominants. The mechanisms by which this occurs are still unclear, but this changeover always parallels soil changes tending to produce brown podzolic profiles in place of podzols. Many of these later colonizers cannot grow, even without competition, in the moorland soils, so the soil changes resulting from the earlier vegetation changes 'facilitate' this phase of the succession, which itself probably operates along the lines of the 'tolerance' model. Finally, as the *Betula* stand dies of old age, *Calluna* begins to re-establish, probably mainly from seed blown in, though small quantities of seed can still occur in the soil seed bank after 90 years. However, if small amounts of *Pteridium aquilinum* are present, they are likely to develop into a dense stand after the release from shading; or if a *Pteridium* stand occurs close by, it can then invade. Either way, a dense stand of *Pteridium* can develop which prevents *Calluna* re-establishment. Seeds or fruits of later successional trees (e.g. *Pinus sylvestris*, *Quercus petraea*) are rarely present, because of the extent of past deforestation. The main sequence of change thus occurs according to, or compatibly with, the sequence of models: initial floristic composition, inhibition, facilitation and tolerance, initial floristic composition or inhibition.

Horn (1976) estimated transition probabilities of tree species in a New Jersey forest formed on abandoned farmland. The inferred replacement series is shown in Fig. 1.5g. He noted that the overall pattern was of a competitive hierarchy (tolerance model), with chronic, patchy disturbance also involved, and 'a hint' of obligatory succession (since *Fagus grandifolia* rarely invades open fields there), plus some self-replacement (i.e. direct succession).

In sum, the models in Fig. 1.5 relate to the succession of particular species in particular situations, and not to vegetation types. This is an unsurprising conclusion, because in succession it is individuals and populations that change. For vegetation to act in successional terms as a unit, and thus to permit useful generalizations about succession at a vegetation level, it would need to be a tightly integrated network of species, probably co-evolved and co-adapted, possessing hypothetical 'emergent' properties, and thus behaving as a vegetational gestalt. There is negligible evidence to support such a view. No properties of vegetation have ever been demonstrated that cannot be explained as the effects of species and of interactions between species. In contrast, much evidence supports Gleason's (1926, 1937) view of the probabilistic nature of vegetation, which predicts that no two patches of vegetation are ever likely to be identical in composition, except by chance. Perusal of the relevés in phytosociological publications supports this idea. Further, the palynological record of invasions and interglacial plant successions, e.g. Fig. 1.8, confirms that vegetation types are not persistent in time, so are unlikely ever to develop the level of integration needed to behave as a unit (West 1964; Watts 1973; Davis, Chapter 18). To quote West (1964): '... we may conclude that our present plant communities have no long history in the Quaternary, but are merely temporary aggregations under given conditions of climate, other environmental factors, and historical factors.'

CURRENT ATTITUDES TO SUCCESSION

I am not aware of any current consensus view about succession, but I shall state what I believe to be the prevalent attitude. Only a tiny proportion of the earth's surface is currently undergoing primary succession as defined by Clements (1916) (although, because the Quaternary climate is never stable, primary succession in theory never ends). Understanding of primary successions has practical use in the reclamation of

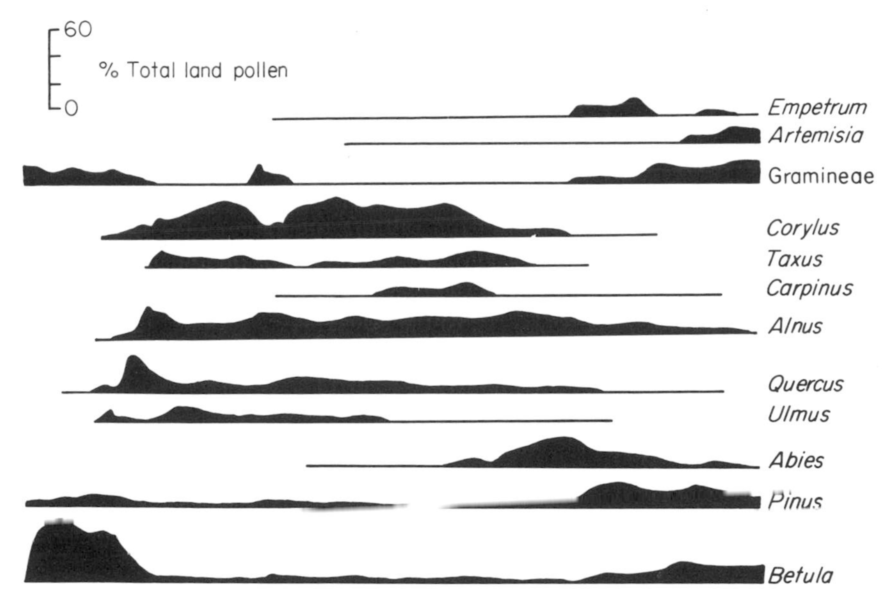

FIG. 1.8. Vegetation changes at Marks Tey, Essex, during the Hoxnian interglacial period. (From Turner 1970, courtesy of the Royal Society.)

quarry spoil and industrial wastelands, while understanding of secondary successions is of direct concern to vegetation managers, farmers and foresters. Hopefully, the quest for unifying, generalized theories of succession has been abandoned. Certainly there is now concentration on three topics: mechanisms of succession, modelling and predicting succession, and understanding the processes that determine the structure and species composition of vegetation.

The first topic, mechanisms of succession, is still poorly understood. More data are needed to test the current speculations and hypotheses, and to develop new hypotheses. This partly accounts for the recent proliferation of studies in permanent plots in continental Europe (Maarel 1984), and it is to be hoped that this may do for vegetation dynamics what the continental Europeans earlier did for static, descriptive vegetation science. There is interest in pursuing the use of 'vital attributes' (Noble & Slatyer 1977, 1979, 1980), but these need to be linked with quantitative methods of predicting change, and to allow for probabilistic events.

Modelling succession continues as a development point. There is a need to be able to model vegetation change as a tool in the use, management and sustaining of biological resources. Particular emphasis has been given to modelling forest succession (Shugart, West & Emanuel 1981; Shugart 1984) because of the economic importance of forests, and because much is known about the life history and behaviour of many common tree species. Markov models have been used increasingly, despite their many limitations and imperfect assumptions (see Usher, Chapter 2). However, their extreme simplicity makes the underlying assumptions crystal clear, an advantage when computers allow the development of increasingly complex models which only a few people may really understand. The use of Markov models can be a useful way of examining largely unknown systems (Miles *et al.* 1985).

Third, there is a valuable concentration on examining the mechanisms determining the structure and species composition of vegetation (Grubb 1977, 1984; Grime 1977, 1979, Chapter 20; Harper 1977; Hubbell 1979; Hubbell & Foster, Chapter 19). This is of crucial importance, because when these mechanisms are modified, by 'natural' events or by management, change in vegetation composition ensues. Equally they can explain how change can, or cannot, occur when new species attempt to invade.

AN EMPIRICAL APPROACH TO SUCCESSION

One aim of succession research should be to predict change in vegetation when or if management alters. Prediction can never be exact or in fine detail. However, predictions that are probabilistic and less detailed are in very many instances all that are needed. Prediction should ideally be made from an understanding of how any particular system functions. Acquiring such knowledge is a long process, and the need for management advice and prediction does not disappear in the meantime. Hence, there is a need for empirical, correlative studies to be pursued concurrently. With time, management advice should steadily improve by a process of successive approximation.

My main interest is in the use and management of semi-natural vegetation in the British uplands, and of similar vegetation in the lowlands. Nine years ago, I reviewed the scanty literature on change in such vegetation, and on its causes, chiefly grazing and fire. This was incorporated into a series of successional diagrams (Miles, Welch & Chapman 1978) which included estimates of rates of change at particular grazing pressures by sheep. These word models were then used to predict the changes in vegetation on part of Dartmoor, in south-west England, that might occur in 25 years given (i) an end to burning, and of grazing by domestic livestock, (ii) a burning frequency and grazing intensity as high as the vegetation could support (Fig. 1.9). If the rules used were formalized,

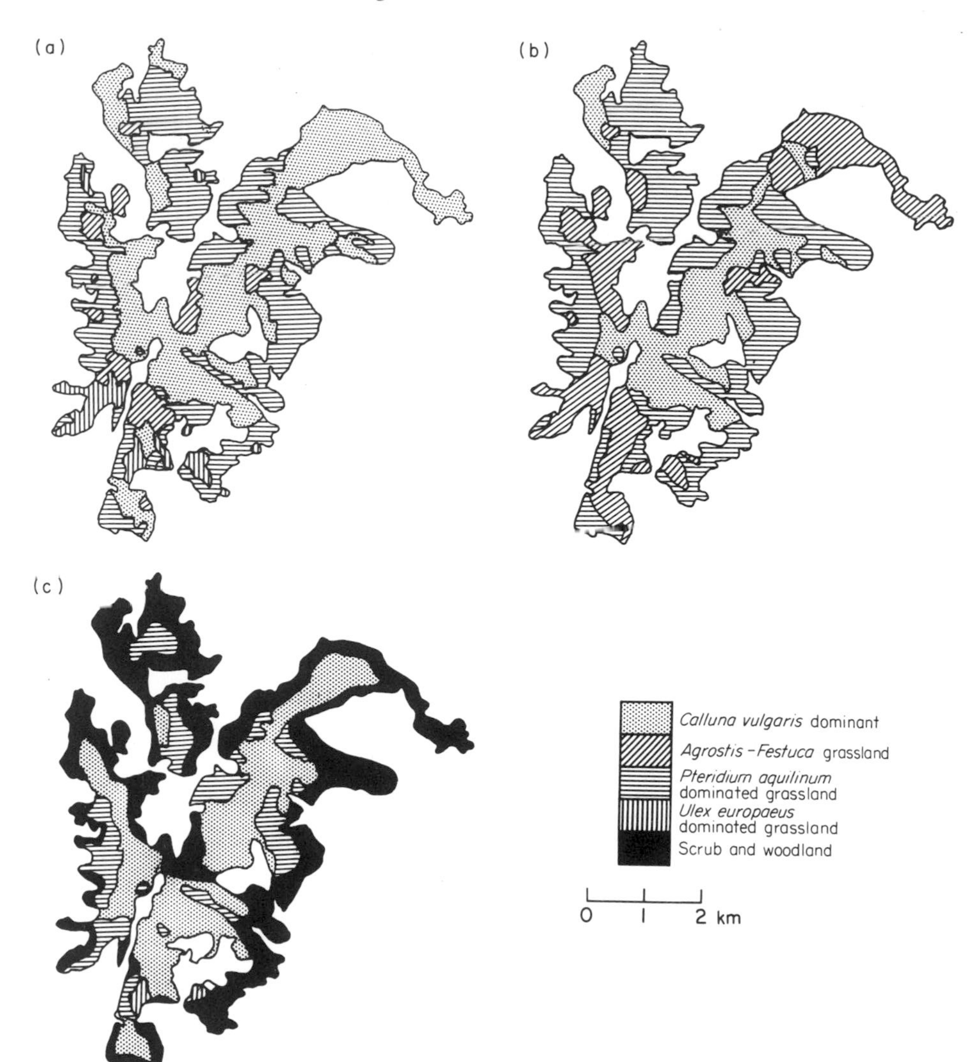

FIG. 1.9. Patterns of vegetation on well-drained soils of the Haytor moorlands, Dartmoor, south-west England: (a) as recorded in 1976; (b) predicted for 2001 if grazing by domestic livestock and burning were at the maximum that could be supported by the vegetation; (c) predicted for 2001 if grazing and burning cease. (From Miles, Welch & Chapman 1978, courtesy of the Countryside Commission.)

such an exercise could be done by an 'expert system' model. The main disadvantages were the hypothetical nature of some of the information used, and the impossibility of making probabilistic predictions.

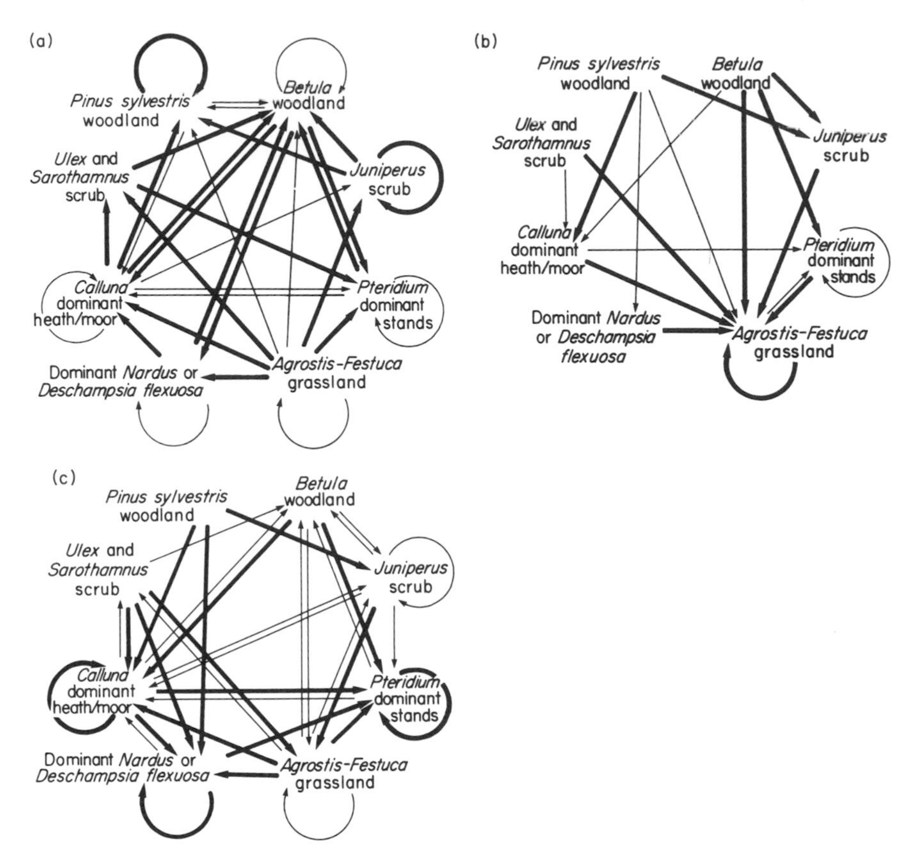

FIG. 1.10. Successional transitions in the British uplands (particularly north-west Scotland) between eight vegetation types given (a) low grazing pressures (<1 sheep equivalent ha^{-1} year^{-1}) and no burning, (b) high grazing pressures (>2–3 sheep equivalents ha^{-1} year^{-1}) and frequent burning, and (c) intermediate levels of grazing (1–2 sheep ha^{-1} year^{-1}) and occasional burning. Broad arrows represent common transitions, thin arrows less frequent transitions, and curved arrows self-replacement. The vegetation types are arranged so that types tending to podzolize and/or acidify soils are on the left, and types with contrasting pedogenic effects are on the right. (From Miles 1985b, courtesy of the British Society of Soil Science.)

We are now trying to collect the data needed to improve such predictions of succession, currently by the analysis of archival aerial photography in north-east Scotland. Formally, the exercise is structured to test the 1978 successional diagrams. Transitions that are known to occur at least sometimes, between various vegetation types on well-drained, mineral soils, and with high, medium and low intensities of grazing and burning, are shown in Fig. 1.10. The problem of species as against vegetation

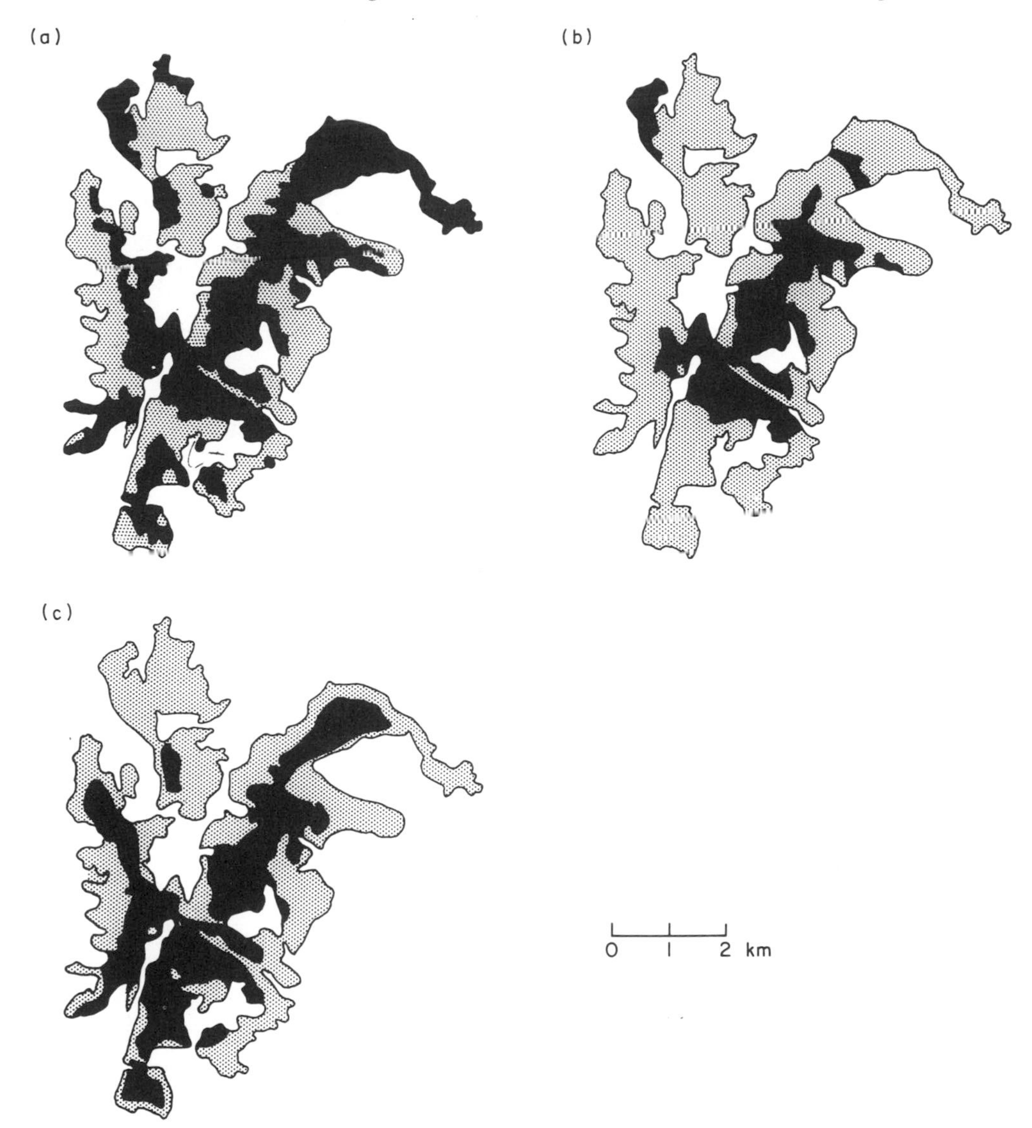

FIG. 1.11. Patterns of vegetation tending to podzolize and/or acidify soils (solid areas), and of vegetation with contrasting pedogenic effects (stippled areas), on well-drained soils of the Haytor moorlands, Dartmoor, south-west England: (a) as recorded in 1976; (b) predicted for 2001 if grazing by domestic livestock and burning were at the maximum that could be supported by the vegetation; (c) predicted for 2001 if grazing and burning ceased.

succession, discussed earlier, is overcome by using a physiognomic classification based on dominance. We believe this is sensible because dominant species *inter alia* give landscapes their characteristic colour and texture, have the greatest ecological effect, especially on the soil (Miles 1985b), and are the main resource for game, domestic livestock, and animals of nature conservation interest. Fig. 1.10 indicates that multiple pathways occur, but the extent to which they occur at a site or local level, or on a larger scale is not yet known. Equally there is little knowledge of variations in rates of change.

Fig. 1.10 is arranged so that species and vegetation known to podzolize or acidify soil are on the left, and those with contrasting tendencies on the right. Inclusion of pedogenic trends permits a new level of prediction. Even without a knowledge of the relative frequencies of the different transitions, inspection of Fig. 1.10a suggests that several vegetation types will tend to coexist within a catchment or landscape (albeit that forest may predominate): effectively the concept of multiple stable points (Sutherland 1974) or of the shifting-mosaic steady state (Bormann & Likens 1979). This would also give a mosaic of contrasting pedogenic trends which, because the mosaic pattern would change with time, might tend to cancel out and give effective equilibrium of soil properties at this scale. Fig. 1.10b suggests a dominant trend away from podzolization, towards brown podzolic soil/brown earth development. Fig. 1.10c suggests even more strongly than Fig. 1.10a that a mosaic of vegetation types and pedogenic trends would exist. Application of pedogenic trends to Fig. 1.9 gives Fig. 1.11. A mosaic effect is seen, shifting as management changes.

In sum, our empirical perspective of succession views a catchment or landscape as a mosaic of sites or ecotesseras (Jenny 1980). This gives the possibility of combining the probabilistic information on ecotesseras (at any scale) with a wider spatial perspective. Thus, we aim to use our knowledge of succession to predict vegetation change both probabilistically and cartographically. It should also be possible to integrate other information (e.g. models predicting element output and changes in faunal populations with changing vegetation) in such landscape-scale models. Seeking to apply ecological knowledge in this way will, on the one hand, expose its deficiencies, and on the other hand may, even if only partly successful, show the practical relevance of such knowledge, and thereby generate continued support for the subject.

ACKNOWLEDGMENT

I thank Professor A.D. Bradshaw for his constructive criticisms of a draft of this paper.

REFERENCES

Abrams, A.D., Sprugal, D.G. & Dickmann, D.I. (1985). Multiple successional pathways on recently disturbed jack pine sites in Michigan. *Forest Ecology and Management*, **10**, 31–48.

Babb, T.A. & Bliss, L.C. (1974). Effects of disturbance on Arctic vegetation in the Queen Elizabeth Islands. *Journal of Applied Ecology*, **11**, 549–562.

Bakker, J.P., Bie, S. de, Dallinga, J.H., Tjaden, P. & Vries, Y. de (1983). Sheep grazing as a management tool for heathland conservation and regeneration in the Netherlands. *Journal of Applied Ecology*, **20**, 541–560.

Bormann, F.H. & Likens, G.E. (1979). *Pattern and Process in a Forested Ecosystem*. Springer-Verlag, New York.

Bradshaw, A.D., Marrs, R.H., Roberts, R.D. & Skeffington, R.A. (1982). The creation of nitrogen cycles in derelict land. *Philosophical Transactions of the Royal Society B*, **296**, 557–561.

Braun–Blanquet, J. (1932). *Plant Sociology. The Study of Plant Communities* (translated by G.D. Fuller & H.S. Conard). McGraw-Hill, New York.

Cattelino, P.J., Noble, I.R., Slatyer, R.O. & Kessel, S.R. (1979). Predicting the multiple pathways of plant succession. *Environmental Management*, **3**, 41–50.

Clements, F.E. (1904). *The Development and Structure of Vegetation*. Botanical Survey of Nebraska, 7. The Botanical Seminar, Lincoln, Nebraska.

Clements, F.E. (1916). Plant succession: an analysis of the development of vegetation. *Carnegie Institute of Washington Publication*, **242**.

Cochon, F. (1846). Alternance des essences dans les forêts. *Annales Forestières*, **5**, 1–13.

Colinvaux, P. (1973). *Introduction to Ecology*. Wiley, New York.

Connel, J.H. & Slatyer, R.O. (1977). Mechanisms of succession in natural communities and their role in community stability and organization. *American Naturalist*, **111**, 1119–1144.

Cooper, W.S. (1926). The fundamentals of vegetation change. *Ecology*, **7**, 391–413.

Cowles, H.C. (1899). The ecological relations of the vegetation on the sand dunes of Lake Michigan. I. Geographical relations of the dune flora. *Botanical Gazette*, **27**, 95–117, 167–202, 281–308, 361–391.

Cowles, H.C. (1901). The physiographic ecology of Chicago and vicinity; a study of the origin, development, and classification of plant communities. *Botanical Gazette*, **31**, 73–108, 145–182.

Cowles, H.C. (1911). The causes of vegetative cycles. *Botanical Gazette*, **51**, 161–183.

Davis, W.M. (1909). *Geographical Essays*. Ginn, Boston.

Dawson, J.W. (1847). On the destruction and partial reproduction of forests in British North America. *American Journal of Science*, Series 2, **4**, 161–170.

Dimbleby, G.W. (1962). The development of British heathlands and their soils. *Oxford Forestry Memoirs*, 23.

Dorp, D. van, Boot, R. & Maarel, E. van der (1985). Vegetation succession on the dunes near Oostvoorne, The Netherlands, since 1934, interpreted from air photographs and vegetation maps. *Vegetatio*, **58**, 123–136.

Douglas, R. (1875). Succession of species in forests. *The Horticulturist and Journal of Rural Art and Taste*, **30**, 138–140.

Douglas, R. (1889). Succession of forest growths. *Garden and Forest*, **2**, 285–286.

Drury, W.H. (1956). Bog flats and physiographic processes in the Upper Kuskokwim River region, Alaska. *Contribution from the Gray Herbarium of Harvard University*, 178.

Drury, W.H. & Nisbet, I.C.T. (1973). Succession. *Journal of the Arnold Arboretum*, **54**, 331–368.

Dureau de la Malle, A.J.C.A. (1825). Memoire sur l'Alternance ou sur ce problème: la succession alternative dans la reproduction des espèces végétales vivant en société, est elle une loi générale de la nature? *Annales des Sciences Naturelles*, **5**, 353–381.

Egler, F.E. (1951). A commentary on American plant ecology, based on the textbooks of 1947–1949. *Ecology*, **32**, 673–695.

Egler, F.E. (1954). Vegetation science concepts. 1. Initial floristic composition: a factor in old-field vegetation development. *Vegetatio*, **4**, 412–417.

Egler, F.E. (1970). *The Way of Science. A Philosophy of Ecology for the Layman.* Hafner, New York.

Finegan, B. (1984). Forest succession. *Nature*, **312**, 109–114.

Gill, A.M. (1975). Fire and the Australian flora: a review. *Australian Forestry*, **38**, 4–25.

Gleason, H.A. (1917). The structure and development of the plant association. *Bulletin of the Torrey Botanical Club*, **44**, 463–481.

Gleason, H.A. (1926). The individualistic concept of the plant association. *Bulletin of the Torrey Botanical Club*, **53**, 7–26.

Gleason, H.A. (1927). Further views on the succession-concept. *Ecology*, **8**, 299–326.

Gleason, H.A. (1937). The individualistic concept of the plant association. *American Midland Naturalist*, **21**, 92–110.

Gleason, H.A. (1953). A letter for the younger generation of ecologists. *Bulletin of the Ecological Society of America*, **34** (2), 40–42.

Gleason, H.A. (1975). Delving into the history of American ecology. *Bulletin of the Ecological Society of America*, **56** (4), 7–10.

Gorham, E., Vitousek, P.M. & Reiners, W.A. (1979). The regulation of chemical budgets over the course of terrestrial ecosystem succession. *Annual Review of Ecology and Systematics*, **10**, 53–84.

Grime, J.P. (1977). Evidence for the existence of three primary strategies in plants and its relevance to ecological and evolutionary theory. *American Naturalist*, **111**, 1169–1194.

Grime, J.P. (1979). *Plant Strategies and Vegetation Processes.* Wiley, Chichester.

Grubb, P.J. (1977). The maintenance of species-richness in plant communities: the importance of the regeneration niche. *Biological Reviews*, **52**, 107–145.

Grubb, P.J. (1984). Some growth points in investigative plant ecology. *Trends in Ecological Research in the 1980s* (Ed. by J.H. Cooley & F.B. Golley), pp. 51–74. Plenum Press, New York.

Harper, J.L. (1977). *Population Biology of Plants.* Academic Press, London.

Harper, R.M. (1914). The 'Pocosin' of Pike County, Alabama, and its bearing on certain problems of succession. *Bulletin of the Torrey Botanical Club*, **41**, 209–220.

Hobbs, R.J., Mallik, A.U. & Gimingham, C.H. (1984). Studies on fire in Scottish heathland communities. III. Vital attributes of the species. *Journal of Ecology*, **72**, 963–976.

Horn, H.S. (1976). Succession. *Theoretical Ecology. Principles and Applications* (Ed. by R.M. May), pp. 187–204. W.B. Saunders, Philadelphia.

Hort, A. (Ed.) **(1916).** Theophrastus. *An Enquiry into Plants.* Book IV. *Of the Trees and Plants Special to Particular Districts and Positions.* Heinemann, London.

Hubbell, S.P. (1979). Tree dispersion, abundance and diversity in a tropical dry forest. *Science*, **203**, 1299–1309.

Jenny, H. (1980). *The Soil Resource. Origin and Behaviour.* Springer-Verlag, New York.

Kellman, M.C. (1970). The influence of accessibility on the composition of vegetation. *Professional Geographer*, **22**, 1–4.

Kerner von Marilaun, A. (1863). *Das Pflanzenleben des Donauländer.* Translated by H.S. Conard in *The Background to Plant Ecology*. Iowa State College Press, Ames.

Lepart, J. & Escarre, J. (1983). La succession végétale, mécanisms et modèles: analyse bibliographique. *Bulletin d'Écologie*, **14**, 133–178.

Londo, G. (1974). Successive mapping of dune slack vegetation. *Vegetatio*, **29**, 51–61.

Maarel, E. van der (1984). Vegetation science in the 1980's. *Trends in Ecological Research in the 1980s* (Ed. by J.H. Cooley & F.B. Golley), pp. 89–110. Plenum Press, New York.

Maarel, E. van der & Werger, M.J.A. (1978). On the treatment of succession data. *Phytocoenosis*, **7**, 257–277.

Margalef, R. (1958). Information theory in ecology. *General Systems*, **3**, 36–71.

Margalef, R. (1963). On certain unifying principles in ecology. *American Naturalist*, **97**, 357–374.

Margalef, R. (1968). *Perspectives in Ecological Theory.* University of Chicago Press, Chicago.

McIntosh, R.P. (1980). The relationship between succession and the recovery process in ecosystems. *The Recovery Process in Damaged Ecosystems* (Ed. by J. Cairns), pp. 11–62.

McIntosh, R.P. (1981). Succession and ecological theory. *Forest Succession. Concepts and Application* (Ed. by D.C. West, H.H. Shugart & D.B. Botkin), pp. 10–23. Springer-Verlag, New York.

Medawar, P.B. (1969). *Induction and Intuition in Scientific Thought.* Methuen, London.

Meeker, D.O. & Merkel, D.L. (1984). Climax theories and a recommendation for vegetation classification: a viewpoint. *Journal of Range Management*, **37**, 427–430.

Miles, J. (1979). *Vegetation Dynamics.* Chapman & Hall, London.

Miles, J. (1981a). Problems in heathland and grassland dynamics. *Vegetatio*, **46**, 61–74.

Miles, J. (1981b). *Effect of Birch on Moorlands.* Institute of Terrestrial Ecology, Cambridge.

Miles, J. (1985a). Soil in the ecosystem. *Ecological Interactions in Soil* (Ed. by A.H. Fitter, D. Atkinson, D.J. Read & M.B. Usher), pp. 407–427. Special Publication of the British Ecological Society, **4**. Blackwell Scientific Publications, Oxford.

Miles, J. (1985b). The pedogenic effects of different species and vegetation types and the implications of succession. *Journal of Soil Science*, **36**, 571–584.

Miles, J., French, D.D., Xu, Z.-B. & Chen, L.-Z. (1985). Transition matrix models of succession in a stand of mixed broadleaved – *Pinus koraiensis* forest in Changbaishan, Kirin Province, north-east China. *Journal of Environmental Management*, **20**, 357–375.

Miles, J., Welch, D. & Chapman, S.B. (1978). Vegetation management in the uplands. *Upland Land Use in England and Wales*, pp. 77–95. Publication CCP111, Countryside Commission, Cheltenham.

Miles, J. & Young, W.F. (1980). The effects on heathland and moorland soils in Scotland and northern England following colonization by birch. *Bulletin d'Écologie*, **11**, 233–242.

Muller, C.H. (1952). Plant succession in arctic heath and tundra in northern Scandinavia. *Bulletin of the Torrey Botanical Club*, **79**, 296–309.

Naveh, Z. (1975). The evolutionary significance of fire in the Mediterranean region. *Vegetatio*, **29**, 199–208.

Naveh, Z. & Dan, J. (1973). The human degradation of Mediterranean landscapes in Israel. *Mediterranean Type Ecosystems. Origin and Structure* (Ed. by F. di Castri & H.A. Mooney), pp. 373–391. Springer-Verlag, Berlin.

Noble, I.R. (1981). Predicting successional change. *Fire Regimes and Ecosystem Properties*

(Ed. by H.A. Mooney), pp. 278–300. Proceedings of the Conference at Honolulu, Hawaii, 1978. *United States Department of Agriculture Forest Service General Technical Report*, WO-26.

Noble, I.R. & Slatyer, R.O. (1977). Post-fire succession of plants in Mediterranean ecosystems. *Environmental Consequences of Fire and Fuel Management in Mediterranean Ecosystems* (Ed. by H.A. Mooney & C.E. Conrad), pp. 27–36. *United States Department of Agriculture Forest Service General Technical Report*, WO-3.

Noble, I.R. & Slatyer, R.O. (1979). The effect of disturbance on plant succession. *Proceedings of the Ecological Society of Australia*, **10**, 135–145.

Noble, I.R. & Slatyer, R.O. (1980). The use of vital attributes to predict successional changes in plant communities subject to recurrent disturbances. *Vegetatio*, **43**, 5–21.

Odum, E.P. (1969). The strategy of ecosystem development. *Science*, **164**, 262–270.

Patten, B.C. (1975). Ecosystem linearization: an evolutionary design problem. *American Naturalist*, **109**, 529–539.

Pound, R. & Clements, F.E. (1898). *Phytogeography of Nebraska.* The Botanical Survey, Lincoln, Nebraska.

Romans, J.C.C. & Robertson, L. (1975). Soils and archaeology in Scotland. *The Effect of Man on the Landscape: the Highland Zone* (Ed. by J.G. Evans, S. Limbrey & H. Cleere), pp. 37–39. *Council for British Archaeology Research Report*, 11.

Shreve, F. (1942). The desert vegetation of North America. *Botanical Review*, **8**, 195–246.

Shugart, H.H. (1984). *A Theory of Forest Dynamics. The Ecological Implications of Forest Succession Models.* Springer-Verlag, New York.

Shugart, H.H., West, D.C. & Emanuel, W.R. (1981). Patterns and dynamics of forests: an application of simulation models. *Forest Succession. Concepts and Application* (Ed. by D.C. West, H.H. Shugart & D.B. Botkin), pp. 74–94. Springer-Verlag, New York.

Smith, A.G. (1970). The influence of Mesolithic and Neolithic man on British vegetation: a discussion. *Studies in the Vegetational History of the British Isles* (Ed. by D. Walker & R.G. West), pp. 81–96. Cambridge University Press, Cambridge.

Spurr, S.H. (1952). Origin of the concept of forest succession. *Ecology*, **33**, 426–427.

Sutherland, J.P. (1974). Multiple stable points in natural communities. *American Naturalist*, **108**, 859–873.

Thoreau, H.D. (1860). The succession of forest trees. Read to the Middlesex Agricultural Society, Concorde, September 1860. Reprinted 1863 in *Excursions*, pp. 135–160. Ticknor & Fields, Boston.

Ugolini, F.C. & Mann, D.H. (1979). Biopedological origin of peatlands in South East Alaska. *Nature*, **281**, 366–368.

Walker, D. (1970). Direction and rate in some British post-glacial hydroseres. *Studies in the Vegetational History of the British Isles* (Ed. by D. Walker & R.G. West), pp. 117–139. Cambridge University Press, Cambridge.

Warming, E. (1895). *Plantesamfund. Grundträk auf den ökologiska Plantegeografi*. German edition published 1896 by Bornträger, Berlin.

Warming, E. & Vahl, M. (1909). *Oecology of Plants. An Introduction to the Study of Plant-Communities.* Oxford University Press, Oxford.

Watt, A.S. (1964). The community and the individual. *Journal of Ecology*, **52** (Supplement), 203–211.

Watts, W.A. (1973). Rates of change and stability in vegetation in the perspective of long periods of time. *Quaternary Plant Ecology* (Ed. by H.J.B. Birks & R.G. West), pp. 195–206. Symposia of the British Ecological Society, **14**. Blackwell Scientific Publications, Oxford.

West, R.G. (1964). Inter-relations of ecology and Quaternary palaeobotany. *Journal of Ecology*, **52** (Supplement), 47–57.

Whittaker, R.H. (1951). A criticism of the plant association and climatic climax concepts. *Northwest Science*, **25**, 17–31.

Whittaker, R.H. (1953). A consideration of climax theory: the climax as population and pattern. *Ecological Monographs*, **23**, 41–78.

Whittaker, R.H. & Levin, S.A. (1977). The role of mosaic phenomena in natural communities. *Theoretical Population Biology*, **12**, 117–139.

Whittaker, R.H. & Woodwell, G.M. (1972). Evolution of natural communities. *Ecosystem Structure and Function* (Ed. by J.A. Wiens), pp. 137–156. Proceedings of the Thirty-first Annual Biology Colloquium. Oregon State University Press, Corvallis.

2. MODELLING SUCCESSIONAL PROCESSES IN ECOSYSTEMS

MICHAEL B. USHER
Department of Biology, University of York, York YO1 5DD

INTRODUCTION

In many respects the subject of modelling cuts across several facets of this symposium. Models are included in many of the other chapters and hence, to avoid duplication of material, this chapter will be selective (with all of the attendant problems of being described as biased). There are four main groups of models.

(a) *Descriptive and compartment models.* During the first two or three decades of this century, observations of succession were being conceptualized, and hence the writings of some of these earlier observers and thinkers can be described as descriptive models (see Miles, Chapter 1). With the advent of computers in the 1950s and 1960s, the descriptions were quantified, and essentially the description has become encapsulated in large sets of computer code rather than in a series of printed or spoken words.

(b) *Models based on population dynamics.* These models are concerned with the organisms involved in any colonization or succession process. In their simplest form they may be concerned with the rate of change of numbers of a single species during time (see Connell, Chapter 16) or with the productivity of a single species during time (see Edwards & Gillman, Chapter 14). Some more complex models, involving the effects of vertical relationships within a food chain (see Lawton, Chapter 11), have been developed. However, this modelling approach is generally limited either to the modelling of a single species or the simultaneous modelling of a small number of species. Modelling of perturbations would clearly fall within this group of models.

(c) *Markovian models.* These models move away from the organisms and concentrate on the more abstract concept of the community, and with the probabilities that certain events will happen. Since probabilities are used in the model formulation, these models are generally viewed as representing stochastic processes.

(d) *Invasion models.* For the process of colonization, the initial step in a successional sequence, the species must have arrived at the new habitat. Although the mechanisms are discussed in a number of chapters of this book, modelling is less well covered. Both Williamson and Crawley (Chapters 17 and 21) are concerned with invasion by alien species, and it is perhaps in the models of the spread of invasive species that one sees the most elegant models of both stochastic and deterministic types.

In the review in the following section, frequent reference will be made to a hypothetical example that is shown in Fig. 2.1. This illustration consists of a 'typical' selection of thirty-two species spread out along a successional sequence in time. In the next section, models of the invasion process will be omitted (they will be covered in a Royal Society volume in 1986), but other methods will be illustrated by reference to Fig. 2.1.

THE MODELS AVAILABLE

Descriptive and compartment models

From the earliest writings of ecologists this century, there has been a division into deterministic and stochastic descriptive models. Clements (1916) was the chief proponent of the former with such assertions as 'The life history of a formation is a complex but definite process ...', whilst Gleason (1926) has become associated with the latter approach with statements such as '... plant associations ... depend solely on the coincidence of environmental selection and migration ...'. Whichever side of the controversy was favoured, succession was seen by McIntosh

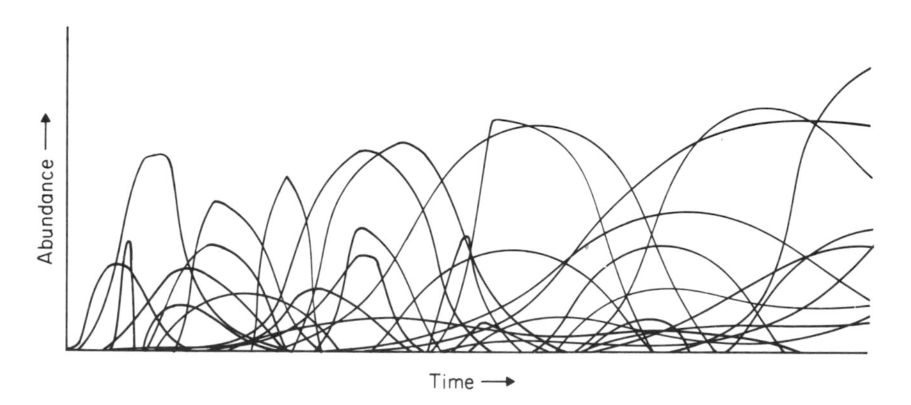

FIG. 2.1. A hypothetical sequence of species during succession. This sequence is used to illustrate various modelling approaches that are discussed in the text.

(1980) as being one of the nine groups of ecological principles that he recognized.

Writings on succession have tended to recognize two forms: *primary*, where a substrate is available to the biota of the geographical area for the first time, and *secondary*, where the substrate has been perturbed, usually massively with the deaths of living organisms, but where organic matter is present. Examples of primary successions include retreating glaciers, coastal accretion and quarries. The most widely quoted examples of secondary successions are abandoned fields, the 'old-fields' of American workers. However, in thinking of models, one needs to ask how the succession starts. What is the probability that a seed will land on a small area of land, and does the number of seeds in that area follow a Poisson distribution? Although Sterling *et al.* (1984) demonstrated that microtopography was important in colonization, some observations (Usher 1967) on the colonization of small ponds at Aberlady Bay Nature Reserve indicated random colonization by aquatic macrophytes, and in quarries (Usher 1979a) that the distribution of propagules was random. Numata (1979) has reviewed the characteristics of many of the species that colonize such new substrates.

There are many descriptions of succession, most of them being based upon spatially separated communities that are taken to represent stages in the succession. Long-term studies of a small area are less common, but van der Maarel (1979) demonstrated the use of phytosociological techniques in the long-term study of a dune grassland. Multivariate methods have been used more recently in an attempt to describe succession in a more quantitative manner: an excellent example is Austin's (1977) study of a secondary succession on a lawn. Orloci (1981) demonstrated the use of other multivariate techniques, as well as time-series analysis, in the interpretation of woodland data. Such statistical models are all essentially concerned either with the description of succession, or with determining whether any trend that could be interpreted as succession exists in a collection of data.

Other descriptive models have endeavoured to incorporate different facets of ecosystems, such as the total number of species or the biomass, into successional theory. Odum (1969) considered there to be an orderly accumulation of biomass in a sigmoid manner. Using the data in Fig. 2.1, fifty random vertical lines have been drawn, and at each the number of species and total abundance have been counted. These data, shown in Fig. 2.2 with the Shannon-Weaver diversity index, show what might be expected from a successional sequence: increases in number of species, abundance and diversity, though with subsequent small decreases. Of

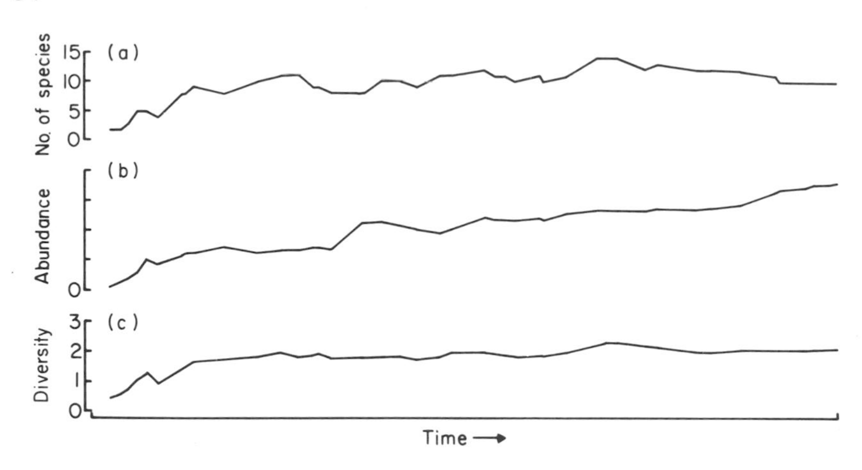

FIG 2.2. Three derived statistics taken from fifty random samples along the sequence shown in Fig. 2.1: (a) the number of species, (b) the total abundance of all species, and (c) the Shannon-Weaver diversity index. The horizontal axis is the same as that in Fig. 2.1.

course, these have to increase since the sequence starts with no species, and species do not all arrive at the same instant of time. Hence, many of these patterns that are included in descriptive models are no more than constraints due to the arrival of propagules over time and their subsequent growth. Similarly speeds of succession are often a feature of the arrival and growth processes. Typically the rate of acquisition of new species slows down, as can be seen in Fig. 2.1. Bornkamm (1981) equated rates of change of less than 5% per year with temporal stability (see also Williamson, Chapter 17).

Compartment models really link the ideas of descriptive or verbal models with the ideas of population dynamic models in the next section. They are generally formulated as extensive sets of computer code, the extensive backing store of the computer being used to simulate every organism or every one of many small spatial areas (generally grid squares) of the system being modelled. Early examples are given by Botkin, Janak & Willis (1972) with the JABOWA program, and Shugart, Crow & Hett (1973), who predicted woodland changes over a 250 year period. Subsequently, Shugart, West & Emanuel (1981) have reviewed the application of simulation models in studies of forest systems, concentrating their review on four models named FORET, BRIND, FORAR and KIAMBRAM (demonstrating the scope for research into model nomenclature!). A full discussion of the programs is beyond the scope of this review.

Population dynamic models

A search for the strategies of species involved in successional sequences has led to a fertile field for modelling. Connell & Slatyer (1977) distinguished three patterns of succession, referred to as the facilitation, tolerance and inhibition models. These depended on whether one set of species facilitated the following set of species, or whether the succession was manifest by the speeds of development of the species so that apparent sets consisted of species maturing at much the same time, or whether there was a set of species that could, as it were, highjack the succession and inhibit its further more normal development. Alongside such discussions of successional strategy were the general discussions of species strategy, from the *r–K* continuum to the *r–K–A* triangle of Greenslade & Greenslade (1983) and the *ruderal–competition–stress* triangle of Grime (1979, see also Chapter 20).

Most applications in succession have been concerned with the population dynamics of single species. Thus, van der Valk (1981) separated out three key life-history traits for succession in freshwater wetlands: these were life-span of the individual, propagule longevity and the requirements for propagule establishment. Pickett (1982) reviewed studies of the secondary succession on old-fields, and again picked out various aspects of the life-histories of the individual species, indicating that many of these species persist in the successional sequence at very low density for long periods of time before or after being abundant. These life-history traits should be included in the population dynamic models.

However, the interactions between species are likely to be as important, or more important, than the life-histories of the individual species (see Lawton, Chapter 11). Noble & Slatyer (1978) realized the need for studying the interactions between pairs of species with different sets of 'vital attributes' (the perceived important life-history traits) so as to predict how species would replace one another in a post-fire succession. Sparkes (1982) went one step further when she took, axiomatically, competition between pairs of species to be important, and then both by experiment and model showed that species later in a succession of soil arthropods tended to be competitively stronger.

The use of population dynamic models is best illustrated in van Hulst's (1979a) investigation of succession in model communities. He experimented with three models. The first considered all S species each to be governed by a logistic equation of the form

$$\mathrm{d}x_i/\mathrm{d}t = r_i x_i (K_i - x_i)/K_i \qquad (1)$$

where r_i is the intrinsic rate of increase of the ith species, x_i is its abundance and K_i is its carrying capacity which is assumed to vary during successional time, and hence should be written $K_i(t)$. In the second model, competition between pairs of species was introduced, so that the Lotka-Volterra equation formed the initial basis of the model,

$$\mathrm{d}x_i/\mathrm{d}t = r_i x_i (K_i - \sum_{j=1}^{S} \alpha_{ij} x_j)/K_i \quad (2)$$

where the coefficient α_{ij} represents the effect of the jth species on the ith species. The third model was similar to the second except that the function $K_i(t)$ was a continuous function rather than an integration of the history of vegetation development. All three models produced the kind of sequence of curves seen in Fig. 2.1, though in general without the long tails that Pickett (1982) noticed. They indicate that such a modelling approach may be useful, but that the difficulty in real communities will lie in the collection of the data. If one takes the artificial series of curves in Fig. 2.1, then with thirty-two species there are thirty-two sets of life-history traits to determine and 496 pairwise interactions to investigate. If there are interesting interactions between groups of three or more species there are a further 4294966766 interactions to consider! Population dynamic models are likely to require too much data for their practical use, even when only the pairwise interactions are considered.

Markovian models

The models discussed in the previous sections are all species-orientated, answering such questions as 'How does the species react in a particular environment?' or 'How does the species interact with other species?'. In formulating Markovian models the species are forgotten and emphasis is placed on recognizing the possible states of the system. Intuitively, this is analagous to many early studies of succession: the biotic community passed through states, referred to as seral stages, until it reached the end-point, the climax. Since Markovian models are concerned with these states, they could be said to be ecosystem-orientated, but the difficulty lies in how to recognize and characterize these states. As seen in Fig. 2.1, there is no obvious division of the sequence into distinct states: the continuum has to be arbitrarily divided into a convenient number of states.

Despite this ecosystem-orientation, states tend to have been designated in terms of single-species dominance (e.g. the applications of these models to forests: Stephens & Waggoner 1970, 1980; Horn 1975; Tucker &

Fitter 1981). Usher's (1975) models of a termite community relied on states defined by single-species occupancy of a baitwood block. However, Hobbs & Legg (1983) defined their states in heathland either as single-species dominance or as co-dominance of pairs of species. Usher (1981) used classification as a means of deriving the states objectively, though because of the restricted nature of the data (only six species) the states tended to reflect dominance by single species. The use of classification and ordination techniques will probably prove to be the best way of defining objectively the states to be used in Markovian models.

The basic assumptions of the use of Markov chain models in ecology have been reviewed by van Hulst (1979b) and Usher (1979b). Bartlett (1955) defined a Markov process as 'a stochastic process for which the values of X_r at any set of times t_r ($r = 1, 2, \ldots, n$) depend on the values X_s at any set of previous times t_s ($s = 0, -1, \ldots, -j$) only through the last available value X_0'. Since the dependence is on the single and last value, X_0, this is the definition of what is termed a first-order process. The majority of applications of Markovian models in the ecological literature assume a first-order process, for which Anderson & Goodman (1957) suggest the use of

$$-2\ln\lambda = 2 \sum_{i=1}^{k} \sum_{j=1}^{k} n_{ij} \ln(p_{ij}/p_j) \qquad (3)$$

as a test criterion. The null hypothesis, that successive steps in the chain are statistically independent (and hence that they *could* represent a first-order Markov chain), is tested against the alternative hypothesis that the steps are not independent. The criterion, $-2\ln\lambda$, is distributed asymptotically as χ^2 with $(k-1)^2$ degrees of freedom, where there are k states in the sytem and where n_{ij} and p_{ij} are derived from tally and probability transition matrices as shown in Table 2.1. The terms p_j are the marginal probabilities of the jth column, and are given by

$$p_j = \sum_{i=1}^{k} n_{ij} \Bigg/ \left(\sum_{i=1}^{k} \sum_{j=1}^{k} n_{ij} \right). \qquad (4)$$

Usher (1979b) has tested many published examples of Markovian models with equation (3), and has found that virtually all have significant values of χ^2.

Bartlett's (1955) definition is much narrower than that used by other statisticians (e.g. Bailey 1964) or by the majority of ecologists. This wider definition allows, for example, for multi-dependence in Markov chains: in ecological terms, this implies that the future state of the system not only depends on the current state of the system but also on the history

Table 2.1. An example of the collection of data for the terms n_{ij} and the estimation of the terms p_{ij} used in equation (3) and subsequently. In the example below, the three states that can be recognized in the system are designated by the letters F, G and M. Out of 2432 observed transitions, the table lists the total in each of the nine classes (F → F, F → G, F → M, etc). This table is referred to as the *tally matrix*, and its elements, which are counts, are designated n_{ij}. If each of the rows is divided by the row total, the results are estimates of the probabilities that the transitions occur. The table is referred to as the *probability transition matrix*, and the elements are designated p_{ij}. Note that, for all i, $\sum_{j=1}^{k} p_{ij} = 1$ (allowing for rounding errors). The data are derived from a series of ungrazed grassland quadrats in Breckland.

	To state			
From state	F	G	M	Total
Counts				
F	602	14	114	730
G	29	185	179	393
M	172	97	1040	1309
Total	803	296	1333	2432
Probabilities				
F	0·825	0·019	0·156	1·000
G	0·074	0·471	0·455	1·000
M	0·131	0·074	0·794	0·999

of that system. A first-order process can be considered as

$$s_0 \rightarrow s_t \tag{5}$$

where s_0 is the current state and s_t is some future state, whereas a second-order process is

$$s_h \rightarrow s_0 \rightarrow s_t \tag{6}$$

where s_h is a historical state (note that, usually, t and h take values of 1 and −1 respectively). Defining marginal probabilities, as in equation (4), as

$$p_{jl} = \sum_{i=1}^{k} n_{ijl} \bigg/ \left(\sum_{i=1}^{k} \sum_{l=1}^{k} n_{ijl} \right) \tag{7}$$

Anderson & Goodman (1957) test for the null hypothesis of a first-order process by

$$-2\ln\lambda = 2 \sum_{i=1}^{k} \sum_{j=1}^{k} \sum_{l=1}^{k} n_{ijl} \ln(p_{ijl}/p_{jl}) \tag{8}$$

which is distributed asymptotically as a χ^2 with $k(k - 1)^2$ degrees of freedom.

Another possible deviation from the narrow definition is for the chain to be non-stationary in time, i.e. that the constant transition probabilities, p_{ij}, are replaced by functions of time, $p_{ij}(t)$. With the null hypothesis that all $p_{ij}(t)$ take a constant value, p_{ij}, Anderson & Goodman's (1957) test is

$$-2\ln\lambda = 2\sum_{i=1}^{k}\sum_{j=1}^{k}\sum_{t=1}^{T} n_{ij}(t)\ln(p_{ij}(t)/p_{ij}) \qquad (9)$$

where measurements are over T time intervals, and $-2\ln\lambda$ is distributed asymptotically as a χ^2 with $k(k - 1)(T - 1)$ degrees of freedom.

AN APPLICATION TO THE GRAZED AND UNGRAZED BRECKLAND GRASSLANDS

The data

The data given by Watt (1960b) are amongst the most detailed ecological information, about a very small spatial area over a long period of time, that has ever been published. The information was collected annually from 1936 until 1957 (except for 1950). Two transects were recorded: one was in an area open to grazing, mainly by rabbits (*Oryctolagus cuniculus*), and the other was in an area that was fenced (the transects will be referred to as grazed and ungrazed respectively). Each transect consisted of 128 quadrats, each of which was subdivided into eight subquadrats, 1·25 cm square. Plant species were scored for presence (1) or absence (0) in each subquadrat, and the scores were added to produce the published scores. Only six taxa of higher plants occurred in these transects: *Agrostis* spp. (both *A. canina* L. and *A. capillaris* L.), *Aira praecox* L., *Festuca ovina* L., *Galium saxatile* L., *Luzula campestris* (L.) DC., and *Rumex tenuifolius* (Wallr.) A. Löve (generic names will be used subsequently). Although not recorded, there was abundant lichen cover in the quadrats. The data matrices for each transect, for each year, can be considered as having 128 rows (the quadrats) each consisting of six columns (the higher plant taxa), each element being an integer between 0 (the taxon absent from all eight subquadrats of a quadrat) and 8 (the taxon present in all eight subquadrats). The broad effects of rabbit grazing, and conversely of enclosure, are discussed by Watt (1960a).

Before a Markov model could be constructed, the data needed to be classified. With 21 years of observations, there were data on 2688 temporally and spatially autocorrelated quadrats in both the grazed and ungrazed grasslands. Reed (1980) used a two-stage relocation method of cluster analysis, using the CLUSTAN package, to produce the two dendrograms in Fig. 2.3. Both dendrograms start at the ten-group level, and show the subsequent fusion of groups. For the grazed grassland two horizontal lines have been drawn arbitrarily to distinguish seven and three groups. One group, F, is dominated by *Festuca*; two groups, G1 and G2, are dominated by *Galium*; and three of the remaining four groups, M1, M2 and M4, are dominated by *Agrostis*, *Luzula* and *Rumex* respectively.

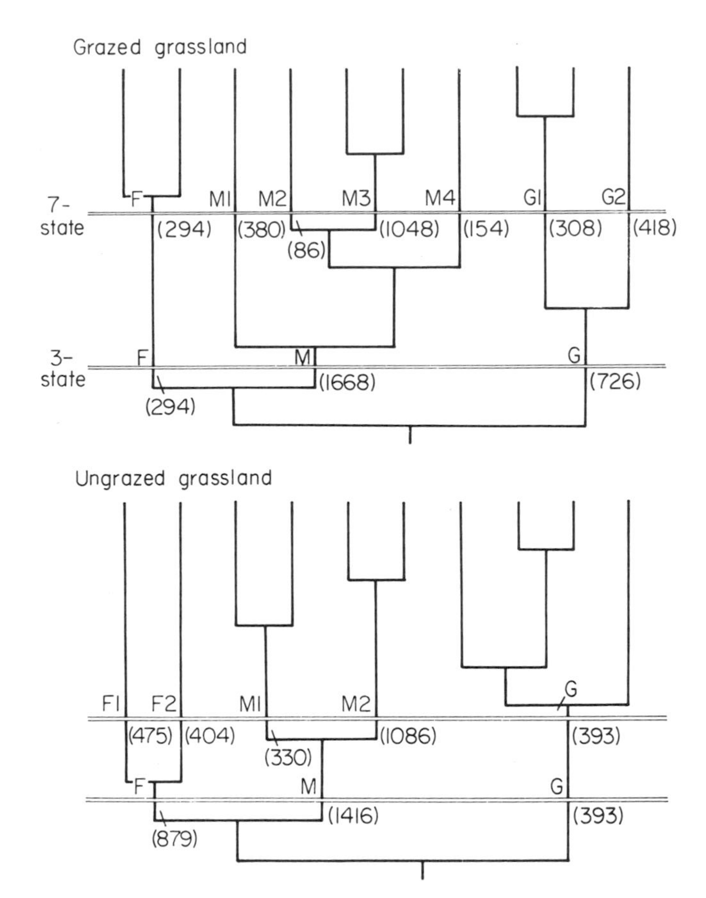

FIG 2.3. Reed's (1980) classification of the grazed and ungrazed Breckland grassland quadrats by a relocation procedure. The horizontal lines divide the grazed grassland data into three- and seven-state systems, and divide the ungrazed grassland data into three- and five-state systems. The states are designated by letters (see Table 2.2). The numbers of quadrats in each state are also shown.

The only group not dominated by higher plant taxa, and hence lichen-dominated, is M3. The abundance of the taxa in these groups is shown in Table 2.2. The Shannon-Weaver diversity indices reflect the dominance by a single taxon: they are 0·77, 1·09 and 1·39 for groups G, F and M respectively. Groups G1 and G2 have indices of 0·48 and 1·09, and groups M1 to M4 have indices of 0·91, 1·09, 1·62 and 1·29 respectively (the maximum index value is $\ln 6 = 1{\cdot}79$).

The analysis of the ungrazed system also yielded groups dominated by *Festuca*, *Galium* and *Agrostis*: the first arbitrary division in the dendrogram is at the five group level. The second division, at the three group level, yields groups F (*Festuca*), G (*Galium*) and M (without dominance) in a similar manner to the grazed grassland. The data, in Table 2.2, also show a similar diversity with the Shannon-Weaver index taking values 0·30, 0·91 and 1·36 for groups F, G and M respectively. Groups F1 and

TABLE 2.2. The results of Reed's (1980) classification of Watt's Breckland grassland data into three and seven clusters for the rabbit-grazed grassland and into three and five clusters for the ungrazed grassland. The naming of the clusters (states) is the same as in Fig. 2.3. The table lists the mean frequency of each of the six higher plant taxa in all quadrats included within a particular state

State	*Festuca*	*Agrostis*	*Aira*	*Luzula*	*Galium*	*Rumex*
Grazed						
F/(F)	3·5	0·5	0·1	0·4	0·4	0·2
M1	0·1	3·9	0·3	0·2	0·3	0·3
M2	0·1	0·5	0	2·6	0·7	0·2
M3	0·1	0·6	0·2	0·2	0·2	0·2
M4	0·2	1·1	0·3	0·3	0·3	2·8
(M)	0·1	1·4	0·2	0·3	0·2	0·4
G1	0·1	0·4	0	0·2	6·5	0·1
G2	0·2	0·7	0·1	0·3	3·1	0·2
(G)	0·1	0·6	0	0·3	4·5	0·2
Ungrazed						
F1	7·2	0·1	0	0	0·1	0
F2	3·7	0·1	0	0·1	0·3	0
(F)	6·0	0·1	0	0·1	0·2	0
M1	0·6	2·8	0	0·2	0·4	0·2
M2	0·5	0·2	0	0·2	0·1	0·1
(M)	0·6	0·8	0	0·2	0·2	0·1
(G)	0·4	0·7	0·1	0·2	4·4	0·1

F2 have low diversity with indices of 0·14 and 0·48, and the M1 and M2 groups have indices of 1·06 and 1·41 respectively.

In the discussion of Markov models in the following sections the three- and seven-state systems will be used for the grazed grassland, and the three- and five-state systems for the ungrazed grassland. An earlier classification of these data into eight states (Usher 1981) is not used here.

Simple Markovian models

The tally matrix for the three-state system of the grazed grassland system is

$$\begin{bmatrix} 179 & 11 & 24 \\ 32 & 404 & 264 \\ 61 & 198 & 1250 \end{bmatrix}$$

where the order of the states is F, G and M. Dividing by row totals yields the matrix of transition probabilities.

$$\mathbf{P}_3 = \begin{bmatrix} 0{\cdot}84 & 0{\cdot}05 & 0{\cdot}11 \\ 0{\cdot}05 & 0{\cdot}58 & 0{\cdot}38 \\ 0{\cdot}04 & 0{\cdot}13 & 0{\cdot}83 \end{bmatrix} \qquad (10)$$

and, for the seven-state system,

$$\mathbf{P}_7 = \begin{bmatrix} 0{\cdot}84 & 0{\cdot}03 & 0{\cdot}02 & 0{\cdot}01 & 0{\cdot}01 & 0{\cdot}09 & 0 \\ 0{\cdot}04 & 0{\cdot}37 & 0{\cdot}27 & 0{\cdot}07 & 0{\cdot}05 & 0{\cdot}19 & 0{\cdot}03 \\ 0{\cdot}05 & 0{\cdot}18 & 0{\cdot}35 & 0{\cdot}08 & 0{\cdot}03 & 0{\cdot}30 & 0{\cdot}01 \\ 0{\cdot}04 & 0{\cdot}01 & 0{\cdot}09 & 0{\cdot}51 & 0{\cdot}01 & 0{\cdot}27 & 0{\cdot}07 \\ 0{\cdot}06 & 0{\cdot}01 & 0{\cdot}10 & 0 & 0{\cdot}36 & 0{\cdot}42 & 0{\cdot}04 \\ 0{\cdot}04 & 0{\cdot}02 & 0{\cdot}09 & 0{\cdot}13 & 0{\cdot}02 & 0{\cdot}66 & 0{\cdot}05 \\ 0{\cdot}04 & 0{\cdot}08 & 0{\cdot}24 & 0{\cdot}09 & 0{\cdot}05 & 0{\cdot}27 & 0{\cdot}24 \end{bmatrix}, \qquad (11)$$

where the order of states is F, G1, G2, M1, M2, M3 and M4. Using (10) and (11) in a first-order Markov chain yields eigenvectors of

$$\mathbf{P}_3 = [20 \quad 20 \quad 60]$$

and

$$\mathbf{P}_7 = [20 \quad 7 \quad 13 \quad 14 \quad 3 \quad 38 \quad 5].$$

The elements of these vectors have been adjusted so that they represent the percentage of quadrats in each of the states. Tests for departure from randomness, using equation (3), give χ^2 of 1795 for the seven-state system and 1220 for the three-state system: in both cases the null hypothesis of randomness would be rejected. Mean recurrence times, as shown by Reed (1980), can be represented by the vectors

$$[4{\cdot}9 \quad 4{\cdot}8 \quad 1{\cdot}7]$$

and

$$[4{\cdot}9 \quad 14{\cdot}9 \quad 7{\cdot}7 \quad 7{\cdot}1 \quad 30{\cdot}5 \quad 2{\cdot}6 \quad 22{\cdot}0]$$

where the vector elements are numbers of years and are arranged in the same order as in equations (10) and (11).

Similar statistics for the ungrazed grassland give the tally and transition probability matrices in Table 2.1 for the three-state system, and

$$\mathbf{P}_5 = \begin{bmatrix} 0{\cdot}77 & 0{\cdot}11 & 0 & 0 & 0{\cdot}12 \\ 0{\cdot}26 & 0{\cdot}51 & 0{\cdot}04 & 0{\cdot}01 & 0{\cdot}19 \\ 0{\cdot}02 & 0{\cdot}05 & 0{\cdot}47 & 0{\cdot}20 & 0{\cdot}25 \\ 0{\cdot}02 & 0{\cdot}04 & 0{\cdot}05 & 0{\cdot}46 & 0{\cdot}43 \\ 0{\cdot}04 & 0{\cdot}12 & 0{\cdot}08 & 0{\cdot}09 & 0{\cdot}67 \end{bmatrix} \qquad (12)$$

for the five-state system where the order is F1, F2, G, M1 and M2. The dominant eigenvectors are

$$\mathbf{P}_3 = [42 \quad 8 \quad 50]$$

and

$$\mathbf{P}_5 = [27 \quad 17 \quad 8 \quad 10 \quad 38].$$

Use of equation (3) gave χ^2 values of 1470 and 2079 for the three- and five-state systems respectively. Mean recurrence times, in years, are given by the vectors

$$[2{\cdot}4 \quad 11{\cdot}8 \quad 2{\cdot}0]$$

and

$$[3{\cdot}8 \quad 6{\cdot}0 \quad 12{\cdot}2 \quad 10{\cdot}0 \quad 2{\cdot}6].$$

There are two conclusions to be drawn from the analyses of these two sets of data. First, the tests for randomness yield very large values of χ^2, and hence the null hypothesis of randomness can be rejected conclusively. The implication of these tests, together with others listed by Usher (1979b, in press), is that ecological succession is not a random process (though, in the Breckland data, this may be an artefact of the use of a 1-year time step for mostly perennial species). Second, no elements in the eigenvectors are approaching zero, or, conversely, no mean recurrence times are approaching infinity. This implies that there are both 'forward' and 'backward' movements in the grassland, and that all of the defined states will remain in the grassland. For conservation purposes, this model does give an assurance that all the types of grassland community, represented by the states in the model, will persist in perpetuity.

Diagrams showing the main features of the probability transition matrices (11) and (12) are shown in Fig. 2.4. A study of the matrices indicates that virtually all transitions are possible as there are only four zeros in the seventy-four elements in the two matrices. The diagrams in Fig. 2.4 show some differences between the grasslands that are not apparent in the eigenvectors. In the grazed grassland the *Festuca*-dominated state, F, is not linked to any other states, and there is generally an attraction towards state M3. In the ungrazed grassland the two *Festuca*-dominated states are linked to the remainder of the diagram, unlike the grazed grassland but, like the grazed grassland, state M2 is characterized by having the maximum diversity and tends to attract quadrats.

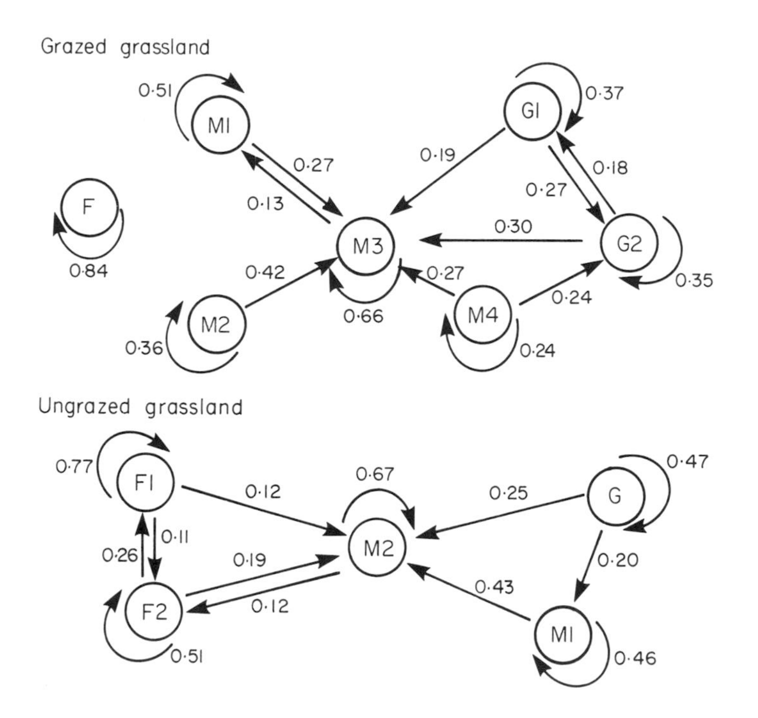

FIG. 2.4. Diagramatic representations of the transition probabilities in equations (11) and (12) for the grazed and ungrazed grasslands respectively. Only transition probabilities greater than 0·10 are included as arrows in the diagrams.

Indirect probability estimates for Markovian models

The difficulty of collecting suitable data for Markov chain models was discussed by Usher (1981). In order to estimate transition probabilities in matrices such as (10)–(12), a large number of permanently marked quadrats have to be recorded at least twice. Such data are rare in the ecological literature. An approach to using more generally available data was described by Cooke (1981), who used time series of frequency data. He illustrated his method by using data from a sheep-grazed sward in North Wales: there were no permanently marked quadrats, but there were estimates of frequencies of occurrence of each of the six states for twelve consecutive years.

Using the Breckland grassland data, Cooke's (1981) method of estimating probabilities can be compared with the true probabilities of transition. Table 2.3 shows how the frequencies in the various states vary from year to year on the grazed grassland. The method, which is similar to that used in 'ridge regression', gives the estimated matrix of transition probabilities, $\widehat{\mathbf{P}}$, by

$$(\mathbf{X}^{\mathrm{T}}\mathbf{X} + c\mathbf{I})\widehat{\mathbf{P}} = \mathbf{X}^{\mathrm{T}}\mathbf{Y} + (c/k)\mathbf{l}\mathbf{l}^{\mathrm{T}} \qquad (13)$$

where $\mathbf{X}$ and $\mathbf{Y}$ are the matrix in Table 2.3 except that the last and first rows are omitted respectively, $\mathbf{l}$ is a vector whose elements are all equal to unity, c is an unknown constant, and k is the number of states in the system. The main problem in estimating the transition probability matrix is the unknown constant c for which Cooke recommends 'Initially c is set equal to zero ... usually some of the estimates are out of range so c is increased by steps of 0·01 until the estimates of all the probabilities are within range'. The problem of estimates within range is due to the constraint that

$$0 \leqslant \widehat{p}_{ij} \leqslant 1$$

for all elements of $\widehat{\mathbf{P}}$. It frequently happens that some estimates, when c is small, are less than zero, whilst others exceed one. Cooke continues '... the estimation procedure is stopped unless any estimates are changing substantially at each iteration, in which case c is stepped on until the estimates change only slightly from one iteration to the next'. Usher (in press) has discussed the problem of when to stop incrementing c; the estimates of p_{ij} all converge on $1/k$ for very large values of c.

Using the three-state system in Table 2.3, and estimating the 1950 missing data as the mean values between the 1949 and 1951 data, gives estimates in relation to values of c shown in Fig. 2.5. All of the probabilities are

TABLE 2.3. The frequency of occurrence of the states recognized in the three- and seven-state systems of the grazed grassland, using the letters designating states shown in Fig. 2.3. Note that the rows sum to one if the F, G and M or the F, G1, G2, M1, M2, M3 and M4 entries are summed. Data were not collected in 1950, and for modelling purposes the means between 1949 and 1951 have been assumed

Year	State								
	F	G1	G2	G	M1	M2	M3	M4	M
1936	0	0·64	0·20	0·84	0	0	0·16	0	0·16
1937	0	0·48	0·27	0·75	0	0	0·25	0	0·25
1938	0	0·14	0·44	0·58	0·05	0·03	0·34	0	0·42
1939	0	0·04	0·11	0·15	0·19	0·05	0·61	0	0·85
1940	0	0	0	0	0·30	0·01	0·66	0·03	1
1941	0	0	0	0	0·42	0·01	0·57	0	1
1942	0	0	0	0	0·35	0·02	0·62	0·01	1
1943	0	0	0·02	0·02	0·30	0·02	0·42	0·24	0·98
1944	0	0·01	0·08	0·09	0·17	0·01	0·39	0·34	0·91
1945	0·01	0·12	0·27	0·39	0·14	0·02	0·27	0·17	0·60
1946	0·06	0·23	0·26	0·49	0·06	0·01	0·28	0·10	0·45
1947	0·19	0·06	0·16	0·22	0·10	0·03	0·46	0	0·59
1948	0·21	0·02	0·09	0·11	0·29	0·02	0·37	0	0·68
1949	0·26	0·01	0·15	0·16	0·23	0·05	0·30	0	0·58
1950	No data								
1951	0·17	0	0·05	0·05	0·15	0·04	0·37	0·22	0·78
1952	0·17	0·04	0·27	0·31	0·05	0·06	0·41	0	0·52
1953	0·16	0·02	0·34	0·36	0·05	0·02	0·39	0·02	0·48
1954	0·16	0·26	0·22	0·48	0·01	0·03	0·31	0·01	0·36
1955	0·19	0·35	0·22	0·57	0·02	0·02	0·20	0	0·24
1956	0·35	0	0·08	0·08	0·05	0·11	0·36	0·05	0·57
1957	0·37	0	0·05	0·05	0·04	0·08	0·46	0	0·58

within range when $c = 0{\cdot}02$, for which the matrix of estimated probabilities is

$$\widehat{\mathbf{P}}_3 = \begin{bmatrix} 0{\cdot}99 & 0 & 0{\cdot}01 \\ 0{\cdot}07 & 0{\cdot}68 & 0{\cdot}25 \\ 0 & 0{\cdot}08 & 0{\cdot}92 \end{bmatrix}. \qquad (14)$$

However, reference to Fig. 2.5 shows that p_{FF} was decreasing rapidly, and both p_{FG} and p_{FM} were increasing rapidly for values of c about 0·02. As the 'true' matrix $\mathbf{P}_3$ was known (equation 10), a matching coefficient was defined as

$$M = \sum_{i=1}^{k} \sum_{j=1}^{k} (p_{ij} - \widehat{p}_{ij})^2, \qquad (15)$$

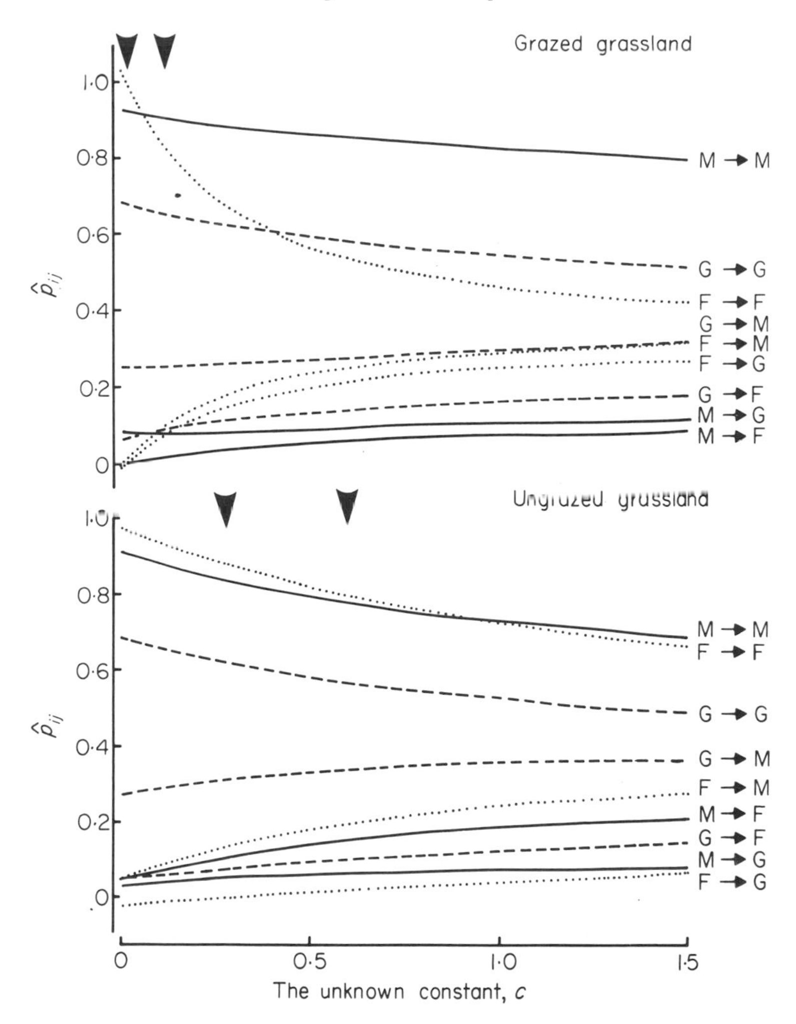

FIG. 2.5. Estimating the matrix of transition probabilities, $\widehat{\mathbf{P}}$, using Cooke's method. The lines indicate the values taken by the estimated probabilities as the value of the unknown constant, c, is increased from zero. The arrows indicate the values of c for which all estimates of $\widehat{p}_{ij}$ are within range (left arrows) and for which the criterion M in equation (15) is minimized.

and the value of c that gave min (M) was accepted as the best estimate of c. The minimum, $M = 0{\cdot}0301$, was achieved with $c = 0{\cdot}12$, when

$$\widehat{\mathbf{P}}_3 = \begin{bmatrix} 0{\cdot}82 & 0{\cdot}08 & 0{\cdot}10 \\ 0{\cdot}09 & 0{\cdot}65 & 0{\cdot}26 \\ 0{\cdot}02 & 0{\cdot}08 & 0{\cdot}90 \end{bmatrix}. \qquad (16)$$

Reference to Fig. 2.5 does not indicate that, in view of the rates of change of some of the probabilities, a value for $c = 0{\cdot}12$ would be preferable to any other value of c less than about 0·3 or 0·4.

The estimate for the seven-state grazed system (probabilities within range) is

$$\widehat{\mathbf{P}}_7 = \begin{bmatrix} 0{\cdot}28 & 0{\cdot}09 & 0{\cdot}14 & 0{\cdot}08 & 0{\cdot}12 & 0{\cdot}20 & 0{\cdot}09 \\ 0{\cdot}11 & 0{\cdot}25 & 0{\cdot}23 & 0{\cdot}06 & 0{\cdot}09 & 0{\cdot}20 & 0{\cdot}06 \\ 0{\cdot}15 & 0{\cdot}16 & 0{\cdot}22 & 0{\cdot}07 & 0{\cdot}08 & 0{\cdot}27 & 0{\cdot}05 \\ 0{\cdot}09 & 0{\cdot}06 & 0{\cdot}08 & 0{\cdot}23 & 0{\cdot}09 & 0{\cdot}28 & 0{\cdot}16 \\ 0{\cdot}16 & 0{\cdot}13 & 0{\cdot}14 & 0{\cdot}13 & 0{\cdot}13 & 0{\cdot}18 & 0{\cdot}13 \\ 0{\cdot}08 & 0{\cdot}03 & 0{\cdot}10 & 0{\cdot}23 & 0 & 0{\cdot}52 & 0{\cdot}05 \\ 0{\cdot}12 & 0{\cdot}14 & 0{\cdot}19 & 0{\cdot}10 & 0{\cdot}11 & 0{\cdot}17 & 0{\cdot}18 \end{bmatrix} \qquad (17)$$

with $c = 1{\cdot}44$. For the ungrazed grassland, the estimates are

$$\widehat{\mathbf{P}}_3 = \begin{bmatrix} 0{\cdot}87 & 0 & 0{\cdot}13 \\ 0{\cdot}07 & 0{\cdot}62 & 0{\cdot}31 \\ 0{\cdot}11 & 0{\cdot}05 & 0{\cdot}84 \end{bmatrix} \qquad (18)$$

and

$$\widehat{\mathbf{P}}_5 = \begin{bmatrix} 0{\cdot}44 & 0{\cdot}23 & 0 & 0{\cdot}02 & 0{\cdot}31 \\ 0{\cdot}29 & 0{\cdot}23 & 0{\cdot}09 & 0{\cdot}09 & 0{\cdot}31 \\ 0{\cdot}04 & 0{\cdot}07 & 0{\cdot}49 & 0{\cdot}24 & 0{\cdot}16 \\ 0{\cdot}01 & 0{\cdot}10 & 0{\cdot}07 & 0{\cdot}46 & 0{\cdot}36 \\ 0{\cdot}17 & 0{\cdot}18 & 0{\cdot}07 & 0{\cdot}04 & 0{\cdot}54 \end{bmatrix} \qquad (19)$$

with values of c of 0·28 and 0·85 respectively.

Two general conclusions can be drawn from this comparison of estimated and 'true' transition probability matrices. First, visual comparison of the appropriate pairs of matrices indicates rather large discrepancies in the estimates of a few of the transition probabilities. Using equation (16) rather than (14) for comparison with (10) shows that probability estimates relating to entry into the first state, or exit from the first state, are good, whilst estimates elsewhere in the matrix are poor. A generally similar pattern can be seen when Table 2.1 and equation (18) are compared. Second, using the criterion M in equation (15), the closest approach of the estimated matrix $\widehat{\mathbf{P}}$ to the 'true' matrix $\mathbf{P}$ is when c takes a value substantially above the value for which all probabilities are within range (in Fig. 2.5, c values of 0·12 and 0·02 for the grazed grassland and 0·60 and 0·28 for the ungrazed grassland). The problems of accepting the estimate, $\widehat{\mathbf{P}}$, when all of the probabilities are within range can be seen when the eigenvectors of equations (10), (14) and (16) are compared. They are

$$[20 \quad 20 \quad 60],$$

$$[58{\cdot}33 \quad 8{\cdot}33 \quad 33{\cdot}33],$$

and

$$[16 \quad 19 \quad 65]$$

respectively. A totally erroneous prediction of the stable structure of the grazed grassland would have been gained by use of equation (14), whereas all predictions are within 5% of the 'true' prediction by use of equation (16), as indeed they would be with the use of c between 0·04 and 0·14.

More elaborate models

All of the models so far discussed are first order Markov chains, and one can ask how appropriate such models really are. The changes of the probabilities p_{ij} during time (see Fig. 2.6) suggest that ecological processes may not be stationary in time, and there is evidence in Watt's research to suggest that the processes may not be stationary in space either. Using the annual year-to-year transition probability matrices, Reed (1980) obtained χ^2 values of 752 and 692 (both with 108 degrees of freedom) using equation (9) for the three-state systems of grazed and ungrazed grassland respectively. With the seven-state system for the grazed grassland, and the five-state system for the ungrazed grassland, the values of χ^2 were also significant ($P \ll 0{\cdot}001$).

Watt's data can also be used to investigate the order of a successional process. Assuming a second-order process with double-dependence and a single time step (i.e. based on data collected in three consecutive years), one can obtain k transition matrices, as shown for the three-state, grazed grassland in Table 2.4. Certain differences become apparent when the three matrices are compared: the greatest differences are in p_{GF} and p_{MF}. When, 1 year earlier, the vegetation of a quadrat was in state G or M, these probabilities are small (0·07 or less). However, when the system had been in state F, the two probabilities are larger, indicating the much increased probability of returning to that state. Using the test in equation (8) gave $\chi^2 = 84$ (12 d.f., $P \ll 0{\cdot}001$), indicating that the null hypothesis that the three matrices are identical could be rejected. This second-order Markov chain gives a stable vector

$$[28 \quad 16 \quad 56]$$

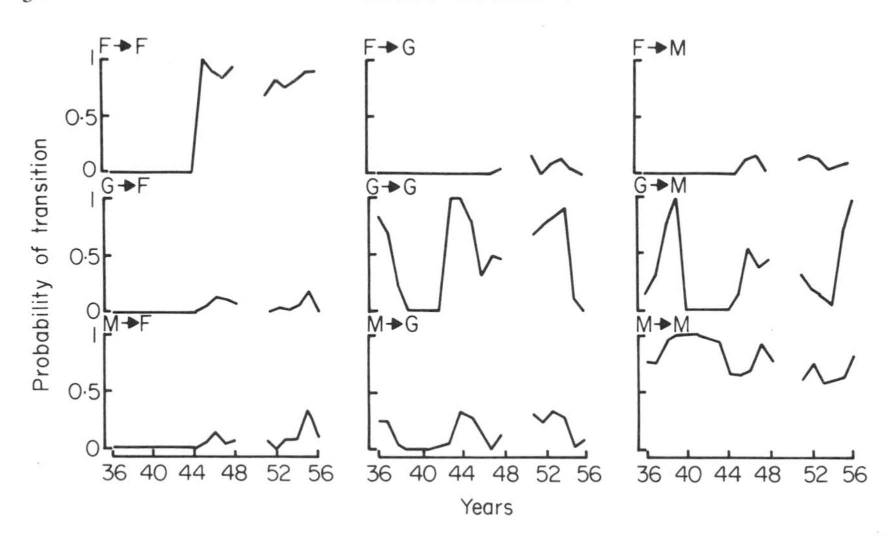

FIG. 2.6. The transition probabilities of the three-state system, for the grazed Breckland grassland, plotted on a year-to-year basis. Thus, the first probability plotted is for transitions observed between 1936 and 1937, and the last plotted is for 1956 to 1957. The records are broken since no observation was made in 1950 (hence 1949–50 and 1950–51 values are omitted). The three states are those shown in Table 2.2 and Fig. 2.3. Undefined probabilities are given the value zero.

in which the *Festuca*-dominated state is much more frequent (28% of quadrats) than in the first-order Markov chain (20% of quadrats).

For the grazed system, using a similar second-order model gave $\chi^2 = 48$ (12 d.f., $P < 0{\cdot}001$) and a stable vector of

$$[47 \quad 6 \quad 47]$$

which again predicts an increase in the *Festuca*-dominated state (from 42 to 47% of the quadrats).

Investigation of the two Breckland grasslands, together with other studies reviewed by Usher (1979b), suggests the following three general rules for ecological communities.

(a) Successional processes are not random (χ^2 in equation (3) is significant). A Markov chain model may, therefore, be a reasonable model of the process.

(b) Successional processes are not first-order (χ^2 in equation (8) is significant). Complicated Markov chain models will be needed for realism.

(c) Successional processes are not stationary in time (χ^2 in equation (9) is significant). There are insufficient studies to know whether successional processes are also non-stationary in space.

TABLE 2.4. The three matrices of transition probabilities used in a second-order Markov chain model of the grazed grassland community. For this model, transitions are recorded between three, equally-spaced (1 year) observations. States are named as in Table 2.2

	1st time								
	F			G			M		
2nd	3rd time								
time	F	G	M	F	G	M	F	G	M
F	0·88	0·04	0·08	0·90	0·03	0·06	0·77	0·02	0·21
G	0·20	0·70	0·10	0·05	0·48	0·47	0·07	0·61	0·31
M	0·39	0·06	0·56	0·03	0·13	0·84	0·03	0·11	0·85

DISCUSSION

This discussion of the modelling of ecological succession focuses attention on four main areas of the modelling process: the experimental collection of data, the kinds of data, the adequacy of the models and, finally, on whether models have contributed anything to an understanding of succession.

The data that have been collected are virtually all based on sampling rather than actual experimental manipulation. It is, therefore, important to question the basis of that sampling to determine whether the data are as good as they appear to be. Two particular problems can be seen in much ecological data: autocorrelation due to repeated sampling in time of the same quadrat (and this is true of the Breckland grassland data) and lack of true replication. Ecologists have tended, particularly in areas where fire is not a common occurrence, to avoid perturbation experiments. These, by and large, can be classified into pulse and press types (Bender, Case & Gilpin 1984). A pulse, removal experiment, which is described by Hils & Vankat (1982), shows many of the problems faced by ecologists using perturbation experiments. The experiment was carried out to test Connell & Slatyer's (1977) models of succession on an old-field system. One difficulty was to know exactly what test criterion could be used to discriminate between the models or, conversely, what null hypothesis to test. The results reflected this since it was concluded that 'more than one model of succession may apply in the same field at the same time . . .'. There is clearly a much greater need for thinking about the design of perturbation experiments, and the appropriate null hypotheses to test, if this potentially powerful experimental technique is to feature in the description, quantification and modelling of succession.

Perhaps one can echo Bender *et al's* (1984) sentiment 'the problems with such studies are formidable'.

The majority of data that have been collected relate to the plants. There are a few studies on soil microbes; these are reviewed by Cromack (1981). There is increasing evidence that animals have an influence on succession (Edwards & Gillman, Chapter 14), though this seems to be related more to plant architecture than to the plant species (MacMahon 1981). During decomposition, animal associations can both be recognized and modelled by Markov chains (Usher & Parr 1977), though competition between pairs of species (Sparkes 1982) is likely to be an important mechanism for changes in the soil animal abundances. Although the plants provide the largest amount of biomass and the greatest energy flow, the animals should not be forgotten as they can modify the processes considerably. Successional studies should integrate the data on plant species change with changes in the grazing and predatory animals as well as with changes in the decomposers.

The adequacy of the model is a feature that has often not been explored. Sensitivity analyses have rarely been performed, usually due to the massive amounts of computer time that would be required, and validation has also often been omitted because of the expense (in time and resources) of collecting an independent set of data. Van Hulst (1978, 1980) has explored some aspects of model adequacy, saying that reasonable predictions are not sufficient evidence of adequacy. Perhaps it was rather tongue-in-cheek for him to suggest that 'students of vegetational succession should either leave the narrow domain of vegetation dynamics in order to study ecosystem dynamics, or explore different succession models'.

Finally, have models contributed to an understanding of succession? On a practical level this question can be answered by taking three examples. Usher (1975) wished to develop a termite test site, and wanted a degree of certainty that, with changing conditions, a wood-feeding termite community would persist. The predictions of a Markov chain model give the assurance required and proved to be surprisingly accurate. Models can help in choosing between two or three possible mechanisms of succession. Usher (1981) has shown how the patterns of probabilities in a Markovian matrix may be related to Connell & Slatyer's three models of succession. Models can also help in answering the kind of questions that ecologists ask: McIntosh (1981) ends his review of succession and ecological theory by posing a set of seven questions. Clearly the behaviour of models, and the results of modelling, could be used to provide answers to some of those questions.

The greatest role of models is, however, in clarifying the concepts which are included in those models. The model provides a discipline within which conceptual advances can be made. The investigations of Markovian models, discussed in relation to the Breckland grasslands, have led to the triple suggestion that succession is not random, that history is important, and that non-stationarity in time (and probably also in space) is the norm. The feedbacks between model, generality (or theoretical development) and experiment (or sampling) can provide the impetus for further research and understanding.

ACKNOWLEDGMENTS

I should like to thank John Reed for permission to quote from his M.Sc. thesis, and Professor Mark Williamson for commenting on a draft manuscript.

REFERENCES

Anderson, T.W. & Goodman, L.A. (1957). Statistical inference about Markov chains. *Annals of Mathematical Statistics*, **28**, 89–110.

Austin, M.P. (1977). Use of ordination and other multivariate descriptive methods to study succession. *Vegetatio*, **35**, 165–175.

Bailey, N.T.J. (1964). *The Elements of Stochastic Processes, with Applications to the Natural Sciences*. Wiley, New York.

Bartlett, M.S. (1955). *An Introduction to Stochastic Processes, with Special Reference to Methods and Applications*. Cambridge University Press, Cambridge.

Bender, E.A., Case, T.J. & Gilpin, M.E. (1984). Perturbation experiments in community ecology: theory and practice. *Ecology*, **65**, 1–13.

Bornkamm, R. (1981). Rates of change in vegetation during secondary succession. *Vegetatio*, **47**, 213–220.

Botkin, D.B., Janak, J.F. & Willis, J.R. (1972). Some ecological consequences of a computer model of forest growth. *Journal of Ecology*, **60**, 849–872.

Clements, F.E. (1916). *Plant Succession: an Analysis of the Development of Vegetation*. Carnegie Institute, Washington, Publication No. **242**.

Connell, J.H. & Slatyer, R.O. (1977). Mechanisms of succession in natural communities and their role in community stability and organisation. *American Naturalist*, **111**, 1119–1144.

Cooke, D. (1981). A Markov chain model of plant succession. *The Mathematical Theory of the Dynamics of Biological Populations, II* (Ed. by R.W. Hiorns & D. Cooke), pp. 231–247. Academic Press, London.

Cromack, K. (1981). Below-ground processes in forest succession. *Forest Succession: Concepts and Application* (Ed. by D.C. West, H.H. Shugart & D.B. Botkin), pp. 361–373. Springer-Verlag, Berlin.

Gleason, H.A. (1926). The individualistic concept of the plant association. *Torrey Botanical Club Bulletin*, **53**, 7–26.

Greenslade, P.J.M. & Greenslade, P. (1983). Ecology of soil invertebrates. *Soils: an Australian Viewpoint* (Ed. by Division of Soils, CSIRO), pp. 645–669. CSIRO, Melbourne & Academic Press, London.

Grime, J.P. (1979). *Plant Strategies and Vegetation Processes*. Wiley, Chichester.

Hils, M.H. & Vankat, J.L. (1982). Species removals from a first-year old-field plant community. *Ecology*, **63**, 705–711.

Hobbs, R.J. & Legg, C.J. (1983). Markov models and initial floristic composition in heathland vegetation dynamics. *Vegetatio*, **56**, 31–43.

Horn, H.S. (1975). Forest succession. *Scientific American*, **232**, 90–98.

Hulst, R. van (1978). On the dynamics of vegetation: patterns of environmental and vegetational change. *Vegetatio*, **38**, 65–75.

Hulst, R. van (1979a). On the dynamics of vegetation: succession in model communities. *Vegetatio*, **39**, 85–96.

Hulst, R. van (1979b). On the dynamics of vegetation: Markov chains as models of succession. *Vegetatio*, **40**, 3–14.

Hulst, R. van (1980). Vegetation dynamics or ecosystem dynamics: dynamic sufficiency in succession theory. *Vegetatio*, **43**, 147–151.

Maarel, E. van der (1979). Experimental succession research in a coastal dune grassland, a preliminary report. *Vegetatio*, **38**, 21–28.

MacMahon, J.A. (1981). Successional processes: comparisons amongst biomes with special reference to probable roles of and influences on animals. *Forest Succession: Concepts and Application* (Ed. by D.C. West, H.H. Shugart & D.B. Botkin), pp. 277–304. Springer-Verlag, Berlin.

McIntosh, R.P. (1980). The background and some current problems of theoretical ecology. *Synthese*, **43**, 195–255.

McIntosh, R.P. (1981). Succession and ecological theory. *Forest Succession: Concepts and Application* (Ed. by D.C. West, H.H. Shugart & D.B. Botkin), pp. 10–23. Springer–Verlag, Berlin.

Noble, I.R. & Slatyer, R.O. (1978). The effect of disturbance on plant succession. *Proceedings of the Ecological Society of Australia*, **10**, 135–145.

Numata, M. (1979). Facts, causal analyses and theoretical considerations on plant succession. *Vegetation und Landschaft Japans (Bulletin of the Yokohama Phytosociological Society*), **16**, 71–91.

Odum, E.P. (1969). The strategy of ecosystem development. *Science*, **164**, 262–270.

Orloci, L. (1981). Probing time series vegetation data for evidence of succession. *Vegetatio*, **46**, 31–35.

Pickett, S.T.A. (1982). Population patterns through twenty years of oldfield succession. *Vegetatio*, **49**, 45–59.

Reed, J. (1980). *Markov models of succession, with reference to data on the East Anglian Brecklands*. M.Sc. thesis, University of York.

Shugart, H.H. Crow, T.R. & Hett, J.M. (1973). Forest succession models: a rationale and methodology for modelling forest succession over large regions. *Forest Science*, **19**, 203–212.

Shugart, H.H., West, D.C. & Emanuel, W.R. (1981). Patterns and dynamics of forests: an application of simulation models. *Forest Succession: Concepts and Applications* (Ed. by D.C. West, H.H. Shugart & D.B. Botkin), pp. 74–94. Springer–Verlag, Berlin.

Sparkes, K.E. (1982). *Studies on ecological succession with particular reference to soil microarthropods*. D.Phil. thesis, University of York.

Stephens, G.R. & Waggoner, P.E. (1970). The forests anticipated from 40 years of natural transitions in mixed hardwoods. *Bulletin of the Connecticut Agricultural Experiment Station, New Haven*, No. 707.

Stephens, G.R. & Waggoner, P.E. (1980). A half century of natural transitions in mixed hardwood forests. *Bulletin of the Connecticut Agricultural Experiment Station, New Haven*, No. 783.

Sterling, A., Pero, B., Casado, M.A., Galiano, E.F. & Pineda, F.D. (1984). Influence of microtopography on floristic variation in the ecological succession in grassland. *Oikos*, **42**, 334–342.

Tucker, J.J. & Fitter, A.H. (1981). Ecological studies at Askham Bog nature reserve. 2. The tree population of Far Wood. *The Naturalist*, **106**, 3–14.

Usher, M.B. (1967). *Aberlady Bay Local Nature Reserve: Description and Management Plan*. East Lothian County Council, County Planning Department, Haddington.

Usher, M.B. (1975). Studies on a wood-feeding termite community in Ghana, West Africa. *Biotropica*, **7**, 217–233.

Usher, M.B. (1979a). Natural communities of plants and animals in disused quarries. *Journal of Environmental Management*, **8**, 223–236.

Usher, M.B. (1979b). Markovian approaches to ecological succession. *Journal of Animal Ecology*, **48**, 413–426.

Usher, M.B. (1981). Modelling ecological succession, with particular reference to Markovian models. *Vegetatio*, **46**, 11–18.

Usher, M.B. (in press). Statistical models of succession. *Succession* (Ed. by D.C. Glen-Lewin). Chapman & Hall, London & New York.

Usher, M.B. & Parr, T.W. (1977). Are there successional changes in arthropod decomposer communities? *Journal of Environmental Management*, **5**, 151–160.

Valk, A.G. van der (1981). Succession in wetlands: a Gleasonian approach. *Ecology*, **62**, 688–696.

Watt, A.S. (1960a). The effect of excluding rabbits from acidiphilous grassland in Breckland. *Journal of Ecology*, **48**, 601–604.

Watt, A.S. (1960b). Population changes in acidiphilous grass-heath in Breckland, 1936–1957. *Journal of Ecology*, **48**, 605–629.

3. CONTRIBUTIONS OF PLANT POPULATION DYNAMICS TO UNDERSTANDING EARLY SUCCESSION

A.M. MORTIMER
Department of Botany, University of Liverpool, P.O. Box 147, Liverpool, L69 3BX

INTRODUCTION

Succession in plant communities has been the subject of considerable ecological debate for much of this century. Miles' (1979, Chapter 1) analysis of this, and of the varied perceptions of succession by ecologists, leads to two inescapable conclusions. The first is simply this. The desire to be able to explain (and ultimately predict) vegetation development led almost at its outset to the hoisting of sets of 'rules' into opposing theories of plant community development. Such as they are, these sets have been summarized as either deterministic (Clements 1916) or stochastic (Gleason 1926). Yet in both, the mechanism(s) by which change arose—the actual 'assembly rules' for a given community—were primarily illusions or, at best, inferences resulting from comparative autecological assessment with scant reference to demographic events or measurement. The second, and not novel, conclusion is that pervading all discussion of succession is the necessity for consideration of events relative to a defined point on a frame of reference whose axes are space and time (Horn 1976).

Succession as defined by many authors (e.g. Miles, Chapter 1; Horn 1974) involves species addition and extinction in the changing composition of the vegetation of a landscape. Whatever the ultimate *direction of change*, the initiation of the process is the colonization of bare ground. Moreover, the absence of vegetation usually indicates the past and often relatively recent occurrence of a disturbance causing plant mortality, the likelihood of recurrence being either periodic or irregular.

Table 3.1. lists the elements determining the rate and pattern of re-establishment of vegetation on a patch following a disturbance as envisaged by Sousa (1984). Disturbance here is defined as 'a discrete, punctuated killing, displacement or damaging of one or more individuals that directly creates an opportunity for new individuals to be established'. Of considerable significance to subsequent vegetation change is the size of the bare patch and the species in last occupancy. This may determine not only the most likely sequence of arrival of species—the last occupants

Table 3.1. Elements determining the rate and pattern of re-establishment of populations following a disturbance. (After Sousa 1984.)

1 Morphological and reproductive traits of species present when disturbance occurs.
2 Reproductive biology of species not present but within dispersal vicinity.
3 Characteristics of the disturbed habitat:
  (a) intensity and severity of the disturbance itself;
  (b) size and shape;
  (c) location and degree of isolation from sources of colonists;
  (d) heterogeneity of its internal environment;
  (e) the time of creation.

may persist at the patch periphery, or in a propagule bank beneath or near the patch—but also the size of the immigrant population. Traditionally, primary successional species are defined as those that appear on 'primary', virgin or new surfaces that are relatively large in area. Yet, as Miles (1979) forcibly argues, the adjective primary is strictly relative to the existing plant community within the dispersal vicinity. Cooper's (1923, 1939) and Lawrence's (1958) observations on the sequence of vegetation development on glacial moraine are examples that attest to this.

Many of the features in Table 3.1 may be incorporated onto the nodes of the time–space continuum envisaged in Southwood's (1977) habitat templet. Whilst this offers a classification from which generalizations on population behaviour and succession may emerge, any attempt must recognize the limitations of existing schemes of life-history analysis (e.g. Stearns 1977; Begon 1984) and of successions themselves (Drury & Nesbit 1973). Equally, simple classification of 'primary successional species' on the basis of presumed adaptive traits (for dispersal, see, for example, Herrera 1985) or simple habitat characteristics risk ecological naivity. Rather it is life-history processes that should bear cogent examination. Colonization in the face of disturbance becomes one focus of demographic interest. A second is the mechanism(s) by which species persist (or become locally extinct) in the face of species immigration, increasing diversity and changing relative abundance, i.e. processes of 'secondary' succession. Studying species population behaviour under a wide range of disturbance regimes and elucidating the factors that regulate population abundance may then lead to causal explanations of species persistence and/or succession. However, measurement must be on temporal and spatial frames of reference appropriate to individual plant species, to generation times and to population flux.

The temporal characteristics of disturbance regimes, as the progenitors of successional events, are of considerable significance. In particular, the predictability and frequency of such disruptions are important forces moulding species' life-histories. Although the response of a species to 'predictable' disturbance regimes over evolutionary time and the exact processes of selection are debatable (Levin 1976; Harper 1977), distinct life-cycles and concomitant life-forms are clearly recognizable. However, it is not my intention to review life-history strategies but to follow an empirical reductionist view of the way in which plant populations respond under regimes of disturbance. The focus of this paper is the dynamics of population renewal after disturbance and its importance in understanding early successional events. As a baseline, my definition of succession remains broad: the pattern of changes in population abundance after disturbance and the opening of a new patch in the physical environment for plant colonization. It encompasses succession in the sense of species renewal in response to secular changes (*sensu* Horn 1974) in the environment and does not presume directionality. A working tenet is that the explanation of population abundance under disturbance regimes can only arise through comprehension of species dynamics and the factors that regulate population size.

GROWTH FORM AND POPULATION RENEWAL

Whilst possibly waiting 'to be counted by ecologists', a more probable consequence of being sessile is that terrestrial plants await disturbance. Not only is the time at which this occurs during the life-cycle important, but also the plant part or module on which it falls.

Population renewal by 'birth' of new individuals in modular organisms may arise either through clonal vegetative growth and disarticulation of ramets (a process which White (1984) has argued is asexual reproduction) or through repeated recruitment from new zygotes (seed). Many species display both, reflecting evolutionary compromises to past disturbance regimes (Armstrong 1982). A convenient comparison, however, is between those species in which adult plant abundance is by recruitment from seed (e.g. annual plants) and those in which it is from ramets alone (those clonal organisms whose fragmented parts retain totipotency).

Besides the evolutionary consequences (Schaal & Leverich 1981), the possession of a two stage life-cycle (seed and adult) has obvious direct demographic consequences. The dormant seed state presents a refuge against extreme climatic conditions. Predictive and consequential dormancy enables synchronization of the passive seed state with secular phases in a regime of recurring disturbance where severe seedling mortality might otherwise ensue. Characteristically, many populations experience a constant death risk during the seed phase. Conversely, in the adult phase the time at which disturbance to a population occurs may have substantial effect on seed production and population abundance of subsequent adult generations.

The punctuated dimorphism typified by annual plants contrasts with the form of renewal in clonal organisms in which disarticulation of modules is the principal means of recruitment to populations. In many species, e.g. grasses, physical disjunction of ramets may arise by a variety of biotic and abiotic means. Disturbance regimes may cause the death of an individual in the sense that fragmentation of an intact individual may be observed but, in turn, this act precipitates the 'birth' of new ramets. The size of the ramet at disjunction will depend on the precise structure of the clonal organism—the position of shoot complexes and 'spacer' internodes (Bell & Tomlinson 1980)—interacting with the actual means by which fragmentation of the genet occurs under disturbance.

Both means of reproduction afford the potential for large numbers of 'offspring' to be produced and both occur among colonists of bare patches. However, although pulsed release of large numbers of seeds is common, rarely do clonal forms display programmed decay causing simultaneous fragmentation of many ramets. Rather our perception must be of an individual with a modular growth form fluctuating in size as ramet disarticulation occurs under disturbance. We may even envisage the equivalent of a fecundity schedule in terms of size-specific disarticulation in a clonal organism under a given disturbance regime and a realized regeneration niche being a function of the size and number of extant ramets.

EXAMPLES OF RENEWAL FROM THE FIELD

Population renewal from seed

Annual plant species provide unambiguous examples of populations which persist by regular recruitment of adults from seed. In habitats characterized by highly predictable regimes of disturbance with a seasonal period of adverse physical conditions for adult plant growth, populations may renew themselves with precision.

Two striking examples of this come from the meticulous studies of Symonides (1979, 1983a, b) on *Androsace septentrionalis* and *Erophila verna* in sand dunes in Poland. Both species germinate when soil temperatures rise (usually in March) and complete their life-cycle in the following

8–10 weeks, successive generations arising from seed produced in the previous season. Over eight generations, populations of *A. septentrionalis* counted as seed immediately after seed dispersal were remarkably constant, despite the fact that the risk of mortality during the time from seed dispersal to seed germination (May–March) increased over the period of study (Fig. 3.1). *k*- factor analysis by Silvertown (1982) suggested that the only component of population regulation which was density-dependent was during seedling establishment. In this example, the primary recurring disturbance regime comprises the cumulative environmental influences of the sand dune environment, which result in seed loss in the dormant state and which present a limited number of microsites for germination and establishment.

Erophila verna illustrates with particular clarity the role of density-dependent regulation and the influence of the environment on the maintenance of population abundance (Fig. 3.2). The pattern of adult survivorship is strongly dependent on seedling density, l_x curves progressively changing from Deevey Type I to Type III (Fig. 3.2a). This, together with an overall plastic response in individual plant fecundity (Fig. 3.2d) leads to spring seedling abundance varying by no more than 8% of the mean over 5 years (1968–72). During these years, variation in rainfall frequency (between March and May) was similarly small. By contrast, in 1973 germination occurred earlier and the habitat was wetter than in past seasons. In 1974 conditions were relatively dry. This resource (water) fluctuation was directly reflected in altered seedling population sizes, survivorship to flowering and mean plant fecundity (Fig. 3.2c). Yet within these fluctuations the influence of intraspecific regulation remains

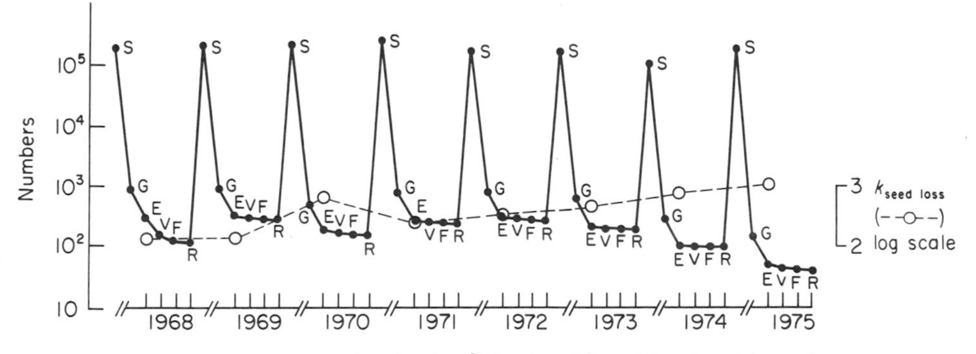

FIG. 3.1. Fluctuations in population size (m^{-2}) (——) and *k* seed loss (– – –) in *Androsace septentrionalis*. S, seed; G, seedling; E, established plant; V, vegetative mature plants; F, flowering plants; R, fruiting plants. Breaks in horizontal axis represent period from May to March (10 months). Data from Symonides (1979), after Begon & Mortimer (1986).

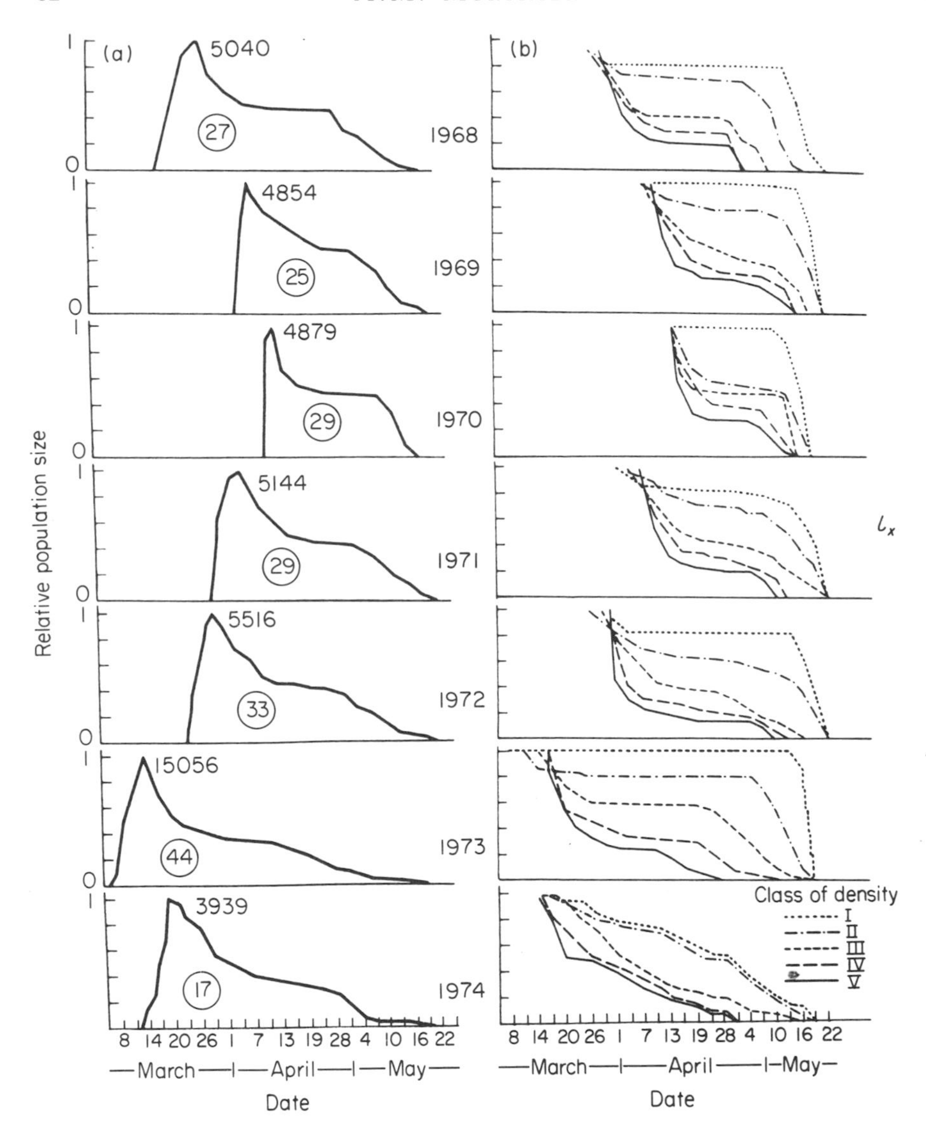

abundantly clear. It is notable also that during drought (1974) the pattern of survivorship was markedly altered in all but the highest density classes.

Fig. 3.2d suggests that on average the ecological neighbourhood area (Antonovics & Levin 1980) of *E. verna* was approximately 100 cm^2. *At this scale* there was considerable heterogeneity in population abundance. It was noted (see Symonides 1983b, Fig. 2) that it was rare for seedling

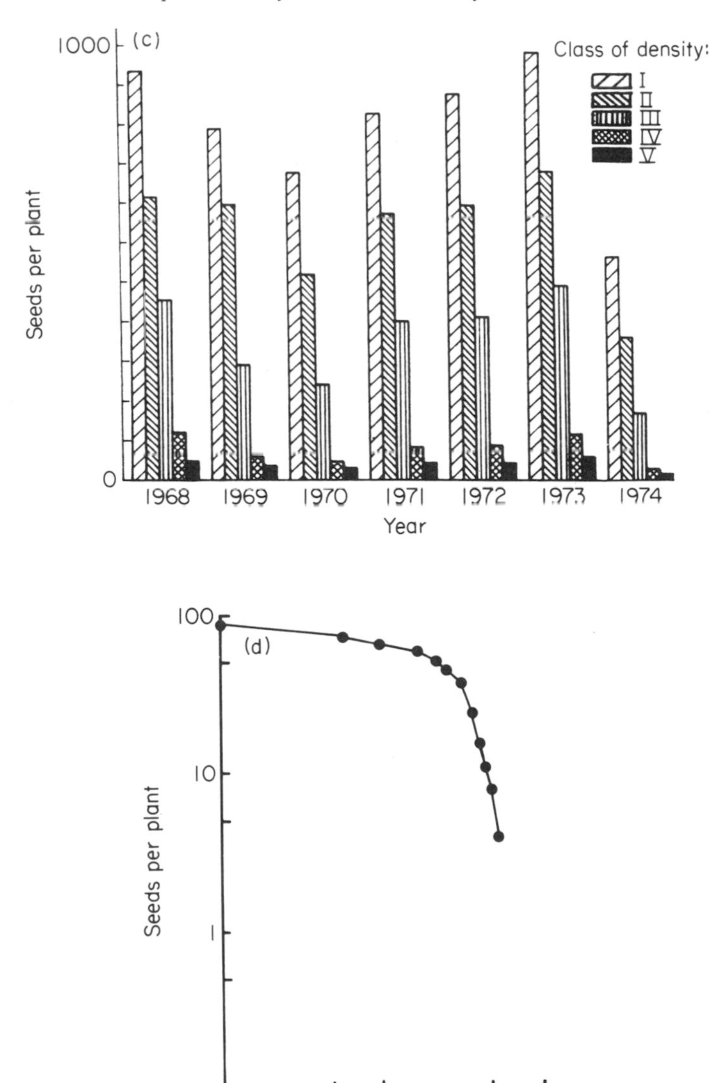

FIG. 3.2. Population responses in *Erophila verna*. (a) *opposite* Relative population sizes over seven consecutive years. Peak sizes given are seedlings m $^{-2}$; data within circles are numbers of days of rainfall in the growing season. (b) *opposite* Survivorship of seedlings at different densities. Numbers (0.01 m $^{-2}$) in each class are I, 1–2; II, 5–10; III, 15–30; IV, 35–50; and V, >55. (c) Seed production per plant according to density class over the study period. (d) Mean plant fecundity averaged over the study period in relation to density. (From Symonides 1983b).

population sizes to remain consistently high over consecutive seasons but rather to oscillate across the density range. The most likely explanation of this is in the manner of seed dispersal and movement on the surface, but the exact mechanism remains unclear.

A quite different picture of population renewal under disturbance is given by Mack & Pyke's (1983) study of *Bromus tectorum*. This relatively recent introduction to the steppe flora of N. America (Mack 1981) expanded its range considerably in the early part of this century, being preadapted (*sensu* Grant 1977) to existing agroecosystems. Extreme variation in disturbance regimes (climate and predator activity) produced considerable amplitude in population attributes. From a comparative study in the same site this species was found to behave simultaneously as an ephemeral monocarpic, annual monocarpic, or winter annual monocarpic species, this ability to display multiple fecundity schedules (*sensu* Harper 1977) precluding its extinction.

Population renewal from ramets

A fundamental problem confronting all investigations of clonal plants in the field is exact definition of the 'individual'. Harper (1985); states 'it is not clear whether (such) regulation (as to avoid self-thinning amongst *plant parts*) can occur among only the connected parts of a clone and it remains an extremely interesting question how far such regulation is lost when a clone fragments through decay or damage'. A tenable hypothesis is that populations of fragmented clones may display conventional demographic processes of regulation. Certainly in clonal plants such as *Ranunculus repens* where recruitment of new individuals is pulsed, there follows a period of density-dependent mortality (Lovett-Doust 1981).

On the other hand in grasses which are regularly disturbed by herbivores, disjunction may occur more or less continuously, detached plant modules (tillers) being frequently established in the population. A comparative study of the demography of *Holcus lanatus* (Weir 1985) in three permanent pastures receiving different disturbance regimes by grazing illustrates this (Fig. 3.3). The life expectancies of detached shoot complexes (individuals) was rarely greater than 3 years and was most commonly around 15 months; throughout this flux equilibrium populations were maintained.

Whilst some evidence of density-dependent regulation was obtained in both birth and death rates, most noticeable was the localized environmental heterogeneity and the differing response of clones. A reciprocal transplant experiment by Billington (1985) showed that the significant sources

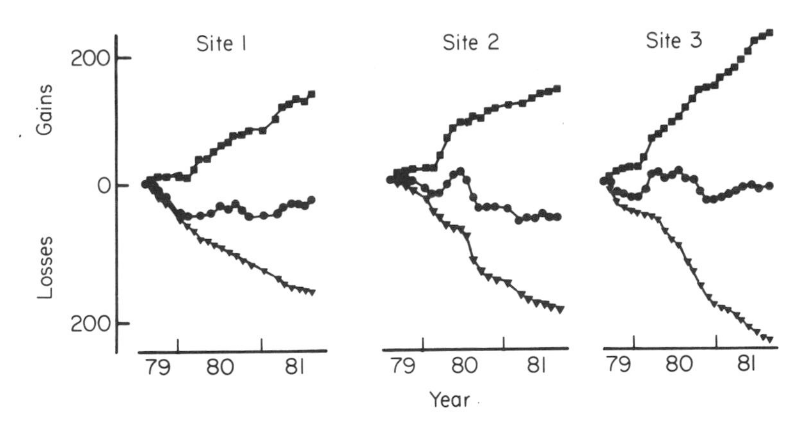

FIG. 3.3. Flux in populations of shoot complexes of *Holcus lanatus* under different disturbance regimes of pasture management: ■ gains; ▼ losses; ● net size. A shoot complex is defined as a morphologically discrete individual comprising one or more tillers. Site 1: cut twice a year, grazed by sheep and cattle between cuts, inorganic fertilizer applied. Site 2: cut once a year in late summer, grazed by cattle in autumn only. Site 3: intensive cattle grazing, high rate of inorganic fertilizer application. Mean population sizes (shoot complexes m^{-2}) Site 1, 309; Site 2, 194; Site 3, 277. Data from Weir (1985).

of variation in the size of surviving transplants in these pastures were genotype and size of recruited shoot complex, area within pasture (56 m^2) and quadrat (690 cm^2) within areas. These factors interacted in a complex and time-dependent way (Table 3.2). Moreover, the survivorship patterns of introduced ramet populations illustrate the importance of genotype–environment interactions in determining clonal fitness (Fig. 3.4).

DEMOGRAPHIC CHARACTERIZATION OF DISTURBANCE REGIMES

The examples just presented provide a backdrop against which the objectives of demographic studies of successional events may be defined. They provide examples of near equilibrium populations in which flux of individuals of differing growth forms is prevalent. It is equally clear that intrinsic population regulation and spatial heterogeneity within the habitat play a major role in determining abundance. As Watkinson (1985) re-asserts, such regulation must involve some component of density-dependent regulation.

Disturbance regimes dramatically alter population densities and alter the balance of resources available for growth. The size of the immigrant or residual population from which renewal occurs may be determined by

TABLE 3.2. The significant sources of variation in the size of surviving shoot complexes of *Holcus lanatus* reciprocally transplanted amongst two grassland sites (Sites 1 and 2, Fig. 3.3) as shown by analysis of variance. Data represent significant ($P \leqslant 0{\cdot}05$) variance components (excluding residual) according to origin of population (Site 1 or 2); size of recruit (1 or 4 tillers per shoot complex); area (block) of transplantation 7 × 8 m; and quadrats (690 cm^2) within area. From Billington (1985)

	1983 Jun	Jul	Aug	Sept	Oct	1984 Feb	Apr	May	Jun	Jul	Aug	Oct	1985 Mar
Site 1													
Population (*P*)	—	—	—	—	—	—	—	—	—	—	—	—	—
Size of recruit (*S*)	7·2	3·7	4·6	3·4	—	—	—	—	—	0·42	—	0·89	0·56
Block (*B*)	—	—	—	—	—	—	—	—	—	—	—	2·7	—
Clone in *P* (*C*)	—	—	3·0	—	—	—	—	—	—	—	11	3·5	—
Quadrat in *B* (*Q*)	0·82	1·4	—	—	—	—	—	—	—	—	—	—	—
P × *S*	—	—	—	—	—	—	—	—	—	—	—	—	—
P × *B*	—	—	—	—	—	—	—	—	—	—	—	—	—
P × *Q*	—	—	—	—	—	—	—	—	—	—	—	—	—
S × *B*	0·39	0·10	—	—	—	—	—	—	—	—	—	—	—
S × *C*	0·63	0·20	—	—	—	—	—	—	—	—	—	—	7·1
S × *Q*	—	—	3·3	2·7	7·1	7·7	—	20	—	—	—	—	—
B × *C*	—	—	—	—	—	—	—	—	—	—	—	—	—
C × *Q*	—	—	—	—	14	—	12	56	70	—	—	—	17
P × *B* × *S*	—	—	—	—	—	—	—	—	—	—	—	—	—
P × *Q* × *S*	—	—	—	—	—	—	—	—	—	—	34	—	—
C × *B* × *S*	—	—	—	—	—	—	127	45	—	—	—	—	—
Residual	3·8	3·4	13	16	60	43	37	39	68	15	29	49	12

Site 2													
Population (P)	—	—	—	—	—	—	—	—	—	—	—	—	—
Size of recruit (S)	2·9	1·1	2·8	—	—	—	0·90	—	—	—	—	—	—
Block (B)	—	—	—	—	—	2·8	0·51	—	—	—	—	—	—
Clone in P (C)	—	—	—	1·1	1·3	—	—	—	—	—	—	—	—
Quadrat in B (Q)	0·08	—	—	—	—	—	—	—	—	—	—	—	—
$P \times S$	—	—	—	—	—	5·2	—	—	—	—	—	—	—
$P \times B$	—	—	—	—	—	—	—	—	—	—	—	—	—
$P \times Q$	—	—	—	—	—	—	—	—	—	—	—	—	—
$S \times B$	—	0·33	0·75	2·6	9·4	—	—	—	—	—	—	—	—
$S \times C$	—	—	—	—	—	—	—	—	—	—	—	87	27
$S \times Q$	0·26	—	—	—	—	—	—	—	—	—	48	—	—
$B \times C$	—	—	—	—	—	0·32	—	3·0	—	—	—	156	57
$C \times Q$	—	—	—	—	—	—	—	—	—	60	66	—	—
$P \times B \times S$	—	—	—	—	—	—	—	—	—	—	—	—	—
$P \times Q \times S$	—	—	—	—	—	—	—	14	—	—	—	—	—
$C \times B \times S$	0·50	0·38	—	—	—	—	—	—	—	211	230	—	—
Residual	1·9	1·3	9·2	33	46	19	22	15	15	66	43	146	203

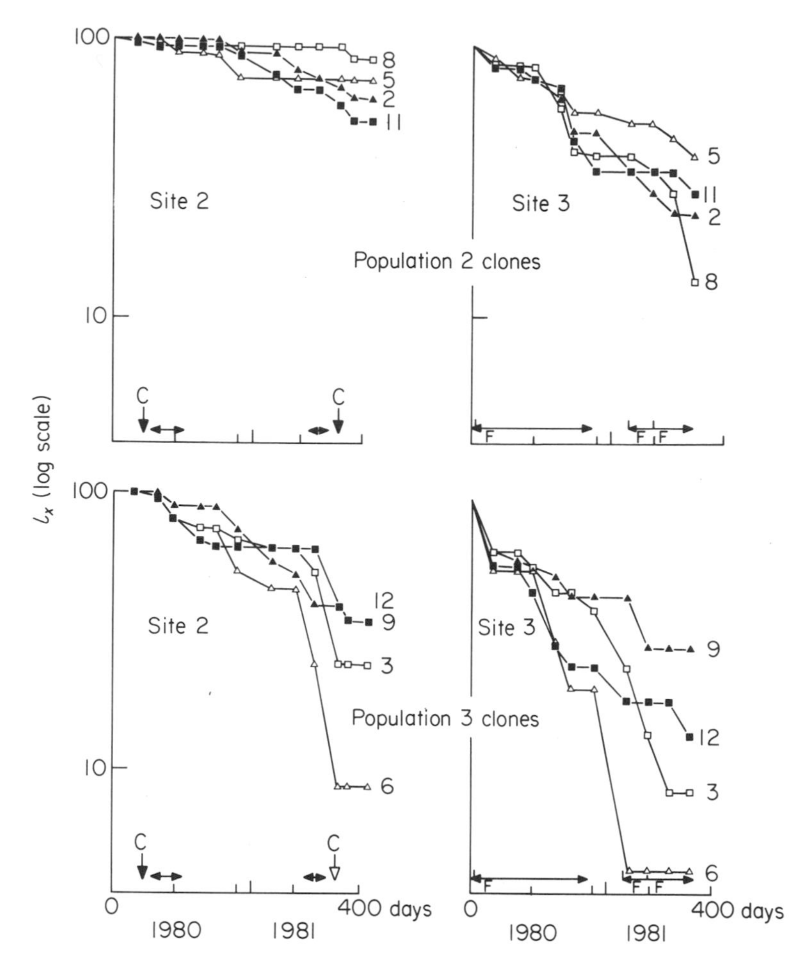

FIG. 3.4. Patterns of survivorship in reciprocally transplanted clones (numbered) of *Holcus lanatus* (see Fig. 3.3 for site details). C, time of cut; F, time of fertilizer application; horizontal arrow, duration of grazing. (From Weir 1985).

chance or be some function of the original population size. Plant populations that subsequently arise will be displaced at various points away from an equilibrium that the habitat might support in the absence of succession. The trajectory of approach of a population towards this equilibrium after a disturbance episode reflects the response of the population. These trajectories described by the map of one generation onto the next provide an objective measure of the effects of disturbance regimes.

For species with discrete generations such as annual plants this focuses onto an understanding of the factors influencing population size from one generation to the next. For those with overlapping generations a useful measure is provided by the reproductive values of different classes within a population. Such an analysis needs to be pursued in a defined series of disturbance regimes over a wide range of population abundance to prove of worth. There have been few attempts to address this approach systematically during the last decade, although much effort has been devoted to describing the dynamics of plant populations.

The dynamical behaviour of single-species populations provides a useful touchstone. May (1975) has advanced the utility of difference equations of the form $N_{t+1} = F(N_t)$, where $F(N)$ is a non-linear (density-dependent) function of N, and exposed the array of dynamical states that theoretically may occur. However, Watkinson (1980) has argued that at least for annual plant populations experiencing density-dependent regulation, single equilibrium points will be approached either monotonically or by damped oscillations, and that they are unlikely to show the more complex dynamical behaviour that is possible (May & Oster 1976).

INTERACTION AMONGST DENSITY-DEPENDENT AND DENSITY-INDEPENDENT FACTORS: TWO EXAMPLES

Central to an understanding of the effects of disturbance regimes is the interaction of density-dependent and independent birth and death rates that determine actual species abundance. The relative importance of each may vary considerably and it is important to distinguish the stages during the life-cycle at which components of regulation act and interact.

One such analysis is that undertaken by Watkinson (1983, 1985) on the sand dune annual *Vulpia fasciculata*. For this species a curvilinear relationship exists between population density and finite rate of increase. Mortality occurs in natural populations during the seed and seedling state in a density-independent manner. Density-dependent regulation is confined to seed fecundity of adult plants. The overall consequences are firstly that in the face of recurring disturbance causing density-independent mortality on seed and seedlings, the equilibrium population density becomes increasingly sensitive to the severity of disturbance-induced mortality; and secondly that equilibrium population densities may only be maintained within defined ranges of mortality in the seed and seedling stage of the life-cycle (Fig. 3.5a).

Where mortality occurs during the vegetative phase of plant growth, the stage of plant development has an important bearing on the outcome

(Fig. 3.5b). Individual yield components (fertile tillers per plant, spikelets per flower and seeds per spikelet) are sequentially determined in accordance with their density at the time of their formation. In consequence, compensatory responses to density reductions may be limited by the yield components already formed. Hence, as the time of disturbance (in this case 50% thinning) was successively delayed during the vegetative phase, the attainable equilibrium population density fell as the compensatory response became increasingly muted. However, the pattern in which this occurred was dependent on the extent to which competition had occurred prior to the time of thinning. Thus, younger plants showed an increased capacity to compensate for reduction during the early vegetative phase of life.

Plant populations are highly likely to experience mortality through disturbance more than once within a lifetime. Recruitment from a persistent seed bank provides an obvious buffer against this, as examplified by many arable weeds. In *Avena fatua*, a persistent component of arable communities, this tactic enables survival in the face of intense (man-induced) mortality. In this monocarpic species, recruitment into seedling populations may occur from September through to June in the UK, except at times of low soil temperatures. All individuals, regardless of age, flower in the subsequent July and August. Monoculture populations

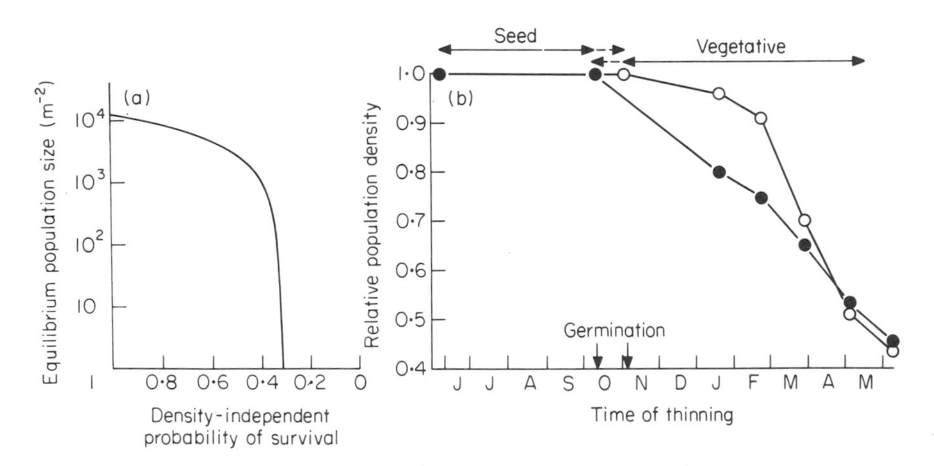

FIG. 3.5. (a) The influence of density-independent mortality in the seed and seedling phase on population abundance in *Vulpia fasiculata*. (From Watkinson 1985.) (b) The influence of time of disturbance (50% thinning) on equilibrium populations in *Vulpia fasiculata*. Population sizes are expressed relative to sizes after thinning in the seed stage in June. Closed circles represent populations germinating early (October) and open ones those germinating late (November). (From Watkinson 1983.)

of *A. fatua* are commonly perturbed in two ways: by invasion of a second species in the form of a crop (usually at high density) and by herbicide application. Crop sowing is usually preceded by destruction of existing cohorts of *A. fatua* when soil disturbance occurs during preparation of the seedbed. Herbicide applications represent agents of mortality that may be applied throughout the life-cycle of *Avena*.

Processes of population regulation in *A. fatua* comprise density-dependent regulation (i) at the seed/seedling transition and (ii) in seed production per plant (Manlove 1985). The stabilizing effects of this can be seen in Fig. 3.6. Net reproductive rate declined monotonically with density to a projected equilibrium of 34 600 seeds m^{-2} counted after dispersal. Both the presence of the companion crop (winter wheat) and the action of a selective herbicide against *A. fatua* (causing seed loss through flower abortion) acted in a density-independent manner. Thus, projected equilibrium densities of *A. fatua* populations became successively reduced in the presence of the second species and further reduced when sprayed with herbicide.

Whilst illustrative of the overall compensatory responses to density, maps of generation changes in *Avena fatua* also reflect the trajectories towards equilibrium displayed in two-species assemblages (Fig. 3.7a).

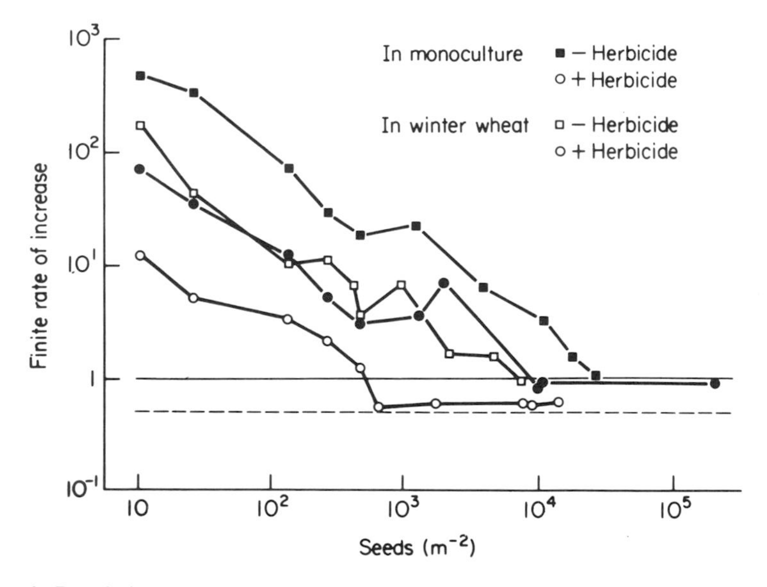

FIG. 3.6. Population regulation in *Avena fatua*. The influence of the herbicide (1-flamprop-isopropyl) serves to reduce seed production and augment the competitive effect of the associated crop species. Census point is immediately post seed dispersal (late August). Dotted line indicates the decay rate of buried seed populations. (Data from Manlove 1985.)

Whilst of the general form exhibited by single species populations, populations of *A. fatua* in winter wheat and spring-sown barley approach the equilibrium locus and then apparently track along it to an equilibrium displayed by monocultures. This pattern arises because of the contributions that late emerging cohorts of *A. fatua* in wheat and barley make to population increase in comparison to those in monoculture (Fig. 3.7b). Field observations suggest that in these two crops these cohorts are able to exploit resources (principally light) released as changes in crop canopy structure occur and plants successfully flower. On the other hand, in monocultures intraspecific competition from larger and older plants renders younger plants barren. Such competition is likely to be for soil resources since canopies in monoculture are far less dense for much of the season.

A THIRD EXAMPLE: POSITIVE DENSITY DEPENDENCE

There is a strong tendency to presume that density-dependent processes generally act on populations in a negative, stabilizing manner. However, some population processes are positively density-dependent, e.g. seed germination (Linhart 1976) and seed predation (Greenwood 1985) and disturbance regimes that precipitate such responses are of considerable interest. An example comes from a study of weed control.

Bromus sterilis is an annual weed of low tillage arable ecosystems, exhibiting discrete generations and a simple population behaviour (Firbank *et al.* 1984). Both seedling survivorship and seed production are negatively density-dependent, resulting in maps of generation change of the form illustrated in Fig. 3.8. The model suggests that two generations are required for equilibrium populations to be approached from an initial density of 1 m^{-1}, in part reflecting the high reproductive capacity of this species. Seedling mortality, which is not noticeably evident in natural populations of densities less than 1000 m^{-2}, may be enhanced by the application of pre-emergence herbicides.

Fig. 3.9 illustrates seedling survival in response to triallate applied to the soil surface (22·5 kg ha^{-1}). This chemical persists in the upper surface layers of the soil and prohibits meristematic activity in *Bromus*. It results in total kill of seedlings below a threshold density of 44 m^{-2}. Above this, survivorship is positively density-dependent, dosage per individual seedling failing to be lethal and diminishing in effectiveness with increasing density. This relationship may be readily described by a power function, the terms of which describe the approach to the locus of density-independent survivorship. Its exact form influences the pattern of finite

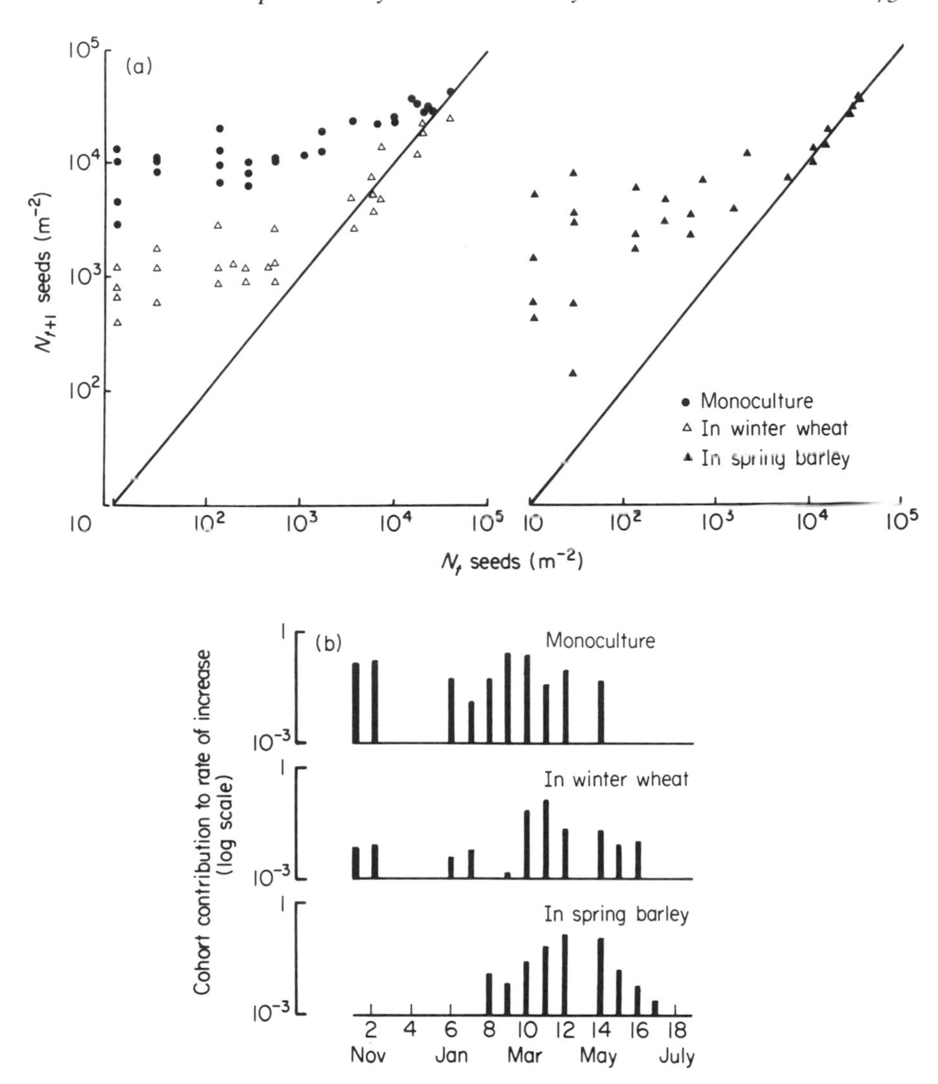

FIG. 3.7. (a) Generation maps of *Avena fatua* from field populations in monoculture, in winter wheat and in spring barley. Census point is immediately post seed dispersal. (b) Contributions to population increase (following Mortimer 1983) by cohorts of *Avena fatua* at high density. Cohorts were recorded approximately every 2 weeks.

rates of increase over successive generations and the approach to equilibrium density (Fig. 3.10). In monoculture the high seed fecundity (1757 seeds in isolated individuals) of adult *B. sterilis* plants enables the population to rapidly achieve densities that nullify the influence of the herbicide. In the presence of a companion crop species, seed production by *Bromus*

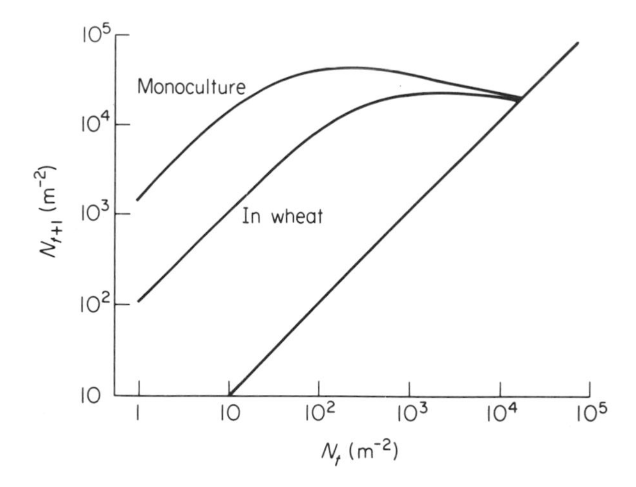

FIG. 3.8. Generation maps for *Bromus sterilis* in field populations in monoculture and in winter wheat. See Firbank *et al.* (1984) for details of construction.

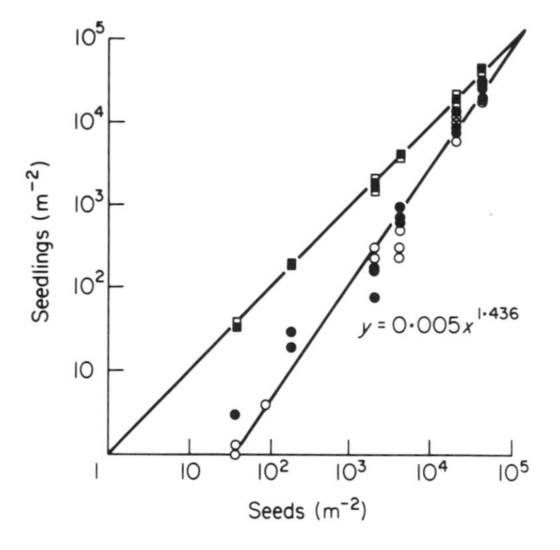

FIG. 3.9. Positive density dependence in seedling survivorship of *Bromus sterilis* under control by a pre-emergence chemical herbicide triallate (22·5 kg ha^{-1}): circles + triallate; squares control; open symbols monocultures; closed symbols in winter wheat. Regression significant at $P < 0{\cdot}001$. $R^2 = 0{\cdot}923$.

is reduced and the population takes longer to achieve equilibrium. Moreover, the population trajectory to equilibrium is determined by the form of the seedling survivorship function. As the degree of control is increased, more generations of the population are required to ascend the seedling survivorship curve (Fig. 3.9). Initially, populations at relatively low densities experience low intraspecific regulation of seed production, and net reproductive rate rises (simulations c–f). Subsequently, populations reach sizes when seed production is strongly regulated and population growth rates fall. Thus, a disturbance regime and its density dependence interacting with natural regulation in *Bromus* yields a complex pattern of growth rates.

DISCUSSION

In this chapter I have raised the proposition that if successional changes are the outcome of disturbance regimes then population trajectories described by generation maps provide a means of describing the consequences of that regime. This proposal is not novel in itself, being in part a reflection of earlier views that an organism's life-history response is a useful measure of environmental heterogeneity (Clements & Goldsmith 1924; Lacey *et al.* 1983), life-histories themselves being viewed as the consequences of genotypic (adaptive?) response, environmental effects and demographic constraints (Law, Bradshaw & Putwain 1977). Yet such maps have been constructed for remarkably few species.

This may be a consequence of (i) over-emphasis on the study of communities which experience constancy in the effects of recurring disturbance regimes (see Grubb, Kelly & Mitchley 1982); (ii) the dominance of the alternative approach of assessing life-history strategies on the basis of measured traits which are *indirect* parameters of actual fitness (survivorship and fecundity); (iii) simply that insufficient time has elapsed for the testing and implementation of tractable models of plant population dynamics (see Pacala & Silander 1985).

The utility of this approach, whilst clearly limited to species with discrete generations is at least threefold.

(a) If disturbance regimes may be sufficiently well circumscribed then population trajectories over successive generations offer a comparative measure of population renewal and relative abundance. For disturbance regimes that result in population replacement (as for the sand dune annuals described earlier) with no change in species diversity, then it offers promise in forecasting plant population sizes (see Firbank, Mortimer & Putwain 1985).

(b) Analysis of the fitness components underlying rates of increase exposes the nature of interactions that may occur amongst population process. The importance of interacting density-dependent and independent processes is well exemplified by the studies of Carter & Prince (1981) and Watkinson (1985).

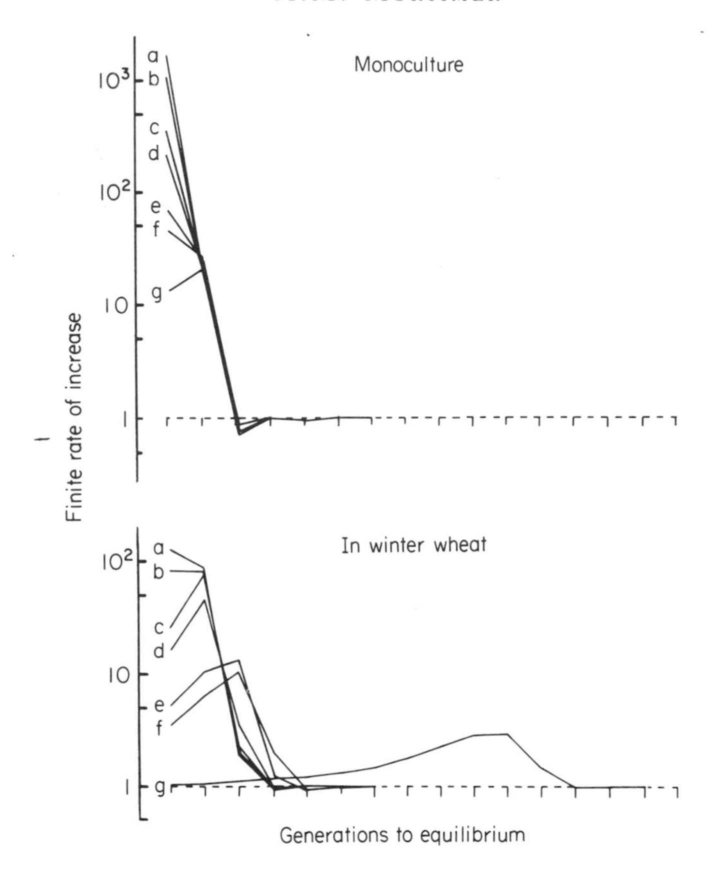

FIG. 3.10. Simulations of the pattern of finite rates of increase in *Bromus sterilis* experiencing positive density-dependent seedling survivorship (Fig. 3.9) and negative density-dependent regulation of seed production. Simulations (a)–(g) represent increasing intensity of control by deliberate choice of the constant in the equation (given in Fig. 3.9) in the range 0·95–0·001, power term remaining 1·436. Each simulation is started with a surviving seedling density of 1 m^{-2}.

(c) Coupled to analytical models of the form proposed by May (1975), the dynamical properties of populations may be examined. The utility of non-linear difference equations that underlie the behaviour of annual plants with discrete generations is clearly illustrated in Watkinson's (1985) instructive re-analysis of the abundance of *Cakile edentula* on an environmental gradient.

To the experimental demographer this approach offers one way to investigate community interactions. The task involves not only fulfilling Harper's (1982) requirement of examining populations far outside their natural habitat and abundance range, but also recognizing that species may switch amongst trajectories with the imposition of different disturbance regimes. It remains a substantial experimental task not least compounded by the difficulty of defining appropriate disturbance regimes. Yet the astute student may capitalize on an extreme disturbance event (see Waite 1984), overlay a range of population densities into the habitat, and chronicle subsequent events.

Of major importance in understanding the shape of population trajectories is the examination of the time at which mortality falls in relation to the development of the population. Both the examples discussed earlier for *Bromus sterilis* and *Avena fatua* point to this. Moreover, such studies highlight the compensatory responses to mortality. For example, populations of *A. fatua* at high density may overcome the loss of cohorts emerging prior to the sowing of spring barley through the contributions of later ones *and* in so doing achieve equivalent abundance as monocultures (Fig. 3.7). Additionally, the time at which mortality is visited upon individuals within populations is crucial (Fig. 3.5b). Whilst there is a substantial literature on yield component analysis for agronomic species, this subject has hardly been touched in investigations of species in 'natural' communities.

The discussion above is clearly inappropriate to clonal organisms with overlapping generations of individuals that persist under regimes of disturbance characterized by herbivory (e.g. *Holcus lanatus*, Table 3.2) in which localized spatial heterogeneity plays a key role in species persistence (Aarssen & Turkington 1985). An appropriate and testable model for the analysis of these populations is that proposed by Noy-Meir (1975). Recently Billington (1985) has demonstrated experimentally that clonal variation in rates of tiller production does indeed interact with grazing intensity and produce the range of steady-state conditions predicted by the model.

Successional events involve species replacement and addition on ecological time-scales that are measured by generation times. I am left with the thought that, to date, the contributions of field studies of plant population dynamics to successional processes have been limited. However, the generation times of those species that are considered early successional—pioneers and colonists—are much shorter than those of the ecologists who study them! And that brings hope.

ACKNOWLEDGMENTS

I am extremely grateful to Dr P.D. Putwain for discussions on this topic and to Drs Helen Billington, Ric Manlove and David Weir for data published in thesis form.

REFERENCES

Aarssen, L.W. & Turkington, R. (1985). Vegetation dynamics and neighbour associations in pasture community evolution. *Journal of Ecology*, **73**, 585–603.

Antonovics, J. & Levin, D.A. (1980). The ecological and genetic consequences of density-dependent regualtion in plants. *Annual Reviews of Ecology and Systematics*, **11**, 411–452.

Armstrong, R.A. (1982). A quantitative theory of reproductive effort in rhizomatous perennial plants. *Ecology*, **63**, 679–686.

Begon, M. (1984). A general theory of life history variation. *Behavioural Ecology* (Ed. by R.M. Sibly & R.H. Smith), pp. 91–97. British Ecological Society Symposium Vol, No. 25. Blackwell Scientific Publications, Oxford.

Begon, M. & Mortimer, M. (1986). *Population Ecology: A Unified Study of Animals and Plants*, 2nd Edn. Blackwell Scientific Publications, Oxford.

Bell, A.D. & Tomlinson, P.B. (1980). Adaptive architecture in rhizomatous plants. *Botanical Journal of the Linnean Society*, **80**, 125–160.

Billington, H.L. (1985). *Population ecology and genetics of* Holcus lanatus L. Unpublished Ph.D. thesis, University of Liverpool.

Carter, R.N. & Prince, S.D. (1981). Epidemic models used to explain biogeographic models. *Nature (London)*, **293**, 644–645.

Clements, F.E. (1916). *Plant succession: an analysis of the development of vegetation.* Carnegie Institute of Washington Publication 242.

Clements, F.E. & Goldsmith, G.W. (1924). *The phytometer method in ecology.* Carnegie Institute of Washington Publication 356.

Cooper, W.S. (1923). The recent ecological history of Glacier Bay Alaska. II. the present vegetation cycle. *Ecology*, **4**, 223–246.

Cooper, W.S. (1939). A fourth expedition to Glacier Bay Alaska. *Ecology*, **20**, 130–159.

Drury, W.H. & Nesbit, I.C.T. (1973). Succession. *Journal of the Arnold Arboretum*, **54**, 331–368.

Firbank, L.G., Manlove, R.J., Mortimer, A.M. & Putwain, P.D. (1984). The management of grass weeds in cereal crops, a population biology approach. *Weed Biology, Ecology and Systematics*, pp. 375–384. 7th International Symposium of the European Weed Research Society, Paris.

Firbank, L.G., Mortimer, A.M. & Putwain, P.D. (1985). *Bromus sterilis* in winter wheat: a test of a predictive population model. *Aspects of Applied Biology*, **9**, 59–66.

Gleason, H.A. (1926). The individualistic concept of the plant association. *Bulletin of the Torry Botanical Club*, **53**, 7–26.

Grant, V. (1977). *Organismic Evolution.* Freeman, San Francisco.

Greenwood, J.J.D. (1985). Frequency-dependent selection by seed-predators. *Oikos*, **44**, 195–210.

Grubb, P.J., Kelly, D. & Mitchley, J. (1982). The control of relative abundance in communities of herbaceous plants. *The Plant Community as a Working Mechanism* (Ed. by E.I. Newman), pp. 79–98. Blackwell Scientific Publications, Oxford.

Harper, J.L. (1977). *The Population Biology of Plants.* Academic Press, London.

Harper, J.L. (1982). After description. *The Plant Community as a Working Mechanism* (Ed. by E.I. Newman), pp. 11–26. Blackwell Scientific Publications, Oxford.

Harper, J.L. (1985). Modules, branches and the capture of resources. *Population Biology and Evolution of Clonal Organisms* (Ed. by J.B.C. Jackson, L.W. Buss & R.E. Cook), pp. 1–33 Yale University Press, New Haven.

Herrera, C.M. (1985). Determinants of plant–animal coevolution: the case of mutualistic dispersal of seeds by vertebrates. *Oikos*, **44**, 132–141.

Horn, H.S. (1974). The ecology of secondary succession. *Annual Reviews of Ecology and Systematics*, **5**, 25–37.

Horn, H.S. (1976). Succession. *Theoretical Ecology, Principles and Applications* (Ed. by R.M. May), pp. 187–204. Blackwell Scientific Publications, Oxford.

Lacey, E.P., Real, L., Antonovics, J. & Heckel, D.G. (1983). Variance models in the study of life histories. *American Naturalist*, **122**, 114–131.

Law, R., Bradshaw, A.D. & Putwain, P.D. (1977). Life-history variation in *Poa annua*. *Evolution*, **31**, 233–246.

Lawrence, D.B. (1958). Glaciers and vegetation in south-eastern Alaska. *American Scientist*, **46**, 89–122.

Levin, S.A. (1976). Population dynamic models in heterogenous environments. *Annual Reviews of Ecology and Systematics*, **7**, 287–310.

Linhart, Y.B. (1976). Density-dependent seed germination strategies in colonizing versus non-colonizing plant species. *Journal of Ecology*, **64**, 375–380.

Lovett Doust, L. (1981). Population dynamics and local specialization in a clonal perennial (*Ranunculus repens*). I. The dynamics of ramets in contrasting habitats. *Journal of Ecology*, **69**, 743–756.

Mack, R.N. (1981). The invasion of *Bromus tectorum* L. into western North America: an ecological chronicle. *Agro-Ecosystems*, **7**, 145–165.

Mack, R.N. & Pyke, D.A. (1983). The demography of *Bromus tectorum*: variation in time and space. *Journal of Ecology*, **71**, 69–93.

Manlove, R.J. (1985). *On the population ecology of* Avena fatua L. Unpublished Ph.D. thesis, University of Liverpool.

May, R.M. (1975). Biological populations obeying difference equations: stable points, stable cycles and chaos. *Journal of Theoretical Biology*, **49**, 511–524.

May, R.M. & Oster, G.F. (1976). Bifurcations and dynamic complexity in simple ecological models. *American Naturalist*, **110**, 573–599.

Miles, J. (1979). *Vegetation Dynamics*. Chapman & Hall, London.

Mortimer, A.M. (1983). On weed demography. *Recent Advances in Weed Research* (Ed. by W.W. Fletcher), pp. 3–41, Commonwealth Agricultural Bureaux, Farnum Royal.

Noy-Meir, I. (1975). Stability of grazing systems: an application of predator–prey graphs *Journal of Ecology*, **63**, 459–483.

Pacala, S.W. & Silander, J.A. (1985). Neighbourhood models of plant population dynamics. I. Single-species models of annuals. *American Naturalist*, **125**, 385–411.

Schaal, B.A. & Leverich, W.J. (1981). The demographic consequences of two stage life cycles: survivorship and the time of reproduction. *American Naturalist*, **118**, 135–138.

Silvertown, J.W. (1982). *Introduction to Plant Population Ecology*. Longman, London.

Sousa, W.P. (1984). The role of disturbance in natural communities. *Annual Reviews of Ecology and Systematics*, **15**, 353–391.

Southwood, T.R.E. (1977). Habitat, the templet for ecological strategies? *Journal of Animal Ecology*, **46**, 337–365.

Stearns, S.C. (1977). The evolution of life history traits. *Annual Reviews of Ecology and Systematics*, **8**, 145–171.

Symonides, E. (1979). The structure and population dynamics of psammophytes on inland dunes. III. Populations of compact psammophyte communities. *Ekologia Polska*, **27**, 235–257.

Symonides, E. (1983a). Population size regulation as a result of intrapopulation interactions. I. Effect of density on the survival and development of individuals of *Erophila verna* (L.) CAM. *Ekologia Polska*, **31**, 839–881.

Symonides, E. (1983b). Population size regulation as a result of intrapopulation interactions.

II. Effect of density on the growth rate, morphological diversity and fecundity of *Erophila verna* (L.) CAM individuals. *Ekologia Polska*, **31**, 883–912.

Waite, S. (1984). Changes in the demography of *Plantago coronopus* at two coastal sites. *Journal of Ecology*, **72**, 809–826.

Watkinson, A.R. (1980). Density dependence in single species populations of plants. *Journal of Theoretical Biology*, **83**, 345–357.

Watkinson, A.R. (1983). Factors affecting the density response of *Vulpia fasiculata. Journal of Ecology*, **70**, 149–161.

Watkinson, A.R. (1985). On the abundance of plants along an environmental gradient. *Journal of Ecology*, **73**, 569–578.

Weir, D.A. (1985). *The population ecology and clonal structure of two grasses.* Unpublished Ph.D. thesis, University of Liverpool.

White, J. (1984). Plant metamerism. *Perspectives on Plant Population Ecology* (Ed. by R. Dirzo & J. Sarukhan), pp. 15–7. Sinauer Assoc., Massachusetts.

4. SOME GENERALIZING IDEAS ABOUT COLONIZATION AND SUCCESSION IN GREEN PLANTS AND FUNGI

P.J. GRUBB

Botany School, University of Cambridge, CB2 3EA

INTRODUCTION

The first and largest part of this chapter is devoted to a review of the characteristics of colonizing plants and fungi; the chief objective is to show that colonizing species are very various in nature, and that useful generalizations cannot be made about colonizers unless particular types of substratum are specified. The second part of the chapter is devoted to a comparison between successions of green plants and of fungi, and the third part is intended to correct a widespread misunderstanding of the meaning of *r*- and *K*-selection.

CHARACTERISTICS OF COLONIZING PLANTS

What are colonizing species?

In the classic symposium entitled *The Genetics of Colonizing Species* published 20 years ago (Baker & Stebbins 1965), the organisms discussed were those which had proved capable of invading parts of the world where they were not native, or of increasing enormously in abundance under the influence of man where they were native. In the case of plants, most of the species considered were annuals or perniciously weedy perennials with extensive spread by rhizomes or roots and/or the ability to regenerate from vegetative fragments. A similar compass for colonizing species was maintained by Brown & Marshall (1981) in their treatment of evolutionary changes accompanying colonization. In view of the title of the present symposium, *Colonization, Succession and Stability*, I have adopted a wider meaning for 'colonizers', as 'the first organisms to become established in a succession'. During (1979) did the same for the special case of bryophytes.

By succession I mean directional change in species composition, whether in terms of the list of species present or of their relative abundances. This definition generally works well enough, but it is not possible to separate succession from fluctuation over a long time-scale (cf. Davis 1986). Both primary and secondary successions are considered in this volume, and thus by implication the widest possible view of a colonizer is adopted. The colonizers in primary successions form a distinct subset which we may call 'pioneers' (During 1979). They may be relatively short-lived and transitory or long-lived and persistent; their essential feature is their ability to become established before other plants (Grubb 1986a).

In this chapter two kinds of secondary succession are recognized: the 'internal successions' that form part of the natural regeneration process in many types of vegetation (Curtis 1959, p. 292) and 'man-induced successions' on such sites as old-fields and forest clearcuts. In internal successions the species which colonize gaps made by fire, windthrow, flood or animal activity are an integral part of the community. In contrast, the colonizing species in man-induced secondary successions are commonly recruited from a wide range of habitats and have not evolved together to the same degree (Grubb 1985).

The variety of colonizing plants

The growth-form of an effective pioneer plant varies in a predictable way with substratum type, as shown in Fig. 4.1. The evidence for this assertion has been summarized elsewhere (Grubb 1986a) and comes from studies on substrata bared by glacial retreat and landslips, and on surfaces newly created by volcanic activity or the deposition of river-borne materials. It has been argued in the same review that the availability of mineral nutrients is at least as important as the availability of water in determining which growth-form is best suited to a particular substratum type.

In general, the pioneers at stable sites with adequate water and mineral nutrients are relatively short-lived, and mean life-length increases during succession; this represents what we may call the 'conventional' type of succession. In contrast, the pioneers at sites poor in resources are typically long-lived. There is another class of habitats in which the pioneers are characteristically long-lived: those where the substratum is physically unstable as a result of tides and wave action, wind or gravity. This is true whether there is an abundance of water and at least a moderate supply of nutrients at the surface, as in the case of tidal flats, or a marked deficiency in both water and nutrients at the surface, as in the case of sand dunes and screes.

On tidal flats the characteristic pioneers in the tropics are mangrove trees, and in the southern temperate zone in areas with a seasonally dry climate the pioneers are bushes of the genus *Sarcocornia* (formerly

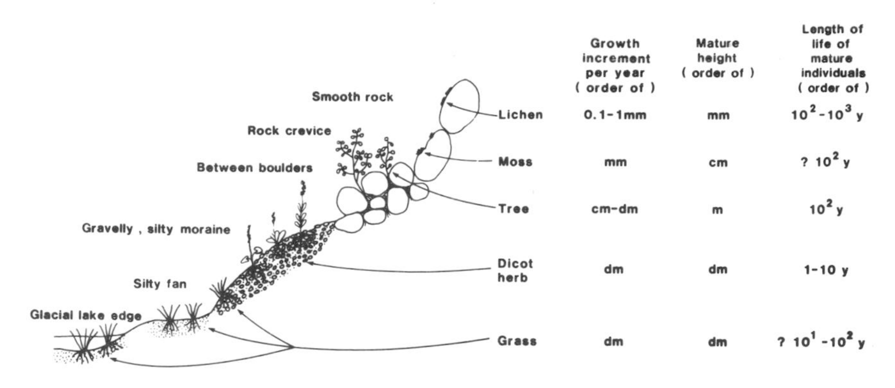

FIG. 4.1. The type of pioneer plant characteristic of different types of substratum in a recently deglaciated area within the range of present-day forest, and selected properties of the plants (reproduced, with permission, from Grubb 1986a). The distribution of pioneers with respect to substratum type is essentially similar on surfaces newly created by landslips, by volcanic activity, and by deposition of river-borne materials. Where conditions are warmer and/or moister, algae or cyanobacteria may replace lichens as the major pioneers on smooth rocks.

Arthrocnemum); in many parts of the northern temperate zone the principal pioneers are perennial grasses of the genera *Puccinellia* and *Spartina* (Chapman 1977). The same phenomenon of long-lived pioneer plants is seen in the perennial 'sea-grasses' (Hydrocharitaceae and Potamogetonaceae; den Hartog 1970) on fine-grained substrata under shallow coastal waters in many parts of the world. A significant exception to the generalization that pioneers on tidal flats and in shallow seas are perennial is *Salicornia dolichostachya*, which is widespread on northern European coasts. This *Salicornia* is found at the level where *Spartina anglica* is often the pioneer, and also a little lower where the 'seagrass' *Zostera noltii* is more usually the pioneer (Ellenberg 1982).

On sand dunes, whether coastal or inland, the pioneers are usually perennial grasses (e.g. species of *Ammophila*, *Elymus*, *Leymus*, *Spartina* or *Sporobolus*) and less often sedges (*Carex*) but on the western North American coast perennial dicotyledons can be equally important (Barbour & Johnson 1977). In all temperate zones there is a marked contrast between the long-lived pioneers on fore-dunes and the annuals which are the characteristic pioneers on the flotsam of the drift-line, such as species of *Atriplex*, *Cakile* and *Salsola*. The key differences between the habitats are that the fore-dunes are relatively nutrient-poor and subject to continual moderate disturbance by sand-blow, while the drift-line materials are nutrient-rich and removed annually (by winter storms) but rarely disturbed after the time of germination of the species whose seeds arrive in the drift.

On screes the principal pioneers are always specialized perennials, as documented for central Europe by Ellenberg (1982), for North America by Thorne (1977), and for New Zealand by Fisher (1952). On the more stable micro-sites on a scree, annuals or biennials are sometimes able to form small pioneer populations, e.g. *Arabis* spp. in Europe, but they contribute little to the total biomass, and are more typical of stable rock-crevice sites.

Where long-lived plants are the pioneers, there are often shorter-lived plants to be found later in succession. For example, where perennial grasses stabilize a tidal mud flat and build up the level of the sediment, annuals such as *Suaeda maritima* and species of *Salicornia* other than *S. dolichostachya* may become established together with the biennial or pauciennial *Aster tripolium* (Ranwell 1972; Gray, Chapter 13). Where mangroves are the pioneers, their crowns may become colonized by three sorts of shorter-lived vascular plant: orchids (*Cymbidium* or *Dendrobium*), myrmecophytes (*Hydnophytum* or *Myrmecodia*) or hemi-parasites in the Loranthaceae (*Amyema*); meanwhile their pneumatophores come to support numerous relatively short-lived cyanobacteria and red and green algae (Saenger *et al.* 1977). On rock outcrops where long-lived desiccation-tolerant cyanobacteria or mosses build up a thin layer of soil we often find drought-avoiding annual flowering plants invading (Oosting & Anderson 1939; Quarterman 1950); succession may not proceed beyond this stage for a very long time if the depressions in the rock are slight.

For the sake of clarity I have presented primary successions as of two basic types: the 'conventional' type where the pioneers are shorter-lived than the later-invading species, and those that are 'back to front' in this respect. In fact, there are many intermediate cases where shorter-lived plants appear soon after the pioneer long-lived plants, or even at the same time, and so the populations of the two groups of species expand together through succession. On the Hawaiian lava-flows described by Vitousek and Walker in Chapter 10 two short-lived herbs (*Epilobium adenocaulon* and *E. cinereum*) become established only a few years after the tree *Metrosideros polymorpha* has begun to ameliorate the microclimatic conditions. On northern European coasts, around low-tide level, the annual *Salicornia dolichostachya* often invades together with one of the two perennials *Spartina anglica* and *Zostera noltii*. Mixtures of longer- and shorter-lived pioneers are also common on shingle banks, which are often very heterogeneous in stability and in the proportion of fine-grained inorganic material and organic detritus present, and therefore the nutrient-supply; examples for Britain are given by Scott (1963).

In most secondary successions, plants of very various growth-forms and life-lengths are present from the earliest stages, ranging from annual herbs to long-lived trees or shrubs, and there is either a progressive total loss of shorter-lived plants (on a local scale) or they persist only in the seed-bank in the soil. This is true both of 'internal successions', e.g. in chaparral after burning (Hanes 1971) and 'man-induced successions', e.g. in old-fields of the eastern USA (Egler 1954).

These observations concerning a wide variety of habitats lead us to question received generalizations about 'colonizing species': not only about their longevity and growth-form but also about their dispersibility, and even more importantly about their breeding systems and levels of heterozygosity. The latter topic is taken up by Brown & Burdon and by Gray in this volume and in Gray (1986).

Significance of dispersibility

A major difference between primary and secondary succession is that many colonizers in secondary succession are dispersed through time rather than through space. Seeds and spores persist in the soil between disturbances, and in some cases (as in many Myrtaceae and Proteaceae in Australia) seeds persist in woody capsules well above ground. In the extreme case, where virtually all dispersal is through time and not space, the vegetation can display a high level of 'resilience *in situ*' (e.g. in the face of fire) but a low level of 'resilience by migration' (e.g. onto extensive bare areas left by large-scale open-cast mining), as documented by Grubb & Hopkins (1986) for Western Australian heathland.

The pioneers of primary succession are commonly supposed to have exceptional powers of dispersal (cf. Fenner, Chapter 5). In those habitats with at least a moderate supply of light, water and mineral nutrients this may be true, as the pioneers often have very tiny seeds or small seeds with plumes, e.g. species of *Epilobium* in many parts of the world. However, where the supply of resources is meagre, the pioneers are not necessarily so dispersible; on talus slopes or landslips the pioneer trees may have relatively light seeds, as in the case of *Betula pubescens* and *B. papyrifera* in Europe and North America respectively, but others have relatively large seeds, e.g. *Picea englemannii* and *P. rubens* in western and eastern North America respectively (Cooper 1916; Flaccus 1959). The oaks (*Quercus* spp.) which are sometimes pioneers on quarry walls, and are dispersed by birds, have notably large seeds. Where the resources are strictly limited, there is, so to speak, less urgency about a plant arriving before its competitors. There is a clear parallel with the importance of maximum relative growth rate in relation to the supply of resources. As argued by Bradshaw *et al.* (1964) and many others later, plants suited to sites with abundant resources have to possess high maximum relative growth rates to maintain their occupancy, while for plants suited to sites with less abundant resources tolerance of unfavourable periods and avoidance of predation are more important than maximum relative growth rate.

At the sites with least resources for plant growth, the pioneers belong to groups of plants in which almost all fertile species produce spores that are minute relative to any seed, i.e. mosses, algae and lichens. However, in both lichens and mosses there is some evidence that pioneers have smaller spores than average. Pioneer lichens, such as species of *Acarospora*, *Buellia* and *Rhizocarpon* tend to have spores of diameter <25 μm, and mean disseminule size (including spores and soredia) is said to increase in most successions of lichens (Topham 1977). Pioneer mosses tend to have spores of diameter <20 μm, where the range for all mosses is 5–200 μm (During 1979); *Andreaea* species, which are widespread pioneers on rocks, are exceptional in having spores 25–40 μm across (Longton & Schuster 1983). In any case, it is likely that in these spore-producing plants an ability to tolerate the conditions on the substrata concerned is at least as important as any ability to arrive quickly in enabling them to be pioneers. This point is brought home by the fact that the effective propagules, especially for short-distance dispersal, are often not spores but shoot-fragments in the case of mosses, and soredia or thallus-fragments in the case of lichens (Keever 1957; Bailey 1976; Lindsay 1977).

Vegetative propagules are also important in habitats which are physically unstable, whether relatively rich in resources or not. For example, shoot-fragments were the sole means of spread for the sterile *Spartina* × *townsendii sensu stricto* (as opposed to the fertile *S. anglica*) on tidal flats, and are important at the present time for *Puccinellia maritima* (Gray & Scott 1977) and *Zostera* species (Tutin 1942). Huge 'viviparous' (i.e. pre-germinated) seeds are characteristic of the pioneer mangrove *Rhizophora* (Walter 1971). In all these latter cases the possession of an adequate size to retain anchorage, and of adequate reserves to grow up through newly deposited sediment, are more important than sheer dispersibility. In an analogous way, invasion by vegetative means is commonly of overwhelming importance in plants of screes, sand dunes and 'underwater dunes' carrying 'sea-grasses' (McComb *et al.* 1981).

CHARACTERISTICS OF COLONIZING FUNGI

Types of fungal succession

Successions among fungi can be studied at two very different levels: that of the individual plant or animal part or product, and that of the plant or animal community. Park (1968) suggested the terms 'substratum succession' and 'seral succession', and they were adopted by Frankland (1981) in her wide-ranging discussion of fungal successions on litter and in the soil. However, these terms are not satisfactory. With respect to the first, it is the succession of fungi with which we are primarily concerned, and not the succession of chemical and physical changes in the substratum. The second term is a tautology, as Clements (1916, p. 4) defined a sere as 'a unit succession'. I suggest as preferable terms 'substratum-level succession' and 'community-level succession'. Little research has been carried out at the community level, and although much has been done at the substratum level there are grave problems in interpreting the results. (i) There is a constant need to consider the relevant scale of study; different successions may be going on in the epidermis, mesophyll and xylem of a fallen leaf, and these are commonly mixed up when conventional methods are used to isolate and identify the fungi present. (ii) It is usually difficult to identify the species which are active at a particular site at a given time, as opposed to those which are merely present as spores or other viable survival structures but able to grow rapidly in laboratory culture. (iii) It is often impossible to define the limits of an individual, and to make confident assertions about longevity of individuals. Despite these problems, many of the major issues in fungal successions have now been clarified.

Colonizing fungi in substratum-level successions

The characteristics of colonizing fungi in seven types of substratum-level succession are summarized in Table 4.1. The following sources of information were used in compiling the analysis. The most up-to-date accounts of succession on the surfaces of living plants and on plant debris are given by Cooke & Rayner (1984) and by Hudson (1986), who cite the original literature in detail. The paper by Willoughby (1974) is especially useful for successions on different kinds of leaves in streams (quoting the work of J.A. Newton), and the paper of Wicklow (1981a) is important as a source of information and ideas on succession in dung. The diversity of pathogens able to pioneer the invasion of healthy plant tissue is well

TABLE 4.1. The chief types of colonizing fungi in various substratum-level successions, and their key characters*

Substratum for succession	Chief colonists	Second- and third-wave invaders	Key characters of colonists
(1) Surfaces of healthy plants			
(a) Shoots	Yeasts, yeast-like fungi and yeast-forms of filamentous types	Filamentous types, some weakly parasitic	Effective growth on slender supply of C and N. Ability to use specialized chemical substrates, e.g. waxes. Tolerance of great variation in temperature and water regimes. Tolerance of bright light, especially ultra-violet. Ability to grow quickly in transiently favourable conditions. Ready local dispersal by wind, rain-splash and insects.
(b) Roots	Fast-growing filamentous types	As above	Rapid germination and growth on moderate concentrations of simple organic compounds. Abundant dormant spores in soil. Small spores (often $<10\ \mu m$) readily dispersed by wind and percolation of soil by water.
(2) Roots forming sheathing mycorrhiza	Basidiomycetes not living on litter	Basidiomycetes living on litter	Responsive to high levels of easily assimilable carbohydrate and probably to nitrogen in mineral form; not demanding of vitamins. Rapid spread by single hyphae. Infection by spores, dormant in soil or blown in.
(3) Diseased plants	Very various parasites	Very various saprophytes	Above all, an ability to overcome host resistance in either intact or slightly damaged organs. Means of dispersal and invasion are suited to scale of infection, ranging from single spores to massive rhizomorphs.

(4) Leaf litter			
(a) Falling to forest floor	Surface-living, filamentous types ('primary saprotrophs'), and some weak parasites (both mainly Asco- and Deuteromycetes).	'Secondary saprotrophs' (mainly Asco-, Deutero- and Basidiomycetes), then 'soil inhabitants' (mainly Zygomycetes and Deuteromycetes)[†].	Like yeasts in (1) but more demanding of nutrients and in some cases able to overcome host resistance. Invasion by spores whereas several later-successional species invade by hyphae or mycelial strands.
(b) Falling into running streams	(Above types quickly replaced)	Deuteromycetes with tetraradiate, sigmoid or multibranched conidia.	Responsive to easily assimilable carbohydrates. May have varying requirements re turbulence. Spread and invasion is by swimming spores or by non-flagellate spores with an increased chance of impaction ('three-point landing').
	Water moulds (Mastigomycetes) and fungi with tetraradiate conidia (mainly Ascomycetes, also Deutero- and Basidiomycetes).		
(5) Faeces of warm-blooded herbivores	Zygomycetes, Ascomycetes and Basidiomycetes all together.	At a late stage 'soil fungi'.	Spores able to withstand gut temperatures, high pH and animal's enzymes. Germination is stimulated by these conditions. Ability to disperse spores effectively from dung to the next generation of leaves to be eaten. Spores or spore masses large, sticky, discharged explosively and phototropically.

* This expression is used in preference to 'adaptation' (cf. Grubb 1985).

[†] 'Primary saprotrophs', 'secondary saprotrophs' and 'soil inhabitants' *sensu* Hudson (1986).

covered by Wheeler (1969) and Manners (1982). An invaluable introduction to the literature on the successions of fungi to be found in root systems with sheathing mycorrhiza is provided by Dighton & Mason (1985).

There is space here to comment on just a few points in Table 4.1. First, there is a marked contrast between leaves and roots in the nature of the colonizing saprotrophic fungi on their surfaces. The fungi on leaves must tolerate conditions analogous to those faced by a lichen on a smooth rock because the cuticle of a newly formed leaf effectively restricts the outward leakage of nutrients, and air-borne organic debris such as pollen and spores takes some time to arrive. In contrast the root cap sloughs off mucilage, and there is a relatively massive leakage of organic compounds of low molecular weight from the extension zone of the young root. The colonizing rhizoplane fungi face conditions analogous to those experienced by short-lived, fast-growing, highly dispersible herbs on sites with moist fine-grained soils (or even organic debris as on drift-lines). One should note that the phylloplane yeasts do not merely tolerate life in their peculiar habitat; they are positively suited to it by their ability to metabolize waxes of the epicuticle, compounds not to be found in the 'more favourable' conditions of the rhizoplane, and their ability (because of their small size) to produce whole new dispersible cells using very limited resources. There is something of a parallel with those lichens which have been found to grow best with a slender nutrient supply, and show a breakdown of the mutualistic symbiosis when the nutrient supply is markedly increased, and therefore 'more favourable' for other plants (Scott 1960; Ahmadjian 1967).

Research into the successions on recently dead plant material, fallen litter and faeces was well under way over 20 years ago (cf. Garrett 1963; Hudson 1968), but the phenomenon of succession in the fungi that form sheathing mycorrhiza has come to be widely appreciated only in the last decade. It is seen most clearly where widely spaced trees are planted at a non-forested site. The fungi forming the first mycorrhiza produce rings of basidiocarps at successively greater distances from the trunk in succeeding years (as the root system grows out) and finally disappear. The fungi next involved behave similarly within the expanding ring of the first species. In this way up to four or five arbitrarily separated phases of succession may be recognized on a single tree species, e.g. *Betula pendula* or *B. pubescens* (Mason *et al.* 1983). Analogous successions in the mycorrhizal fungi forming basidiocarps are seen in closely planted commercial stands of conifers of increasing age, and in naturally regenerated ageing stands of *Picea abies* (references given by Dighton & Mason 1985). It is not a matter of any single absorptive root containing a succession of fungi, but of successive fine roots on any given main root being infected by different associates.

Evidence summarized by Dighton & Mason (1985) strongly suggests that the fungi which form the earliest mycorrhizas infect by spores, and then spread from root to root by rapidly growing hyphae, while the fungi forming later mycorrhiza infect by slower-growing mycelial strands from an inoculum living either saprotrophically in the litter, or in mutualistic symbiosis in the roots of other trees. In many cases the spores of colonizing species will have been dormant in the soil, while others will have blown in. In natural forests these species have their chance when wind or fire opens up the canopy so that pockets of soil are left free of litter and of roots infected with late-successional mycorrhizal fungi. The succession depends, however, not only on differences in modes of infection, but also on nutrient-demand. Dighton & Mason (1985) summarize evidence that the colonizers have a lower carbohydrate demand. They may also have a higher demand for readily assimilable inorganic forms of nitrogen and phosphorus, as opposed to organic forms present in the litter, while being less dependent on complex mixtures of vitamins provided by either the host root or cohabiting micro-organisms in the litter. In the boreal, subalpine and temperate forests in which sheathing mycorrhiza are abundantly developed, we may see that as a regeneration-unit of forest ages after being opened up by wind or fire, the trees become progressively more dependent on internal recycling for the supply of nitrogen and phosphorus to actively growing parts; the litter becomes increasingly poor in these nutrients and richer in phenolics, and the trees devote an increasing proportion of their carbon capital to the support of litter-decomposing mycorrhizal fungi. In this way substratum-level fungal succession merges with the community-level succession discussed in the next section of this paper.

Colonizing fungi in community-level successions

Two types of succession are considered: primary and secondary ('internal'). Attempts to follow the changes in the soil microflora and rhizosphere flora through a primary succession have been made for sand dunes (Webley, Eastwood & Gimingham 1952; Brown 1958), salt marshes (Pugh 1962) and glacial moraines (Cooke & Lawrence 1959). The problems of identifying active, as opposed to dormant, fungi are at their worst in these studies. Nevertheless, the changes can be marked and consistent. For example, Frankland, née Brown (1981, p. 407) wrote concerning the sand dune system described in her 1958 paper 'the center of a seral stage could be recognised as readily by the fungal population of a culture plate, inoculated with only 5 mg of sand, as by a community of higher plants.' The colonizing fungi seem to be characterized more by their tolerance of the relevant soil conditions than by being especially highly dispersible.

The best-studied internal succession is that following forest fire (Gochenaur 1981). The 'pyrophilous fungi', which colonize the newly bared ground and charred plant parts, resemble dung-inhabiting fungi in their ability to grow at high pH, to respond positively to high levels of available nitrogen and phosphorus, and to survive (as spores) conditions which kill other fungi, in this case very high temperatures rather than the conditions in an animal's gut. They are also analogous in the way that spore-germination is triggered by the very conditions which kill other fungi. Spores of at least some species remain dormant for long periods between fires in the soil.

Generalizations about colonizing fungi

The diversity of key characters in the colonizers listed in Table 4.1 is impressive. However, the range of key characters could be made much greater by inclusion of other substratum-level successions, e.g. those on leaf-fragments taken into the nests of ants or termites or mould-building birds (Megapodiidae), the faeces of carnivorous animals, hair falling onto soil, or pollen falling into a water film on, say, *Sphagnum* (Hudson 1986). The range of key characters would be widened again by consideration of the colonizers in community-level successions. The only absolute generalizations that can be made are trivial: the fungi must be able to reach the substratum readily, tolerate the conditions en route and at the site and, if invasion is by spores, these must not germinate until the appropriate time.

The characteristics commonly assigned to colonizing fungi are summarized, together with exceptions, in Table 4.2. The exceptions in respect of life-length are particularly interesting for our discussion of succession as they can initiate 'back-to-front' sequences, e.g. where comparatively slow-growing and long-lived lignin-decomposing basidiomycetes attack micro-sites composed of ligno-cellulose, release sugars, and let in potentially fast-growing and presumably short-lived 'secondary sugar fungi' (Frankland 1981). Several of the 'exceptions' in Table 4.2 parallel the pioneering green plants characteristic of substrata poor in resources.

It is clear from Table 4.1 that there is little or nothing to be gained from recognizing a single category or 'strategy' of colonizer fungus, any

TABLE 4.2. Partial generalizations about colonizing fungi in substratum-level successions

Generalization	Exceptions
(1) Invasion is by spores rather than by hyphae or mycelial strands.	Parasites infecting via bark of roots, and saprophytes growing over unfavourable media to favourable new media (e.g. *Serpula lacrimans*); spread of colonizers by hyphae is probably commonplace in soil, e.g. in fungi forming vesicular-arbuscular mycorrhiza and rhizoplane fungi in densely rooted topsoil.
(2) Dispersal units are very widespread and travel rapidly.	Parasites and saprophytes on various recalcitrant substrata. The issue of scale is important; many commonplace substrata contain cell-types or micro-sites on which establishment is delayed, e.g. vessels, tracheids, lignified fibres and idioblasts in leaf litter.
(3) Germination rate and specific growth rate both high.	As under (2). Growth is slow on recalcitrant micro-sites.
(4) Responsive to, and dependent upon, easily assimilable carbohydrates.	As under (2). Species of recalcitrant micro-sites again do not conform.
(5) Adult mycelial mass and sporulating organs small.	Parasites of roots, trunks and branches of trees.

more than there is in the case of green plants. Nor does it help much to recognize two categories or 'strategies', one for fungi of sites rich in resources and the other for those of sites poor in resources, as is done (in effect) by Cooke & Rayner (1984). Nature has too many dimensions of variation to make such an approach lastingly satisfying. The parallel problems of generalization for green plants are discussed in detail elsewhere (Grubb 1985).

COMPARISON OF SUCCESSIONS OF GREEN PLANTS AND SUCCESSIONS OF FUNGI

Comparison of substratum-level successions of fungi and successions of green plants

In the case of green plants we expect primary succession to lead to a type of vegetation that is relatively constant in composition (integrated over a sufficient area and through sufficient time) with localized 'internal successions' occurring wherever the plant cover is disturbed. Strictly speaking there is no parallel in fungal successions at the substratum level, but the nearest to a parallel is seen in the fungal species forming sheathing mycorrhiza on a long-lived forest tree; in a natural forest the late-successional fungi probably remain in balance with the host for a period of the order of 10^2 years (rarely 10^3 years), and as new roots are formed each year they are occupied by the same fungi. To complete the analogy with succession in green plants we have to swap to the community level, and note the way in which the fungi forming sheathing mycorrhiza on young trees invade from the spore-bank in the soil, or are blown in, just like colonizing green plants in a forest gap.

The next nearest parallel to primary succession among green plants is seen in the fungi on the surfaces of healthy, living plant parts. However, in this case there is no equally long-lived constant terminal state, except perhaps on the bark of the trunk and major roots. Even here there is typically a continuing succession of micro-organisms and associated animals, for example as a kind of soil is built up from the remains of epiphytic green plants on the bark. Certainly on leaves there is a continuing succession until death, although the rate of change may be slow once mature size is reached in long-lived leaves of plants in areas too dry for development of many epiphylls, e.g. leaves of holly (*Ilex aquifolium*) living 5–8 years in eastern England and studied by Mishra & Dickinson (1981).

When plant parts die of old age or disease a new succession of fungi is initiated. This situation is analogous to that where a forest develops in a valley system after deglaciation of a landscape, and is then killed by flooding when a beaver builds its dam downstream, or a landslip blocks the drainage through a gorge, and invasion by quite different species is invoked. The successions of fungi on plant parts recently killed by pathogens or dying of old age have no parallel, strictly speaking, among green plants because the resources in the substratum are eventually all used up, and all species drive themselves (or are driven) to local extinction. They persist indefinitely only at the community level, migrating to new suitable sites like animals that have eaten the whole of an item of prey. The position is the same with the fungi in dung except that the succession there is primarily of the 'initial floristic composition' type, while that on litter follows the 'relay floristics' model of Egler (1954).

Mechanisms of succession

Connell & Slatyer (1977) suggested three models of succession for plants and sessile animals: tolerance, facilitation and inhibition. In fact, their classification is inadequate. As pointed out by Grubb & Hopkins (1986), there are many successions where it is helpful to distinguish between mere facilitation, i.e the phenomenon whereby species B is able to establish more quickly or more widely as a result of the prior occupation of the site by species A, from *enablement* which is the phenomenon whereby species A literally makes it possible for species B to invade. Similarly it is useful to distinguish mere inhibition, i.e. the slowing down of invasion of species B through the activity of species A, from *exclusion*, the phenomenon whereby species A makes it quite impossible for species B to become established. The more extreme phenomena, not considered by Connell & Slatyer, are seen most clearly where plants alter the properties of the soil beneath them dramatically, and examples are given in Table 4 of Grubb & Hopkins (1986). However, enablement may also involve the attraction to a site of specific dispersal agents, e.g. squirrels dispersing seeds of *Carya* and *Quercus* when visiting pines in long-abandoned old-fields in eastern North America (Finegan 1984).

The 'tolerance' model of Connell & Slatyer may explain several features of the fungal successions covered in Table 4.1, such as the change from a predominance of yeasts to a predominance of filamentous types on the surfaces of leaves as these senesce and leak more nutrients. Examples of both facilitation and enablement are also well established. The following cases are quoted from Hudson (1968, 1986). Plant litter is comminuted by animals of very various sizes, and the surface area for enzyme activity is increased; faecal pellets are not only enriched in available nitrogen but also have an improved water-holding capacity. In all these ways development of fungi is facilitated. There is also evidence that bacterial products in faeces stimulate the growth of certain fungi, while the monosaccharides released by cellulolytic fungi certainly increase the populations of 'sugar fungi'. A clear example of the more extreme phenomenon of enablement is provided by *Aspergillus halophilicus* and *A. restrictus* which can grow on stored cereal grain too dry for any other micro-organisms (13·2–13·5% fresh weight as water). Their respiration produces water, and at about 14·0–14·2% water content species of the *A. glaucus* group invade. Their respiration adds more water and thus progressively more and more species are enabled to invade. The accompanying large rise in temperature (up to 50 °C or more) certainly facilitates growth of thermophilic species, and may truly enable the most demanding thermophiles to invade.

Inhibition and exclusion are not clearly distinguished in the mycological literature. This much is clear from the review of Wicklow (1981b), who cites many cases where prior occupation of a substratum by one saprotrophic fungus has been said to 'prevent' invasion by another. Also several parasites seem to be able to delay, though not prevent, the entry of saprotrophs into recently dead tissue. Action at a distance via antibiotic compounds is commonly claimed; if this is true, exclusion rather than inhibition probably occurs. A special form of antagonism is found where the hypha of one fungal species can bring about the death of another on contact (Ikedingwu & Webster 1970). In this case the substratum—a pellet of dung, for example—may be sufficiently heterogeneous for the potential 'loser' to persist in micro-sites not reached by the potential 'winner', and thus inhibition may occur rather than exclusion.

THE TENUOUS CONNECTION BETWEEN *r*-SELECTION AND EARLY-SUCCESSIONAL STATUS

The ideas of *r*- and *K*-selection put forward by MacArthur & Wilson (1967), on the basis of the theory of MacArthur (1962), concerned evolution where selection is respectively density-independent (D-I) and density-dependent (D-D). As Boyce (1984) has pointed out, a most unfortunate confusion has arisen in the literature between the original issue of D-I and D-D selection on the one hand, and the evolution of life-history traits (so-called 'strategies' and 'tactics') on the other. The paper of Pianka (1970) seems to have played a key part in leading people to link the idea of *r*-selection with occupation of transient sites, and the possession of highly dispersible propagules, early reproduction, high allocation of resources to reproduction and large numbers of offspring. (In fact, high values of *r*, the intrinsic rate of increase of the population, may arise from low rates of mortality just as well as from high rates of fecundity, as Boyce points out, but this point has been generally overlooked.) *K*-selection in Pianka's (1970) paper was associated with the opposite characters from those associated with *r*-selection. Very soon the idea of *r*-selection was being associated with short-lived plants of transient sites, and *K*-selection with longer-lived 'more competitive' plants (Gadgil & Solbrig 1972). The correspondence between *r*- and *K*-selection on the one hand, and early- and late-successional plants on the other, was made explicit by Wells (1976), Grime (1977), Harper (1977), Topham (1977) and During (1979) for plants, and by Lacey (1979) for fungi, and has been widely accepted. I have to confess to being at fault too (Grubb 1976, p. 70).

In fact, this supposed correspondence is not real. As Boyce (1984,

p. 433) wrote 'Obviously different life histories are associated with temporary vs. permanent habitats, but the association with *r*- and *K*-selection is not always a clear one.'

I think that the historical root of the confusion may be in the development of the ideas of *r*- and *K*-selection by MacArthur & Wilson (1967) in the context of the emerging study of island biogeography. Clearly it is reasonable to suppose that only one or a very few individuals of a colonizing species will reach a far-away virgin island. Initially, the population of this species may increase in a way that is largely free of D-D effects, provided that its disseminules are mostly very widely dispersed. In contrast, the species arriving later and destined to build up a persistent cover composed mostly of their offspring are likely to experience interspecific D-D effects from the start and in species-poor systems are bound to run into intraspecific D-D effects eventually. A similar position may hold in some primary successions of plants which are not literally on islands but where colonizers are sparse, e.g. on unstable moraines and screes or on some rocks with few fissures.

The position in secondary succession is quite different. In this case the pulses of resources, which are tracked by species so often referred to as '*r*-selected', commonly become available in 'islands' amidst vegetation containing mature, seed-producing individuals of those very species (i.e. on other 'islands' not far away). And in many cases the site of the 'island' of newly available resources will have been in a similar state at some time during the last decade or century, and so have in the soil a bank of dormant seeds of relevant species ready to germinate with appropriate stimuli. It is therefore hardly surprising that the plant cover in the early stages of many secondary successions is typically dense in appearance. Very probably D-D effects on survival and fecundity are the rule (cf. Raynal & Bazzaz 1975). Certainly an elegant and critical analysis of D-D effects has been published for the early-successional tree *Prunus pensylvanica* (Mohler, Marks & Sprugel 1978). In an analogous way D-D effects are probably common in colonizing fungi, whether on the surfaces of living plants, on litter or in faeces. This is true despite the fact that the fungi may cover only about 1–10% of the surface because they are able to act at a distance, both through production of antibiotics and through overlapping zones of absorption.

As far as species-rich late-successional plant communities are concerned, there is evidence that for species which are sparsely distributed intraspecific D-D effects may be of relatively little importance in controlling population size (Grubb 1986b; Hubbell & Foster, Chapter 19). However, the generalized effects of 'crowding' of any one species by all the others, emphasized by Parry (1981), are bound to be important in closed vegetation—whether it is closed above-ground or only below-ground, as in a semi-desert. Presumably the same is true for late-successional fungal communities.

The issue of *scale* is critical for studies on *intra*specific density-dependence. Often a species will occupy only a small fraction of the micro-sites suitable for it in a given community, and D-D effects over substantial areas may not be detectable, while D-D effects within small patches may be important. The 'biennial' plants studied by van der Meijden *et al.* (1985) on dunes in Holland can show just this kind of effect. In runs of dry years the populations crash, and many patches become extinct. When the populations expand in a run of wet years, they never occupy more than a fraction of the physically suitable sites. But within the sites that are occupied some D-D effects have been found, e.g. for *Cynoglossum officinale*; of the species studied by van der Meijden *et al.* this is the one least eaten at the seed stage. Significantly, the same study did not reveal any D-D effects, even in small patches, in *Cirsium vulgare* which suffers heavy predation on its seeds. However, the unit-area of study was 2 dm^2, and D-D effects might have been found if smaller unit-areas had been used. Ideally, studies on D-D effects should be made in terms of the survival or performance of a plant in relation to the numbers of other plants at measured distances from the one in question, i.e. in relation to the neighbourhood characteristics of an individual plant (cf. Mack & Harper 1977).

In future, ecologists should be content to label species as respectively early-, mid- or late-successional, and keep this issue separate from that of *r*- and *K*-selection.

CONCLUSIONS

At sites with a poor supply of water and mineral nutrients the best suited pioneer green plants are relatively slow-growing and long-lived, and are not necessarily highly dispersible. At unstable sites the pioneers are also usually long-lived; some have large disseminules, and others invade primarily by vegetative spread. In many cases shorter-lived species appear once the substratum has become stabilized and/or the supply of resources increased. All these cases contrast with the common type of primary succession seen on stable sites with relatively large supplies of water and nutrients, in which shorter-lived plants are the pioneers, and also contrast with most secondary successions in which shorter- and longer-lived plants appear together as colonizers but the shorter-lived plants are quickly lost or confined to the seed bank.

Among fungi there is a parallel diversity in colonizing species, which can be related to the amounts of water and nutrients available, difficulties in penetrating the substratum and problems involved in reaching the substratum and beginning growth at the appropriate time.

In the past, much of the theory about colonizers has been too narrowly based. New critical work is needed on all kinds of biological characteristics, including breeding systems, heterozygosity and niche breadth in colonizers that are not short-lived herbaceous angiosperms, but trees, ferns, mosses and lichens, and also fungi of the phylloplane, rhizoplane or sheathing mycorrhiza.

The phenomenon of succession among green plants often involves a greater impact of the plants on the site than has been suggested in much recent literature; for example, changes in pH, availability of mineral nutrients, or aeration. Mycologists have paid more attention to the extreme phenomena here called 'enablement' and 'exclusion' but much more research is needed on the precise mechanisms involved.

The difference between colonization of distant virgin sites and reoccupation of newly available patches of resources at sites near to existing populations ought to be accorded more importance. Ecologists working with all groups of organisms should be careful not to confuse the issue of *r*- versus *K*-selection with that of early- versus late-successional status.

ACKNOWLEDGMENTS

I thank all three editors for constructive criticism of an early draft. Drs H.J. During and J. White also made valuable suggestions. The section on fungi benefited greatly from discussions with Dr H.J. Hudson.

REFERENCES

Ahmadjian, V. (1967). *The Lichen Symbiosis*. Blaisdell, Waltham, Mass.

Bailey, R.H. (1976). Ecological aspects of dispersal and establishment in lichens. *Lichenology: Progress and Problems* (Ed. by D.H. Brown, D.L. Hawksworth & R.H. Bailey), pp. 215–247. Academic Press, London.

Baker, H.G. & Stebbins, G.L. (Eds) **(1965).** *The Genetics of Colonizing Species*. Academic Press, New York.

Barbour, M.G. & Johnson, A.F. (1977). Beach and dune. *Terrestrial Vegetation of California* (Ed. by M.G. Barbour & J. Major), pp. 223–261. Wiley, New York.

Boyce, M.S. (1984). Restitution of *r*- and *K*-selection as a model of density-dependent natural selection. *Annual Review of Ecology & Systematics*, **15**, 427–447.

Bradshaw, A.D., Chadwick, M.J., Jowett, D. & Snaydon, R.W. (1964). Experimental investigations in the mineral nutrition of several grass species. IV. Nitrogen level. *Journal of Ecology*, **52**, 665–676.

Brown, J.C. (1958). Soil fungi of some British sand dunes in relation to soil type and succession. *Journal of Ecology*, **46**, 641–664.

Brown, A.H.D. & Marshall, D.R. (1981). Evolutionary changes accompanying colonization in plants. *Evolution Today* (Ed. by G.G.E. Scudder & J.L. Reveal), pp. 351–363. Hunt Institute for Botanical Documentation, Pittsburgh.

Chapman, V.J. (1977). *Ecosystems of the World. I. Wet Coastal Ecosystems.* Elsevier, Amsterdam.

Clements, F.E. (1916). *Plant Succession.* Carnegie Institution, Washington.

Connell, J.H. & Slatyer, R.O. (1977). Mechanisms of succession in natural communities and their role in community stability and organization. *American Naturalist*, **111**, 1119–1144.

Cooke, E.C. & Rayner, A.D.M. (1984). *The Ecology of Saprotrophic Fungi.* Longman, London.

Cooke, W.B. & Lawrence, D.B. (1959). Soil mould fungi isolated from recently glaciated soils in south-eastern Alaska. *Journal of Ecology*, **47**, 529–549.

Cooper, W.S. (1916). Plant successions in the Mount Robson region, British Columbia. *Plant World*, **19**, 211–238.

Curtis, J.T. (1959). *The Vegetation of Wisconsin.* University of Wisconsin Press, Madison.

Davis, M.B. (1986). Climatic instability, time lags and community disequilibrium. *Community Ecology* (Ed. by J. Diamond & T.J. Case), pp. 269–284. Harper & Row, New York.

Dighton, J. & Mason, P.A. (1985). Mycorrhizal dynamics during forest tree development. *Developmental Biology of Higher Fungi* (Ed. by D. Moore, L.A. Casselton, D.A., Wood & J.C. Frankland), pp. 117–139. Cambridge University Press.

During, H.J. (1979). Life strategies of bryophytes: a preliminary review. *Lindbergia*, **5**, 2–18.

Egler, F.E. (1954). Vegetation science concepts: Initial floristic composition—a factor in old-field development. *Vegetatio*, **4**, 412–417.

Ellenberg, H. (1982). *Die Vegetation Mitteleuropas mit den Alpen in ökologischer Sicht*, 3rd edn. Ulmer, Stuttgart.

Finegan, B. (1984). Forest succession. *Nature (London)*, **312**, 109–114.

Fisher, F.J.F. (1952). Observations on the vegetation of screes in Canterbury, New Zealand. *Journal of Ecology*, **40**, 156–167.

Flaccus, E. (1959). Revegetation of landslides in the White Mountains of New Hampshire. *Ecology*, **40**, 692–703.

Frankland, J.C. (1981). Mechanisms in fungal successions. *The Fungal Community: Its Organization and Role in the Ecosystem* (Ed. by D.T. Wicklow & G.C. Carroll), pp. 403–426. Dekker, New York.

Gadgil, M. & Solbrig, O.T. (1972). The concept of *r*- and *K*-selection: evidence from wildflowers and theoretical considerations. *American Naturalist*, **106**, 14–31.

Garrett, S.D. (1963). *Soil Fungi and Soil Fertility.* Pergamon, Oxford.

Gochenaur, S.E. (1981). Response of soil fungi communities to disturbance. *The Fungal Community: Its Organization and Role in the Ecosystem* (Ed. by D.T. Wicklow & G.C. Carroll), pp. 458–479. Dekker, New York.

Gray, A.J. (1986). Do invading species have definable genetic characteristics? *Proceedings of the Royal Society of London. B.* (in press)

Gray, A.J. & Scott, R. (1977). Biological Flora of the British Isles: *Puccinellia maritima* (Huds.) Parl. *Journal of Ecology*, **65**, 699–716.

Grime, J.P. (1977). Evidence for the existence of three primary strategies in plants and its relevance to ecological and evolutionary theory. *American Naturalist*, **111**, 1169–1194.

Grubb, P.J. (1976). A theoretical background to the conservation of ecologically distinct groups of annuals and biennials in the chalk grassland ecosystem. *Biological Conservation*, **10**, 53–76.

Grubb, P.J. (1985). Plant populations and vegetation in relation to habitat, disturbance and competition: problems of generalization. *The Population Structure of Vegetation* (Ed. by J. White), pp. 595–621. Junk, Dordrecht.

Grubb, P.J. (1986a). The ecology of establishment. *Ecology and Landscape Design* (Ed. by A.D. Bradshaw, D.A. Goode & E. Thorp), pp. 83–98. British Ecological Society Symposium Vol. No. **24**. Blackwell Scientific Publications, Oxford.

Grubb, P.J. (1986b). Problems posed by sparse and patchily distributed species in species-rich plant communities. *Community Ecology* (Ed. by J. Diamond & T.J. Case), pp. 207–225. Harper & Row, New York.

Grubb, P.J. & Hopkins, A.J.M. (1986). Resilience at the level of the plant community. *Resilience in Mediterranean-type Ecosystems.* (Ed. by B. Dell, A.J.M. Hopkins & B.B. Lamont), pp. 21–38. Junk, Dordrecht.

Hanes, T.L. (1971). Succession after fire in the chaparral of southern California. *Ecological Monographs*, **41**, 27–52.

Harper, J.L. (1977). *Population Biology of Plants*. Academic Press, London.

Hartog, C. den (1970). *The Sea-grasses of the World*. North-Holland, Amsterdam.

Hudson, H.J. (1968). The ecology of fungi on plant remains above the soil. *New Phytologist*, **67**, 837–874.

Hudson, H.J. (1986). *Fungal Biology*. Edward Arnold, London.

Ikedingwu, F.E.O. & Webster, J. (1970). Antagonism between *Coprinus heptemerus* and other coprophilous fungi. *Transactions of the British Mycological Society*, **54**, 181–204.

Keever, C. (1957). Establishment of *Grimmia laevigata* on bare granite. *Ecology*, **38**, 422–429.

Lacey, J. (1979). Aerial dispersal and the development of microbial successions. *Microbial Ecology: a Conceptual Approach* (Ed. by J.M. Lynch & N.J. Poole), pp. 140–170. Blackwell Scientific Publications, Oxford.

Lindsay, D.C. (1977). Lichens of cold deserts. *Lichen Ecology* (Ed. by M.R.D. Seaward), pp. 183–209. Academic press, London.

Longton, R.E. & Schuster, R.M. (1983). Reproductive biology. *New Manual of Bryology* (Ed. by R.M. Schuster), pp. 386–462. Hattori Botanical Laboratory, Michinan.

MacArthur, R.H. (1962). Some generalized theorems of natural selection. *Proceedings of the National Academy of Sciences*, **48**, 1893–1897.

MacArthur, R.H. & Wilson, E.O. (1967). *The Theory of Island Biogeography*. Princeton University Press, Princeton, New Jersey.

McComb, A.J., Cambridge, M.L., Kirkman, H. & Kuo, J. (1981). The biology of Australian sea grasses. *The Biology of Australian Plants* (Ed. by J.S. Pate & A.J. McComb), pp. 259–294. University of Western Australia Press, Nedlands.

Mack, R. & Harper, J.L. (1977). Interference in dune annuals: spatial pattern and neighbourhood effects. *Journal of Ecology*, **65**, 345–363.

Manners, J.G. (1982). *Principles of Plant Pathology*. Cambridge University Press, Cambridge.

Mason, P.A., Wilson, J., Last, F.T. & Walker, C. (1983). The concept of succession in relation to the spread of sheathing mycorrhizal fungi on inoculated tree seedlings in unsterile soils. *Plant & Soil*, **71**, 247–256.

Meijden, E. van der, De Jong, T.J., Klinkhamer, P.G.L. & Kooi, R.E. (1985). Temporal and spatial dynamics in populations of biennial plants. *Structure and Functioning of Plant Populations* 2 (Ed. by J. Haeck & J.W. Woldendorp), pp. 91–103. North-Holland, Amsterdam.

Mishra, R.R. & Dickinson, C.H. (1981). Phylloplane and litter fungi of *Ilex aquifolium*. *Transactions of the British Mycological Society*, **77**, 329–337.

Mohler, C.L., Marks, P.L. & Sprugel, D.G. (1978). Stand structure and allometry of trees during self-thinning of pure stands. *Journal of Ecology*, **66**, 599–614.

Oosting, H.J. & Anderson, L.E. (1939). Plant succession on granite rock in eastern North Carolina, *Botanical Gazette*, **100**, 750–768.

Park, D. (1968). The ecology of terrestrial fungi. *The Fungi: An Advanced Treatise* (Ed. by G.C. Ainsworth & A.S. Sussman), pp. 5–39. Academic Press, New York.

Parry, G.D. (1981). The meanings of *r*- and *K*-selection. *Oecologia*, **48**, 260–264.

Pianka, E.R. (1970). On *r*- and *K*-selection. *American Naturalist*, **104**, 592–597.

Pugh, G.J.F. (1962). Studies on fungi in coastal soils. II. Fungal ecology in a developing salt marsh. *Transactions of the British Mycological Society*, **45**, 560–566.

Quarterman, E. (1950). Major plant communities of Tennessee cedar glades. *Ecology*, **31**, 234–254.

Ranwell, D.S. (1972). *The Ecology of Salt Marshes and Sand Dunes*. Chapman & Hall, London.

Raynal, D.J. & Bazzaz, F.A. (1975). The contrasting life-cycle strategies of three summer annuals found in abandoned fields in Illinois. *Journal of Ecology*, **63**, 587–596.

Saenger, P., Specht, M.M., Specht, R.L. & Chapman, V.J. (1977). Mangal and coastal salt-marsh communities in Australia. *Ecosystems of the World*. I. *Wet Coastal Ecosystems* (Ed. by V.J. Chapman), pp. 293–345. Elsevier, Amsterdam.

Scott, G.A.M. (1963). The ecology of shingle beach plants. *Journal of Ecology*, **51**, 517–527.

Scott, G.D. (1960). Studies of the lichen symbiosis. I. The relationship between nutrition and moisture content in the maintenance of the symbiotic state. *New Phytologist*, **59**, 374–381.

Thorne, R.F. (1977). Montane and subalpine forests of the Transverse and Peninsula Ranges. *Terrestrial Vegetation of California* (Ed. by M.G. Barbour & J. Major), pp. 537–557. Wiley, New York.

Topham, P.B. (1977). Colonization, growth, succession and competition. *Lichen Ecology* (Ed. by M.R.D. Seaward), pp. 31–68. Academic Press, London.

Tutin, T.G. (1942). Biological Flora of the British Isles: *Zostera*. *Journal of Ecology*, **30**, 217–226.

Walter, H. (1971). *The Ecology of Tropical and Subtropical Vegetation*. Oliver & Boyd, Edinburgh.

Webley D.M., Eastwood, D.J. & Gimingham, C.H. (1952). Development of a soil microflora in relation to plant succession on sand-dunes, including the 'rhizosphere' flora associated with colonizing species. *Journal of Ecology*, **40**, 168–178.

Wells, P.V. (1976). A climax model for broadleaf forest: a *n*-dimensional ecomorphological model of succession. *Central Hardwood Forest Conference* (Ed. by J.S. Fralish, G.J. Weaver & R.C. Schlesinger), pp. 131–176. Department of Forestry, Southern Illinois University, Carbondale, Illinois.

Wheeler, B.E.J. (1969). *An Introduction to Plant Diseases*. Wiley, London.

Wicklow, D.T. (1981a). The coprophilous fungal community: a mycological system for examining ecological ideas. *The Fungal Community: its Organization and Role in the Ecosystem* (Ed. by D.T. Wicklow & G.C. Carroll), pp. 47–76. Dekker, New York.

Wicklow, D.T. (1981b). Interference competition and the organization of fungal communities. *The Fungal Community: Its Organization and Role in the Ecosystem* (Ed. by D.T. Wicklow & G.C. Carroll), pp. 351–375. Dekker, New York.

Willoughby, L.G. (1974). Decomposition of litter in fresh water. *Biology of Plant Litter Decomposition* (Ed. by C.H. Dickinson & G.J.F. Pugh), pp. 659–681. Academic Press, London.

5. SEED CHARACTERISTICS IN RELATION TO SUCCESSION

M. FENNER
Department of Biology, Southampton University, SO9 5NH

INTRODUCTION

During the course of many terrestrial seres the majority of higher plant species enter the succession as seeds. The seeds may arrive at any stage in the sere, but their entry, germination and establishment may be either hindered or facilitated by the vegetation already present. This chapter examines the properties of seeds of species which are dominant at different stages in various terrestrial seres, and attempts to show how their characteristics help to explain the sequence of floristic changes which constitute succession. The focus of this chapter is on plants characteristic of relatively mesic sites with adequate supplies of minerals and water. However, as Grubb (Chapter 4) points out, generalizations derived from such examples do not necessarily apply to all cases of succession. For convenience the sere will be considered to go through three phases: pioneer, intermediate and climax.

PIONEER SPECIES

One of the chief characteristics of a pioneer species, especially in fertile sites, is the ability to arrive at the newly-created open site as quickly as possible. One way of achieving this is by having an effective means of dispersal. In primary successions the pioneer species are predominantly small-seeded and wind dispersed. For example, the first colonizers on the gravel outwash of the Muldrow Glacier, Alaska, are the wind-dispersed *Dryas drummondii*, *Epilobium latifolium*, *Crepis nana* and several *Salix* species. A pioneer legume, *Astragalus nutzotinensis* is curiously well suited for seed dispersal on glacier deposits by possessing a pod which forms a seed-scattering wheel after dehiscence (Viereck 1966).

In many primary habitats (such as glacier gravel and sand dunes) the newly exposed surface appears continuously, and all stages of the sere are represented locally. Seeds of the pioneer species are therefore readily available, often within a few metres of the fresh substrate. In fact, many pioneers of sand dunes are stoloniferous and colonize by vegetative means. However, primary habitats which arise by means of sudden catastrophic events, such as volcanoes, may not have appropriate colonizing species locally available. The colonization of isolated newly-formed volcanic islands such as Surtsey indicates the importance of wind dispersal for pioneers in such habitats (Fridriksson 1975), though a minority of species may be transported by sea. Once the vegetation develops beyond the pioneer stage birds may become the predominant means of seed dispersal to islands (Carlquist 1967).

The dispersal problems faced by colonizing species in secondary successions are essentially similar to those in primary successions. The sites involved are unpredictable in occurrence, dispersed in space, and fleeting in duration. In nature, major disturbances causing vegetation removal over any appreciable area would be caused only occasionally by floods, fires, landslides and hurricanes. In recent times the exposure of large tracts of bare soil by agriculture has resulted in a great expansion of these early-successional species. The seeds of the species which grow in these secondary habitats tend to have a suite of characters which enable them to exploit erratic and ephemeral sites.

Since heavy seeds are in general less easily dispersed by wind than light ones, it is perhaps not surprising to find that colonizing species as a group have notably small seeds. This is well illustrated by the data from Salisbury's (1942) classic study on seed sizes of British plants from various successional stages of maturity (Fig. 5.1). Note that although the most frequent weight category for the open-habitat species is 0·24–0·98 mg, that for mid-successional shrubs 16 times larger, and that for forest trees 256 times larger, there is a very wide range of seed weights amongst the plants at each successional stage.

This size difference might of course be partly a function of the size of the parent plants. A world survey of seed sizes in relation to parent plant sizes indicates that Salisbury's British data is fairly representative in this respect (Levin & Kerster 1974). Nevertheless, there is some evidence that seed size is related to successional status independent of parent size. Examples are provided by species which occur in both early and late stages in a sere. Werner & Platt (1976) recorded the seed weights of some goldenrod (*Solidago*) species which have early- and late-successional populations. Their data (Table 5.1) show that for four out of the five species tested, small seeds are associated with the early successional populations.

Small size alone is not, of course, a guarantee of wide dispersal. The exact mechanism employed by many pioneer species is still unknown. Some species have obvious mechanisms which aid dispersal by wind, e.g.

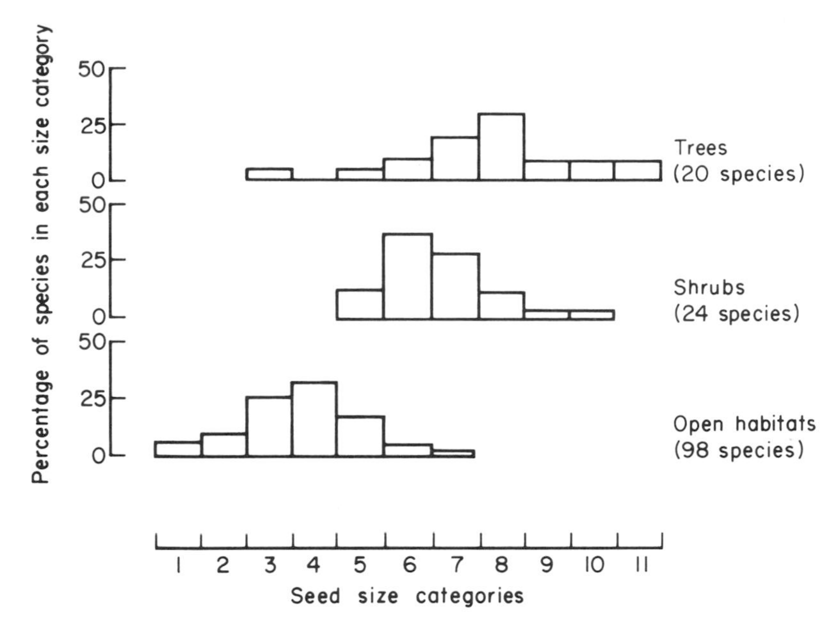

FIG. 5.1. The distribution of seed weights in three habitats of increasing maturity. The frequencies are expressed as percentages of total numbers sampled in each habitat. The size classes increase geometrically, each class being four times that of the preceding one. The lower boundaries of the classes are (g): $3{\cdot}81 \times 10^{-6}$(1), $1{\cdot}53 \times 10^{-5}$(2), $6{\cdot}10 \times 10^{-5}$(3), $2{\cdot}44 \times 10^{-4}$(4), $9{\cdot}77 \times 10^{-4}$(5), $3{\cdot}91 \times 10^{-3}$(6), $1{\cdot}56 \times 10^{-2}$(7), $6{\cdot}25 \times 10^{-2}$(8), 0·25(9), 1(10), 4(11). (From Salisbury 1942.)

TABLE 5.1. Mean weights (μg ± SE) of seeds (achenes) of five species of goldenrod (*Solidago*) which have populations in early and late successional stages in prairie (Werner & Platt 1976)

Species	Old-field	Prairie
S. nemoralis	26·7 ± 2·1	104·0 ± 8·3
S. missouriensis	17·6 ± 0·6	39·3 ± 3·2
S. speciosa	19·5 ± 1·6	146·3 ± 11·7
S. canadensis	27·3 ± 2·3	58·3 ± 11·1
S. graminifolia	24·5 ± 2·7	10·6 ± 1·6

groundsel (*Senecio vulgaris*) and willowherb (*Epilobium montanum*), but most species do not. Open habitats throughout the world are often dominated by members of the Compositae, a family noted for the modification of its calyx into a feathery pappus for wind dispersal of the one-seeded fruits. But even amongst the Compositae a substantial minority of colonizing species lack the familiar pappus, e.g. scentless mayweed (*Tripleurospermum inodorum*), pineapple weed (*Matricaria matricarioides*), corn marigold (*Chrysanthemum segetum*) and even the common daisy itself (*Bellis perennis*). With regard to plants in other families, it is unlikely that wind would normally carry the seeds of species like knotgrass (*Polygonum aviculare*) and poppies (*Papaver rhoeas*) more than a few metres.

The most obvious alternative to wind transport for those species is external carriage by animals, probably in the mud on their feet, or on human footwear. In a survey of seeds carried on footwear (Clifford 1956) recorded forty-three species which were carried in this way, including annual meadow grass (*Poa annua*), daisy (*Bellis perennis*), chickweed (*Stellaria media*), plantain (*Plantago lanceolata*) and nettle (*Urtica dioica*). In nature these species may have been associated with animal tracks and watering places.

Seedlings of *Stellaria media* and *Poa annua* can often be seen in horse dung deposited in heathland in the New Forest (England), arising from seeds ingested elsewhere. Welch (1985) recorded eighty-eight species of plants in dung of cattle, red deer, sheep, grouse, hares and rabbits on moorland. This form of seed dispersal may be more common among early colonizers than is suspected, and its occurrence lends weight to Janzen's idea that the vegetative parts of plants may act as the attractant for herbivorous seed dispersers (Janzen 1984), though Collins & Uno (1985) contest this idea.

By far the best way for a pioneer species of secondary succession to ensure that it is first on the spot when an opening occurs is to be there already in anticipation of the disturbance. A feature of many such species is the possession of a reserve of dormant seeds in the soil waiting to germinate as soon as the opportunity arises. This mechanism of colonization is clearly not available to pioneers of primary seres. Examples of species which characteristically have large soil seed reserves are given in Thompson & Grime (1979). A familiar example is the poppy (*Papaver rhoeas*) which appeared so quickly on the French battlefields after the First World War, and which undoubtedly arose from a soil seed bank suddenly released from dormancy.

Having a reserve of dormant seeds in the soil can be seen as an alternative strategy to long distance dispersal. However, the strategy of waiting for the next episode of disturbance to occur has an important disadvantage: it can only succeed if disturbance occurs within the period that the seeds remain viable. Although some remarkable examples of longevity exist (e.g. Kivilaan & Bandurski 1981), most open-habitat species lose their viability under field conditions within a few decades at most. Roberts & Feast (1973) showed rather neatly that the number of viable seeds of a range of weed species declines exponentially with time in the field, with an overall loss of 12% per year in an uncultivated soil (Fig. 5.2). This gives a half-life of just over 5 years taken over all the species monitored, but some species (e.g. *Polygonum aviculare*, *P. convolvulus* and *Thlaspi arvense*) declined much more slowly (6–8% per year). The potential for long-term viability would be most advantageous in an environment where major disturbance takes place only at long intervals.

Having arrived at the appropriate site (either by distant dispersal or by remaining dormant in the soil) the pioneer seed needs to be able to detect that it is favourably placed for establishment before it germinates. Among open-habitat species there are a variety of gap-detecting mechanisms which prevent their seeds from germinating in the competitive environment of closed vegetation.

One such mechanism is a requirement for fluctuating temperatures to break dormancy. The temperature regime in a gap is much more extreme than that under a closed canopy, so the sudden occurrence of high diurnal

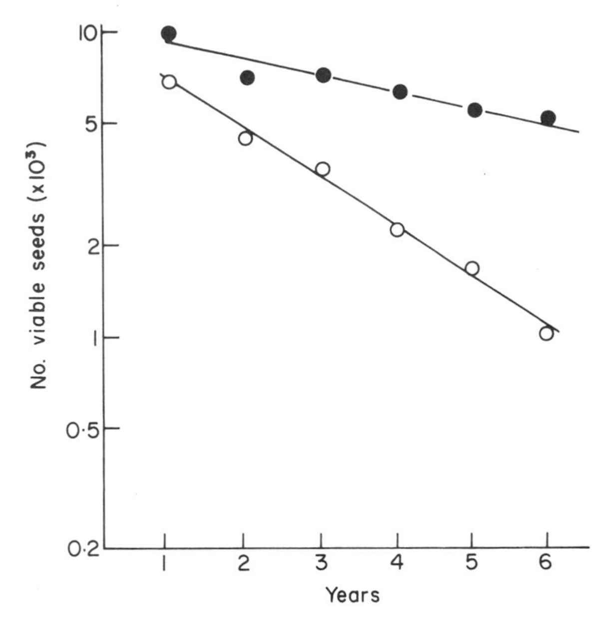

FIG. 5.2. Numbers of viable seeds of twenty species of annual arable weeds remaining after 1–6 years in undisturbed (●) and in cultivated (○) soil. The soil samples containing the seeds were placed in open cyclinders under field conditions. (Roberts & Feast 1973.)

temperature fluctuations can act as a signal to a buried seed that an opening has been created. Sensitivity to temperature fluctuations can also act as a depth-sensing mechanism, preventing seeds from germinating at depths too great for emergence to be possible (see Fenner 1985). Thompson & Grime (1983) quantify the temperature fluctuation requirement as the minimum fluctuation needed to achieve 50% of maximum germination. For example, the early-successional species sowthistle (*Sonchus arvensis*) and persicaria (*Polygonum persicaria*) require temperature fluctuations of 7·5 and 8·0 °C respectively.

Another characteristic which may act as a gap-detecting mechanism in gap colonizers, is sensitivity to light. Germination inhibition by darkness would force any buried seed to remain dormant, whether in open ground or not. As a group, the pioneer species are light-requiring, as shown in the large-scale screening tests of Grime *et al.* (1981). But perhaps more important is the sensitivity of many of these species to the quality of the light. Under a leaf canopy the ratio of red to far-red light is reduced. Most open-habitat species have the ability to differentiate between daylight and leaf-transmitted light. There are many examples in the literature of germination being inhibited in leaf-transmitted light. Gorski, Gorska & Nowicki (1977) tested a wide range of species and showed that the germination of seeds of open-habitat plants were generally inhibited by light with a low red/far-red ratio. Out of eighteen East African weeds tested by Fenner (1980a), only two were indifferent to light quality. This sensitivity to leaf canopy shade can be regarded as another gap-detecting mechanism. The effectiveness of this feature can be inferred from the fact that the pioneers which dominate the community in the first season are often absent from above-ground vegetation in the second season. They appear to be unable to establish a second generation, probably due to an inability to germinate. Even in short grass (15 cm high) germination is markedly inhibited in a sensitive species (Fenner 1980b).

Germination inhibition in leaf-canopy shade is a feature which would only be of benefit to species in a community subject to regular disturbance causing canopy removal on a large scale. Without frequent removal of at least part of the canopy the seeds would remain ungerminated. In communities subjected only to small-scale disturbance, such as that caused by the death of individuals, the proportion of canopy-sensitive species may be reduced. In a permanent chalk grassland only seventeen out of twenty-seven species showed germination inhibition in low red/far-red light (Silvertown 1980). Even within a single species, seeds from arable populations may show a greater degree of canopy inhibition than seeds from grassland sites. This was shown in *Poa trivialis* by Hilton, Froud-Williams & Dixon (1984).

The relevance of these seed characteristics to succession is that they help to explain (a) how the colonizers arrive quickly, (b) how they detect they are in open ground, and (c) why these species disappear so quickly from the vegetation. In addition, the small seed size of at least some of these open-habitat species makes their newly germinated seedlings very sensitive to competition from neighbouring plants, and so may exclude them from rapidly closing vegetation (Fenner 1978). The disadvantage of small seed size is well illustrated in experiments by Gross (1984) in which establishment in *Poa pratensis* turf is compared in six monocarpic perennials. Emergence and subsequent growth were consistently lower in the small seeded plants.

MID-SUCCESSIONAL SPECIES

The mid-successional stages of temperate forest seres are often characterized by a predominance of berry-bearing shrubs whose seeds are dispersed mainly by birds. For example, the scrub stage on chalk in Britain is dominated by species such as hawthorn (*Crataegus monogyna*), privet (*Ligustrum vulgare*), blackthorn (*Prunus spinosa*), purging buckthorn (*Rhamnus catharticus*), dogwood (*Thelycrania sanguinea*), wayfaring tree (*Viburnum lantana*), spindle (*Euonymus europaeus*), etc. In spite of their great prominence in mid-successional stages (and their abundance in hedgerows) surprisingly little is known about the reproductive ecology of these species. Their germination rates seem to be very low if tested straight from the fruits, though in some cases this increases with the passage of time or with chilling. (W.G. Lee, pers. comm; Table 5.2). Their germination rate may increase after passage through the digestive system of a bird, but few experiments have been done on such pre-treated seeds.

A substantial proportion of ingested seeds may be digested or at least rendered inviable (Krefting & Roe 1949). The necessity for the seed to survive passage through the bird's digestive tract may have led to the evolution of the thick endocarp seen in many species. The difficulties of absorbing water through this impermeable seed covering may at least partly account for the characteristically delayed germination.

Unlike seeds of the early successional species, seeds of berried plants are likely to be dispersed very unevenly, perhaps in concentrated groups and often with a mixture of other berried species. The accompanying

TABLE 5.2. Percentage germination in seeds of common British shrubs in chalk grassland soil. The seeds were planted on 4 December 1984 and left out of doors (W.G. Lee, unpublished)

Species	No. of seeds planted	15 April (132 days)	16 May (163 days)	16 June (194 days)	14 July (222 days)
Thelycrania sanguinea	960	0	2·3	6·0	6·7
Crataegus monogyna	320	0	0	0	0
Ligustrum vulgare	800	0	5·1	31·6	36·5
Rhamnus catharticus	640	0	4·5	16·7	17·3
Rosa canina	480	0	1·9	1·9	1·9
Viburnum lantana	480	0	0·8	39·4	60·6

faeces may have an important fertilizing effect for the newly emerged seedlings, though some seeds are merely regurgitated from the crop rather than defecated. The effect of the clumped pattern of deposition may well account for the fact that mature plants of these species so often occur in clumps with other bird-dispersed species.

Some relevant observations on the feeding of a range of berry-eating birds have been made by Sorensen (1981) in scrub near Oxford. She found that different species of birds not only choose characteristic selections of fruits, but also have different types of perching sites where they are presumed to defecate the ingested seeds. Tits tended to take the smaller fruits, e.g. elderberries (*Sambucus nigra*), honeysuckle (*Lonicera periclymenum*), bittersweet (*Solanum dulcamara*), and then to retire into the undergrowth. Thrushes, in contrast, tended to choose larger fruits, e.g. haws (*Crataegus monogyna*), sloes (*Prunus spinosa*), rosehips (*Rosa canina*), and to eat them on the spot. The effect of these differences in behaviour on the distribution pattern of the seeds was not investigated but it may well result in quite distinct differences in the patterns of deposition of thrush and tit-dispersed seeds, and ultimately affect the pattern of the vegetation.

Another feature which has a bearing on the course of succession is the *rate* at which seeds of these species are deposited. In a secondary succession bird-dispersed seeds will enter the habitat in very low numbers until the birds have somewhere to roost. This was shown very neatly by McDonnell & Stiles (1983) by comparing seed deposition in traps along transects in two abandoned fields, one of 2 years old and one of 12 years old. The younger field had vegetation 0·5–1·2 m high of simple structure; the older field had a more complex vegetation structure with scattered shrubs. Overall, seed deposition by birds was 3·7 times greater in the older field (Table 5·3). The fact that this was related to the existing woody plants in the vegetation was demonstrated by an experiment in which artificial 'trees' of various shapes were set up. There is little doubt that the appearance of a few pioneer trees or shrubs (in some cases perhaps wind-dispersed) which act as roosts greatly accelerates the rate of succession to the main shrub stage. In the terminology of Connell & Slatyer (1977) the existing vegetation 'facilitates' the arrival of subsequent species into the same sere.

LATE SUCCESSIONAL SPECIES

Rather more is known about the seeds of the later successional species such as forest trees. This is partly because of their commercial interest. The seeds tend to have a cluster of characteristics which contrast with those of the pioneers. As we have seen from Salisbury's data they are generally larger, but this cannot be taken to mean they are necessarily poorly dispersed. Many European and North American forest trees have wind-dispersed seed, e.g. elm (*Ulmus* spp.), maple (*Acer* spp.), hornbeam (*Carpinus* spp.), lime (*Tilia* spp.), ash (*Fraxinus* spp.), and many pines (*Pinus* spp.). Those with larger seeds, such as oak (*Quercus* spp.), beech (*Fagus* spp.), and chestnut (*Castanea* spp.), are dispersed by cache-hoarding squirrels and jays. Here again succession is probably facilitated by the action of animals and birds. A single blue jay, (*Cyanocitta cristata*)

TABLE 5·3. Numbers of seeds deposited in eleven 1 m^2 seed traps along transects in a 2-year-old and a 12-year-old field in New Jersey, USA between 7 July 1977 and 16 March 1978 (McDonnell & Stiles 1983)

	2-year-old field	12-year-old field
Cornus florida	8	95
Juniperus virginiana	13	393
Phytolacca americana	—	32
Prunus serotina	10	72
Rosa multiflora	113	61
Rubus spp.	50	14
Toxicodendron radicans	6	48
Vitus spp.	6	49
Others	2	6
	208	770

is said to bury 4600 acorns per season (Darley-Hill & Johnson 1981). Although most of the acorns cached are of course subsequently eaten, the few that remain are very favourably sited for establishment. In a very thorough study of jays and oaks, Bossema (1979) found that even though only a small proportion of the acorns available were cached by jays, over half the first-year oak saplings were derived from jay-planted acorns. A recent study of dispersal of beech nuts by blue jays in Wisconsin indicates that the birds are highly selective in their choice of beech mast. Although all but 11% of the beech nuts on the trees contained no kernel, all of those recovered from jays were sound (Johnson & Adkisson 1985) (Table 5.4).

The effectiveness of jays in dispersing these heavy-seeded species may be inferred from the fact that these birds have been recorded as caching beech nuts at a distance of 4 km from the parent tree (Johnson & Adkisson 1985). It has been calculated that the recolonization of Britain by oak (*Quercus robur*) in the late Boreal took place at an average of 4·8 km per year (Webb 1966). This was presumably due to jays or squirrels, or both, depositing acorns at considerable distances from the edge of the oak forest and failing to retrieve them. Webb's figure for the spread of oak contrasts with that of Davis (Chapter 18) for beech (*Fagus grandifolia*) in North America. In the case of the latter species, the northward spread took place at the rate of only 0·13 km per year, and was probably brought about mainly by blue jays. Since the *Fagus* was spreading through established forest the rate may have been limited by lack of opportunities for establishment rather than by dispersal.

The seeds of late-successional or climax forest trees are often rather short-lived, and so these species do not form banks of dormant seeds in the soil. As a group, climax forest trees have seeds that lose their viability quickly, often within one year. It might at first sight seem disadvantageous in nature for any species to have such short-term viability, but the

TABLE 5.4. Germinability of beech nuts (a) collected from trees by investigators and (b) selected by jays. Sound nuts had a kernel, unsound ones only an empty pericarp (Johnson & Adkisson 1985)

Source of seed	Number collected	Number sound	Percentage sound	Percentage sound nuts germinated	Percentage all nuts germinated
Trees	1100	121	11	88	10
Blue jays	40	40	100	88	88

lack of dormancy may simply be a result of their vulnerability to predation. If all the seeds which fail to germinate immediately are invariably eaten within a short time, long-term dormancy becomes irrelevant and will not be selected for.

In any case these climax species often have an alternative strategy to the maintenance of soil seed banks, i.e. the maintenance of a population of saplings whose growth is suppressed. Having germinated and established, often in deep shade, many tree seedlings can survive for years in a state of suspended development. These saplings may be less vulnerable to attack by predators than the seeds are, possibly because a well-rooted sapling is less easily carried away by small animals and birds. It is also probable that a bank of established seedlings can react to the presence of a small gap much more quickly than ungerminated seeds could. When these quiescent saplings are released from shade they are known to grow exceptionally quickly but may need several periods of gap creation before reaching the canopy (Canham 1985). The bank of suppressed seedlings may thus be seen as a mechanism for ensuring rapid responses to the occurrence of quickly-closing gaps in the canopy.

REFERENCES

Bossema, I. (1979). Jays and oaks: an eco-ethological study of a symbiosis. *Behaviour*, **70**, 1–117.

Canham, C.D. (1985). Suppression and release during canopy recruitment in *Acer saccharum*. *Bulletin of the Torrey Botanical Club*, **112**, 134–145.

Carlquist, S. (1967). The biota of long-distance dispersal. V. Plant dispersal to Pacific Islands. *Bulletin of the Torrey Botanical Club*, **94**, 129–162.

Clifford, H.T. (1956). Seed dispersal on footwear. *Proceedings of the Botanical Society of the British Isles*, **2**, 129–131.

Collins, S.L. & Uno, G.E. (1985). Seed predation, seed dispersal, and disturbance in grasslands: a comment. *American Naturalist*, **125**, 866–872.

Connell, J.H. & Slatyer, R.O. (1977). Mechanisms of succession in natural communities and their role in community stability and organisation. *American Naturalist*, **111**, 119–144.

Darley-Hill, S. & Johnson, W.C. (1981). Acorn dispersal by the blue jay (*Cyanocitta cristata*). *Oecologia*, **50**, 231–232.

Fenner, M. (1978). A comparison of the abilities of colonizers and closed-turf species to establish from seed in artificial swards. *Journal of Ecology*, **66**, 953–963.

Fenner, M. (1980a). Germination tests on thirty-two East African weed species. *Weed Research*, **20**, 135–138.

Fenner, M. (1980b). The inhibition of germination of *Bidens pilosa* seeds by leaf canopy shade in some natural vegetation types. *New Phytologist*, **84**, 95–101.

Fenner, M. (1985). *Seed Ecology*. Chapman and Hall, London.

Fridriksson, S. (1975). *Surtsey: Evolution of Life on a Volcanic Island.* Butterworths, London.

Gorski, T., Gorska, K. & Nowicki, J. (1977). Germination of seeds of various herbaceous species under leaf canopy. *Flora*, **166**, 249–259.

Grime, J.P., Mason, G., Curtis, A.V., Rodman, J., Band, S.R., Mowforth, M.A.G., Neal, A.M. & Shaw, S. (1981). A comparative study of germination characteristics in a local flora. *Journal of Ecology*, **69**, 1017–1059.

Gross, K.L. (1984). Effects of seed size and growth form on seedling establishment of six monocarpic perennial plants. *Journal of Ecology*, **72**, 369–387.

Hilton, J.R., Froud-Williams, R.J. & Dixon, J. (1984). A relationship between phytochrome photoequilibrium and germination of seeds of *Poa trivialis* L. from contrasting habitats. *New Phytologist*, **97**, 375–379.

Janzen, D.H. (1984). Dispersal of small seeds by big herbivores: foliage is the fruit. *American Naturalist*, **123**, 338–353.

Johnson, W.C. & Adkisson, C.S. (1985). Dispersal of beech nuts by blue jays in fragmented landscapes. *American Midland Naturalist* **113**, 319–324.

Kivilaan, A. & Bandurski, R.S. (1981). The one hundred-year period for Dr. Beal's seed viability experiment. *American Journal of Botany*, **68**, 1290–1292.

Krefting, L.W. & Roe, E.I. (1949). The role of some birds and mammals in seed germination. *Ecological Monographs*, **19**, 269–286.

Levin, D.A. & Kerster, H.W. (1974). Gene flow in seed plants. *Evolutionary Biology*, **7**, 139–220.

McDonnell, M.J. & Stiles, E.W. (1983). The structural complexity of old field vegetation and the recruitment of bird-dispersed plant species. *Oecologia* (*Berlin*), **56**, 109–116.

Roberts, H.A. & Feast, P.M. (1973). Emergence and longevity of seeds of annual weeds in cultivated and undisturbed soil. *Journal of Applied Ecology*, **10**, 133–143.

Salisbury, E.J. (1942). *The Reproductive Capacity of Plants.* Bell, London.

Silvertown, J.W. (1980). Leaf-canopy-induced seed dormancy in a grassland flora. *New Phytologist*, **85**, 109–118.

Sorensen, A.E. (1981). Interactions between birds and fruit in a temperate woodland. *Oecologia* (*Berlin*), **50**, 242–249.

Thompson, K. & Grime, J.P. (1979). Seasonal variation in the seed banks of herbaceous species in ten contrasting habitats. *Journal of Ecology*, **67**, 893–921.

Thompson, K. & Grime, J.P. (1983). A comparative study of germination responses to diurnally-fluctuating temperature. *Journal of Applied Ecology*, **20**, 141–156.

Viereck, L.A. (1966). Plant succession and soil development on gravel outwash of the Muldrow Glacier, Alaska. *Ecological Monographs*, **36**, 181–199.

Webb, D.A. (1966). Dispersal and establishment: what do we really know? *Reproductive Biology and Taxonomy of Vascular Plants.* Botanical Society of the British Isles Conference Report No. 9, pp. 93–102.

Welch, D. (1985). Studies in the grazing of heather moorland in north-eastern Scotland. IV. Seed dispersal and plant establishment in dung. *Journal of Applied Ecology*, **22**, 461–472.

Werner, P.A. & Platt, W.J. (1976). Ecological relationships of co-occurring goldenrods (Solidago: Compositae). *American Naturalist*, **110**, 959–971.

6. MATING SYSTEMS AND COLONIZING SUCCESS IN PLANTS

A.H.D. BROWN AND J.J. BURDON
Division of Plant Industry CSIRO, P.O. Box 1600, Canberra, ACT 2601, Australia

INTRODUCTION

The process of colonization, which entails the founding of a population in a locality where none recently existed, is a basic one in vegetational history. However, such a general concept encompasses so broad a range of phenomena that it is readily imbued with very different meanings. Thus ecologists, being interested in interspecific interactions, have sharply distinguished primary from secondary colonization. Evolutionary geneticists, on the other hand, being aware of the dynamics of gene flow, have stressed the distinction between two geographic scales. At one extreme is the introduction of exotic organisms from a remote part of the world, while at the other, is the localized process of range expansion. Here we are mainly concerned with genetic aspects of broad-scale, secondary colonization after long-distance dispersal.

The notion of succession in ecology is a classical concept stressing the temporal sequence of communities and the waxing and waning of their constituent species. In direct analogy, it is natural to think of the populations of individual species and the flux of genotypes in time, i.e. the nature and extent of temporal micro-evolutionary changes. Unfortunately, there are very few chronological genetic studies of colonizing plants. Rather than comparing temporal sequences at a single site most studies have compared colonial populations with known or presumed source material. This approach confounds changes accompanying colonization with subsequent local evolution.

Many features of colonizing plants play a role in the founding, and subsequent history, of colonial populations (for recent reviews see Jain 1983; Oka 1983; Clegg & Brown 1983; Barrett & Richardson 1986; Gray 1986). The mating system, in particular, has long been recognized to have a pre-eminent effect upon colonization (Baker 1955) and micro-evolution (Allard 1965, 1975) because it modulates the temporal flux of genotypes.

MATING SYSTEMS

Plant species display a rich diversity of mating systems. Since reproduction is the fundamental challenge in the establishment and perpetuation of a population, the mating systems of colonizing species continue to receive much attention. Individual plants reproduce by one or more of four main modes, namely clonal or vegetative propagation, apomixis (agamospermy), autogamy and outcrossing. The mating system of a population is determined by which of these modes predominates. The question arises as to which modes distinguish invading species (Barrett 1982). Two broad approaches can be taken. The first of these is to consider those features that notably successful colonizers (such as the 'world's worst weeds') have in common. When this was done for the eighteen worst weeds (Brown & Marshall 1981), their capacity for uniparental reproduction was outstanding. Thus, about half the species are self-pollinated, and the remaining outbreeders invest massive resources in vegetative propagules. The second approach is to look for species correlations in a floristic survey. Thus, Price & Jain (1981) classified 400 species growing in the British Isles jointly for their reproductive mode and colonizing ability, and found predominant selfing or apomixis significantly more common among known or assumed colonizers. It seems clear from both approaches that colonizing species form a distinctive yet heterogeneous sample of plant mating systems, i.e. there is a higher representation of species which possess some mode of uniparental reproduction.

Taken as an overall strategy, the mating system must meet four important yet frequently conflicting demands. First, it should assure reproduction, indeed numerical increase, of the individual genotype and indirectly the population. Second, it should allow genetic experimentation, to develop new, better adapted genotypes. Third, it must achieve the fertilization of ovules, and fourth promote the dissemination of pollen.

Each of these functions is clearly at a premium for successful colonization. Thus, when population density is low, or suitable pollinators absent, uniparental reproduction of 'general purpose genotypes' will help the successful founding of a population. Second, as new selection pressures, new competitor species, or new pest or pathogen variation are met, the capacity to produce new genotypes is needed. Third, the struggle to attract and reward pollinators, and in rapidly expanding populations to outdo neighbours in pollen dissemination, would allow scope for sexual selection. Possible examples are the evolution of flower colour polymorphism in *Ipomoea purpurea* in the US (Brown & Clegg 1984), and the brilliant floral display in New South Wales (NSW) populations of *Echium plantagineum*.

As predictable from pluralities of function, the mating system of colonizing species itself may be flexible and indeed may show a mixture of modes. Thus, self-pollinated species are rarely exclusively so, and the rare outcrossed progeny in highly selfing species provide new recombinant genotypes. A possible example of such evolution has occurred in *Trifolium subterraneum* in South Australia, where a large number of novel strains, differing from the introduced cultivars, have arisen (Cocks & Phillips 1979). A shift in mating system—for example, the tenfold increase in outcrossing rate in *Bromus mollis* in Australia over that in England (Brown & Marshall 1981)—is thus a key micro-evolutionary change affecting the succession of genotypes subsequent to colonization.

The dynamics of mating systems that consist of a mixture of apomixis and outbreeding have rarely been studied (see Barrett & Richardson 1986). Obligate apomixis associated with sexual sterility such as in triploids does not provide flexibility within a lineage. However, colonial populations of such species can be genotypically diverse, consisting of a number of races, e.g. *Chondrilla juncea* in Australia (Burdon, Marshall & Groves 1980), *Taraxacum officinale* in California (Lyman & Ellstrand 1984). The source and replenishment of genotypic diversity in such cases is important in their continued colonizing success. It is noteworthy that mating systems of mixed selfing and apomixis, or mixed selfing and clonal propagation are rare in plant species in general and colonizers in particular.

MEASURES OF POPULATION GENETIC STRUCTURE

Genetic polymorphism

How genetically variable are populations of colonizing plant species? Furthermore, does the level of variation differ in colonial versus source populations, and does it alter during succession? Like the ecologist's concern with the species diversity of animal and plant communities, these are basic questions.

Many recent studies of plant populations employing electrophoretic techniques are providing answers to these questions, at least for the structural genes of certain enzymes. Estimates can be made of the proportion of loci that are polymorphic, the number of alleles per locus and the evenness of their frequencies (measured by $h = 1 - \Sigma p_i^2$). In a recent collation of estimates from a wide range of plant species, Hamrick (1983) found that colonizing and/or early-successional species had, on average, lower estimates of genetic variation. This general pattern may be attributed either to the assumed relatively specialized niche or restricted habitats occupied by many of the colonizers that have been studied (disturbed, open habitats with reduced biotic complexity), or to reductions or 'bottlenecks' in population size (the so-called 'founder effect').

Yet not all colonizing species are depauperate in genetic variation. We (Burdon, Marshall & Brown 1983; Brown & Burdon 1983) have been studying *Echium plantagineum*, a highly successful weed in temperate Australia. Remarkably high levels of allozyme variation were found in a recent survey of eight NSW populations (Table 6.1). Average heterozygosity (*H*) in these populations exceeded 30%. *Echium plantagineum* is an outcrossing species, mainly pollinated by *Apis mellifera*. In contrast, most of the colonizing species with low genetic variation within colonial populations are inbreeders, e.g. *Xanthium strumarium*, *Emex spinosa* and *Avena barbata* (Brown & Marshall 1981), and *Hordeum murinum* (Giles 1984).

Clearly there is an all-important interaction between the size of the founding population and its breeding system. The capacity for vigorous uniparental reproduction after long-distance dispersal could allow one individual to colonize successfully (Baker 1955). Hence, we might expect such species to be over-represented in a list of colonizer species. This in turn would result in estimates of polymorphism in colonizers as a class, to be below average. Yet this lower polymorphism is due to the mating system rather than a necessary consequence of being a colonizer (Layton & Ganders 1984).

Heterozygosity

Given the capacity to resolve the products of separate loci at the protein or indeed at the DNA level, the question of heterozygosity comes clearly into focus. In this issue, mating system has a substantial effect. The molecular diversity present in an individual can be divided into segregational heterozygosity and 'fixed' heterozygosity. The former arises from allelic variation at a locus (allozymes in the cases of isozyme loci); the latter arises from duplication of chromosomal segments, or genomes as in polyploids. Elements of the genetic system such as translocation heterozygosity or self-incompatibility genes may curtail or restrict the degree of segregation among progeny.

Self-fertilization rapidly erodes existing segregational heterozygosity. An obvious difference in the genetics of inbreeding as opposed to outcrossing plants is the markedly lower levels of heterozygosity. This does not mean heterozygosity is absent from inbreeders. Indeed, inbreeders tend to have more heterozygosity than expected from their low outcrossing rate, whereas outbreeders tend to have less (Brown 1979). This is

TABLE 6.1. Allozyme variation at sixteen loci in eight New South Wales (NSW) and ACT populations of *Echium plantagineum*. Single buds from fifty mature plants were sampled (Burdon & Brown, in press)

	Populations								Average	Total
	1	2	3	4	5	6	7	8		
Alleles per locus	3·1	2·7	2·7	2·6	2·6	2·6	2·4	3·1	2·72	3·56
Diversity (h)	0·37	0·33	0·35	0·36	0·33	0·28	0·33	0·40	0·343	0·388
Heterozygosity (H)	0·35	0·32	0·34	0·33	0·29	0·26	0·33	0·36	0·323	0·323
Fixation index	0·05	0·0	0·02	0·07	0·11	0·08	0·0	0·10	0·062	
Variance in heterozygosity										
Observed (S_K^2)	2·3	2·5	4·0	3·8	2·8	3·2	2·8	4·0	3·17	
Expected (σ_K^2)	2·9	2·9	3·0	2·9	2·7	2·4	2·8	3·3	2·86	

Populations: (1) Wagga Wagga; (2) Gundagai; (3) Burrinjuck Junction, Hume H'way; (4) Binalong; (5) Canberra; (6) Tharwa; (7) Lake George; (8) Braidwood.

paradoxical as mechanisms promoting outcrossing are presumed to operate through increased heterozygosity. Because genetic variation in inbreeders is correlated among loci (see below), the occasional outcrossed plant has a significance far greater than its rarity would suggest (Stebbins 1957; Brown, Zohary & Nevo 1978). In the genetic demography of inbreeders, all individuals are very far from being equal!

In outbreeding organisms there have been several recent attempts to relate the degree of heterozygosity of an individual (as measured by allozymes) positively with either its growth rate or homeostatic properties. In a recent review, Mitton & Grant (1984) concluded that 'roughly 70–80% of the effects on growth and developmental stability can be attributed to heterozygosity *per se*', and that the 'individual organisms' level of heterozygosity is a major organizing principle in natural populations of both plants and animals'. It is hard to reconcile this conclusion with the remarkable success of inbreeders as colonizing plants. Mitton and Grant discuss three mechanisms which would bring about such a correlation: (i) that isozyme loci mark blocks of the chromosome and variation at nearby loci affects growth and development; (ii) that the level of heterozygosity at isozyme loci is an inverse measure of the degree of inbreeding of the whole genome; and (iii) that enzyme polymorphism *per se* brings about the relationship, as a consequence of biochemical diversity for buffered metabolism. Mitton and Grant favour the last hypothesis although the arguments supporting their choice are questionable.

Marked shifts in the genotypic and demographic constitution of a population can occur during the course of a single generation. In one population of *E. plantagineum* at least seven distinct cohorts germinated in response to seasonal rains; yet only 6% of seedlings survived to produce seed (Burdon, Marshall & Brown 1983). At the same time genetic changes occurred at four isozyme loci, with greater survival of individuals heterozygous for the *Pgi* or *Pgm* loci. This was apparently due in part to heterozygote advantage, or inbreeding depression.

Population divergence

The genetic structure of a series of populations in space deals with the spatial deployment of variation in a species, and the factors which control it. Since the process of colonization emphasizes the transience of populations, colonizing species *a priori* may be assumed to have distinctive patterns. The genetic distinctiveness of individual populations of such species would arise from: (i) a bottleneck in population size at founding, (ii) the absence of genetic enrichment by repeated migration during the lifetime of the local population; and (iii) the possibility of novel selection pressures in the new habitat (Clegg & Brown 1983).

Variation in allele frequency

Measurement of population differentiation in plants has commonly been by Nei's D_{ST} statistic which is based on analogues of Simpson's measure of ecological diversity. The D_{ST} statistics have considerable appeal because values correspond with the amount of heterozygosity expected under random mating. Thus, D_{ST} measures the difference between the 'total' diversity that would obtain in a hypothetical bulk of all populations (h_T) and the average diversities ($\bar{h}$) of all the constituent populations. The ratio:

$$G_{ST} = D_{ST}/h_T = 1 - \bar{h}/h_T$$

allows estimates of differentiation relative to the total diversity. Recently Hamrick (1983) and Loveless & Hamrick (1984) have summarized estimates of these statistics from the plant isozyme literature, and have studied the overall trends in average values of h_T, $\bar{h}$ and G_{ST} for species classified according to their 'species' characteristics. Table 6.2 is an extract of their results for overall estimates with respect to breeding system and successional stage. Loci which were monomorphic within each species were not included in this study. Remarkably, the overall allozyme

TABLE 6.2. Population differentiation in plant species classified as to mating system, and successional stage (from Loveless & Hamrick 1984). The symbol *N* is the number of species studied, h_T and $\bar{h}$ are the total and the within-population gene diversities, and G_{ST} is the proportion of diversity among populations

	N	h_T	$\bar{h}$	G_{ST}
Mating system				
Autogamous	39	0·29	0·13	0·52
Mixed mating	48	0·24	0·17	0·24
Predominantly outcrossed	36	0·25	0·21	0·12
Successional stage				
Early	81	0·28	0·15	0·41
Middle	47	0·23	0·18	0·18
Late	33	0·30	0·26	0·11

diversity within species (h_T) showed little difference among these categories. In contrast, either predominant self-fertilization or colonization (early succession stage), or more likely both, were associated with lower diversity within populations and higher values of G_{ST}.

Table 6.1 gives our estimates for eight NSW populations of the outbreeding *Echium plantagineum*. These estimates indicate above-average values for total diversity (h_T = 0·388) and within population diversity ($\bar{h}$ = 0·343) with G_{ST} = 0·12. This apparently low G_{ST} value is not typical (Table 6.2) of early-successional plants and may point to two problems associated with the G_{ST} analysis. First, gene diversity is a function primarily of the frequency of the more common alleles, and is bounded by 1·0. As $\bar{h}$ increases it becomes numerically more difficult for D_{ST} (= $h_T - \bar{h}$) to increase. Second, the D_{ST} analysis does not indicate whether genetic divergence is spread over all populations, or confined to a few. This requires the full matrix of genetic distances to be computed. Alternatively, the *Echium* results emphasize that the mating system can override the effect of long-distance colonization and continuing bottlenecks, thereby restraining population divergence.

An example of strong population divergence in a predominantly selfing species is found in three closely related species of barnyard grass (*Echinochloa*). The three species are *E. crus-galli*, which is hexaploid and occurs widely throughout the world, and two large-seeded rice mimic forms; *E. oryzoides*, an early flowering hexaploid, and *E. phyllopogon* (= *E. oryzicola*) a late flowering tetraploid (Barrett 1983). Comparable samples of these were made from Californian rice fields and were assayed for variation in the same spectrum of isozymes (S.C.H. Barrett & A.H.D.

Brown, unpublished). The genetic statistics are summarized in Table 6.3. Although the generalist *E. crus-galli* was sampled only from nearby sites on the levy banks, its populations were markedly more variable genetically. Three times as many loci were polymorphic in this race than in the mimics. The other statistics are expressed on the basis of per polymorphic locus and so tend to mask this basic difference. The populations of *E. crus-galli* were differentiated to a level typical of autogamous colonizers. The rice crop mimics, which are normally restricted to the rice fields themselves, show less polymorphism, but this was strongly differentiated, presumably due to restricted migration and continued founder effects.

Variation in gene diversity

A second kind of population divergence, aside from differentiation of allele frequencies, is variation in total levels of diversity among populations. Early electrophoretic studies of plant populations (reviewed in Brown 1978) have shown that the geographic pattern of diversity is markedly affected by breeding system. Outbreeders tend to a fairly uniform spread of diversity, whereas inbreeders have non-uniform patterns. The frequency distribution of populations with respect to diversity levels can be markedly L-shaped (e.g. *Avena barbata* in California, where the majority of populations lack diversity), or bimodal (e.g. *Hordeum spontaneum* in Israel) or trimodal (e.g. *Lycopersicon pimpinellifolium* in Peru). In these studies the genetically homogeneous populations constitute one mode and are presumably recently founded, small, or outlying populations. In *Lycopersicon*, the two non-zero modes correspond to weedy populations outside the main zone of the species on the one hand, and to central populations in more permanent communities on the other.

The major difference between outbreeders and inbreeders in the evenness of diversity has been confirmed in more recently studies. Thus, for example, genetic diversity (h) in *E. plantagineum* varies only between 0·28 and 0·40 (Table 6.1), whereas the ranges of the three *Echinochloa* subspecies are more than threefold (Table 6.3). The variance in gene diversity (h) estimates for the selfing *Plectritis brachystemon* is twice that of its outcrossing relative *P. congesta* (Layton & Ganders 1984). In a survey of eighty-one European populations of *Capsella bursa-pastoris*, a selfing species, Bosbach & Hurka (1981) found a large range in genotypic diversity among the sites. The sites differed for several variables including population size and sample size yet the factor which was strongly related to genotypic diversity was the degree and frequency of disturbance. The number of genotypes was higher in the eighteen most 'highly disturbed'

TABLE 6.3. Genetic variation of *Echinochloa* species from Californian rice-field populations. *E. crus-galli* is a generalist weed, whereas *E. oryzoides* and *E. phyllopogon* are rice mimics (S.C.H. Barrett & A.H.D. Brown, unpublished)

	E. crus-galli (6*x*)	*E. oryzoides* (6*x*)	*E. phyllopogon* (4*x*)
Populations	11	12	12
Total number of allozyme loci screened	31	32	25
Prop. of loci polymorphic	0·52	0·16	0·28
Alleles per locus polymorphic†	2·56	2·0	2·14
Overall diversity (h_T)†	0·168	0·238	0·263
Diversity within populations ($\bar{h}$)†	0·112	0·120	0·051
Range h-minimum	0·05	0·05	0·0
h-maximum	0·17	0·20	0·16
G_{ST}	0·33	0·50	0·81

† Where the average is taken over the number of loci polymorphic over all the populations within each variety, i.e. 16, 5 and 7 respectively.

sites (average 4–6 genotypes) than in eighteen 'less disturbed' (average 1–4 genotypes). This result draws attention to the cryptic gene reservoir present in the dormant seed bank. Frequent disturbance of habitat may allow such variation to be manifest. Preliminary data on apomictic colonizers also indicate an extreme range in levels of genotypic diversity. In Lyman & Ellstrand's (1984) survey of *Taraxacum officinale* the number of clones in twenty-two North American populations ranged from 1 to 13.

Multilocus structure

The final aspect of the genetic structure of populations of colonizing species which is profoundly affected by mating system, is their multilocus structure. Under this heading, we are concerned with the extent to which the frequencies of combinations of alleles at various polymorphic loci depart from those expected when combinations are random. Since few individuals may be involved in founding a new population, especially after intercontinental introduction, a restricted number of multilocus genotypes may be established. For example, *Chondrilla juncea*, an apomictic weed, occurs in Australia as three distinct races. These races differ for several isozyme loci (Burdon, Marshall & Groves 1980) so that, on the basis of single loci, the species has a high level of genetic diversity. Yet its multilocus diversity is severely restricted to three multilocus genotypes. Autogamy also allows the build-up of intense multilocus structure as found in *Avena barbata* in California, either directly (Allard *et al.* 1972), or by 'hitchhiking' (Hedrick & Holden 1979). Subsequent immigrants to the colonial population, if few in number, are likely to increase the level of multilocus association.

In contrast, a mating system of predominant outbreeding is expected to break up combinations of alleles generated by small population sizes (Felsenstein 1974). As a result, the question arises as to which of these forces predominates in outbreeding colonizers. To test the extent of multilocus structure in a population for several loci simultaneously, Brown, Feldman & Nevo (1980) developed a procedure based on the number of heterozygous loci in two randomly chosen gametes (see also Brown 1984). In outbreeding species this distribution can be observed directly as the distribution of the number of heterozygous loci among individuals. The major advantage of this approach is that both the expected distribution assuming *zero* associations among loci, and the distribution assuming *complete* association, can be calculated.

The expected distribution assuming independence among the loci is obtained by convolution of the generating functions for heterozygosity at each single locus. This generating function for m loci (with X as the index variable) is:

$$G(X) = \prod_{j=1}^{m} [1 - H_j + H_j X]$$

The function is expanded numerically and the coefficient of the term X^i is the probability that i loci are heterozygous in an individual. The variance of this distribution is

$$\sigma_K^2 = \Sigma H_j - \Sigma H_j^2$$

The values for σ_K^2 and for the observed variances (S_K^2) of the numbers of heterozygous loci per individual in the populations of *E. plantagineum* are given in Table 6.1. In no single case was there significant difference, and hence no evidence that the high genetic variation in this species is organized into a limited number of genotypes. These results contrast with earlier studies of Israel populations of *Hordeum spontaneum* (Brown, Feldman & Nevo 1980). For this predominantly self-pollinating species, the average value of S_K^2 was 2·80, significantly higher than that of the variance (σ_K^2) which was 1·56. The data for *E. plantagineum* agree with that obtained in a previous study (Brown & Burdon 1983) and clearly indicate that outbreeding systems encourage sufficient recombination to overcome the effects of bottlenecks in population size on multilocus structure.

One problem of this variance-in-heterozygosity procedure concerns its statistical power. Chakraborty (1984) compared the procedure with the earlier method of estimating individual coefficients of linkage disequilibrium. For the 2 locus–2 allele situation, the variance-in-heterozygosity is a statistically inefficient method of detecting disequilibrium. However, simulation studies of the multilocus situation have yet to be carried out.

ALLOZYME VARIATION AND PHENOTYPIC VARIATION

Since its introduction in the 1960s, the technique of protein electrophoresis has come to dominate experimental plant population genetics. Much of the above discussion in consequence relies on electrophoretic evidence. This raises the obvious question, how representative of the genome is this evidence? From the molecular standpoint, it is clear that allozyme variation gives a biased picture of genetic variation. It studies only certain classes of mutation in the coding region of the structural genes for a very small subset of proteins. At present, the relative ease and low cost are its major advantages over the more fundamental approach of DNA sequencing. For this imperfect tool we can ask then a parallel question: what is the relationship between allozyme variation and genetical variation for ecological or fitness-related (Jain 1983) characters? This has been a major question during the electrophoretic era, and most of the literature bears either directly or implicitly on this question. As a result several opinions have been expressed on the relationship between isozymes and 'real' characters. In the context of this chapter, we want to stress that the mating system is a critical variable in discussing the question. This point has been made before (e.g. Price *et al.* 1984) but is frequently overlooked.

Effects of mating system

As noted in the section above on heterozygosity, three potential pathways can link variation detectable by isozyme techniques with other kinds of genetic variation. These are:

(a) genetic linkage of allozyme alleles that are non-randomly associated with variation at other loci which affects the character;
(b) variation among populations or individuals in co-ancestry or degree of inbreeding;
(c) direct phenotypic manifestation of the allozyme difference itself.

The first pathway refers to correlations which would decay with recombination at rates inversely related to the degree of linkage. In principle such associations are separable and, in practice, the rates of decay may be much faster than predicted by simple models (Clegg, Kidwell & Horch 1980). However, a tightly linked block of genes can be very resilient to decay (Avery & Hill 1979). Examples of such blocks are loci affecting leaf and inflorescence morphology and the *Pgm1* locus in the self-fertilizing species *Plantago major* (van Dijk 1984), and loci affecting germination and the *Idh–Est5* group in the outcrossing species *Plantago lanceolata* (van Dijk 1985).

The second pathway may be evident in, for example, the outcrossed individuals of predominantly selfing populations where allozyme heterozygotes may show heterosis for growth characters (Brown & Marshall 1981); or it may be evident in the phenotypic variation found in hybrid zones between two formerly isolated populations. In these scenarios the isozyme loci are acting as markers for whole genome effects. Correlations between allozyme genotypes and morphological phenotype which stem from this cause would decrease rapidly with random mating.

Under the third pathway we include all differences between the coding sequence for two alleles as well as differences in the specific controlling sequences upstream from the structural gene. This includes any effect stemming from the expression of the gene in question, but not from adjacent structural genes.

Clearly the breeding system affects the likelihood that each of these pathways is involved. Predominant self-fertilization retards the decay of linkage disequilibrium, increases the variance of co-ancestry, and increases the level of homozygosity (and hence gene expression). Other factors may complicate the relation between isozymes and phenotypes. These include the degree of isozyme multiplicity, the degree of polymorphism, and the level of allelic multiplicity.

The degree of isozyme multiplicity is affected by a change in level of ploidy, or by gene duplication. Duplication provides opportunity for gene divergence or gene silencing. In general, inbreeders and apomictic species are more likely to be polyploids, thus providing another route by which breeding system affects the relationship between isozymes and 'real' characters of ecological interest.

The proportion of polymorphic loci will affect the amount of variation at loci other than the specific few loci under study. Here again there is a major difference between inbreeders and outcrossers. In highly polymorphic outbreeding species such as *Echium plantagineum*, it requires only a modest sample of loci to prove that each individual in a population is genetically unique. Chakraborty (1981) has considered the question of the adequacy of a sample of independent isozyme loci for ranking individuals for their genome mean heterozygosity. He found that the correlation of ranking was approximately the square root of the proportion of polymorphic loci examined. This is potentially a very weak correlation. In inbreeders, however, it may be much higher because of departure from the basic assumption of independence of loci. In similar vein, Lewontin (1984) has pointed to contrasts in statistical power in that it is more difficult to detect differences in allele frequencies than differences in the metric traits they control.

Finally, high allelic multiplicity itself may hinder the detection of a relationship. For example, if there are two alleles (giving three genotypes at a marker locus) a relationship may be more apparent than if there are five alleles (fifteen genotypes at a locus). A related problem is that concerned with the extent of so-called 'hidden heterogeneity'. As inbreeders tend to have fewer alleles per locus, again, mating system may affect the strength of the relationship.

Kinds of relationship

More precision is needed when considering the relationship between allozyme variation and phenotypes. Several distinct yet overlapping categories of questions can be specified. In each category the relationship of allozyme variation could be studied with any of the following types of phenotypic variation.

(a) Other simply inherited traits such as morphological polymorphisms or disease resistance.
(b) Measurement traits in material grown *ex situ* in 'common garden' controlled environments. Variation expressed in such experiments can be partitioned into genetic and environmental sources.
(c) Measurement traits in natural populations.

The four kinds of questions are as follows.

1 At the *individual* level: are there correlations observable between allozyme genotypes (single-locus, multilocus, heterozygosity) and morphological phenotypes? For example, Hamrick & Allard (1975) showed that two multilocus genotypes of *Avena barbata* differed for morphological and life-history traits when grown in a common environment. In *Bromus mollis*, variants at an alcohol dehydrogenase locus differed in their germination at low temperatures and response to waterlogging (Brown, Marshall & Munday 1976). Mitton & Grant (1984) have argued

for a general relationship between heterozygosity and growth (see above).

2 For *pairwise divergence*: is the extent of divergence for allozymes (genetic distance) between populations related to phenotypic divergence? Positive examples include those of Price *et al.* (1984) for *ex situ* studies of *Avena barbata*, *Hordeum vulgare*, and *H. jubatum*, and of Brown, Marshall & Albrecht (1974) for *in situ* traits in *Bromus mollis*. Apparent negative examples include morphological races of the outcrossing roadside weed *Gaillardia pulchella* in Texas (Heywood & Levin 1984).

3 For *amount of variation*: is there a correlation between the level of allozyme diversity within a population or species and the level of phenotypic variation? A positive example is that of *Avena barbata* (Marshall & Allard 1970). At present the reported lack of relationship in two inbreeding colonizers, namely *Xanthium strumarium* (Moran, Marshall & Müller 1981) and *Hordeum murinum* (Giles 1984), point to the need for further studies, because these examples revealed little allozyme polymorphism anyway.

4 For *hierarchical sources of variation* (individuals, families, population, species): are the patterns of variance components comparable? Are morphological characters more differentiated among populations than are allozymes? For spikelet characters in *Hordeum spontaneum*, this would appear to be the case (Brown *et al.* 1978).

Most of the published examples in the above four categories with positive outcomes are for inbreeding species. However, the paucity of experimental studies indicates that generalizations may be premature. Clearly allozymes may serve adequately as neutral markers to index variation and relationship, without necessarily being causative agents. Indeed, because of the cohesive nature of the genomes in inbreeders and apomicts, this should be the standard expectation.

CONCLUSIONS

In outbreeding populations, individual genotypes are transient. Even in colonizing situations, there is apparently sufficient variation in the genome to ensure that recombination provides new genotypes. For colonizing success, an outbreeder must at least tolerate, if not positively utilize this genotype flux. Recent evidence suggests that such genotypic variation is important for exotics in facing the considerable biological barrier to invasion (Scorza 1983). Thus, sexually reproducing exotic weeds have proven to be more difficult to control with biological agents, than have asexually producing weeds (Burdon & Marshall 1981).

Yet a strong bias toward uniparental mating systems among successful plant colonizers is evident. Part of this bias may be accidental, in the sense that species which possess a capacity for uniparental mating are assured of some reproduction after dispersal away from their home range. The second advantage is the capacity to exploit successful genotypes in 'predictable' habitats or those recently opened up by human disturbance. Here continued colonizing success is achieved by a degree of stability in the genome, and suppression of genotypic flux.

Mating system is clearly a fundamental parameter in colonizing success. The founding and perpetuation of a population entails new and at times conflicting demands. There is need for high reproductive rate and yet genetic experimentation. It is no wonder that several flexible solutions are evident among colonizing plants.

REFERENCES

Allard, R.W. (1965). Genetic systems associated with colonizing ability in predominantly self-pollinated species. *The Genetics of Colonizing Species* (Ed. by H.G. Baker & G.L. Stebbins), pp. 49–76. Academic Press, New York.

Allard, R.W. (1975). The mating system and microevolution. *Genetics*, **70**, 115–126.

Allard, R.W., Babbel, G.R., Clegg, M.T. & Kahler, A.L. (1972). Evidence for coadaptation in *Avena barbata*. *Proceedings, National Academy Sciences, USA*, **69**, 3043–3048.

Avery, P.J. & Hill, W.G. (1979). Distribution of linkage disequilibrium with selection and finite population size. *Genetical Research, Cambridge*, **33**, 29–48.

Baker, H.G. (1955). Self-compatibility and establishment after 'long distance' dispersal. *Evolution*, **9**, 347–349.

Barrett, S.C.H. (1982). Genetic variation in weeds. *Biological Control of Weeds with Plant Pathogens* (Ed. by R. Charaduttan & H.L. Walker), pp. 73–97. Wiley, New York.

Barrett, S.C.H. (1983). Crop mimicry in weeds. *Economic Botany*, **37**, 255–282.

Barrett, S.C.H. & Richardson, B.J. (1986). Genetic attributes of invading species. *The Ecology of Biological Invasions: an Australian Perspective* (Ed. by R.H. Groves & J.J. Burdon), pp. 21–33. Australian Academy of Science, Canberra.

Bosbach, K. & Hurka, H. (1981). Biosystematic studies on *Capsella bursa-pastoris* (Brassicaceae): enzyme polymorphism in natural populations. *Plant Systematics & Evolution*, **137**, 73–94.

Brown, A.H.D. (1978). Isozymes, plant population genetic structrue and genetic conservation. *Theoretical & Applied Genetics*, **52**, 145–157.

Brown, A.H.D. (1979). Enzyme polymorphism in plant populations. *Theoretical Population Biology*, **15**, 1–42.

Brown, A.H.D. (1984). Multilocus organization of plant populations. *Population Biology and Evolution* (Ed. by K. Wöhrmann & V. Loeschcke), pp. 159–169. Springer–Verlag, Berlin.

Brown, A.H.D. & Burdon, J.J. (1983). Multilocus diversity in an outbreeding weed, *Echium plantagineum* L. *Australian Journal of Biological Sciences*, **36**, 503–509.

Brown, A.H.D., Feldman, M.W. & Nevo, E. (1980). Multilocus structure of natural populations of *Hordeum spontaneum*. *Genetics*, **96**, 523–536.

Brown, A.H.D. & Marshall, D.R. (1981). Evolutionary changes accompanying colonization in plants. *Evolution Today* (Ed. by G.G.E. Scudder & J.L. Reveal), pp. 351–363. *Proceedings, Second International Congress of Evolutionary Biology.*

Brown, A.H.D., Marshall, D.R. & Albrecht, L. (1974). The maintenance of alcohol dehydrogenase polymorphism in *Bromus mollis* L. *Australian Journal of Biological Sciences*, **27**, 545–559.

Brown, A.H.D., Marshall, D.R. & Munday, J. (1976). Adaptedness of variants at an alcohol dehydrogenase locus in *Bromus mollis. Australian Journal of Biological Sciences*, **29**, 389–396.

Brown, A.H.D., Nevo, E., Zohary, D. & Dagan, O. (1978). Genetic variation in natural populations of wild barley (*Hordeum spontaneum*). *Genetica*, **49**, 97–108.

Brown, A.H.D., Zohary, D. & Nevo, E. (1978). Outcrossing rates and heterozygosity in natural populations of *Hordeum spontaneum* Koch in Israel. *Heredity*, **41**, 49–62.

Brown, B.A. & Clegg, M.T. (1984). Influence of flower color polymorphism on genetic transmission in a natural population of the common morning glory, *Ipomoea purpurea. Evolution*, **38**, 796–803.

Burdon, J.J. & Brown, A.H.D. (1986). Population genetics of *Echium plantagineum*—a target weed for biological control. *Australian Journal of Biological Sciences*, **39** (in press).

Burdon, J.J. & Marshall, D.R. (1981). Biological control and the reproductive mode of weeds. *Journal of Applied Ecology*, **18**, 649–658.

Burdon, J.J., Marshall, D.R. & Brown, A.H.D. (1983). Demographic and genetic changes in populations of *Echium plantagineum. Journal of Ecology*, **71**, 667–679.

Burdon, J.J., Marshall, D.R. & Groves, R.H. (1980). Isozyme variation in *Chondrilla juncea* L. in Australia. *Australian Journal of Botany*, **28**, 193–198.

Chakraborty, R. (1981). The distribution of the number of heterozygous loci in an individual in natural populations. *Genetics*, **98**, 461–466.

Chakraborty, R. (1984). Detection of nonrandom association of alleles from the distribution of the number of heterozygous loci in a sample. *Genetics*, **108**, 719–731.

Clegg, M.T., Kidwell, J.F. & Horch, C.R. (1980). Dynamics of correlated genetic systems. V. Rates of decay of linkage disequilibria in experimental populations of *Drosophila melanogaster. Genetics*, **94**, 217–234.

Clegg, M.T. & Brown, A.H.D. (1983). The founding of plant populations. *Genetics and Conservation* (Ed. by C.M. Schonewald–Cox, S.M. Chambers, B. MacBryde, & L. Thomas), pp. 216–228. Benjamin/Cummings, Menlo Park, California.

Cocks, P.S. & Phillips, J.R. (1979). Evolution of subterranean clover in South Australia. I. The strains and their distribution. *Australian Journal of Agricultural Research*, **30**, 1035–1052.

Felsenstein, J. (1974). The evolutionary advantage of recombination. *Genetics*, **78**, 737–756.

Giles, B.E. (1984). A comparison between quantitative and biochemical variation in the wild barley *Hordeum murinum. Evolution*, **38**, 34–41.

Gray, A.J. (1986). Do invading species have definable genetic characteristics? *Proceedings of the Royal Society of London B*, (in press).

Hamrick, J.L. (1983). The distribution of genetic variation within and among natural plant populations. *Genetics and Conservation* (Ed. by C.M. Schonewald–Cox, S.M. Chambers, B. MacBryde, & L. Thomas), pp. 335–348. Benjamin/Cummings, Menlo Park, California.

Hamrick, J.L. & Allard, R.W. (1975). Correlation between quantitative characters and enzyme genotypes in *Avena barbata. Evolution*, **29**, 438–442.

Hedrick, P.W. & Holden, L. (1979). Hitch-hiking: an alternative to coadaptation for the barley and slender wild oat examples. *Heredity*, **43**, 79–86.

Heywood, J.S. & Levin, D.A. (1984). Allozyme variation in *Gaillardia pulchella* and *G. amblyodon* (Compositae). Relation to morphological and chromosomal variation and to geographic isolation. *Systematic Botany*, **9**, 448–457.

Jain, S.K. (1983). Genetic characteristics of populations. *Ecological Studies: Analysis and Synthesis* 44 (Ed. by H.A. Mooney & M. Godron), pp. 240–258. Springer, Berlin.

Layton, C.R. & Ganders, F.R. (1984). The genetic consequences of contrasting breeding systems in *Plectritis* (Valerianaceae). *Evolution*, **38**, 1308–1325.

Lewontin, R.C. (1984). Detecting population differences in quantitative characters as opposed to gene frequencies. *American Naturalist*, **123**, 115–124.

Loveless, M.D. & Hamrick, J.L. (1984). Ecological determinants of genetic structure in plant populations. *Annual Review of Ecology and Systematics*, **15**, 65–95.

Lyman, J.C. & Ellstrand, N.C. (1984). Clonal diversity in *Taraxacum officinale* (Compositae), an apomict. *Heredity*, **53**, 1–10.

Marshall, D.R. & Allard, R.W. (1970). Isozyme polymorphisms in natural populations of *Avena fatua* and *A. barbata*. *Heredity*, **25**, 373–382.

Mitton, J.B. & Grant, M.C. (1984). Associations among protein heterozygosity, growth rate, and developmental homeostasis. *Annual Review of Ecology and Systematics*, **15**, 479–499.

Moran, G.F. & Marshall, D.R. (1978). Allozyme uniformity within and variation between races of the colonizing species *Xanthium strumarium* L. (Noogoora Burr). *Australian Journal of Biological Sciences*, **31**, 283–291.

Moran, G.F., Marshall, D.R. & Müller, W.J. (1981). Phenotypic variation and plasticity in the colonizing species *Xanthium strumarium* L. (Noogoora Burr). *Australian Journal of Biological Sciences*, **34**, 639–648.

Oka, H.I. (1983). Life-history characteristics and colonizing success in plants. *American Zoologist*, **23**, 99–109.

Price, S.C. & Jain, S.K. (1981). Are inbreeders better colonizers? *Oecologia*, **49**, 283–286.

Price, S.C., Shumaker, K.M., Kahler, A.L., Allard, R.W. & Hill, J.E. (1984). Estimates of population differentiation obtained from enzyme polymorphisms and quantitative characters. *Journal of Heredity*, **75**, 141–142.

Scorza, R. (1983). Ecology and genetics of exotics. *Exotic Plant Pests and North American Agriculture* (Ed. by C.L. Wilson & C.L. Graham), pp. 229–238. Academic Press, New York.

Stebbins, G. (1957). Self fertilization and population variability in the higher plants. *American Naturalist*, **91**, 337–354.

van Dijk, H. (1984). Genetic variability in *Plantago* species in relation to their ecology. 2. Quantitative characters and allozyme loci in *P. major*. *Theoretical & Applied Genetics*, **68**, 43–52.

van Dijk, H. (1985). Allozyme genetics, self incompatibility and male sterility in *Plantago lanceolata*. *Heredity*, **54**, 53–63.

7. FEATURES OF COLONIZING ANIMALS: PHENOTYPES AND GENOTYPES

P.A. PARSONS
Department of Genetics and Human Variation, La Trobe University, Bundoora, Victoria, 3083 Australia

INTRODUCTION

'Natural selection which guides evolutionary change acts primarily on phenotypes, and only secondarily on genotypes' (Waddington 1965). In order to establish patterns of variation, population geneticists typically follow visible polymorphisms, chromosome rearrangements, electrophoretic variants, or lethal genes. A search for the ecological significance of these patterns follows a posteriori, as documented in the ever-expanding literature on geographical patterns and environmental gradients in central–marginal populations (see Brussard 1984). In parallel, there have been substantial developments in theoretical population genetics, but contact with empirical studies is generally indirect (see Nevo 1983).

In contrast, ecologists tend to emphasize quantitative traits of obvious adaptive significance in determining the distribution and abundance of organisms. The normally unstated assumption is that the organism is the unit of selection. This follows from the conventional understanding that natural selection acts primarily at the organismic level. Individuals with certain phenotypic characteristics are favoured at the expense of others. Genotypic changes follow phenotypic changes, but the connections between genotype and phenotype range from direct, where clear-cut polymorphisms are involved, to indirect for most quantitative traits.

The $r-K$ continuum of life-history characteristics (see Stearns 1976; Southwood 1977) embodies a phenotypic approach. However, estimating r is too complex to handle for anything but the simplest of genetic analyses. In recent years, therefore, there has been emphasis on the study in natural populations of quantitative traits such as development time, survival, and fertility; the aim has been to dissect life histories into their components (see Istock 1983), and to investigate selection in natural populations (Arnold & Wade 1984).

The question of which of many possible traits should be considered in detail is extremely difficult. Andrewartha & Birch (1954), in their classic book *The Distribution and Abundance of Animals*, considered four major components of the environment: weather, resources, other organisms and habitats. Following their approach, Parsons (1983a) identified certain ecologically-important characteristics, which he defined as 'ecological phenotypes', such as tolerance to extreme physical environments, development time, and certain aspects of resource utilization. Their significance in determining the distribution and abundance of animals was examined using *Drosophila melanogaster* as a case study.

An understanding of the interface between ecology and genetics is more likely using quantitative genetic methods because these are concerned with the phenotypic level (in contrast to population genetics, which operates at the genotypic level). Proceeding via quantitative genetics, it should be possible to encourage more effective collaboration between ecologists and geneticists. One object of this chapter is to develop this theme using colonizing species. The arguments will be principally illustrated with the genus *Drosophila*, in which there are many species occurring in or near human habitations, gardens, orchards, and garbage dumps, and hence which have spread with the creation of such habitats.

For the purposes of this discussion, a colonizing species is defined as one in which range expansions are occurring or have been documented in historic time, so that species spreading with man are, together with man, colonists.

EXTREME STRESSES AND ECOLOGY

'Climate plays an important rôle in determining the average numbers of a species, and periodical seasons of extreme cold or drought seem to be the most effective of all checks' (Darwin 1859). Ehrlich *et al.* (1980) investigated the dynamics of two species of the checkerspot butterfly, *Euphydryas*, during a major Californian drought in 1975–77. Several populations became extinct, some were dramatically reduced, others remained stable, and at least one increased. Ehrlich *et al.* suggested that severe climatic stresses may be effective at this level once only in 50–100 generations in California. Such 'ecological crises' are therefore unpredictable events of short duration relative to the total time under consideration, but such disturbances may lead to fundamental changes in the natural order of a system (Raup 1981; Sousa 1984). The major problem in studying ecological crises is that they can be regarded as random events of low frequency, meaning that observations over long periods of time are mandatory. Studies over shorter periods of time may lead to erroneous conclusions concerning factors determining distribution and abundance;

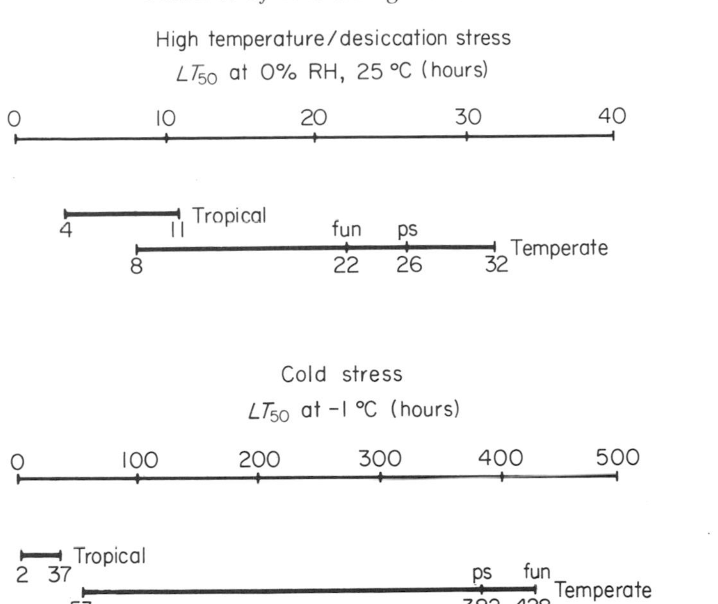

FIG. 7.1. Range of LT_{50} values (number of hours at which 50% of flies die from stress) for nine tropical (including the cosmopolitan species, *D. ananassae*) and twelve temperate-zone (including seven cosmopolitan species) *Drosophila* species. For cold tolerance, there is no overlap across zones and for high temperature/desiccation, the overlap is minimal, comprising only *Drosophila simulans* which is a cosmopolitan species distributed across both zones. Results for a population of *D. funebris* (fun) and for a New Zealand population of *D. pseudoobscura* (ps) are individually plotted.

conversely, it follows that all biological groups will be subjected to crises, which means that steady states in evolutionary time are not expected (Raup 1981).

Following an ecological crisis, rapid population growth will be likely, whether by recolonization or by colonization of a new habitat. The ecology of the reoccupied habitat may differ from that before the crisis, and similarly a colonizing population is likely to encounter habitats differing ecologically in some way from the source population. Far more common than actual crises will be periods of ecological stress: indeed, stress periods on a diurnal and seasonal time-scale are characteristic features of ephemeral habitats where physical conditions and resource availability are highly variable. While the distinction between ecological stress and crisis is arbitrary, for the purposes of this discussion an ecological crisis is defined as likely to lead to species extinctions, while an ecological stress is likely to lead to adaptation and eventual intraspecific climatic race formation.

In spreading from the humid tropics to the temperate zone which is characterized by a seasonal pattern of cool, wet winters and warm, dry summers, species of the genus *Drosophila* (Fig. 7.1) have evolved substantial tolerance of climatic extremes. These may be assessed in the laboratory by testing their resistance to stresses imposed by temperature and desiccation (Parsons 1983a). More generally, extremes of temperature correlate well with floral type in the Australian flora (Nix 1981). This is important for *Drosophila* since species of the genus predominantly depend upon plant degradation products for resources. The biogeographic importance of extreme temperatures is therefore clear.

The greatest diversity of *Drosophila* species in Australia occurs in the mesotherm (thermal optimum in the 18–22 °C range) element of the flora, especially in the complex mesophyll vine forests of the humid tropics of north Queensland, where nearly half of the described Australian *Drosophila* species have been found (Bock 1982; Parsons 1982). These forests contain fungus-feeding species of the subgenus *Hirtodrosophila* which are very sensitive to the extreme physical stresses of desiccation at 25 °C and −1 °C (Table 7.1). In contrast, the Northern Territory is in the megatherm (thermal optimum around 28 °C) range where there is a desicccation-resistant, depauperate fauna consisting of a few mainly widespread species of the dominant Australian subgenus *Scaptodrosophila*; almost all of these species also occur in lowland north Queensland habitats where the heat stress is substantially higher than in the complex, mesophyll vine forests. Biologically, the 'heat–stress status' of the Northern Territory is confirmed from comparative analyses of ecological phenotypes of *D. melanogaster* populations from various Australian localities (Parsons 1983a).

Widespread and/or colonizing *Drosophila* species therefore tend to be resistant to environmental stresses. The depauperate New Zealand *Drosophila* fauna consists of seven of the eight cosmopolitan (colonizing) *Drosophila* species (the eighth, *D. ananassae*, being restricted to the tropics), two endemic *Scaptodrosophila* species, and two recent colonists, namely *D.* (*Scaptodrosophila*) *enigma*, a colonist of orchards in Australia, and the North American species *D. pseudoobscura* (Parsons 1982, 1983a), which is more cold-resistant than all the species in Fig. 7.1 except *D. funebris*, a species of high latitudes in both hemispheres. The cold-resistance of *D. pseudoobscura* is consistent with its native habitat in the forests of western North America, although range expansions have occurred into orchards (see Brussard 1984). In summary, these and other data

TABLE 7.1. LT_{50} values (h) for field-collected flies (sexes combined) compared with *D. melanogaster*

Subgenus and species	Desiccation (25 °C, O% RH)	−1 °C cold stress
Hirtodrosophila		
*mycetophaga**	3·8	15
*polypori**	2·7	24
Sophophora		
melanogaster†	16	70

* As the LT_{50}s are from field-collected flies, sample sizes are generally low and highly variable, so that means only are given. However, individual LT_{50}s do not overlap across subgenera.

† Results vary somewhat according to climate of habitat where flies collected.

(Parsons & Stanley 1981) suggest that the resistance of *Drosophila* species to the above environmental stresses can be related broadly to their distribution patterns. This analysis must not, however, be applied too rigorously because of genetic variation based upon geographic races within species (Stanley & Parsons 1981; Coyne, Bundgaard & Prout 1983).

Under less ephemeral and more benign circumstances than would be typical of the habitats of colonizing *Drosophila* species, environmental stability tends towards *K*-selection, when phenonomena such as density and frequency-dependent selection and competition have been argued to be powerful forces in determining community structure, and at the intraspecific level in maintaining genetic diversity (see Futuyma 1979). This implies some sort of equilibrium between populations and resources which certainly is not a feature of those colonizing species discussed so far. Genetic analysis of colonizing insect species at the individual and population levels should therefore be simplified, since 'second-order' phenomena can be largely discounted. Indeed, assuming that all populations are subject at some time to ecological crises, this conclusion may be of wider application in evolutionary biology (see Raup 1981; Connell 1983).

While not wishing to enter the wide-ranging debate in the literature on competition, it is worth repeating from Strong (1983) that much of orthodox competition theory refers to birds, lizards and other vertebrates, which are only a small component of biotic diversity. Indeed, he considers that for decomposer insects (including *Drosophila*) 'the theoretical model that most accurately describes these organisms is not one of homogeneous niches at some equilibrium but one with highly dissected, rapidly changing resource patches'. Since *Drosophila* is involved at the initial stage of microbial degradation, and indeed is often attracted to resources (e.g. flowers) before breakdown is obvious, Strong's (1983) comment on herbivorous insects is relevant: 'Interspecific competition is probably not commonly important for herbivorous insects because autecology, vertical food-web factors, and the weather normally serve to maintain populations below densities that would deplete resources'. In summary, interactions among *Drosophila* species utilizing similar resources may be relatively unimportant. This is consistent with Atkinson's (1985) observations on field-collected rain forest flies, where he found no evidence for interspecific competition, even though there is demonstrable larval competition within species. In addition, Shorrocks *et al.* (1984) have recently argued against interspecific competition as a major organizing force in many insect communities, as have Parsons & Hoffmann (1985), based upon habitat marking experiments in *Drosophila*.

GENETIC VARIABILITY AND ENVIRONMENTAL STRESSES

'It is unlikely that any of the problems and contradictions of evolutionary genetics can be resolved by even more arcane models of single loci, or of constant environments' (Lewontin 1973).

Field populations: electrophoretic traits

A large literature has been accumulated on molecular polymorphisms in nature since the advent of electrophoretic techniques. This has led to a debate as to whether the molecular variation is adaptive, irrelevant to natural selection, or whether both interpretations apply (see, for example, Nevo 1978; Fitch 1982; McDonald 1983). Discrimination between the different models is frequently ambiguous. This unsatisfactory situation largely follows from a posteriori attempts at providing causal interpretations of data from natural populations. Allele frequencies of electrophoretic polymorphisms often tend to be correlated directly or indirectly with geographical variations in temperature. There are many examples of this, such as the alcohol dehydrogenase (ADH) polymorphism in *D. melanogaster* (Alahiotis 1982) and the majority of enzyme or protein polymorphisms in humans (Piazza, Menozzi & Cavalli-Sforza 1981). Such associations follow from the importance of temperature in all biological processes, and are quite usual if adequate data are collected.

Using data from 1111 species for an average of 23 loci each, Nevo, Beiles & Ben-Shlomo (1984) studied the genetic diversity of enzymes and proteins among populations, species and higher taxa in relation to ecological parameters (life-zone, geographical range, habitat type and range, and climatic region), demographic parameters (species size and population structure, gene flow and sociality), and a series of life-history characteristics (longevity, generation length, fecundity, origin, and parameters related to the mating system and mode of reproduction). Fig. 7.2a shows that ecological factors explained by far the highest proportion of the genetic diversity as compared with demographic or life-history factors (90%, 39% and 3·5%, respectively). While this is an a posteriori approach, it is based on all readily available published data, and as such is a most valuable compilation and analysis. The unequivocal conclusion is that the major force producing evolutionary adjustment at the protein level is at the level of ecological parameters, characterized for each category into the following subdivisions:

(a) life zone into Arctic, Temperate, Tropical, Temperate and Tropical, and Cosmopolitan;
(b) geographical range into endemic or relict, narrow, regional and widespread;
(c) habitat type into underground, overground, arboreal, air and terrestrial, aquatic, aquatic–littoral, aquatic and terrestrial, and aquatic and air, while habitat range refers to specialists vs. generalists;
(d) climatic range into arid, sub-arid, sub-humid, mesic, extra mesic, and mesic and arid.

In a biogeographical context, these habitat subdivisions can, in many cases, be reduced to Andrewartha & Birch's (1954) major components of the environment. Therefore, climate is clearly important at both the habitat and geographical level, indicating agreement with Nix's (1981) conclusions for the biogeographical importance of climatic extremes based upon Australian floral types.

Fig. 7.2a shows that about 80% of the overall genetic diversity is open to further analysis and interpretation, even though the 20% that can be explained is primarily accounted for by ecological factors. The inefficiency of this a posteriori approach is to be expected, and it is noteworthy that the explained proportion of genetic diversity increases within closer groupings. Fig. 7.2b shows that 66% of genetic diversity can be explained in Drosophila mainly by ecological factors. Even though the approach uses electrophoretic variation, it emphasizes the important relationship between ecological phenotypes and habitat.

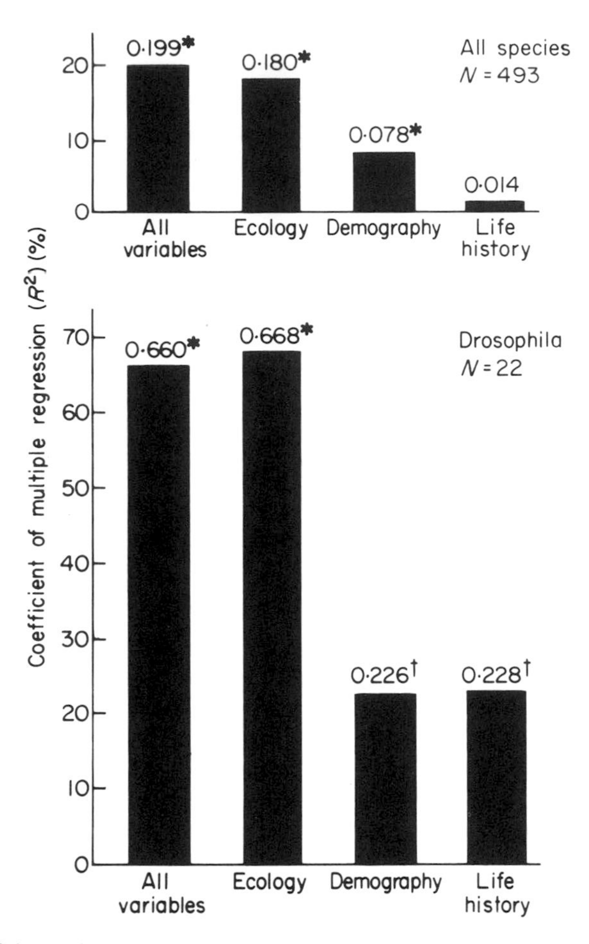

FIG. 7.2. Coefficients of multiple regression with dependent value heterozygosity, and as independent variables (a) fifteen biotic variables, and (b) a subset of six ecological, (c) five demographic and (d) four life-history variables for all species and for the genus *Drosophila*. The coefficients represent the proportion of genetic variance that can be explained by biotic variables. Significance: $^{*}P < 0{\cdot}01$, $^{\dagger}P < 0{\cdot}05$ (plotted from Nevo, Beiles & Ben-Shlomo 1984).

Predictably, over a wide array of taxa, the coefficient of multiple regression for life-history factors is very low (Fig. 7.2a); in the genus *Drosophila* this coefficient is higher and is significant at $P < 0{\cdot}05$, although small compared with the value for ecological factors (Fig. 7.2b). The lack of success of Taylor & Condra (1980) in selecting for rapid generation cycles under uncrowded conditions (*r* selection) and ability to withstand crowding (*K* selection) in *D. pseudoobscura* is consistent with relatively low genetic variability for life-history characteristics. There is, however, sufficient additive genetic variance for life-history traits, such as fecundity, in *D. melanogaster* to permit some responses to selection (Charlesworth 1984; Rose 1983), although heritabilities tend to be $<0{\cdot}5$ with values in the region of $0{\cdot}2-0{\cdot}3$ being typical for such traits (Falconer 1960). Especially for resistance to such climatic extremes as desiccation and mortality after exposure to extreme temperatures, there is some limited evidence for a higher genetic component (Parsons 1970, 1983a), which would be reflected in a higher coefficient of multiple regression for ecological compared with life-history factors. Simultaneous studies of life-history factors and ecological phenotypes in the same population under optimal and stressful environments are needed.

Field populations: ecological traits

Organismic selection at the interspecific level (Fig. 7.1) provides a model for studies at the intraspecific level. For example, tropical populations of the sibling species *D. melanogaster* and *D. simulans*, and of the Queensland fruit fly *Dacus tryoni*, are more sensitive to high temperature/desiccation and cold stresses than those from the temperate zone (Lewontin & Birch 1966; Parsons 1983a). More generally, ecological phenotypes related to factors determining distribution and abundance tend to vary clinally among *D. melanogaster* populations in a manner predicted by geographical variation in climatic factors (Stanley & Parsons 1981). Coyne, Bundgaard & Prout (1983) found geographical variation for three environmental factors (heat, cold and desiccation) in seven *D. pseudoobscura* populations at the larval and pupal stages. Based upon the a priori prediction that survivorship reflects local adaptations to climate, they found the pupal results to be closer to those expected than were the adult results—perhaps because pupae cannot respond behaviourally to extreme climates.

Data on three traits from three populations of *Drosophila melanogaster* ranging from the temperate zone to two contrasting tropical climates (humid and wet-dry) are given in Table 7.2. For cold tolerance, the Melbourne and Darwin population means differ by more than seven standard deviations in the expected direction. In contrast, there is overlap between populations for development time which is consistent with directional selection (Lewontin 1965). Ethanol tolerance is relatable to habitat and is apparently an ecologically-important phenotype (Parsons 1983a); for this, the Melbourne and Townsville populations differ by more than eight standard deviations.

TABLE 7.2. Means and standard deviations for three ecological traits in three Australian populations of *Drosophila melanogaster*. Values are based on seven to ten isofemale strains per population. Data are from Stanley & Parsons (1981)

	Latitude (°S)	Climate	Ethanol tolerance* $\bar{x} \pm$ S.D.	Development time† $\bar{x} \pm$ S.D.	Cold tolerance‡ Females $\bar{x} \pm$ S.D.	Cold tolerance‡ Males $\bar{x} \pm$ S.D.
Melbourne	37	Temperate	57·8±6·3	13·8±0·2	6·6 ± 0·7	10·3 ± 1·7
Townsville	20	Humid tropics	7·9±6·2	14·0±0·2	–	–
Darwin	12	Wet-dry tropics (stressful)	34·5±6·9	13·7±0·2	19·2±1·7	23·6±0·4

* Time taken in hours for half of 20 flies to die when exposed to atmospheric ethanol vapour at a concentration of 12% in a sealed system.
† Mean time in days from hatching to emergence at 20 °C.
‡ Number dead out of 25 after 40 hours exposure to −1 °C.

Non-overlapping distributions for quantitative traits frequently occur; e.g. morphology and ethanol tolerance in European and tropical African strains of *D. melanogaster* (David & Bocquet 1975). The ethanol-tolerance of *D. melanogaster* from a wine cellar was found to be twice that of a nearby urban population, with means separated by two to four standard deviations (McKenzie & Parsons 1974a). At this local level there is a clear separation between populations for ethanol-tolerance, although they are indistinguishable for allele frequencies at the *Adh* locus. Indeed, natural populations often show no detectable correlation between ethanol-associated phenotypes and variation at the *Adh* structural locus (McKenzie & McKechnie 1978; Middleton & Kaiser 1983). Inferences based upon ecological phenotypes such as ethanol-tolerance are more directly relatable to habitat than variants at the electrophoretic locus. Lewontin (1984) has calculated that it is very much more difficult to show a detectably-significant difference between gene frequencies than between the means of metrical characters in comparisons of populations, when the loci are equally differentiated. Calculations by Hoffmann, Nielsen & Parsons (1984) agree with this conclusion.

Electrophoretic variability was investigated for three loci (Hoffmann, Nielsen & Parsons 1984) at three levels of spatial heterogeneity in *D. melanogaster*:

(a) variation among breeding sites (single fruits in an orchard);
(b) within populations (among banana-baited traps attracting adults in the orchard);
(c) among populations at the geographical level (east coast of Australia).

The variation among breeding sites, (a), was greater than variation within the orchard, (b), in all cases (Table 7.3). Furthermore, for the *Gpdh* and *Tpi* loci the variation, (a), was greater than geographic variation, (c), although this ranking is reversed for the *Adh* locus. It is this pattern of variation, largest at the microhabitat level, which argues against its involvement in evolutionary shifts at the primary level, including colonization events. The study of ecological phenotypes is a more direct approach than that via electrophoresis.

Optimal and extreme environments

There is increasing emphasis on studying the genotypes of a polymorphic system under optimal and extreme environments (Parsons 1974, 1983a). Intertidal marine invertebrates form good material for such studies, since stress arising from thermal, chemical, and heavy metal pollution tends to emphasize fitness differences among allozymes, as do parallel laboratory experiments simulating the field situation (Nevo, Lavie & Ben-Shlomo 1983). On the island of Skokholm, Berry (1979) found that the mortality of house mouse populations in winter varied between 40 and 90%. In mild winters, deaths were independent of haemoglobin genotype, but in cold winters animals carrying the *s* allele survived better. Parasites can be an ecological stress, as shown by the advantage of the sickle-cell gene heterozygote in man under the influence of the malarial parasite. Berry (1979) summarized other similar examples involving environmental stresses of various types, such as temperature extremes, pesticides, predation, food availability, and population density.

For a random set of polymorphic loci, there is usually an association between genetic variability and fitness, which is maximized under extreme environments. Examples include the oyster, *Crassostrea virginica*, the coot clam, *Mulinia lateralis*, the oldfield mouse, *Peromyscus polionotus*, the side-blotched lizard, *Uta stansburiana*, and of course *Drosophila*

TABLE 7.3. Standardized variances F_{st} at three electrophoretic loci in *Drosophila melanogaster* (simplified from Hoffmann, Nielsen & Parsons 1984)

Locus	Triosephosphate isomerase	Glycerophosphate isomerase	Alcohol dehydrogenase
Among breeding sites (individual fallen ripe fruits)	0·046	0·039	0·030
Among adults (baited traps)	0·013	0·012	0·013
Among geographic localities	0·028	0·023	0·256

(Garten 1976; Soulé 1979; Zouros, Singh & Miles 1980; Parsons 1983a; Garton, Koehn & Scott 1984). These results are consistent with theoretical considerations (Turelli & Ginzburg 1983), and recent experimental data suggest an association between routine metabolic efficiency and heterozygosity (Garton, Koehn & Scott 1984).

Turning to quantitative characters, the utilization of ethanol and acetic acid as resources has been studied in populations of *D. melanogaster* consisting of isofemale strains (strains set up from single inseminated founder females) under optimal and stressful situations, whereby 3% ethanol and 6% acetic acid were optimal and utilized as resources, and 12% ethanol and 12% acetic acid were stressful, since exposure to the metabolites at these concentrations shortened longevity (Parsons 1983a). Intraclass correlations were higher for the stressful concentrations than for the optimal concentrations. The two highest correlations were for the tropical Townsville populations, where longevities were reduced most of all under the stress used, indicating that the level of variation among isofemale strains depends on the environment because the expressed variation increases with stress. In mice, an analysis of the weights of endocrine glands subjected to chronic stress during pregnancy showed an increase in the genetic components of the phenotypic variation for such characters as pre-implantation loss, litter size and relative weights of the thymus and adrenals (Belyaev & Borodin 1982).

In summary, these and many other examples show that phenotypic variability, genetic variability and heterozygote advantage are maximized under severe stress conditions. This implies that stress can increase the range of genetic variation and so affect the rate and direction of evolutionary changes. This conclusion follows directly from Fisher's (1930) Fundamental Theorem of Natural Selection, whereby as an evolutionary equilibrium is approached in stable environments, the additive genetic variance for fitness traits should tend towards zero.

Mutational responses

In adjustments to new adaptive zones, it can be asked whether there is any *de novo* variation provided by mutational events of any type, especially as the genomes of most organisms contain DNA several times in excess of their apparent coding requirements. Recent findings in molecular biology indicate that many mutational events may have a significant environmental component and, in particular, there are suggestions that both mutational responses and recombinations may increase under stress conditions (Belyaev & Borodin 1982; Echols 1982; Wills 1983). Therefore, genetic variability of all kinds, including *de novo* variation, could be highest under stress conditions, depending on the magnitude of the environmental challenge. *Escherichia coli*, for example, displays a series of inducible responses (the SOS response) to a variety of environmental stresses including inhibition of cell division, prophage induction, elevated capacity for DNA repair, and increased rates of mutagenesis which include point mutations, general and site-specific recombination and genetic rearrangements (Little & Mount 1982). Interestingly, observations of this nature were anticipated by McClintock (1951, 1978), who argued for the importance of both genomic and environmental stress as triggers for rapid genomic reorganization via controlling elements, which are now known to be one example of a broad class of movable or transposable elements present within perhaps all prokaryotic and eukaryotic genomes (Green 1978; Echols 1982; Little & Mount 1982; Wills 1983). Following the groundwork laid by McClintock in maize, and by Green (1967) in *Drosophila melanogaster* in particular, such fragments of DNA have been intensively studied in the last few years.

Transposons therefore appear to provide a mechanism whereby mutational events increase dramatically during periods of environmental challenge, and the consequent periods of organismic stress. Does it follow that precisely at those challenging moments in evolutionary history when there is a premium on major adaptive shifts, the probability that the appropriate variants will be produced is increased? Much more needs to be known about the nature and rôle of environmentally-induced responses, and detailed studies of laboratory stresses simulating field stresses are needed.

ECOLOGICAL PHENOTYPES AND GENETIC ANALYSIS

'Intra-population genetic variation has been largely ignored by ecologists, probably because they had no easy means of recognizing the effects of segregating genes' (Berry 1979).

Since colonists may arise from populations at the margins of the distribution of species, populations from central and marginal habitats must be compared. Much information has been obtained on gene and chromosome variation of central and marginal populations defined mainly in a geographical sense. This has turned out to be remarkably difficult to interpret (Parsons 1983a; Brussard 1984), but in view of the variability among breeding sites presented in Table 7.3, this is not surprising.

Populations can also be collected as a set of single inseminated females taken at random to enable population phenotypes to be described by variability among isofemale strains. Even assuming multiple insemination, an isofemale strain goes through a bottleneck when establishing, since it is unlikely that the effective population size $N > 3$. Different isofemale strains should possess their own unique phenotypic characteristics over many generations, on the assumption that relatively few genes contribute most variability to a quantitative trait; data consistent with this prediction have been published for morphological, ecological, behavioural and molecular traits (see Parsons 1983a, b). In addition, there is a steadily accumulating body of evidence from directional selection experiments suggesting that major genes often underlie quantitative traits (Thompson & Thoday 1979). This genetic architecture permits a rapid response to new environments during range expansions.

The isofemale-strain approach to quantitative inheritance takes some of the simpler features of the biometrical and polygene-location approaches in order to concentrate upon the range and nature of variation in natural populations and is of particular value for traits of ecological importance. Comparisons among populations each based upon several isofemale strains permit the genetic dissection of populations from ecologically-central and marginal habitats based upon ecological phenotypes as outlined in a case study by Parsons (1980).

There has been substantial laboratory research on acute stresses of an ecological nature, in particular resistance to temperature extremes and desiccation. Crosses among isofemale strains from natural Australian populations of *D. melanogaster* and *D. simulans* have shown that desiccation resistance is largely under additive genetic control, but with a significant but lesser dominance component in most cases (McKenzie & Parsons 1974b). Among eighteen isofemale strains from a habitat characterized by long, hot, dry summers, a correlation coefficient between mortality following desiccation (0% RH at 25 °C) and following high temperature of 30·5 °C came to 0·52 ($P < 0{\cdot}05$). This result suggests that the physiological basis of resistance to high temperature and desiccation may be similar (Parsons 1980). However, differing results are possible from other populations, since some *Drosophila* populations also occur in areas of high temperature and humidity. Morrison & Milkman (1978) successfully selected for decreased resistance to heat shock (25–30 min at 40 °C) in one of forty isofemale strains; a major locus on chromosome 2 was largely responsible. Stephanou & Alahiotis (1983) found substantial variability in a natural population for heat shock resistance (25 min at 40 °C), based upon fifty isofemale strains. In this case, selection experiments produced strains exhibiting a twentyfold difference when subjected to heat shock, which was ascribed mainly to quantitative differences transmitted through the maternal cytoplasm, together with nuclear modifier genes as a minor component.

While much more needs to be done in order to understand the genetic basis of resistance to extreme stresses, the studies cited collectively demonstrate the usefulness of isofemale strains in analyses at the population level. In particular, they permit the testing of hypotheses as to predicted genetic architectures using simple, one-generation procedures such as diallel crosses among strains. A significant advantage is that genetic information is rapidly available, enabling comparative assessments of the evolutionary significance of quantitative variation in species where genetic components of variation cannot be accurately estimated or genes specifically located. Indeed, at this level, isofemale strains can readily provide genetic information on populations for an array of ecologically-important traits incorporating environmental stress when additivity tends to be maximized.

DISCUSSION

Can the approach to variability in natural populations outlined in this chapter be generalized to incorporate life-history classifications such as the $r-K$ continuum? The electrophoretic data analysis of Nevo, Beiles and Ben-Shlomo (1984), as summarized in Fig. 7.2, would argue for substantial difficulties. Boulétreau-Moule *et al.* (1982) showed that tropical (African) populations of *D. melanogaster* have lower fecundity than temperate (French) populations, indicating that temperate populations are closer to the r end of the $r-K$ continuum, but other life-history characteristics do not support this interpretation. Atkinson (1979) compared reproductive strategies of seven domestic species of *Drosophila* and found that species from similar breeding sites tend to have similar reproductive strategies, which in themselves are inconsistent with $r-K$ selection theory. In agreement with Kambysellis & Heed (1971), reproductive strategies apparently evolved as adaptations to the frequency of finding breeding sites, so that the larval niches of *Drosophila* are of primary concern. Similarly, Lachaise (1983) found a wide array of reproductive patterns among twenty-three species of tropical African *Drosophila* associated with varying degrees of specialization in the field, and concluded that the frequency of lifetime oviposition opportunities is a major determinant of interspecific variation in reproductive traits. In other words, rather than being bound by the $r-K$ continuum, an understanding of organisms in relation to their habitat is more fundamental—as stressed by Southwood (1977). An example of this is the analysis of ecological phenotypes in *D. melanogaster*, including desiccation resistance, development time, ethanol tolerance and larval preference for ethanol for populations from climatically-benign and stressful habitats (Parsons 1980, 1983a).

What are the criteria for identifying potential colonists? Rather than attempting to answer this in a general sense, one approach is to compare closely-related colonist and non-colonist species, such as the *melanogaster* subgroup of *Drosophila* species which contains two cosmopolitan (colonizing) species (*melanogaster*, *simulans*) and at least six species restricted to the tropics. Comparative studies show the following:

(a) Adults of the two cosmopolitan species are substantially more resistant to the physical stresses of desiccation at 25 °C and −1 °C than tropical species, in particular *D. erecta* (Stanley *et al.* 1980).

(b) Larvae give parallel results for resistance to temperature extremes, albeit with some variation among developmental stages, indicating that all life-cycle stages must show adaptations for range expansions into temperate zones (Parsons 1983a).

(c) Considering resources, ethanol and acetic acid utilization falls into three groups (Parsons 1983a), i.e.

D. melanogaster > *D. simulans* > *D. yakuba* > *D. erecta*
D. tiessieri *D. mauritiana*

Tropical populations of *D. melanogaster* are almost indistinguishable from *D. simulans*, with a cline of increasing ethanol-tolerance/utilization towards the temperate zone (David & Bocquet 1975), which may be consistent with the characteristics of resources (Stanley & Parsons 1981). *D. tiessieri* and *D. yakuba* are widespread in tropical Africa, and so are almost certainly generalists compared with *D. erecta*—an apparent specialist almost exclusively associated with the fruits of *Pandanus candelabrum* in tropical Africa (Lemeunier & Ashburner 1976). *D. mauritiana* is a sibling species of *D. simulans* on the tropical island of Mauritius, where it could have evolved a low level of ethanol-utilization in response to the specific resources of this island habitat.

(d) In Africa, the number of host-plant families (genera) utilized by four of the species (Lachaise 1983) has been found to be *D. melanogaster* 26 (27)), *D. yakuba* 12 (13–16), *D. tiessieri* 12 (13) and *D. erecta* 2 (2), which parallels the sequence of sensitivities to environmental stresses and ethanol-utilization above (Lachaise 1983).

From a consideration of items (a)–(d), there emerges in terms of environmental stresses tolerated and resources utilized a broad distinction between cosmopolitan species and those of more restricted distribution.

In a study of twenty-three African species of Drosophilidae, Lachaise

(1983) found that for the array of larval food items utilized, generalists tend to have short pre-reproductive periods as adults and to produce more offspring than specialists. This is in accord with the $r-K$ theory, but there is some evidence that colonizing (actually expanding) species tend to fall between the generalist and specialist categories. He argued that colonists have evolved phenotypes permitting colonization into new habitats; hence, some delays in terms of maturity would be advantageous compared with generalist species. Once again, therefore, we see a dependence of reproductive characteristics upon habitat.

Can these conclusions be extended more widely? Certainly there are parallels between *Drosophila* and the tephritid genus, *Dacus*, where a combination of physiological, ecological and climatological data have been used to ascertain the bioclimatic limits to the spread of the Queensland fruit fly, *D. tryoni* (Bateman 1967; Meats 1981). Furthermore, since there are about 100 known species of *Dacus* in Australia, mostly specialists colonizing very few (and often only one) host-plant species (Drew 1982), comparative studies among species should assist in defining more accurately the features of colonizing *Dacus* species.

Among other recent colonists in Australia, the giant toad, *Bufo marinus*, introduced in 1935, has now spread over most of Queensland. It has been predicted that the southern distributional limits will be mainly determined by climate, while spread into the arid interior will probably be restricted by insufficient breeding sites (Sabath, Boughton & Easteal 1981). Detailed work on climate-related ecological phenotypes will be important, as emphasized by studies on the northern limits of the wild rabbit in Australia (Cooke 1977). Extension of this approach to species such as the house mouse, *Mus musculus*, and its various relatives would be of substantial interest in view of the environmental challenges to which this species has been exposed (Berry, Peters & van Aarde 1978).

At the phenotypic level, therefore, some progress has been made towards unravelling the features of colonizing animal species based upon an understanding of the biology of organisms in their habitats and concentrating upon ecological phenotypes important in determining distribution and abundance.

No unique features of colonizing animals have been proposed at the genetic level, since the discussions on genetic architectures are not restricted to colonizing species. Perhaps detailed studies of the genomes of closely-related species may ultimately be informative, but the amount of work needed is enormous. Such studies would need to include investigations of mutational response, and its molecular basis, under stress conditions, and of the connection between the molecular level and the potential for the rapid development of races in an array of habitats (defined principally in terms of climate and resources). The *melanogaster* subgroup of *Drosophila* species may provide good material. Laboratory investigations might usefully attempt to see whether artificial selection can simulate the natural situation for tolerances to physical conditions, and move an apparent non-colonist into the range characteristic of closely-related colonizing species. For such experiments, stresses approaching the ecological crisis level should be used. Responses might then be analysed in terms of whether a reorganization of already-present variability has occurred and/or whether *de novo* variability has played a significant rôle. In other words, comparisons among closely-related colonists and non-colonists are needed where populations are severely challenged by relevant stresses as, arguably, these are the circumstances when variability is maximized and there is the greatest potential for promoting change. Such an approach may also provide insights into discontinuities in the fossil record where there is increasing evidence for ecological change as a primary factor (Turner 1983). Therefore, an understanding of the circumstances under which a species can be converted to a colonist may facilitate interpretation of evolutionary rates and shifts over much longer time-spans.

REFERENCES

Alahiotis, S.N. (1982). Adaptation of *Drosophila* enzymes to temperature, IV. Natural selection at the alcohol-dehydrogenase locus. *Genetica*, **59**, 81–87.

Andrewartha, H.G. & Birch, L.C. (1954). *The Distribution and Abundance of Animals.* University of Chicago Press, Chicago.

Arnold, S.J. & Wade, M.J. (1984). On the measurement of natural and sexual selection: applications. *Evolution*, **38**, 720–734.

Atkinson, W.D. (1979). A comparison of the reproductive strategies of domestic species of *Drosophila. Journal of Animal Ecology*, **48**, 53–64.

Atkinson, W.D. (1985). Coexistence of Australian rainforest Diptera breeding in fallen fruits. *Journal of Animal Ecology*, **54**, 507–518.

Bateman, M.A. (1967). Adaptations to temperature in geographic races of the Queensland fruit fly, *Dacus (Strumeta) tryoni. Australian Journal of Zoology*, **15**, 1141–1161.

Belyaev, D.K. & Borodin, P.M. (1982). The influence of stress on variation and its role in evolution. *Biologisches Zentralblatt*, **100**, 705–714.

Berry, R.J. (1979). Genetical factors in animal population dynamics. *Population Dynamics* (Ed. by R.M. Anderson, B.D. Turner & L.R. Taylor), pp. 53–80. Blackwell Scientific Publications, Oxford.

Berry, R.J., Peters, J. & van Aarde, R.J. (1978). Sub-antarctic house mice: colonization, survival and selection. *Journal of Zoology (London)*, **184**, 127–141.

Bock, I.R. (1982). Drosophilidae of Australia. V. Remaining genera and synopsis (Insecta: Diptera). *Australian Journal of Zoology Supplementary Series*, **89**, 1–164.

Boulétreau-Moule, J., Allemand, R., Cohet, Y. & David, J.R. (1982). Reproductive strategy in *Drosophila melanogaster*: significance of a genetic divergence between temperate and tropical populations. *Oecologia* (*Berlin*), **53**, 323–329.

Brussard, P.F. (1984). Geographic patterns and environmental gradients: The central-marginal model in *Drosophila* revisited. *Annual Review of Ecology and Systematics*, **15**, 25–64.

Charlesworth, B. (1984). The evolutionary genetics of life histories. *Evolutionary Ecology* (Ed. by B. Shorrocks), pp. 117–133. Blackwell Scientific Publications, Oxford.

Connell, J.H. (1983). On the prevalance and relative importance of interspecific competition: evidence from field experiments. *American Naturalist*, **122**, 661–696.

Cooke, B.D. (1977). Factors limiting the distribution of the wild rabbit in Australia. *Proceedings of the Ecological Society of Australia*, **10**, 113–120.

Coyne, J.A., Bundgaard, J. & Prout, T. (1983). Geographic variation of tolerance to environmental stress in *Drosophila pseudoobscura*. *American Naturalist*, **122**, 474–488.

Darwin, C. (1859). *On the Origin of Species by Means of Natural Selection*. Murray, London.

David, J. & Bocquet, C. (1975). Similarities and differences in latitudinal adaptation of two *Drosophila* sibling species. *Nature* (*London*), **257**, 588–590.

Drew, R.A.I. (1982). Taxonomy. *Economic Fruit Flies in the South Pacific Region*, 2nd edn (Ed. by R.A.I. Drew, G.H.S. Hooper & M.A. Bateman), pp. 2–97. Queensland Department of Primary Industries, Brisbane.

Echols, H. (1982). Mutation rate: some biological and biochemical considerations. *Biochimie*, **64**, 571–575.

Ehrlich, P.R., Murphy, D.D., Singer, M.C., Sherwood, C.B., White, R.R. & Brown, I.L. (1980). Extinction, reduction, stability and increase: the responses of checkerspot butterfly (*Euphydryas*) populations to the California drought. *Oecologia*, **46**, 101–105.

Falconer, D.S. (1960). *Introduction to Quantitative Genetics*, 1st edn. Oliver and Boyd, Edinburgh.

Fisher, R.A. (1930). *The Genetical Theory of Natural Selection*. Clarendon Press, Oxford.

Fitch, W.M. (1982). The challenge to Darwinism since the last centennial and the impact of molecular studies. *Evolution*, **36**, 1133–1143.

Futuyma, D.J. (1979). *Evolutionary Biology*. Sinauer Associates, Sunderland, Mass.

Garten, C.J. (1976). Relationships between aggressive behavior and genic heterozygosity in the Oldfield mouse, *Peromyscus polionotus*. *Evolution*, **30**, 59–72.

Garton, D.W., Koehn, R.K. & Scott, T.M. (1984). Multiple-locus heterozygosity and the physiological energetics of growth in the coot clam, *Mulinia lateralis*, from a natural population. *Genetics*, **108**, 445–455.

Green, M.M. (1967). The genetics of a mutable gene at the white locus of *Drosophila melanogaster*. *Genetics*, **56**, 467–482.

Green, M.M. (1978). The genetic control of mutation in *Drosophila*. *Stadler Symposium*, **10**, 95–104.

Hoffmann, A.A., Nielsen, K.M. & Parsons, P.A. (1984). Spatial variation of biochemical and ecological phenotypes in *Drosophila*—electrophoretic and quantitative variation. *Developmental Genetics*, **4**, 439–450.

Istock, C.A. (1983). The extent and consequences of heritable variation for fitness characters. *Population Biology: Retrospect and Prospect* (Ed. by C.E. King & P.S. Dawson), pp. 61–96. Columbia University Press, New York.

Kambysellis, M.P. & Heed, W.B. (1971). Studies of oogenesis in natural populations of Drosophilidae. I. Relation of ovarian development and ecological habitats of the Hawaiian species. *American Naturalist*, **105**, 31–49.

Lachaise, D. (1983). Reproductive allocation in tropical Drosophilidae: further evidence on the role of breeding-site choice. *American Naturalist*, **122**, 132–146.

Lemeunier, F. & Ashburner, M. (1976). Relationships within the *melanogaster* species subgroup of the genus *Drosophila* (*Sophophora*). II. Phylogenetic relationships between six species based upon polytene chromosome banding sequences. *Proceedings of the Royal Society of London, Series B*, **193**, 275–294.

Lewontin, R.C. (1965). Selection for colonizing ability. *The Genetics of Colonizing Species* (Ed. by H.G. Baker & G.L. Stebbins), pp. 77–94. Academic Press, New York.

Lewontin R.C. (1973). Population genetics. *Annual Review of Genetics*, **7**, 1–17.

Lewontin, R.C. (1984). Detecting population differences in quantitative characters as opposed to gene frequencies. *American Naturalist*, **123**, 115–124.

Lewontin, R.C. & Birch, L.C. (1966). Hybridization as a source of variation for adaptation to new environments. *Evolution*, **20**, 315–36.

Little, J.W. & Mount, D.W. (1982). The SOS regulatory system of *Escherichia coli*. *Cell*, **29**, 11–22.

McClintock, B. (1951). Chromosome organization and genic expression. *Cold Spring Harbor Symposium in Quantitative Biology*, **16**, 13–47.

McClintock, B. (1978). Mechanisms that rapidly reorganize the genome. *Stadler Symposium*, **10**, 25–47.

McDonald, J.F. (1983). The molecular basis of adaptation: a critical review of relevant ideas and observations. *Annual Review of Ecology and Systematics*, **14**, 77–102.

McKenzie, J.A. & McKechnie, S.W. (1978). Ethanol tolerance and the *Adh* polymorphism in a natural population of *Drosophila melanogaster*. *Nature* (*London*), **272**, 75–76.

McKenzie, J.A. & Parsons, P.A. (1974a). Microdifferentiation in a natural population of *Drosophila melanogaster* to alcohol in the environment. *Genetics*, **77**, 385–394.

McKenzie, J.A. & Parsons, P.A. (1974b). The genetic architecture of resistance to desiccation in populations of *Drosophila melanogaster* and *D. simulans*. *Australian Journal of Biological Sciences*, **27**, 441–456.

Meats, A. (1981). The bioclimatic potential of the Queensland fruit fly, *Dacus tryoni*, in Australia. *Proceedings of the Ecological Society of Australia*, **11**, 151–161.

Middleton, R.J. & Kaiser, H. (1983). Enzyme variation, metabolic flux and fitness: alcohol dehydrogenase in *Drosophila melanogaster*. *Genetics*, **105**, 633–650.

Morrison, W.W. & Milkman, R. (1978). Modification of heat resistance in *Drosophila* by selection. *Nature* (*London*), **273**, 49–50.

Nevo, E. (1978). Genetic variation in natural populations: patterns and theory. *Theoretical Population Biology*, **13**, 121–177.

Nevo, E. (1983). Population genetics and ecology: the interface. *Evolution from Molecules to Man* (Ed. by D.S. Bendall), pp. 287–321. Cambridge University Press, Cambridge.

Nevo, E., Beiles, A. & Ben-Shlomo, R. (1984). The evolutionary significance of genetic diversity: ecological, demographic and life-history correlates. *Evolutionary Dynamics of Genetic Diversity* (Ed. by G.S. Mani), pp. 13–213. Springer Verlag, Berlin.

Nevo, E., Lavie, B. & Ben-Shlomo, R. (1983). Selection of allelic isozyme polymorphisms in marine organisms: pattern, theory and application. *Isozymes: Current Topics in Biological and Medical Research*, Vol. 10, *Genetics and Evolution*, pp. 69–92.

Nix, H.A. (1981). The environment of *Terra Australis*. *Ecological Biogeography of Australia* (Ed. by A. Keast), pp. 103–133. W.W. Junk, The Hague.

Parsons, P.A. (1970). Genetic heterogeneity in natural populations of *Drosophila melanogaster* for ability to withstand desiccation. *Theoretical and Applied Genetics*, **40**, 261–266.

Parsons, P.A. (1974). Genetics of resistance to environmental stresses in *Drosophila* populations. *Annual Review of Genetics*, **7**, 239–265.

Parsons, P.A. (1980). Isofemale strains and evolutionary strategies in natural populations. *Evolutionary Biology*, **13**, 175–217.

Parsons, P.A. (1982). Evolutionary ecology of Australian *Drosophila*: a species analysis. *Evolutionary Biology*, **14**, 297–350.

Parsons, P.A. (1983a). *The Evolutionary Biology of Colonizing Species*. Cambridge University Press, New York.

Parsons, P.A. (1983b). The genetic basis of quantitative traits: evidence for punctuational evolutionary transitions at the intraspecific level. *Evolutionary Theory*, **6**, 175–184.

Parsons, P.A. & Hoffmann, A.A. (1985). Ecobehavioral genetics: habitat preference in *Drosophila*. *Evolutionary Processes and Theory* (Ed. by S. Karlin & E. Nevo), pp. 535–559. Academic Press, London.

Parsons, P.A. & Stanley, S.M. (1981). Domesticated and widespread species. *The Genetics and Biology of Drosophila* (Ed. by M. Ashburner, H.L. Carson & J.N. Thompson, Jr), Vol. 3a, pp. 349–393. Academic Press, New York.

Piazza, A. Menozzi, P. & Cavalli-Sforza, L.L. (1981). Synthetic gene frequency maps of man and selective effects of climate. *Proceedings of the National Academy of Sciences*, USA, **78**, 2638–2642.

Raup, D.M. (1981). Introduction: What is a crisis? *Biotic Crises in Ecological and Evolutionary Time* (Ed. by M.H. Nitecki), pp. 1–12. Academic Press, New York.

Rose, M.R. (1983). Theories of life-history evolution. *American Zoologist*, **23**, 15–23.

Sabath, M.D., Boughton, W.C. & Easteal, S. (1981). Expansion of the range of the introduced toad *Bufo marinus* in Australia from 1935 to 1974. *Copeia*, **3**, 676–680.

Shorrocks, B., Rosewell, J., Edwards, K. & Atkinson, W.D. (1984). Interspecific competition is not a major organizing force in many insect communities. *Nature*, **310**, 310–312.

Soulé, M.E. (1979). Heterozygosity and development stability: another look. *Evolution*, **33**, 396–401.

Sousa, W.P. (1984). The role of disturbance in natural communities. *Annual Review of Ecology and Systematics*, **15**, 353–391.

Southwood, T.R.E. (1977). Habitat, the templet for ecological strategies? *Journal of Animal Ecology*, **46**, 337–365.

Stanley, S.M. & Parsons, P.A. (1981). The response of the cosmopolitan species, *Drosophila melanogaster*, to ecological gradients. *Proceedings of the Ecological Society of Australia*, **11**, 121–130.

Stanley, S.M., Parsons, P.A., Spence, G.E. & Weber, L. (1980). Resistance of species of the *Drosophila melanogaster* subgroup to environmental extremes. *Australian Journal of Zoology*, **28**, 413–421.

Stearns, S.C. (1976). Life history tactics: a review of the ideas. *Quarterly Review of Biology*, **51**, 3–47.

Stephanou, G. & Alahiotis, S.N. (1983). Non-Mendelian inheritance of 'heat-sensitivity' in *Drosophila melanogaster*. *Genetics*, **103**, 93–107.

Strong, D.S. (1983). Natural variability and the manifold mechanisms of ecological communities. *American Naturalist*, **122**, 636–660.

Taylor, C.E. & Condra, C. (1980). *r*- and *K*-selection in *Drosophila pseuboodscura*. *Evolution*, **34**, 1183–1195.

Thompson, J.N. Jr & Thoday, J.M. (1979). *Quantitative Genetic Variation*. pp. 1–305. Academic Press, New York.

Turelli, M. & Ginzburg, L.R. (1983). Should individual fitness increase with heterozygosity? *Genetics*, **104**, 191–209.

Turner, J.R.G. (1983). Mimetic butterflies and punctuated equilibrium: some old light on a new paradigm. *Biological Journal of the Linnean Society*, **20**, 277–300.

Waddington, C.H.(1965). Introduction to the symposium. *The Genetics of Colonizing Species* (Ed. by H.G. Baker & G.L. Stebbins), pp. 1–6. Academic Press, New York.
Wills, C. (1983). The possibility of stress-triggered evolution. *Evolutionary Dynamics of Genetic Diversity* (Ed. by G.S. Mani), pp. 299–312. Springer Verlag, Berlin.
Zouros, E., Singh, S.M. & Miles, H.E. (1980). Growth rate in oysters: an overdominant phenotype and its possible explanations. *Evolution*, **34**, 856–867.

8. COLONIZATION OF EPHEMERAL HABITATS

ILKKA HANSKI
Department of Zoology, University of Helsinki, P. Rautatiekatu 13, SF-00100 Helsinki, Finland

INTRODUCTION

Ephemeral habitats can be colonized by species that are mobile enough, when rare, to establish more local populations than are lost through the inevitable local extinctions. It is now recognized (Andrewartha & Birch 1954; Southwood 1977) that most patchy habitats are more or less temporary, hence greater dispersal is expected (Den Boer 1971; Gadgil 1971; Comins, Hamilton & May 1980; Levin, Cohen & Hastings 1984) and observed (Southwood 1960, 1962; Johnson 1969) in species living in them. Colonization—establishment of new populations—even if on a small spatial scale, is commonplace and a necessity for species inhabiting ephemeral and patchy environments.

Standard examples of ephemeral habitats are dung, carrion, fungal fruiting bodies, early successional plants, dead wood, fruits, seeds, etc., which are patchily distributed in both time and space. This chapter focuses on such habitats, and especially on insects colonizing them. Typically, one generation develops in each habitat patch, and many species often coexist.

Another class of examples consists of patches not defined by the habitat *per se*; e.g. small areas of space that may support individual plants or sessile animals, or slightly larger areas defended as adult territories as by coral reef fishes. Empty sites are created by chronic and patchy disturbances (Grubb 1977; Horn 1981; Dethier 1984). Naturally, quite different models are best suited to describe the two kinds of situations, though it is interesting to ask what the similarities are.

Succession of species may occur in ephemeral habitats, and it has frequently been observed that more specialized species arriving early are replaced in time by generalists (Mohr 1943; Elton 1966; Beaver 1984; Koskela & Hanski 1977). By the nature of the process, such succession will not lead to a stable community structure in individual habitat patches but causes and/or reflects their absorption by the surrounding environment. Questions about stability must be asked about larger spatial scales and systems of habitat patches. Regions may be invaded, and regional stability re-established after invasion by exotic species. An example is the recent invasion of South America by four species of *Chrysomya* blowflies from the Old World tropics (Baumgartner & Greenberg 1984). These colonizations—attributable to increased chances of transatlantic dispersal with ships and airplanes—have caused dramatic declines in the numbers of native species (Guimarães, Prado & Buralli 1979), e.g. the relative abundance of *Cochliomyia macellaria* plummeted from 90% to less than 1% in 18 months in a locality in Peru following the invasion of the region by *Chrysomya putoria* and *C. albiceps* (Baumgartner & Greenberg 1984).

Ephemeral habitats—and more generally patchy environments—have wide ecological interest for two reasons. First, as suggested above, systems may be unstable at one spatial scale but stable on another scale. Paradoxically, the very instability of populations in habitat patches may contribute to stability on a larger scale in competitive and predator–prey systems. One can argue that any species living in an environment subject to local disturbances occurs as temporary local populations (Levin & Paine 1974) hence the conceptual framework erected for species inhabiting ephemeral habitats may find wide application. Second, ephemeral and patchy environments are often exceptionally rich in species, and interest in these habitats touches the general problem of maintenance of species diversity. Well studied examples include tree species diversity in tropical forests, where regeneration sites occur patchily in time and space (Connell 1978; Hubbell & Foster, Chapter 19), intertidal sessile animal and plant communities that experience frequent disturbances (Paine 1966; Paine & Levin 1981; Dethier 1984; Connell, Chapter 16), coral reef fish assemblages (Sale 1977) and, in ephemeral habitats proper, insect assemblages in carrion (Payne 1965) or droppings (Hanski 1986a). Host individuals are ephemeral habitat patches for parasites, but these systems are only touched on here as there are many excellent recent reviews (Price 1980; Anderson 1981, 1982 and references therein), and a more specific approach would be needed.

The next two sections outline some of the basic theoretical ideas and simple models about survival in ephemeral and patchy habitats, with special emphasis on competition between the early colonizers. The fourth section focuses on flies colonizing carrion as a paradigm of organisms living in ephemeral habitats. As pointed out above, carrion and like habitats have short durational stability (Southwood 1977), usually one generation for insects. But as there is no practical way of delimiting all ephemeral habitats from other patchy habitats, both single- and multi-generation patchy habitats will be considered.

ESTABLISHMENT AND EXTINCTION IN PATCHY ENVIRONMENTS

Efficient exploitation of resources requires that one is in the right place at the right time (Andrewartha & Birch 1954). This is especially true for species that are dependent upon patchy and ephemeral resources. Local populations go extinct relatively rapidly and, for the species to survive regionally, empty habitat patches must be colonized fast enough to compensate for extinctions. The prototype model of single-species dynamics in a patchy environment may be used to make the point. In this model within-patch dynamics are ignored and instead changes in the number of local populations Q are modelled:

$$\frac{dQ}{dt} = mQ(S - Q) - eQ, \qquad (1)$$

where S is the number of patches suitable for colonization, and m and e are dispersal and extinction parameters, respectively. In this deterministic model, a rare plant or animal may spread into the system of patches if

$$mS/e > 1. \qquad (2)$$

This condition is analogous to the condition of an infectious disease spreading into a host population (e.g. Anderson 1981). The rate at which patches are colonized (mS) times the expected lifetime of a local population ($1/e$) must exceed unity when the species is rare ($Q \approx 1$).

The oversimple assumptions made in this model about dispersal (the spatial locations of the patches are ignored) and extinction (constant extinction probability) may seem to make suspect any conclusions based on it, but simulation studies by Zeigler (1977), Gurney & Nisbet (1978) and Vance (1984) suggest that the model is structurally robust.

From condition (2) it immediately follows that if S is less than a threshold number of patches, the species cannot establish itself in the patchy environment. This simple result is most important; it parallels and may in fact help to explain the increase in species number with increasing area (MacArthur & Wilson 1967). In patchy environments, species are expected to accumulate with increasing numbers of patches. Surprisingly, although anecdotal examples are numerous, rigorous tests of this prediction seem to be lacking.

Many plants occur as isolated local populations in particular habitats, especially near their geographic range limits (Carter & Prince 1981). Linkola (1916; see also Hanski 1982b) compared species richness of anthropochorous plants (species confined to man-made habitats) in villages in two neighbouring regions in eastern Finland. After correction had been made for village size, there were more species in villages in the region where distances between the villages were smaller, and where dispersal hence was assumed to be greater (Linkola 1916).

Herbivorous insects often occur as more or less discrete local populations because of the patchy distribution of their host-plants, and one would expect a particular plant species to have more herbivores in a region where its density is high than where the density is low. Unfortunately, regression studies on herbivore species richness have shown that host abundance usually correlates well with host range (e.g. Neuvonen & Niemelä 1981; Fowler & Lawton 1982), and it is difficult to disentangle the effect of host density when the scale of the study is large. It would be instructive to have similar studies for relatively small regions, say 100 km^2.

More generally, geographic distribution (the number of sites occupied) and average local abundance have been found to be positively correlated in many groups of animals and plants and on many spatial scales (Hanski 1982a; Bock & Ricklefs 1983; Brown 1984), suggesting that there are no distinct time-scales of 'local' and 'regional' dynamics (Hanski 1983). If one makes the realistic assumption that the probability of local extinction increases with decreasing population size (MacArthur & Wilson 1967; Leigh 1981; Nisbet & Gurney 1982; Toft & Schoener 1983; Diamond 1984), the constant e in the prototype model should be a decreasing function of Q (Hanski 1982a). The consequences of this observation are discussed below.

Human activity, among other things, alters the structure of the environment, and often decreases the density of certain types of habitat patches. Such changes may cause the regional extinction of specialist species before the last patches have been destroyed by decreasing the number of patches below the threshold level (Hanski 1985). Relevant examples include woodland ground beetles in the Netherlands (Den Boer 1979), insects associated with juniper in England (Ward 1977; Ward & Lakhani 1977), and *Pieris virginiensis*, a rare butterfly, feeding on *Dentaria diphylla* in scattered woodlands in North America (Cappuccino & Kareiva 1985).

On longer time-scales, we may enquire about the numbers of species specializing in an ephemeral habitat in different geographical regions; or about the numbers specializing in different kinds of habitats in one region. Thus, the numbers of insects colonizing the pitchers of *Nepenthes* are highest in the geographic distribution centre of the genus *Nepenthes*

(Beaver 1983); probably because there the density of pitchers as well as the diversity of forms is highest. An example of different numbers of species specializing in different microhabitats is found in fungivorous insects. *Leccinum scabrum*, *L. versipelle* and *Boletus edulis* are three common edible fungi, with the fruit body production decreasing in this order in southern Finland (E. Ohenoja, pers. comm.). Of the seventeen fungivorous *Pegomya* flies in Finland, ten have been recorded from *L. scabrum*, five from *L. versipelle* and two from *B. edulis* (Hackman 1979; Hackman & Meinander 1979). The rate of ecophysiological adaptation to different sorts of ephemeral habitats may be expected to increase with their density, as proposed in Southwood's 'encounter-frequency' hypothesis (Southwood 1961; Strong, Lawton & Southwood 1984).

A corollary is that the degree of specialization is expected to be higher where densities of habitat patches are greater. Carrion, dung and dead wood are consumed faster, causing presumably a lower patch density in tropical rain forests than in temperate habitats, and it is interesting to note that, contrary to the accepted rule (MacArthur 1972; Pianka 1974; Pielou 1975), the insects living in these habitats in tropical forests are less microhabitat specialists than their temperate relatives (Beaver 1979, 1984; Hanski 1986a, b).

Stochastic models

In the real world the colonization and extinction rates of model (1) correspond to colonization and extinction probabilities in an infinitesimally small time interval dt: $mQ(S - Q)\mathrm{d}t$ and $eQ\mathrm{d}t$, respectively (Nisbet & Gurney 1982). Thus, if the deterministic model (1) predicts a low stable state (a small number of occupied patches at any one time), by chance the species may go extinct from the system of patches. What is the time to regional extinction? Nisbet & Gurney (1982) give the approximation,

$$T_{\text{ext}} \approx \frac{1}{e}\exp\left\{\frac{Q^{*2}}{2(S - Q^*)}\right\}, \qquad (3)$$

where $Q^* = S - e/m$ is the steady-state number of occupied patches. A further approximation (for reasonably large S) leads to the conclusion that long-term regional persistence, say 1000 times longer than the average lifetime of a patch population, is only possible if the average proportion of occupied patches is greater than a critical value of order of magnitude of $3S^{-1/2}$ (Nisbet & Gurney 1982). For a species to persist in a system of 100 patches 100 times longer than a local population persists in a single habitat patch, Q^*/S should not be less than 0·26 (from eqn 3).

The kind of stochasticity discussed above is analogous to demographic stochasticity (May 1973) in local populations; it should certainly be considered when the number of patches is small. Environmental stochasticity, variation in the extrinsic conditions, may affect the immigration and extinction rates and thus affect regional dynamics in a patchy environment even if the number of patches is large.

If parameter e in model (1) is assumed to be a random variable instead of constant, but variation is not excessive, a distribution of Q values in time may be expected with a mode less than the deterministic equilibrium (Levins 1970). If, however, model (1) is modified along the lines suggested above, by letting parameter e decrease towards zero with increasing Q/S, substantial environmental stochasticity tends to push the species either to superabundance ($Q/S \approx 1$, the deterministic equilibrium) or to rarity (Hanski 1982a, b). Rare species may go extinct, or they may survive in the best habitat patches ('safe sites'), or they may survive because of dispersal from outside the system.

Fig. 8.1 gives a botanical example that appears to support this prediction. Other mechanisms than stochasticity may produce this dichotomy between 'core' (regionally common) and 'satellite' (regionally rare) species, e.g. density-dependent dispersal (Horn 1975, 1981; Acevedo 1981) and a local Allee effect (Hanski 1983). In any case the dynamics in a system of habitat patches may be more interesting than just the sum of local dynamics.

Perennial plants are probably less affected by environmental stochasticity (succession of bad and good years) than are annuals, and perennials thus have a better chance of spreading in a patchy environment, other things being equal (Carter & Prince 1981). In support of this conjecture, one observes that perennials were significantly more frequent than annuals amongst the widespread (core) species in the material in Fig. 8.1 ($P < 0{\cdot}001$; Hanski 1982b).

COEXISTENCE OF COMPETITORS

Competition for ephemeral and patchy resources is frequently severe. Space is the limiting resource in many communities of plants and sessile animals (Paine 1984). High quality, defenceless resources such as carrion and dung are often consumed by insects to the last piece (Hanski 1986a, b), and competition has been demonstrated for insects breeding in fungal fruiting bodies (Grimaldi & Jaenike 1984) and fallen fruits (Atkinson 1985). Nevertheless, in spite of competition, many ephemeral habitats are colonized by tens or hundreds of species (see Hanski 1986a for examples of dung and carrion feeders).

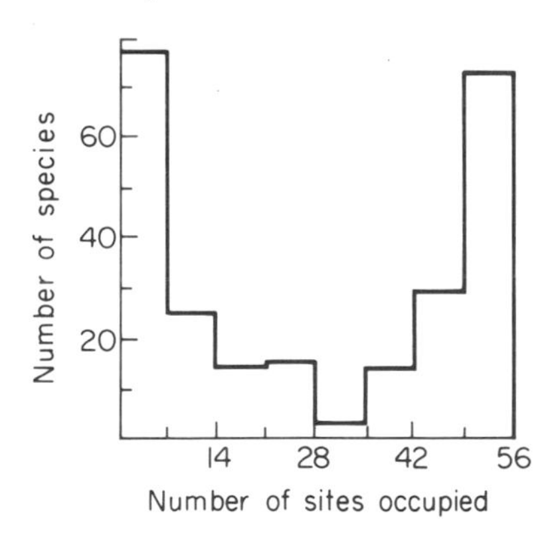

FIG. 8.1. Distribution of the site occurrence frequency of 250 anthropochorous plants in villages surrounded by forest: most species are either widespread or scarce (data from Linkola 1916; see also Hanski 1982b).

This section discusses three kinds of models of competing populations in patchy environments. In the first model, a system of relatively stable sites is assumed, which may support local populations for many generations. The second type of model focuses on situations where the habitat patches (usually arbitrary, not defined by the habitat) may support only one individual, namely the first one to arrive. Plants and sessile animals competing for space may fit in here. The third kind of model has applicability especially for insects living in ephemeral habitats that last only for one generation; but perhaps also more widely whenever processes exist that create substantial spatial variance in density not related to variation in resource availability.

Model A: Immigration–extinction competition

This class of models, extensions of the prototype model (1), offers a macroscopic description of population dynamics in patchy environments by ignoring local dynamics (numbers) and focusing instead on the proportion of habitat patches that are occupied (Horn & MacArthur 1972; Slatkin 1974; Hanski 1983). (Note that the number of patches S in model (1) is now absorbed in the immigration parameter m.) The processes modelled are immigration to, and extinction from, habitat patches. Both may be affected by interspecific competition (Levins & Culver 1971). These models are meant as phenomenological descriptions of competition in patchy environments, but the predictions have been tested in a laboratory study on fungi (Armstrong 1976), and field observations on water fleas in rock pools (Hanski & Ranta 1983) and on sea birds on islands (Caraco & Whitham 1984) have been interpreted within this framework.

Hanski (1983) gives the sufficient condition for stability of a one-species regional community: a second species cannot invade a system of patches in which its competitor is present at its equilibrium patch-occurrence frequency p_1^* if

$$p_2^* < \alpha_{12} p_1^*. \quad (4)$$

Here $\alpha_{12} = \mu_{12}/m_2$ models the reduction in the immigration rate of the invading species to patches occupied by the competitor (thus $0 \leqslant \mu_{12} \leqslant m_2$). Note that the same resistance to invasion by a second species may be caused by a widespread, weak competitor and a rare, strong competitor (αp_1^* constant). This is analogous to the condition of stability in the Lotka–Volterra model, where invasion becomes increasingly difficult with the product αK, the resident species' competitive effect times its carrying capacity.

More generally, in Model A species may survive competition either because of good dispersal or good competitive abilities. Such differences are often reported in the literature (e.g. Hutchinson 1951; Armstrong 1976; Vepsäläinen 1978; Paine 1979; Paine & Levin 1981; Platt & Weis 1985), and are embodied in the $r-K$ selection concept of MacArthur & Wilson (1967) and in the tolerance model of succession of Connell & Slatyer (1977). Because density of sites affects immigration rate, the relative merits of good dispersal and competitive abilities change along a gradient of site density (Hanski & Ranta 1983). Thus, spatial heterogeneity in density of sites is likely to contribute to coexistence of competitors (Platt & Weis 1985).

Other, less obvious conclusions from these models are that similar species may coexist (Slatkin 1974) and that, especially if competition greatly increases local extinction rates, alternative stable states are possible: the initially more widespread species has an advantage and may exclude its competitor from a patchy environment (Hanski 1983). While I do not know of examples of the latter phenomenon (but see Hanski & Ranta 1983), the coexistence of similar competitors predicted by the model is in agreement with the often large number of coexisting species in patchy habitats.

Model B: Lottery competition

This class of models applies to species with a sessile adult stage and a mobile propagule stage. Consider an imaginary set of sites that may be occupied by one adult individual. An established adult cannot be replaced, hence all competition occurs in the immigration phase; this is an extreme example of the inhibition model of succession (Connell & Slatyer 1977). Sites become vacant through deaths of their occupants. Offspring are produced to a common 'pool', thus spatial locations of the sites are ignored as in Model A.

The probability that species *i* colonizes a site is simply the proportion of species *i* in the offspring pool. Ågren & Fagerström (1984) use a slightly different colonization scheme, with the propagules first being distributed randomly among the sites, then the winner of the local competition being selected with equal or biased probability in order to model differences in competitive ability.

The model is originally due to Sale (1977, 1979, 1982), who proposed that the lottery mechanism would allow coexistence in communities of territorial coral reef fishes. However, with constant parameters the model is essentially the same as Model A with no extinction competition but infinitely strong immigration competition ($m_2 - \mu_{12} = 0$). Thus, inequality (4), which in the case of pure immigration competition is both a necessary and a sufficient condition for the stability of the equilibrium (Hanski 1983), reduces to

$$p_2^* < p_1^*, \tag{5}$$

in which p_2^* is the equilibrium patch-occurrence frequency of the invading species when alone. The second species can only invade if p_2^* is greater than p_1^*; reversing the subscripts shows that two species cannot coexist (cannot both invade when rare) except in the extremely unlikely case $p_1^* = p_2^*$. Lottery competition itself is insufficient to allow coexistence in equilibrium systems, a conclusion worth emphasizing as there appears to be much misunderstanding on this question in the literature. Coexistence may be attained without any form of resource partitioning if there is, e.g. environmentally-caused variation in birth rates (Chesson & Warren 1981; Ågren & Fagerström 1985), but it should be noted that such variability promotes coexistence in nonlottery models as well (Abrams 1984a; see discussion in Abrams 1984b).

In the enforced absence of predation and disturbance, single sessile species in marine communities often occur as monocultures (Paine 1984). This suggests, in agreement with the theory, that processes other than the simple lottery mechanism are required for multispecies coexistence. Lottery models with extra assumptions based on Janzen's (1970) and Connell's (1971) seed predation hypothesis have been developed by Hubbell (1979, 1980) and Becker *et al.* (1985) to explain high tree species richness in tropical forests.

Model C: Competition and spatial variance in density

Assume that, as almost always is the case in nature, species' instantaneous spatial distributions are clumped in habitat patches (for examples dealing with ephemeral habitats see Hanski 1980a; Kuusela & Hanski 1982; Atkinson & Shorrocks 1984). If the species' abundance maxima do not coincide, most individuals experience most competition from conspecifics. As spatial variance tends to increase rapidly with mean density (Taylor, Woiwood & Perry 1979), intraspecific competition increases faster with a species' abundance than does interspecific competition, and regional coexistence is possible in spite of severe competition (Shorrocks, Atkinson & Charlesworth 1979; Lloyd & White 1980; Hanski 1981; Atkinson & Shorrocks 1981; de Jong 1982; Ives & May 1985).

The key assumption of Model C is independently clumped distributions. Little or no spatial covariance is reported for coexisting species of coral-associated decapod crustaceans (Gotelli & Abele 1983), flies on fallen fruits (Atkinson & Shorrocks 1984), dung beetles (below) and carrion flies (next section), but this is something of an enigma, as the likeliest mechanism causing aggregation—variable habitat patch quality—might be expected to produce positive covariance. This observation is discussed below, but meanwhile it may be noted that several mechanisms may be important in nature, as long as they produce independently aggregated distributions *and* the degree of intraspecific crowding increases with density faster than density (Hanski 1981; Green 1986). Atkinson & Shorrocks (1984) propose a model in which ovipositing females visit the breeding sites randomly, and have a constant probability of finishing the egg-laying and leaving. This behaviour would lead to independently clumped distributions of eggs in different species, but the degree of crowding decreases with density, and hence this mechanism does not guarantee coexistence (Green 1986).

In the majority of cases movements between breeding sites are non-random, however, presumably because subtle differences usually exist in the attractiveness of the sites, e.g. owing to their microlocation. The next few paragraphs discuss these questions in some detail, because Model C represents the most promising approach for ephemeral habitats, and because the spatial pattern of colonization is its key ingredient.

In a study of dung beetle colonization (Hanski 1979), a set of five large pitfalls were baited with a fresh cow pat (*c.* 1·5 kg) in four pastures fifteen times between April and October, and the beetles were collected after 5–10 days. The twelve most abundant species, representing three genera—*Aphodius* (Scarabaeidae), *Cercyon* and *Sphaeridium* (Hydrophilidae)—were selected for a two-way analysis of variance (trap position × season).

The trap position had a significant effect in eleven of the forty ANOVAS run (there was insufficient material for eight species × pasture combinations). Ten of the significant results came from two of the four pastures, indicating that in these areas the trap positions were most unequal (though no differences were evident to a human eye). Nonetheless, the results also indicate that different microlocations tended to be most favourable for colonization on different days. This is not surprising, as merely the direction from which the wind blows must affect the attractiveness of the baits.

Differences between habitat patches may be expected to create positive spatial correlations between pairs of species. To test this hypothesis the same pitfall material (Hanski 1979) was used to calculate spatial correlations between pairs of colonizing beetles. These data reflect immigration to pats, and are unaffected by possible density-dependent emigration (cf. Holter 1979).

If one omits cases with fewer than ten individuals in the pooled sample from five pitfalls, there remain 795 pair-wise correlations. The distribution of the correlation coefficients is strongly skewed to the left, and positive correlations are definitely more numerous than negative ones (Fig. 8.2). This pattern suggests that differences between the pitfalls were perceived to some extent in the same way by the twelve species.

One could expect that the more similar two species are in physiology, behaviour and ecology, the more similar they would perceive any differences between habitat patches, and hence the greater would be their spatial correlation. To test this hypothesis, non-congeneric and congeneric pairs of species were compared, and the latter were further divided into three genera, which clearly differ in the degree of ecological similarity between the species: the six species of *Aphodius* form a heterogeneous group of species (Hanski 1980a); the three *Cercyon* species resemble each other both taxonomically (Vogt 1968) and ecologically (Hanski 1980b); while the two abundant *Sphaeridium* species have very similar physiological responses (Landin 1967), morphologies and ecologies (Otronen & Hanski 1983). The results show that positive spatial correlations within the genera increase in this order (Fig. 8.2), and the difference from the non-congeneric pairs is significant in *Cercyon* ($P < 0{\cdot}01$) and *Sphaeridium* ($P < 0{\cdot}001$).

In conclusion, coexisting species tend to be more or less independently clumped in habitat patches because the many physiological, behavioural and ecological differences between the species that almost always exist,

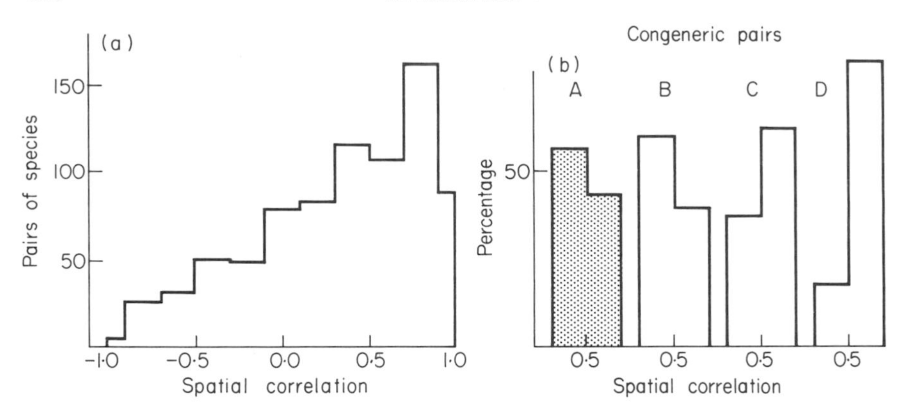

FIG. 8.2. (a) Pair-wise correlation coefficients for twelve dung-breeding beetles trapped in four pastures with five pitfalls in each for one summer. Populations with less than ten individuals in the five traps have been omitted, leaving 795 pairs of species. (b) Comparison between non-congeneric (shaded histogram A) and congeneric species pairs, with the data divided into two classes (correlation smaller or greater than 0·5). The genera and the results of χ^2 tests are: B *Aphodius* (6 spp.), $\chi^2 = 0{\cdot}59$, NS; C *Cercyon* (3 spp.), $\chi^2 = 7{\cdot}96$, $P < 0{\cdot}01$; and D *Sphaeridium* (2 spp.), $\chi^2 = 10{\cdot}09$, $P < 0{\cdot}001$. The more similar two species are in their behaviour and ecology, the greater is their spatial correlation.

though they may not be great, nonetheless suffice to make the species perceive their environment somewhat differently. An important point is that practically *any* interspecific differences tend to decrease spatial covariance, not only differences directly affecting resource use. In particular, differences in foraging behaviour, which are commonplace, can be important. One example is Vet & van Alphen's (1985) study on differences in host detection behaviour in fifty-seven species of Eucoilidae and Alysiinae, parasitic wasps.

The basis for coexistence

Table 8.1 summarizes the assumptions, domains of application, and the basis for coexistence in the three models. As discussed above, Model B is a limiting case of Model A when there is a complete priority effect and local population size is 1. In a constant environment, the species with smaller pre-competitive frequency of occupied patches will be excluded, even if each of two species would occupy only a small fraction of patches available when on its own. However, such competitive exclusion between rare species depends critically on the probably unrealistic assumption that occupied sites are strictly uninvadable *and* that the excluded individuals must die. The first assumption is unrealistic for plants: seeds may germinate close to each other, and though the developing individuals compete,

TABLE 8.1. Comparison of the assumptions, domains of application, and basis of coexistence in the competition models A–C

	Assumptions			Domain of application	Coexistence based on	Example
	Individual movements	Distribution of pop. sizes	Parameters			
Model A	Random, via pool	Not considered	Constant	Local populations may survive many generations	Temporal refugia and random colonization	Mammals on small islands
Model B	Random, via pool	$n = 0$ or 1, no local coexistence	Constant or varying	'Local sites' support one adult	Uncorrelated temporal variation in birth rates	Plants
Model C	Not specified	$\mathrm{Var}(n) \approx a\bar{n}^2$ $\mathrm{cov}(n_1, n_2) \approx 0$	Constant	Ephemeral habitats, one generation	Independently clumped spatial distributions	Flies colonizing carrion

both may survive to reproduce (Harper 1977; Platt & Weis 1985). The second assumption is unrealistic for many animals: if a territory owner excludes an intruder, the latter may try its luck elsewhere.

Model C is well suited to describe interactions in ephemeral habitats, but a challenging task remains in the study of its more general applicability. In all species in all environments spatial density is variable, and this variation leads to spatial variation in the chances of survival and reproduction. Localized (inhomogeneous) interactions are the rule in nature, at least in terrestrial environments, and should be included in all population models (Chesson 1981).

A MODEL SYSTEM: FLIES ON CARRION

The fly assemblage colonizing vertebrate carcasses is a prime example of a guild of competitors exploiting an ephemeral and patchy resource. Any one locality has up to fifteen species, though most are rare (Table 8.4; Nuorteva 1970; Beaver 1977; Palmer 1980; Kuusela & Hanski 1982). Exploitative competition is generally severe (Hanski 1986a and references therein). Carrion flies form an important part of a more comprehensive food web: in temperate regions vertebrate scavengers and burying beetles *Nicrophorus* remove many large and small carcasses, respectively, before flies have had time to colonize (Hanski 1986a), and in tropical forests beetles (Scarabaeidae) and ants compete with the flies (Hanski 1986b).

An appropriate theoretical framework is Model C, though applicability of Model A has been discussed by Hanski (1981). A complete test of Model C would require data on the distribution of eggs of different species in carcasses in the field, knowledge of the type of interactions between developing larvae, and of mortality and pupal size (adult fecundity) at different larval densities. With some exceptions (below), the larvae are engaged in exploitative, scramble competition (Ullyett 1950). Insufficient resource availability decreases first pupal weight and hence adult fecundity (Waterhouse 1947; Ullyett 1950; Valiela 1969a, b), and only severe overcrowding, which nonetheless occurs regularly in the field (Hanski 1986a), increases mortality. We unfortunately lack data on egg distributions—the spatial pattern of colonization of a set of habitat patches in a multispecies community. Looking instead at the numbers of emerging adults in field experiments, we observe more or less independent clumping (Table 8.2; Beaver 1977), which is the situation favourable for coexistence in spite of competition. But distributions of the emerging flies may have been affected by competition, predation, and factors other than egg distributions.

TABLE 8.2. Means, standard deviations and correlations in the emerging individuals of the three most abundant species of blowflies in a rearing experiment (the first column of Table 8.4; $n = 20$)

Species	Mean	S.D.	Correlation (r)	
Lucilia illustris	229	155		
L. silvarum	23	38	0·007	
Muscina assimilis	11	30	−0·173	−0·174

To further test whether larval resource patchiness facilitates coexistence, some experiments have been conducted.

Cage experiment

Large samples of the local fly community in southern Finland were enclosed in July 1982, in 1 m^3 outdoor cages, constructed of nylon mesh. The cages have a natural soil floor in which the populations overwinter. Larvae are fed with 50 g of cow liver once or twice a week from May to September. The larval food is cut into 1, 2, 4, 8 or 16 pieces, placed in random positions on the cage floor. Each treatment has three replicates.

The starting community had at least fourteen species, mostly the same as in Table 8.4. Nine rare species were lost during the first or second summer, or populations declined to such small size that the species were not represented in the samples taken from the cages. Three species were common in the starting community, *Lucilia illustris*, *L. silvarum* and *Muscina assimilis*, and accounted for 84, 5 and 8% of the emerging flies in the first generation (Table 8.3).

Muscina assimilis was lost from all cages during the second generation (first winter; Table 8.3). *Lucilia silvarum* has disappeared from thirteen cages and declined in the remaining ones during 3 years. The winner has consistently been *L. illustris*, the species that is by far the most abundant one outside the cages (Table 8.4), and has been so for the past 10 years (Hanski & Kuusela 1980). Abundance relations in the field are thus at least partly determined by some deterministic process(es): *Lucilia illustris* is the best competitor, and it is the most abundant species.

Two initially uncommon *Sarcophaga* species have become more abundant in some cages (Table 8.3). Significantly, the cages in which they have been successful are the ones that receive the larval food in many pieces (Fig. 8.3). In this experiment *Sarcophaga* appears to have benefited from resource patchiness though *M. assimilis* and *L. silvarum* did not survive.

TABLE 8.3. The five most abundant species in the cage experiment: their pooled numbers in the samples taken from the fifteen cages. The figure in brackets is the number of cages from which the species was collected (this is the minimum number of cages in which the species was present)

	1982	1983	1984
Lucilia illustris	2718 (15)	7917 (15)	3109 (15)
Lucilia silvarum	173 (14)	1608 (9)	169 (5)
Muscina assimilis	250 (13)	— (—)	— (—)
Sarcophaga spp.*	23 (5)	70 (14)	284 (10)

* *S. scoparia* and *S. aratrix*.

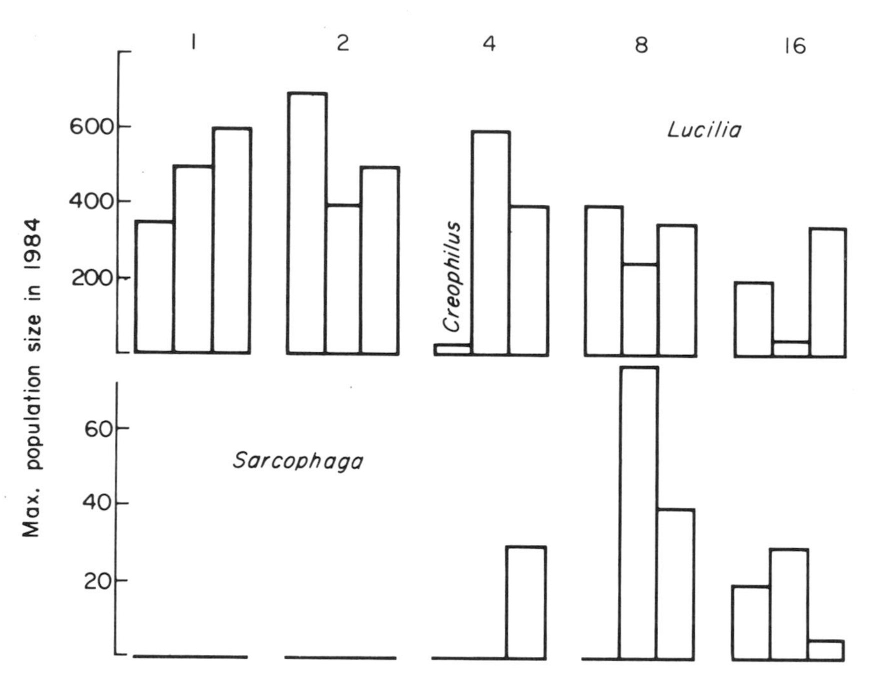

FIG. 8.3. Maximum population sizes of *Lucilia* spp. (mostly *L. illustris*) and *Sarcophaga* spp. (*S. scoparia* and *S. aratrix*) in the cages in summer 1984. The fifteen cages have been arranged in five groups on the basis of the number of pieces into which the larval food (50 g of liver) is cut. The six *Sarcophaga* populations are present in cages with many pieces (P = 0·002; one or two individuals were collected in 1984 from four other cages, in which *Sarcophaga* seem to go extinct). *Creophilus maxillosus*, a large predatory staphylinid, has managed to colonize one of the four-piece cages through the nylon mesh, and has much suppressed the *Lucilia* population.

The explanation is straightforward. *Sarcophaga* females lay 1st instar larvae instead of large clusters of eggs like *Lucilia*. Ovoviviparity gives a rapid start to a developing larva (the egg stage is omitted and *c.* 24 hours saved), but equally, or more importantly, ovoviviparity is a constraint that has important consequences: *Sarcophaga* cannot lay many larvae at one time, and as females fly between bouts of larviposition, the larvae are likely to become especially well spread between carcasses. *Sarcophaga* is well adapted to exploit spatial variance in resource availability, and does well on those pieces of carrion that by chance receive the smallest number of *Lucilia* eggs.

Does the difference in the clutch size between *Sarcophaga* and blowflies make a difference in nature? *Sarcophaga* are usually so uncommon (Table 8.4) that the question is difficult to answer. In a rearing experiment started in late summer 1976 *Sarcophaga scoparia* was, however, the dominant species (Hanski & Kuusela 1980). Eighty rearing pots were placed in the field in eight groups of ten pots each, the groups being situated 15 m from each other. The emerging *Lucilia* and *Calliphora* showed a markedly patchy distribution between the groups, 50% or more of the individuals in each species emerging from a single group. The corresponding figure was only 21% for *S. scoparia*, and there was no statistically significant variation in its numbers between the groups (F = 1·48, df = 7, 72; Hanski & Kuusela 1980). Thus, *Sarcophaga* appear to be good at exploiting spatial variance in resource availability also in the field.

By the above reasoning, any species would be selected to spread its eggs among a large number of carcasses. There is little data for species other than *Lucilia*, which lay large clusters of eggs (Mackerras 1933; Hobson 1938), and *Calliphora*, many of which lay many eggs per carcass but singly (I. Hanski, pers. obs.). Some density dependence is indicated by an experiment in which egg removal increased the total number of eggs laid by *Lucilia* on carrion (Kuusela 1984). Availability of carrion, which affects the optimal egg-laying behaviour (Parker & Courtney 1984), is difficult to assess in the field (Hanski 1986a).

Field experiment

To study more closely the significance of resource patchiness in natural populations, another rearing experiment was conducted (J. Kouki & I. Hanski, unpubl.). Small plastic pots, half-filled with soil, were placed in a large field, in a regular grid, the pots being 15 m apart. Fifty grams of cow liver was placed in each pot, and the pots were left in the field for 3 days

Table 8.4. The flies that emerged from the twenty control rearings and from two sets of experimental rearings, with either two or four pieces of liver (50 g) combined after egg-laying. The figures are relative abundances

	Control	Experiment	
		Two pieces	Four pieces
Lucilia illustris	81·8	95·9	94·2
L. silvarum	8·3	2·1	3·4
L. caesar	1·2	0·2	0·7
L. richardsi	0·5	0·2	0·9
L. sericata	<0·1	0·1	0·1
Calliphora vicina	2·8	0·1	0·2
Cynomyia mortuorum	0·2	<0·1	—
Sarcophaga scoparia	0·6	<0·1	0·2
S. aratrix	0·2	—	—
S. subvicina	<0·1	—	—
Muscina assimilis	4·0	1·0	0·2
Hydrotaea dentipes	0·2	0·1	—
Nemopoda cylindrica	0·2	0·2	0·2
Number of flies	5601	7401	6449
Number of rearings	20	11	9
Total resource (kg)	1·0	1·1	1·8
Flies per gram of resource	5·6	6·7	3·6

to let flies oviposit. After 3 days the pieces of liver with eggs and larvae from twenty randomly selected pots were moved to twenty larger rearing pots, which were closed with nylon mesh (for the method see Nuorteva 1970; Hanski 1976). Larvae were allowed to develop and eventually to pupate in the soil in the pots. These pots were also kept in the field, in one group, to guarantee natural and identical rearing conditions. The emerging flies were collected with a net. This set of pots is the control.

In the experiment, the pieces of liver with eggs and larvae from two or four randomly selected pots were placed into one rearing pot, and the larvae were allowed to develop as in the control rearings. The only difference between the control and the experimental rearings is that in the latter larvae could easily move from one piece of liver to another within the sets of two or four pieces. If this does not make a difference, the experimental results should be indistinguishable from results obtained by pooling the numbers of emerging flies from two or four randomly selected control rearings.

Simple analyses prove that resource patchiness makes a great difference in larval competition. Removing patchiness causes a decline in the number of emerging flies per rearing (Fig. 8.4), and leads to an increase

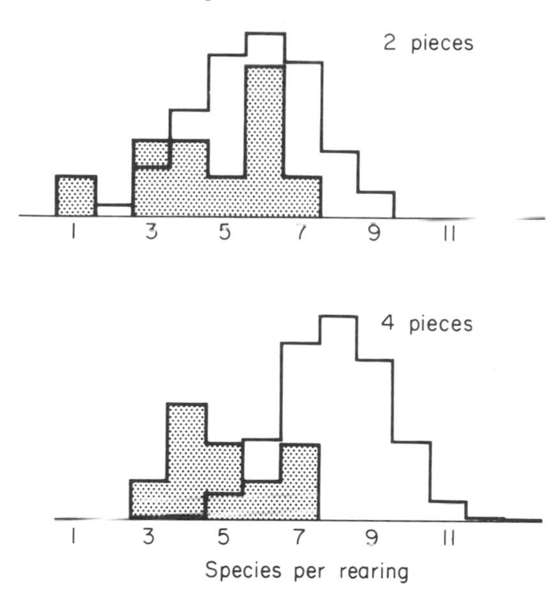

FIG. 8.4. Comparison between the control and experimental rearings in species number. The open histogram is the null hypothesis, derived by pooling flies from each combination of two and four of the twenty control rearings (n = 190 and 4845, respectively). The number of experimental rearings was eleven (two pieces) and nine (four pieces). Removing resource patchiness increases the dominance of the best competitor, *L. illustris* (Table 8.4), and decreases species richness.

in the dominance of the best competitor, *Lucilia illustris* (Table 8.4). Patchiness of the larval resource thus appears to maintain species diversity, most probably because of the mechanism described in Model C.

What makes *Lucilia illustris* a superior competitor, both in the field (Table 8.4) and in enclosed cage populations (Table 8.3)? To answer this question we need to consider the kinds of interspecific interactions taking place in the community. Although the larvae of some species are reported to engage in interspecific interference (e.g. *Sarcophaga aratrix*; Blackith & Blackith 1984) or are facultative predators (e.g. *Hydrotaea dentipes*; Skidmore 1973), most species including *Lucilia* interact via simple resource competition (Ullyett 1950). As the food is often depleted before every larva has completed development, it is important to develop as quickly as possible (Hanski 1986a). Further evidence for strong selection pressure towards short larval development comes from Collins' (1980) 'stress response' analysis: of four kinds of flies, only blowflies exhibit no extension of larval development time under levels of intraspecific competition that reduces their pupal weight by 60–70% (data from Ullyett

1950). This implies that extensions of larval development time result in greater sacrifice to fitness for blowflies than for flies such as *Musca* or *Drosophila* (Collins 1980).

No comparative studies on the larval growth rates of the species involved here have yet been made, but it is suggestive that there is a negative relationship between the (modal) development time until emergence and abundance in the field (Fig. 8.5). I predict that of the species of *Lucilia* in Table 8.4, *L. illustris* has the fastest larval growth rate under conditions prevailing in the study area in southern Finland.

In summary, spatial variance in the numbers of the larvae of the best competitor, *Lucilia illustris*, seems to be sufficient to allow other species to survive regionally, though they go extinct when the community is enclosed in a cage and the number of resource units provided is small. Much remains to be done, however. We need to understand better the behaviour that leads to the observed variance–covariance structure in the larval populations. We need to analyse which other factors affect the numbers of the species (predatory staphylinids are common). And we need a theory capable of predicting species richness and the abundance distribution in the community (Table 8.4).

HETEROTROPHIC SUCCESSION IN EPHEMERAL HABITATS

Small carcasses are consumed by fly maggots so fast that little or no

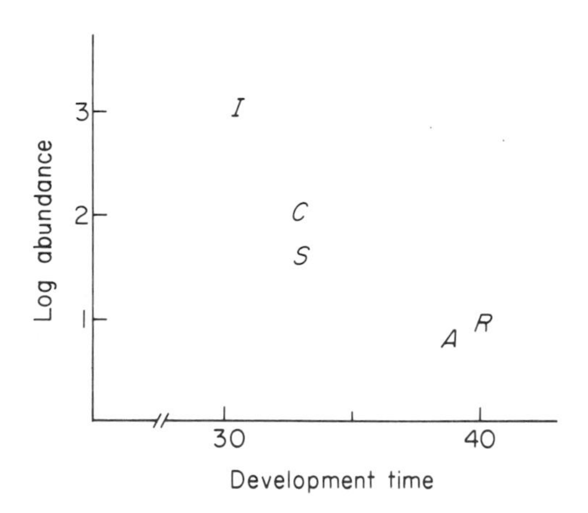

FIG. 8.5. Relationship between development time until emergence and abundance in a field experiment (Kuusela & Hanski 1982). The species are: I = *Lucilia illustris*, C = *L. caesar*, S = *L. silvarum*, R = *L. richardsi* and A = *Muscina assimilis*.

changes in species composition take place. A definite succession of species occurs in larger carcasses (Fuller 1934; Payne 1965; Lane 1975) if these are not removed by vertebrate scavangers. Beetles colonizing cow pats provide an example for this section, though it must be remembered that they are only a part of the dung community: fly larvae especially are rich in species and individuals (Laurence 1954; Valiela 1969a; and references in Valiela 1969b; Legner & Poorbaugh 1972; Hanski 1986a), and form the main prey items of predatory beetles.

The number of species first increases, then declines, on a logarithmic time axis (Fig. 8.6). Two trends in the species composition deserve attention: the proportion of predatory species increases from 20 to 80%, while the proportion of microhabitat (dung) specialists decreases with time (Fig. 8.6). These trends are generally true for heterotrophic succession in ephemeral habitats (Mohr 1943; Elton 1966; Beaver 1984).

Ephemeral habitats are generally maximally distinct from their surroundings at the moment of their origin, and it is not surprising that guilds of specialist species form the bulk of the first colonizers. With time the habitat patches lose their distinctness, and finally merge into the environmental matrix. The facilitation model of succession (Connell & Slatyer 1977) is supported by ephemeral habitats, as many later arriving species seem to benefit from the activity of the early arrivals (Savely 1939;

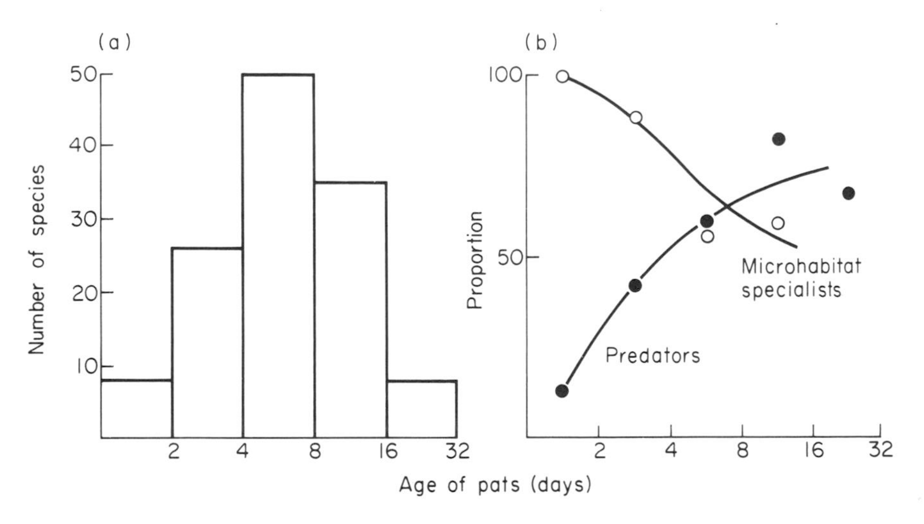

FIG. 8.6. (a) The numbers of beetle species in cow pats varying in age (note logarithmic time axis). (b) The proportion of predatory and microhabitat generalist species during succession. The latter were defined as species also present in carrion (data from Kuusisto 1933; Hanski & Koskela 1977).

Payne 1965; Valiela 1974). As prey populations accumulate and grow in the habitat patches, the role of predation increases. It is probable that similar trends occur in many other habitats where succession is initiated by local disturbances.

Nonetheless, although at the scale of a single habitat patch a heterotrophic succession is usually evident (Fuller 1934; Mohr 1943; Payne 1965; Elton 1966; Beaver 1977, 1983; Hanski 1986a), it is not really sufficient to know the average pattern of species occurrence in patches—one should also know the variance between patches. This is important because populations in ephemeral habitats, and to a lesser extent in all patchy environments, are temporary, and provide the source of colonizers of other habitat patches.

Predators that occur in ephemeral habitats tend to be generalists, feeding indiscriminately on the (often many) coexisting prey species (carrion: Payne 1965; dung: Mohr 1943; *Nepenthes*: Beaver 1983). The rates of immigration to and/or emigration from patches are likely to be non-random, predators and parasitoids generally congregating in the patches where the prey density is highest (see examples in Hassell & May 1974; Hassell 1982). What are the regional consequences of mortality inflicted by such natural enemies? Assuming that prey species tend to be independently clumped in habitat patches, and that each species' spatial variance increases rapidly with its mean abundance (Model C in Table 8.1), it may easily be shown that non-discriminating predators that congregate in the patches with highest pooled prey density cause the greatest regional mortality on the prey species that happens to be regionally most abundant, though rates of local mortality are by assumption frequency-independent (Hanski 1981). This is another example of the pervasive stabilizing influence of non-random predator search on clumped prey (Hassell & May 1974, 1985, and references therein).

The picture that emerges of succession in ephemeral habitats therefore has an important between-patch component. Although in the beginning each habitat unit has a more or less independently developing community of specialists at the second trophic level, the appearance of their mobile predators builds links between the patches, and succession develops as a complex spatio-temporal process. Although studies on this process are lacking, it seems probable that the exceptionally high numbers of species in many ephemeral habitats are partly due to habitat patchiness (localized interactions), and partly to the role of generalist natural enemies, which because of their non-random movements between patches are likely to inflict the heaviest regional mortality on the most abundant prey species.

SPECIES DIVERSITY

Consider an ephemeral habitat in which the resource is consumed or which is saturated by the colonizing species. If one species is given a sufficient lead before the next species arrives, the first species will have had time to use up all the resource, or to establish itself in a way (e.g. all vacant space used) that prevents other species from immigrating. The advantage accruing to the first colonizer(s) is called the priority effect.

For a given rate of between-patch dispersal, the number of species that may develop in a single habitat patch depends on three factors: within-patch rate of development, size of the habitat patch, and its isolation. Specifically, the number of species that may develop in a habitat patch decreases with increasing within-patch rate of development, decreasing patch size, and increasing isolation. The size effect is most easily documented (below).

The above trends suggest that local species richness is highest in large patches situated close to other patches, and it is lowest in small, isolated patches. The equilibrium theory of island biogeography (MacArthur & Wilson 1967) generates the same predictions, though the mechanisms are different: ephemeral habitats do not support stable populations, and the area and isolation effects on species number stem from immigration-related processes.

The increase in species number with increasing patch size has been observed for parasites and herbivores (Price 1980; Kuris, Blaustein & Alió 1980 and references therein). Studies on sessile marine invertebrates show an increase in species number with increasing size of the patch to be colonized (Mook 1976; Abele 1976; Day 1977; Osman 1978; Sousa 1979; Schoener & Schoener 1981), but in some cases important life-history differences between species have been identified, the best competitors tending to exclude other species from large patches, while the latter species—usually good dispersers—survive in small patches (Jackson 1977; Keough 1984). The same result has been reported for carrion flies: the highest number of species emerged from intermediate sized carcasses (Kuusela & Hanski 1982), while in large carcasses the best competitor was disproportionately abundant. Competition is less severe and species richness is higher in the fly assemblage breeding in snails than in vertebrate carcasses (Beaver 1977; Kneidel 1984). The general hypothesis may be advanced that high regional species richness is maintained, in many systems of ephemeral habitats, by small habitat patches in which species composition is variable and interspecific competition weak.

A study by Edler & Mrciak (1975) on Gamasina mites parasitizing small mammals allows us to test this idea on yet another system. Frequency of infestation (by one or more species) and the average number of mites per infested host increase, but species richness of mites decreases with host size (Fig. 8.7), supporting the above hypothesis. The largest small mammals had exceptionally abundant and presumably competitively superior mite species: *Hirstionyssus isabellinus* and *Laelaps hilaris* on *Microtus agrestis* (the two species accounting for 85% of all mites on this host), *L. hilaris* and *Hyperlaelaps microti* on *M. oeconomus* (83%), *H. isabellinus* and *Laelaps clethrionomydis* on *Clethrionomys rufocanus* (78%), and *H. isabellinus* and *Laelaps lemmi* on *Lemmus lemmus* (84%).

Changes in mean distance between patches may have similar effects on regional species richness. Price (1973) has observed that the diversity in the parasite guild of *Neodiprion swainei* first increases but then declines with increasing host density.

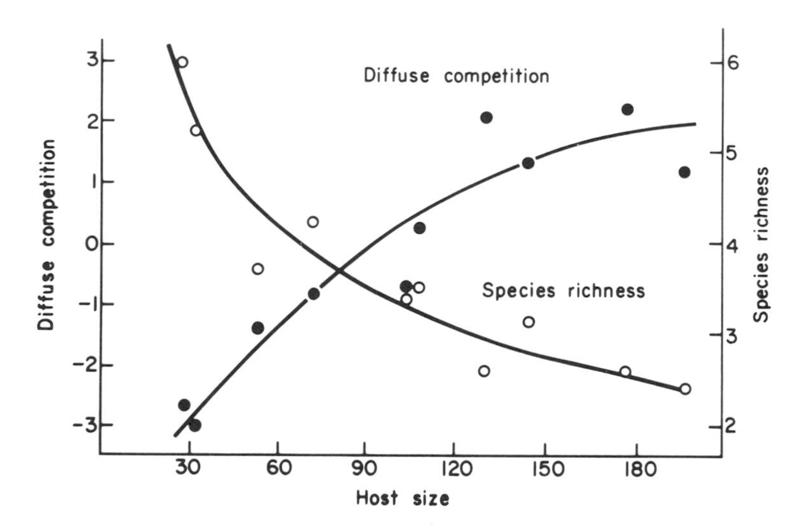

FIG. 8.7. The frequency of infestation (by one or more species) p and the average number of Gamasina mites per infested host n increase, but species richness of mites decreases with increasing host size. p and n are combined in a measure of 'diffuse competition' as $\ln(pn)$ (this is well correlated with p and n, $r = 0{\cdot}963$ and $0{\cdot}908$, respectively). Species richness is the number of species in a random sample of six mites from each host (rarefaction; Simberloff 1979). The lines have been drawn by eye. Data: 17879 Gamasina mites of thirty species on ten species of small mammals from northern Fennoscandia (from Edler & Mrciak 1975).

CONCLUSIONS

Although they stem from a discussion of ephemeral habitats, the findings of this analysis have wider interest in population ecology.

Colonization. Many communities experience disturbances that maintain a spatial mosaic of habitat patches, and small scale extinction–immigration dynamics contribute to spatial variation in density. An example from competition in ephemeral habitats demonstrates that spatial variation in density, density-dependent dynamics, and localized interactions combine to produce regional dynamics not predictable from average spatial density. The general point has been made and developed by Chesson (1981).

Succession. Where local succession takes place in relatively small habitat patches, the between-patch dimension is likely to be significant. Heterotrophic succession in ephemeral habitats often involves generalist predators, which through their non-random spatial distribution have a direct impact on regional dynamics, as the mortality they cause affects the numbers of colonizers of new patches.

Stability. The example of ephemeral habitats forces us to recognize the role of spatial scale: populations may be numerically unstable on one scale while approaching numerical stability on another scale. Paradoxically, instability at a small scale, because it contributes to within-patch variability, may be *necessary* for stability on a larger scale.

ACKNOWLEDGMENT

I am grateful to Dan Simberloff for comments on the manuscript.

REFERENCES

Abele, L.G. (1976). Comparative species richness in fluctuating and constant environments: coral-associated decapod crustaceans. *Science*, **192**, 461–463.

Abrams, P.A. (1984a). Variability in resource consumption rates and the coexistence of competing species. *Theoretical Population Biology*, **25**, 106–124.

Abrams, P.A. (1984b). Recruitment, lotteries, and coexistence in coral reef fish. *American Naturalist*, **123**, 44–55.

Acevedo, L.M.F. (1981). On Horn's Markovian model of forest dynamics with particular reference to tropical forests. *Theoretical Population Biology*, **19**, 230–250.

Ågren, G.I. & Fagerström, T. (1984). Limiting dissimilarity in plants: randomness prevents exclusion of species with similar competitive abilities. *Oikos*, **43**, 369–375.

Anderson, R.M. (1981). Population ecology of infectious diseases. *Theoretical Ecology* (Ed. by R.M. May), pp. 318–355. Blackwell Scientific Publications, Oxford.

Anderson, R.M. (1982). *Population Dynamics of Infectious Diseases*. Chapman and Hall, London.

Andrewartha, H.G. & Birch, L.C. (1954). *The Distribution and Abundance of Animals*. University of Chicago Press, Chicago.

Armstrong, R.A. (1976). Fugitive species: experiments with fungi and some theoretical considerations. *Ecology*, **57**, 953–963.

Atkinson, W.D. (1985). Coexistence of Australian rain forest Diptera breeding in fallen fruits. *Journal of Animal Ecology*, **54**, 507–518.

Atkinson, W.D. & Shorrocks, B. (1981). Competition on a divided and ephemeral resource: a simulation model. *Journal of Animal Ecology*, **50**, 461–471.

Atkinson, W.D. & Shorrocks, B. (1984). Aggregation of larval Diptera over discrete and ephemeral breeding sites: the implications for coexistence. *American Naturalist*, **124**, 336–351.

Baumgartner, D.L. & Greenberg, B. (1984). The genus Chrysomya (Diptera: Calliphoridae) in the New World. *Journal of Medical Entomology*, **21**, 105–113.

Beaver, R.A. (1977). Non-equilibrium 'island' communities: Diptera breeding in dead snails. *Journal of Animal Ecology*, **46**, 783–798.

Beaver, R.A. (1979). Host specificity of temperate and tropical animals. *Nature*, **281**, 139–141.

Beaver, R.A. (1983). The communities living in Nepenthes pitcher plants: fauna and food webs. *Phytotelmata: Terrestrial Plants as Hosts for Aquatic Insect Communities* (Ed. by L.P. Lounibos & J.H. Frank), pp. 129–159. Plexus Publ., New Jersey.

Beaver, R.A. (1984). Insect exploitation of ephemeral habitats. *South Pacific Journal of Natural Sciences*, **6**, 3–47.

Becker, P., Lee, L.W., Rothman, E.D. & Hamilton, W.D. (1985). Seed predation and the coexistence of tree species: Hubbell's models revisited. *Oikos*, **44**, 382–390.

Blackith, R. & Blackith, R. (1984). Larval aggression in Irish flesh-flies (Diptera: Sarcophagidae). *Irish Nature Journal*, **21**, 255–257.

Bock, C.E. & Ricklefs, R.E. (1983). Range size and local abundance of some north American songbirds: a positive correlation. *American Naturalist*, **122**, 295–299.

Brown, J.H. (1984). On the relationship between abundance and distribution of species. *American Naturalist*, **124**, 253–279.

Cappuccino, N. & Kareiva, P. (1985). Coping with a capricious environment: a population study of a rare pierid butterfly. *Ecology*, **66**, 152–161.

Caraco, T. & Whitham, T.S. (1984). Immigration–extinction competition on islands: associations among three species. *Journal of Theoretical Biology*, **110**, 241–252.

Carter, R.N. & Prince, S.D. (1981). Epidemic models used to explain biogeographical distribution limits. *Nature*, **293**, 644–645.

Chesson, P.L. (1981). Models for spatially distributed populations: the effect of within-patch variability. *Theoretical Population Biology*, **19**, 288–325.

Chesson, P.L. & Warren, R.R. (1981). Environmental variability promotes coexistence in lottery competitive systems. *American Naturalist*, **117**, 923–943.

Collins, N.C. (1980). Developmental responses to food limitation as indicators of environmental conditions for Ephydra cinerea Jones (Diptera). *Ecology*, **61**, 650–661.

Comins, H.N., Hamilton, W.D. & May, R.M. (1980). Evolutionary stable dispersal strategies. *Journal of Theoretical Biology*, **82**, 205–230.

Connell, J.H. (1971). On the role of natural enemies in preventing competitive exclusion in some marine animals and in rain forest trees. *Dynamics of Populations* (Ed. by P.J. Den Boer & G.R. Gradwell), pp. 298–312. Pudoc, Wageningen.

Connell, J.H. (1978). Diversity in tropical forests and coral reefs. *Science*, **199**, 1302–1310.

Connell, J.H. & Slatyer, R.O. (1977). Mechanisms of succession in natural communities and their role in community stability and organization. *American Naturalist*, **111**, 1119–1144.

Day, R.W. (1977). Two contrasting effects of predation on species richness in coral reef habitats. *Marine Biology*, **44**, 1–6.

Den Boer, P.J. (1971). Stabilization of animal numbers and the heterogeneity of the environment: the problem of persistence of sparse populations. *Dynamics of Populations* (Ed. by P.J. Den Boer & G.R. Gradwell), pp. 77–97. Centre for Agricultural Publishing and Documentation, Wageningen.

Den Boer, P.J. (1979). Dispersal power and survival. Carabids in a cultivated countryside. *Miscellaneous papers* 14, Landbouwhogeschool, Wageningen.

Dethier, M.N. (1984). Disturbance and recovery in intertidal pools: maintenance of mosaic patterns. *Ecological Monographs*, **54**, 99–118.

Diamond, J.M. (1984). 'Normal' extinctions of isolated populations. *Extinctions* (Ed. by M.H. Nitecki), pp. 191–246. University of Chicago Press, Chicago.

Edler, A. & Mrciak, M. (1975). Gamasina mites (Acari: Parasitiformes) on small mammals in northernmost Fennoscandia. *Entomologisk Tidskrift*, **96**, 167–177.

Elton, C. (1966). *The Pattern of Animal Communities*. Methuen, London.

Fowler, S.V. & Lawton, J.H. (1982). The effects of host-plant distribution and local abundance on the species richness of agromyzid flies attacking British Umbellifers. *Ecological Entomology*, **7**, 257–265.

Fuller, M.E. (1934). The insect inhabitants of carrion: a study in animal ecology. *Bulletin. Council of Scientific and Industrial Research. Australia*, **82**, 1–62.

Gadgil, M. (1971). Dispersal: population consequences and evolution. *Ecology*, **52**, 253–260.

Gotelli, N.J. & Abele, L.G. (1983). Community patterns of coral-associated decapods. *Marine Ecology*, **13**, 131–139.

Green, R.F. (1986). Does aggregation prevent competitive exclusion? A response to Atkinson and Shorrocks. *American Naturalist*, **128**, 301–304.

Grimaldi, D. & Jaenike, J. (1984). Competition in natural populations of mycophagous Drosophila. *Ecology*, **65**, 1113–1120.

Grubb, P.J. (1977). The maintenance of species-richness in plant communities: the importance of the regeneration niche. *Biological Reviews*, **52**, 107–145.

Guimarães, J.H., Prado, A.P. & Buralli, G.M. (1979). Dispersal and distribution of three newly introduced species of Chrysomya Robineau-Desvoidy in Brazil (Diptera, Calliphoridae). *Revista Brasileira de Entomologia*, **23**, 245–255.

Gurney, W.S.C. & Nisbet, R.M. (1978). Single species population fluctuations in patchy environments. *American Naturalist*, **112**, 1075–1090.

Hackman, W. (1979). Reproductive and developmental strategies of fungivorous Pegomya species (Diptera, Anthomyiidae). *Aquilo Series Zoologica*, **20**, 62–64.

Hackman, W. & Meinander, M. (1979). Diptera feeding as larvae on macrofungi in Finland. *Annales Zoologici Fennici*, **16**, 50–83.

Hanski, I. (1976). Breeding experiments with carrion flies (Diptera) in natural conditions. *Annales Entomologici Fennici*, **42**, 113–121.

Hanski, I. (1979). *The community of coprophagous beetles*. D.Phil. thesis, University of Oxford.

Hanski, I. (1980a). Spatial patterns and movements in coprophagous beetles. *Oikos*, **34**, 311–321.

Hanski, I. (1980b). The community of coprophagous beetles (Coleoptera, Scarabaeidae and Hydrophilidae) in northern Europe. *Annales Entomologici Fennici*, **46**, 57–74.

Hanski, I. (1980c). Migration to and from cow droppings by coprophagous beetles. *Annales Zoologici Fennici*, **17**, 11–16.

Hanski, I. (1981). Coexistence of competitors in patchy environments with and without predation. *Oikos*, **37**, 306–312.

Hanski, I. (1982a). Dynamics of regional distribution: the core and satellite species hypothesis. *Oikos*, **38**, 210–221.

Hanski, I. (1982b). Distributional ecology of anthropochorous plants in villages surrounded by forest. *Annales Botanici Fennici*, **19**, 1–15.

Hanski, I. (1983). Coexistence of competitors in patchy environment. *Ecology*, **64**, 493–500.

Hanski, I. (1985). Single-species spatial dynamics may contribute to long-term rarity and commonness. *Ecology*, **66**, 335–343.

Hanski, I. (1986a). Nutritional ecology of dung- and carrion-feeding insects. *Nutritional Ecology of Insects, Mites, and Spiders* (Ed. by F. Slansky, Jr & J.G. Rodriguez), John Wiley, New York.

Hanski, I. (1986b). Beetles associated with dung and carrion in tropical forests. *Ecosystems of the World, 14b* (Ed. H. Lieth & M.J.A. Werger), Elsevier Publishing Company, Amsterdam.

Hanski, I. & Koskela, H. (1977). Niche relations among dung-inhabiting beetles. *Oecologia*, **28**, 203–231.

Hanski, I. & Kuusela, S. (1980). The structure of carrion fly communities: differences in breeding seasons. *Annales Zoologici Fennici*, **17**, 185–190.

Hanski, I. & Ranta, E. (1983). Coexistence in a patchy environment: three species of Daphnia in rock pools. *Journal of Animal Ecology*, **52**, 263–279.

Harper, J.L. (1977). *Population Biology of Plants*. Academic Press, London.

Hassell, M.P. (1982). Patterns of parasitism by insect parasitoids in patchy environments. *Ecological Entomology*, **7**, 365–377.

Hassell, M.P. & May, R.M. (1973). Stability in insect host--parasite models. *Journal of Animal Ecology*, **42**, 693–726.

Hassell, M.P. & May, R.M. (1974). Aggregation in predators and insect parasites and its effect on stability. *Journal of Animal Ecology*, **43**, 567–594.

Hassell, M.P. & May, R.M. (1985). From individual behaviour to population dynamics. *Behavioural Ecology* (Ed. by R.M. Sibly & R.H. Smith), pp. 3–32. Blackwell Scientific Publications, Oxford.

Hobson, R.P. (1938). Sheep blowfly investigations. VII. Observations on the development of eggs and oviposition in the sheep blowfly, Lucilia sericata (Mg.). *Annals of Applied Biology*, **25**, 573–582.

Holter, P. (1979). Abundance and reproductive strategy of the dung beetle *Aphodius rufipes* (Scarabaeidae). *Ecological Entomology*, **4**, 317–326.

Horn, H.S. (1975). Markovian properties of forest succession. *Ecology and Evolution of Communities* (Ed. by M. Cody & J. Diamond), pp. 196–211. Harvard University Press, Cambridge, Mass.

Horn, H.S. (1981). Succession. *Theoretical Ecology* (Ed. by R.M. May), pp. 253–271. 2nd edn. Blackwell Scientific Publications, Oxford.

Horn, H.S. & MacArthur, R.H. (1972). Competition among fugitive species in a harlequin environment. *Ecology*, **53**, 749–752.

Hubbell, S.P. (1979). Tree dispersion, abundance, and diversity in a tropical dry forest. *Science*, **203**, 1299–1309.

Hubbell, S.P. (1980). Seed predation and the coexistence of tree species in tropical forests. *Oikos*, **35**, 214–229.

Hutchinson, G.E. (1951). Copepodology for the ornithologist. *Ecology*, **32**, 571–577.

Ives, A.R. & May, R.M. (1985). Competition within and between species in a patchy environment: relations between microscopic and macroscopic models. *Journal of Theoretical Biology*, **115**, 65–92.

Jackson, J.B.C. (1977). Habitat area, colonization, and development in a marine epifaunal community. *Biology of Benthic Organisms* (Ed. by B.F. Keegan *et al.*), pp. 349–358. Pergamon Press, Oxford.

Janzen, D.H. (1970). Herbivores and the number of tree species in tropical forests. *American Naturalist*, **104**, 501–528.

Johnson, C.G. (1969). *Migration and Dispersal of Insects by Flight*. Methuen, London.

de Jong, G. (1982). The influence of dispersal pattern on the evolution of fecundity. *Netherlands Journal of Zoology*, **32**, 1–30.

Keough, M.J. (1984). Effects of patch size on the abundance of sessile marine invertebrates. *Ecology*, **65**, 423–437.

Kneidel, K.A. (1984). Competition and disturbance in communities of carrion-breeding Diptera. *Journal of Animal Ecology*, **53**, 849–866.

Koskela, H. & Hanski, I. (1977). Structure and succession in a beetle community inhabiting cow dung. *Annales Zoologici Fennici*, **14**, 204–223.

Kuris, A.M., Blaustein, A.R. & Alió, J.J. (1980). Hosts as islands. *American Naturalist*, **116**, 570–586.

Kuusela, S. (1984). Suitability of carrion flies for field experiments on reproductive behaviour. *Annales Entomologici Fennici*, **50**, 1–6.

Kuusela, S. & Hanski, I. (1982). The structure of carrion fly communities: the size and the type of carrion. *Holarctic Ecology*, **5**, 337–348.

Kuusisto, I. (1933). *Tutkimuksia raadoilla elävistä kovakuoriaisista Turun Ruissalossa*. M.S. Thesis, University of Turku, Finland (unpubl.)

Landin, J. (1967). On the relationship between the microclimate in cow droppings and some species of Sphaeridium (Col. Hydrophilidae). *Opuscula Entomologica*, **32**, 207–212.

Lane, R.P. (1975). An investigation into blowfly (Diptera: Calliphoridae) succession on corpses. *Journal of Natural History*, **9**, 581–588.

Laurence, B.R. (1954). The larval inhabitants of cow pats. *Journal of Animal Ecology*, **23**, 234–260.

Legner, E.F. & Poorbaugh, J.H. (1972). Biological control of vector and noxious synanthropic flies: a review. *Californian Vector Views*, **19**, 81–100.

Leigh, E.G. Jr (1981). The average lifetime of a population in a varying environment. *Journal of Theoretical Biology*, **90**, 213–239.

Levin, S.A., Cohen, D. & Hastings, A. (1984). Dispersal strategies in patchy environments. *Theoretical Population Biology*, **26**, 165–191.

Levin, S.A. & Paine, R.T. (1974). Disturbance, patch formation, and community structure. *Proceedings of the National Academy of Sciences, USA*, **71**, 2744–2747.

Levins, R. (1970). Extinction. *Some Mathematical Problems in Biology* (Ed. by M. Gerstenhaber), pp. 77–107. American Mathematical Society, Providence, R.I.

Levins, R. & Culver, D. (1971). Regional coexistence of species and competition between rare species. *Proceedings of the National Academy of Sciences, USA*, **68**, 1246–1248.

Linkola, K. (1916). Studien über den Einfluss der Kultur auf die Flora in den Gegenden nördlich von Ladogasee. I. Allgemeiner Teil. *Acta Societas Fauna Flora Fennica*, **45**, 1–432.

Lloyd, M. & White, J. (1980). On reconciling patchy microspatial distributions with competition models. *American Naturalist*, **115**, 29–44.

MacArthur, R.H. (1972). *Geographical Ecology*. Harper & Row, New York.

MacArthur, R.H. & Wilson, E.O. (1967). *The Theory of Island Biogeography*. Princeton University Press, Princeton.

Mackerras, M.J. (1933). Observations on the life-histories, nutritional requirements and fecundity of blowflies. *Bulletin of Entomological Research*, **24**, 353–361.

May, R.M. (1973). *Complexity and Stability in Model Ecosystems*. Princeton University Press, Princeton.

Mohr, C.O. (1943). Cattle droppings as ecological units. *Ecological Monographs*, **13**, 275–309.

Mook, D. (1976). Studies of fouling invertebrates in the Indian River. *Bulletin of Marine Science*, **26**, 610–615.

Neuvonen, S. & Niemelä, P. (1981). Species richness of macrolepidoptera on Finnish deciduous trees and shrubs. *Oecologia*, **51**, 364–370.

Nisbet, R.M. & Gurney, W.S.C. (1982). *Modelling Fluctuating Populations.* John Wiley, New York.

Nuorteva, P. (1970). Histerid beetles as predators of blowflies (Diptera, Calliphoridae) in Finland. *Annales Zoologici Fennici*, **7**, 195–198.

Osman, R.W. (1978). The influence of seasonality and stability on the species equilibrium. *Ecology*, **59**, 383–399.

Otronen, M. & Hanski, I. (1983). Movement patterns in Sphaeridium: differences between species, sexes, and feeding and breeding individuals. *Journal of Animal Ecology*, **52**, 663–680.

Paine, R.T. (1966). Food web complexity and species diversity. *American Naturalist*, **100**, 65–75.

Paine, R.T. (1979). Disaster, catastrophe, and local persistence of the sea palm Postelsia palmaeformis. *Science*, **205**, 685–687.

Paine, R.T. (1984). Ecological determinism in the competition for space. *Ecology*, **65**, 1339–1348.

Paine, R.T. & Levin, S.A. (1981). Intertidal landscapes: disturbance and the dynamics of pattern. *Ecological Monographs*, **51**, 145–178.

Palmer D.H. (1980). *Partitioning of the carrion resource by sympatric Calliphoridae (Diptera) near Melbourne.* Ph.D. thesis, La Trobe University, Melbourne.

Parker, G.A. & Courtney, S.P. (1984). Models of clutch size in insect oviposition. *Theoretical Population Biology*, **26**, 27–48.

Payne, J.A. (1965). A summer carrion study of the baby big *Sus scrofa*. *Ecology*, **46**, 592–602.

Pianka, E.R. (1974). *Evolutionary Ecology*. Harper & Row, New York.

Pielou, E.C. (1975). *Ecological Diversity*. John Wiley, New York.

Platt, W.J. & Weis, I.M. (1985). An experimental study of competition among fugitive prairie plants. *Ecology*, **66**, 708–720.

Price, P.W. (1973). The development of parasitoid communities. *Proceedings of the Northeastern Forest Insect Work Conference*, **5**, 29–41.

Price, P.W. (1980). *Evolutionary Biology of Parasites.* Princeton University Press, Princeton.

Sale, P.F. (1977). Maintenance of high diversity in coral reef fish communities. *American Naturalist*, **111**, 337–359.

Sale, P.F. (1979). Recruitment, loss, and coexistence in a guild of territorial coral reef fishes. *Oecologia*, **42**, 159–177.

Sale, P.F. (1982). Stock-recruitment relationships and regional coexistence in a lottery competitive system: a simulation study. *American Naturalist*, **120**, 139–159.

Savely, H.E. (1939). Ecological relations of certain animals in dead pine and oak logs. *Ecological Monographs*, **9**, 321–385.

Schoener, A. & Schoener, T.W. (1981). The dynamics of the species–area relation in marine fouling systems. 1. Biological correlates of changes in the species–area slope. *American Naturalist*, **118**, 339–360.

Shorrocks, B., Atkinson, W.D. & Charlesworth, P. (1979). Competition on a divided and ephemeral resource. *Journal of Animal Ecology*, **48**, 899–908.

Simberloff, D. (1979). Rarefaction as a distribution-free method of expressing and estimating diversity. *Ecological Diversity in Theory and Practice* (Ed. by J.F. Grassle, G.P. Patil, W. Smith & C. Taillie), pp. 159–176. International Cooperating Publishing House, Maryland.

Skidmore, P. (1973). Notes on the biology of palaearctic muscids. *Entomologist*, **106**, 25–59.

Slatkin, M. (1974). Competition and regional coexistence. *Ecology*, **55**, 128–134.

Sousa, W.P. (1979). Experimental investigations of disturbance and ecological succession in a rocky intertidal algal community. *Ecological Monographs*, **49**, 227–254.

Southwood, T.R.E. (1960). The flight activity of Hemiptera. *Transactions of the Royal Entomological Society, London*, **112**, 173–220.

Southwood, T.R.E. (1961). The evolution of the insect-host tree relationship—a new approach. *Proceedings of the XIth International Congress on Entomology, Vienna*, pp. 651–654.

Southwood, T.R.E. (1962). Migration of terrestrial arthropods in relation to habitat. *Biological Review*, **37**, 171–214.

Southwood, T.R.E. (1977). Habitat, the templet for ecological strategies? *Journal of Animal Ecology*, **46**, 337–365.

Strong, D.R., Lawton, J.H. & Southwood, T.R.E. (1984). *Insects on Plants*. Blackwell Scientific Publications, Oxford.

Taylor, L.R., Woiwood, I.P. & Perry, J.N. (1979). The density-dependence of spatial behaviour and the rarity of randomness. *Journal of Animal Ecology*, **47**, 383–406.

Toft, C.A. & Schoener, T.W. (1983). Abundance and diversity of orb spiders on 106 Bahamian islands: biogeography at an intermediate trophic level. *Oikos*, **41**, 411–426.

Ullyett, G.C. (1950). Competition of food and allied phenomena in sheep-blowfly populations. *Philosophical Transactions of the Royal Society, Series B*, **234**, 77–174.

Valiela, I. (1969a). An experimental study of the mortality factors of larval Musca autumnalis DeGeer. *Ecological Monographs*, **39**, 199–220.

Valiela, I. (1969b). The arthropod fauna of bovine dung in central New York and sources on its natural history. *Journal of New York Entomological Society*, **77**, 210–220.

Valiela, I. (1974). Composition, food webs and population limitation in dung arthropod communities during invasion and succession. *American Middland Naturalist*, **92**, 370–385.

Vance, R.R. (1984). The effect of dispersal on population stability in one-species, discrete-space population growth models. *American Naturalist*, **123**, 230–254.

Vepsäläinen, K. (1978). Coexistence of two competing corixid species (Heteroptera) in an archipelago of temporary rock pools. *Oecologia*, **37**, 177–182.

Vet, L.E.M. & van Alphen, J.J.M. (1985). A comparative functional approach to the host detection behaviour of parasitic wasps. 1. A qualitative study on Eucoilidae and Alysiinae. *Oikos*, **44**, 478–486.

Vogt, H. (1968). Cercyon–Studien. *Entomological Bulletin*, **64**, 172–191.

Ward, L.K. (1977). The conservation of juniper: the associated fauna with special reference to southern England. *Journal of Applied Ecology*, **14**, 81–120.

Ward, L.K. & Lakhani, K.H. (1977). The conservation of juniper: the fauna of food-plant island sites in southern England. *Journal of Applied Ecology*, **14**, 121–135.

Waterhouse, D.F. (1947). The relative importance of live sheep and of carrion as breeding grounds for the Australian sheep blowfly *Lucilia cuprina*. *Bulletin. Council of Scientific and Industrial Research. Australia*, **217**, 1–31.

Zeigler, B.P. (1977). Persistence and patchiness of predator–prey systems induced by discrete event population exchange mechanisms. *Journal of Theoretical Biology*, **67**, 687–713.

9. COLONIZATION AND SPECIATION

HAMPTON L. CARSON
Department of Genetics, University of Hawaii, Honolulu, Hawaii 96822 USA

'The natural history of this archipelago is very remarkable: it seems to be a little world within itself; the greater number of its inhabitants, both vegetable and animal, being found nowhere else.'

CHARLES DARWIN
written in 1837

INTRODUCTION

Colonization of an area by an animal or plant species may be either primary or secondary. The former may be defined as the establishment of a population in an area not previously occupied by the species (Safriel & Ritte 1983). Secondary colonization is more characteristic of small locally-confined areas that are usually less drastically altered and less isolated from colonizing propagules (see Miles Chapter 1). Areas open to primary colonization may lack a species because of recent origin (e.g. volcanic islands or mountain peaks) or recent catastrophies (e.g. earthquakes, floods or lava flows). Examples of secondary colonization are such things as forest gaps, old-field successions, emergent strands or the various environments provided by the retreat of a glacier. The source of propagules, unlike primary colonization, is often close to the altered area so that normal preadaptive dispersal accounts for the easy arrival of propagules.

Such classifications as the above, in view of the diversity of both species and the world environment, are quite imprecise and a continuum of conditions probably exists. Nevertheless, the distinction is useful in approaching the subject of this paper, namely, to determine what sort of genetic changes may occur in populations descended from the propagules in the new site, especially after primary colonization.

Colonization, either primary or secondary, may occur without significant genetic change. In the presence of preadaptation and efficient dispersal, the normal variable gene pool of the species is essentially transported without alteration to the new site. On the other hand, modified environmental conditions may elicit, through natural selection, relatively minor intraspecific phyletic change. Such changes may be manifested in altered frequencies of genes, minor local adaptations or even racial or subspecific differentiation. Primary colonization, especially when it is accompanied by long-distance dispersal, may involve a change of much greater magnitude. Thus, the new colony may give evidence that a newly integrated, novel gene pool, unique in nature, has been formed. Such a population may then be recognized as a new species.

Both primary colonization and the dynamics of speciation involve an intricate set of genetic and environmental processes that are difficult, if not impossible, to observe directly. The scientist is therefore reduced to making nebulous inferences about historic events. As everybody knows, however, it was Darwin who first realized the inferential simplifications that could be afforded by a consideration of the terrestrial biota of oceanic islands.

In the past 20 years, a considerable amount of new information has been gathered on the inhabitants of the terrestrial ecosystems of the Hawaiian archipelago (Fig. 9.1), which consists of a chronologically and spacially linear series of highly isolated oceanic islands in the central Pacific Ocean (Mueller-Dombois, Bridges & Carson 1981). Parallel to the ecological studies, and to some degree integrated with them, has been a multidisciplinary investigation of the Drosophilidae of Hawaii (Carson *et al.* 1970; Carson & Kaneshiro 1976; Carson & Yoon 1982). A crucial dimension added by these latter studies are chromosomal and other genetic analyses. The methods employed have made possible the tracing of both the continental and the inter-island origin of certain populations and species by the use of giant polytene chromosomes and DNA analysis (Carson 1983; Hunt & Carson 1983).

To this broad phylogenetic approach has been added a concentrated study of the population biology and genetics of *Drosophila silvestris*, a species endemic to the high-altitude rain forests of the newest island (see review in Carson 1982a). Accordingly it is possible to consider Hawaiian examples for the discussion of primary colonization and speciation at three levels: the archipelago as a whole, single islands within the archipelago and, indeed, very recent colonizations and incipient speciation at the population level within a single island.

Recent information on the geological history of the islands has added greatly to the precision with which matters pertaining to colonization and speciation can be approached. Specifically, each island appears to have been successively formed over a plume of mantle (a 'hot spot') that appears to be fixed in position under the Pacific tectonic plate in the central Pacific Ocean (Dalrymple, Silver & Jackson 1973). This plume

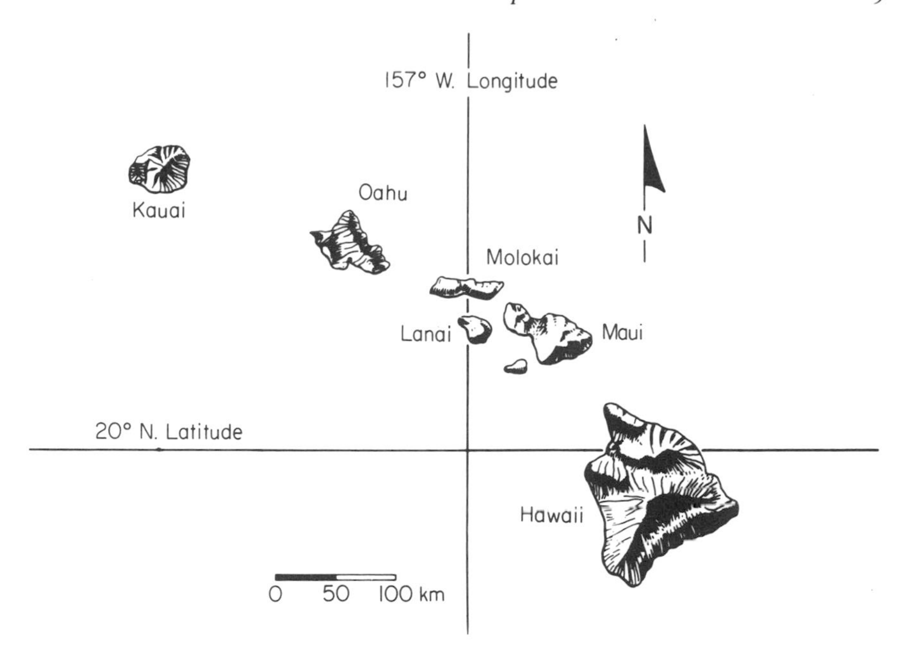

FIG. 9.1. The Hawaiian islands.

appears to accomplish periodic perforations of the plate as the latter moves in a north-westerly direction at the rate of about 10 cm per year. Potassium–argon and magnetic declination measurements indicate the age of the island of Hawaii to be less than 0·4 million years. This island is the newest in the chain; two of the volcanoes at its south-east end show currently active vulcanism. In contrast, as one proceeds north-westward, each island shows currently inactive volcanoes; they are successively older as one proceeds to Kauai, on which the lava flows are not less than 5 million years old. Kauai is now about 500 km north-east of the current zone of active vulcanism. The result is a linear chain of high volcanic islands of successively younger ages as one proceeds to the south-east. This permits special attention to be given to the island of Hawaii (called 'The Big Island' to distinguish it from the archipelago as a whole). Colonization and speciation events involving this island can be inferred as being especially recent ones. This island and its biota may thus be viewed as a natural laboratory for evolutionary studies whether one is interested in the genetical properties of the evolution of species at the population level or the novel arrivals and integrations of species that are implicated in the development of ecosystems.

Using examples from the Hawaiian Islands, I will review the manner in which these new geologial and biological data impinge on the twin subjects of primary colonization and speciation and will attempt to chart the directions that further research might take.

The manner in which the inferred colonization events may be related to the process of speciation are outlined in Table 9.1. For certain groups of organisms found in Hawaii, biogeographical information is sufficient to classify each species as to whether it is indigenous (Table 9.1, Ia), exotic (Table 9.1, Ib) or endemic (Table 9.1, IIa and b). Species in the latter category are of particular interest since the remoteness of the archipelago and the uniqueness of the species enables the investigator to propose the past occurrence of a founder event whereby the ancestor of the species concerned can be inferred to have colonized one of the islands. The first attempt to do this on a large scale was that of Zimmerman (1948). As summarized by Hubbell (1968, Table 1), 6753 endemic species and subspecies of insects, land snails, land birds, ferns and fern-allies and flowering plants could be related to between 677 and 700 immigrant ancestral forms arriving by long-distance dispersal. Since these data were compiled, a large amount of new information has become available, but the general picture has not essentially changed. Rather than present a modern revision of all of these data, I will concentrate in this chapter on the problem of why some lineages established from an ancestral immigrant are species-poor (Table 9.1, IIa), and others are species rich (Table 9.1, IIb). In particular, an attempt will be made to consider to what extent the genetic system of the species can be related to the demonstrated capacity to proliferate species. Knowledge of the genetic properties of the exuberantly speciose Drosophilidae will be invoked in this discussion.

Preadaptive colonization, an event not conducive to speciation

The strand areas of oceanic islands such as Hawaii illustrate a very important principle of colonization, i.e. many species appear to be preadapted to movement over long distances and to quite literally establish beachheads on isolated and distant islands in the course of their normal dispersal. A simple example is the coconut palm, *Cocos nucifera*. This plant is virtually pantropical in distribution and there is little question that its seeds, with their corky floating husks and abundant stored nutrition, have basically evolved as a dispersal mechanism. In the various places where it is found around the world, genetic variation in this species appears to be minor. Accordingly, this plant seems to offer a good

TABLE 9.1. Colonization and speciation on a remote oceanic archipelago: selected examples from Hawaii

I. Colonizers that have *not* undergone speciation.

(a) *Indigenous species*. Single species that are native not only to Hawaii but whose normal range of distribution includes a wider geographical area:

- 54 species of seed plants, the majority from strand habitats (Table 9.2)
- about 27 species of birds, predominantly sea birds
- one species of mammal (*Lasiurus cinereus*, a bat)

(b) *Exotic species*. Single species introduced by man and now naturalized:

- over 600 species of seed plants
- about 25 species of land birds
- about 26 species of land mammals
- over 2000 species of insects
- 26 species of Drosophilidae (Table 9.3)

II. Colonizers that have undergone speciation after arrival (endemic species).

(a) Colonizers forming *one and only one* endemic species:

- 58 species of seed plants (Table 9.4)
- 42 species of insects (Table 9.5)
- 6 species of birds
- 1 species of mammal (*Monachus schaninslandi*, a seal)

(b) Colonizers forming speciose lineages:

- 22 seed plant genera having 15 or more endemic species (Table 9.6)
- 23 insect genera having 35 or more endemic species (Table 9.6)
- 25 species (10 genera) of the Family Drepanididae (Hawaiian finches)

Excluded from the above are the land snails, certain freshwater fauna, ferns, fern-allies and mosses. Endemic genera, subspecies, varieties and forms have not been considered.

example of preadaptation for dispersal and colonization of the vast areas of its distribution. The hypothesis that the coconut is basically preadapted for dispersal through flotation of seeds is difficult to test in the modern era, however, in the light of the fact that the plant has been cultivated by island peoples around the tropical world and thus owes much of its modern distribution to this fact.

Table 9.2 lists a series of plants that appear to have colonized the Hawaiian islands long before human intervention could possibly have affected their distribution. Listed are fifty-four species of seed plants found as indigenous members of the Hawaiian flora. Each of these species, however, is widespread, sometimes over large tropical areas, yet each is at the same time integrated into a natural Hawaiian ecosystem. None of these appears to have made its way to Hawaii under the influence of man; it is the uncertainty about this that has led to the omission of coconut palm from this list.

Table 9.2. Non-speciose, wide-ranging indigenous seed plants of Hawaii

Family	Species	Family	Species
Pandanaceae	*Pandanus tectorius*	Malvaceae	*Abutilon incanum*
Potamogetonaceae	*Ruppia maritima*		*Hibiscus furcellatus*
Gramineae	*Eragrostis whitneyi*		*Hibiscus tiliaceus*
	Heteropogon contortus		*Sida fallax*
	Lepturus repens	Sterculiaceae	*Waltheria indica*
	Microlaena stipoides	Lythraceae	*Lythrum maritimum*
Cyperaceae	*Uncinia uncinata*	Myrtaceae	*Eugenia reinwardtiana*
Piperaceae	*Peperomia leptostachya*	Epacridaceae	*Styphelia tameiameiae*
	Peperomia tetraphylla	Primulaceae	*Lysinmachia mauritania*
Moraceae	*Streblus pendulinus*	Plumbaginaceae	*Plumbago zeylandica*
Urticaceae	*Pilea peploides*	Convolvulaceae	*Cressa truxillensis*
Viscaceae	*Korthasella complanata*		*Ipomoea congesta*
	Korthasella platycaula		*Ipomoea gracilis*
	Korthasella rubescens		*Ipomoea pes-caprae*
Aizoaceae	*Sesuvium portulacastrum*		*Ipomoea imperati*
Menispermaceae	*Cocculus trilobus*		*Jaquemontia ovalifolia*
Lauraceae	*Cassytha filiformis*	Boraginaceae	*Heliotropium anomalum*
Droseraceae	*Drosera anglica*		*Heliotropium curassavicum*
Rosaceae	*Fragaria chiloensis*	Verbenaceae	*Vitex ovata*
	Osteomeles anthyllidifolia	Labiatae	*Lepechinia hastata*
Leguminosae	*Caesalpinia major*		*Plectranthus parviflorus*
	Mucuna gigantea	Scrophulariaceae	*Bacopa monnieria*
	Vigna marina	Myoporaceae	*Myoporum sandwicense*
Zygophyllaceae	*Tribulus cistoides*	Rubiaceae	*Canthium odoratum*
Aquifoliaceae	*Ilex anomala*		*Nertera granadensis*
Sapindaceae	*Dodonaea viscosa*	Goodeniaceae	*Scaevola sericea*
Rhamnaceae	*Colubrina asiatica*	Compositae	*Adenostemma lavenia*

Listed are fifty-four single species that are integrated into Hawaiian ecosystems but are native to one or more areas in addition to the Hawaiian Islands. Subspecies and varieties as well as species introduced since human arrival in the islands are excluded. Data from St. John (1973), and unpublished data from *Manual of the Flowering Plants of Hawaii* by W.L. Wagner, D. Herbst & S.H. Sohmer (in prep.).

Examination of the ecological context of these species shows that a large number (more than half) of them are found on or close to the coastal strand so that long-distance dispersal of propagules floating in oceanic waters suggests itself. Fitting this notion in particular are the five extremely widespread species of the morning-glory genus, *Ipomoea*, one of which, *I. pes-caprae* is virtually world-wide in the tropics.

In the emphasis supplied by the present chapter, it is noteworthy that the species listed in Table 9.2 have been accorded little or no subspecific or other systematic differentiation throughout their enormous geographical ranges. This suggests that there is likewise little recombinable free genetic variation present in these species. Primary colonization has not elicited change. Direct study of both genetic variation and the genetic system (e.g. breeding structure) has not been done on most of these species and is badly needed. The suggestion may be made, following Baker (1965), that these species are basically preadapted for these wide dispersals and their genetic systems, like those of many weedy plants around the world, consist of a complex coadapted heterotic system that enables them to perform these feats of colonization ('general purpose genotypes'). Even though some may be self-fertilized, they may still retain blocks of genes that provide a basis for their ability to colonize. Self-fertilization and inbreeding do not necessarily always override stable polymorphisms (Allard, Kahler & Clegg 1975). These obligatory general-purpose adaptations appear to be inhibitory to founder effects followed by genomic disorganizations that might be expected to ensue if speciation events were to follow colonization.

Each of the plant species listed in Table 9.2 represents a special set of circumstances but it is interesting to note that at least two temperate boreal upland species (*Drosera anglica* and *Fragaria chiloensis*) probably owe their wide distribution to transportation by birds into high-altitude areas on oceanic islands like Hawaii (Carlquist 1970). No attempt has been made to be inclusive in the formation of the lists in Tables 9.1 and 9.2. In the animal kingdom, the analogue of the strand plants are about twenty-seven species of sea birds, most of which have very wide distributions in the Pacific area; all also reproduce within the Hawaiian archipelago. Table 9.1 also recognizes the extraordinary fact that the only clearly native terrestrial mammal known in Hawaii is a slightly variant subspecies of the Hoary bat, a migratory species widespread in the Western Hemisphere, including Bermuda and Galápagos.

In Table 9.1, Ib, a listing has been attempted for large numbers of clearly exotic species found in Hawaii. These differ from the indigenous category in giving every evidence of having been brought to the islands under the influence of man, starting with the arrival of the Polynesians, possibly as early as 200 A.D. Excluding the ferns, mosses and land snails, over 2500 species have been so introduced, including twenty-six species of Drosophilidae (Titus, Carson & Wisotzkey 1985). Recent introductions represent extraordinary natural experiments in the genetics of colonization but the situations have unfortunately not been exploited. The origin of some purposely introduced organisms is precisely known, for example, Hawaiian populations of the small Indian mongoose originated from East Indian stocks brought originally to Jamaica (Tomich 1969).

Perhaps the best-studied colonizing species around the world are the domesticated and widespread species of Drosophila (Carson 1965). Parsons & Stanley (1981) have reviewed the various studies that attempt to characterize different populations of some of these species. Although they find evidence that some species (e.g. *D. melanogaster*) are characterized by local temperature adaptation, the differences are not large. Indeed, like other colonizing species, the Drosophila species do not appear to be undergoing systematic differentiation as they are spread over the globe. Some of the basic facts are given in Table 9.3, wherein most of the 'weed' species of Drosophila are listed. Omitted are a number of species showing similar tendencies but which have not yet been the object of genetic investigations.

Like the strand plants, there is no evidence that the great number of isolated populations, and the founder effects that could be inferred as important in their origin, have produced anything resembling subspecific or specific differentiation. Like all cross-fertilizing diploid species, these

TABLE 9.3. Some widespread species of Drosophila that manifest only minor genetic variation after colonization

Species	Distribution	Genetic differentiation
melanogaster	World-wide in temperate and tropical zones	Clines in inversion and alcohol dehydrogenase gene frequencies.
simulans	Same as above	Chromosomally monomorphic; no clinal frequencies.
ananassae	World-wide in tropics	Ubiquitous inversions that change little in geographic frequency.
busckii	World-wide in temperate and tropical zones	Genetic variation tends to be ubiquitous with minor frequency differences.
kikkawai	World-wide in tropics	Colour polymorphism and inversions tend not to be locally differentiated.
virilis	World-wide	No significant amount.
hydei	World-wide	One ubiquitous inversion.
repleta	World-wide	No significant variation.
buzzatii	Virtually ubiquitous where *Opuntia* cactus has been transported	Complex balanced polymorphisms; minor local variation.
m. mercatorum	Has colonized upland and cooler tropics	Chromosomally monomorphic.
immigrans	World-wide in temperate and many tropical areas	Balanced inversion polymorphisms not much affected by environment.
funebris	Temperate zones	Colonizes without significant genetic change.

manifest a system of extensive genetic variability based on mutation and meiotic recombination but there is no evidence that significant incipient speciation changes have occurred. They are characterized largely by heterotic balance.

Many of the Drosophila species, like *melanogaster*, *simulans*, *hydei* and *immigrans* are ovipositional generalists; in this regard they resemble the weed species in plants and appear to depend on some sort of general-purpose genotype as suggested by Baker (1965) and Carson (1965). The species differ from some well-known cases like, for example, the wide-spread apomictic clones of the dandelion plant *Taraxacum*. The latter are, like strand plants, very efficient in dispersal yet no directional genetic variation is capable of being generated by selection in local situations. One may suggest that many of the plants listed in Table 9.2 may have various kinds of fixed genotypic situations that preadapt them to dispersal and colonization, in particular widely dispersed habitats.

Drosophila buzzatii (Table 9.3) represents a somewhat different case of preadaptive colonization. This species is adapted to breed in a specific ecological site, namely, the cladodes of species of the prickly-pear cactus, *Opuntia*. Both the host-plant and the Drosophila species originated in the New World; in this case, South America is indicated as the probable site of origin for *D. buzzatii* (see Sene, Pereira & Vilela 1982). As the cactus has been introduced by man into the rest of the world (Europe, Africa, Australia), *D. buzzatii* has apparently been carried along with the plant host and the weedy cactus invasions continue to support large populations (Barker 1982). There is no evidence that *D. buzzatii* is able to maintain populations on any plant host other than cactus. Accordingly, its wide-spread colonization is made possible only because of the invasive and colonizing nature of the plant host. *D. buzzatii* displays phyletic evolution based on complex balanced chromosomal and genic polymorphisms in the colonized areas (Barker 1982). Nevertheless, like the rest of the species listed in Table 9.3, it provides no evidence for incipient speciation; it is apparently a preadapted colonizing species but one whose colonizing ability is based on adaptation to a particular breeding niche. Its colonizing ability, therefore, is somewhat different from that of *melanogaster*, *simulans* and *hydei* which breed in many kinds of fermenting wastes resulting from human activity.

As Baker (1965) has emphasized, colonizing species of the two types mentioned above may manifest a number of different kinds of genetic systems, such as polyploidy, self-fertilization or apomixis. If they are diploid and cross-fertilizing, as the *Drosophila* species are, they may manifest fixed or balanced heterotic genetic systems that are not easily perturbed by inbreeding or recombination. The net result that ensues is that the status quo between ancestral and derived population is basically maintained.

Preadapted colonizers accomplish the expansion of their territories due to efficient adaptations for dispersal and, upon arrival, to the establishment of the organism at the new site. These adaptations are frequently such that repeated colonizations into the same site very probably occur. This may enhance the gene pool of cross-fertilizing species but would be of only minor importance for forms with apomixis or fixed heterotic systems.

The conventional explanation of why preadapted colonizers do not appear to speciate suggests that the efficiency of colonization effectively prevents the accumulation of genetic differences from the source. This view should be tempered by the realization that many genetic systems tend to lock up variation into polymorphisms of various kinds. Accordingly, new data on the genetic systems of colonizers are much to be desired.

Stochastic colonization by an outcrossed founder propagule with only minor preadaptation to the new site: an event conducive to speciation

In contrast to all the cases considered so far are those cases of colonization that appear to spawn the formation of new species endemic to the colonized area. It may again be instructive to consider the biota of isolated oceanic islands, such as the Hawaiian archipelago. Two categories of endemic species are recognizable that might briefly be called non-speciose and speciose (Table 9.1, IIa and b). I will discuss these categories by contrasting two somewhat arbitrary states: some ancestral colonizers have produced only a single endemic species whereas others have undergone exuberant speciation that is also nevertheless clearly traceable to a single original immigrant ancestral species. As will be discussed below, in certain groups the diversification from a single ancestral introduction in Hawaii has been so great that some of the forms have been judged as new endemic genera or even families. A striking case of the latter is the Drepanididae, the extraordinary Hawaiian honeycreepers; these are now known to be derived from the cardueline finches (Sibley & Ahlquist 1983).

Before proceeding to a consideration of exuberant speciating lineages, I wish to discuss the case of the colonizers that appear to form one and only one endemic species. In Table 9.4, I present a list of fifty-eight endemic Hawaiian seed plants that apparently have produced only a

TABLE 9.4. Non-speciose endemic seed plants of the Hawaiian islands

Family	Species	Family	Species
Pandanaceae	*Freycinetia arborea*	Leguminosae	*Sophora chrysophylla*
Gramineae	*Dissochondrus* biflorus*		*Strongylodon ruber*
	Festuca hawaiiensis		*Vicia menziesii*
	Garnotia sandwicensis		*Vigna owahuensis*
	Ischaemum byrone	Euphorbiaceae	*Claoxylon sandwicense*
Cyperaceae	*Cladium leptostachyum*		*Columbrina oppositifolia*
	Oreobolus furcatus		*Neowawraea* phyllanthoides*
	Scleria testacea		*Phyllanthus distichus*
Junaceae	*Luzula hawaiiensis*	Anacardiaceae	*Rhus sandwicensis*
Joinvilleaceae	*Joinvillea ascendens*	Celastraceae	*Perrottetia sandwicensis*
Liliaceae	*Smilax melastomifolia*	Sapindaceae	*Sapindus oahuensis*
Iridaceae	*Sisyrinchium acre*	Rhamnaceae	*Alphitonia ponderosa*
Orchidaceae	*Anoectochilus sandwicensis*	Tiliaceae	*Elaeocarpus bifidus*
	Liparis hawaiiensis	Malvaceae	*Gossypium sandvicense*
	Platanthera holochila	Theaceae	*Eurya sandwicensis*
Urticaceae	*Boehmeria grandis*	Guttiferae	*Hypericum degeneri*
	Hesperocnide sandwicensis	Flacourtiaceae	*Xylosoma hawaiiense*
	Touchardia latifolia	Begoniaceae	*Hillebrandia* sandwicensis*
Chenopodiaceae	*Chenopodium oahuense*		*Peucedanum sandwicense*
Amaranthaceae	*Amaranthus brownei*	Umbelliferae	*Spermolepis hawaiiensis*
Phytolaccaceae	*Phytolacca sandwicensis*	Myrsinaceae	*Embelia pacifica*
Papaveraceae	*Argemone glauca*	Oleaceae	*Osmanthus sandwicensis*
Hydrangeaceae	*Broussaisia* arguta*	Gentianaceae	*Centaurium sabaeoides*
Rosaceae	*Acaena exigua*	Apocynaceae	*Rauvolfia sandwicensis*
Leguminosae	*Caesalpinia kavaiense*	Convolvulaceae	*Bonamia menziesii*
	Canavalia galeata	Cuscutaceae	*Cuscuta sandwichiana*
	Erythrina sandwicensis	Hydrophyllaceae	*Nama sandwicensis*
	Senna gaudichaudii	Cucurbitaceae	*Sarx alba*
	Sesbania tomentosa	Compositae	*Gnaphalium sandwicensium*

Listed are fifty-eight endemic Hawaiian species. The ancestor of each apparently arrived by long-distance dispersal, formed a single species and has remained as a single species integrated into native Hawaiian ecosystems. Endemic subspecies and varieties as well as species introduced by man are omitted. Asterisks(*) indicate a monotypic endemic genus. Data from St. John (1973) and unpublished list derived from *Manual of The Flowering Plants of Hawaii* by W.L. Wagner, D. Herbst and S.H. Sohmer (in prep.).

single species after colonization. In these cases, the genus is generally widespread in the Pacific if not in the world at large. In four cases (marked with asterisks in Table 9.4), the divergence from the putative ancestor has resulted in the erection of an endemic genus which, of course, is monotypic.

All of the plants listed appear to have arrived in the islands by natural, long-distance dispersal. All are integrated into native Hawaiian ecosystems and quite a few are found on several islands of different ages in the archipelago. Others are confined to single islands. Some of the species have been split into subspecies, varieties or forms. In this and the speciose plant list in Table 9.6, I have not taken such variations into account. As a geneticist, I am not surprised at this manifestation of intraspecific variation and I have taken the systematicist's judgement as a guide to whether or not the species level has been reached. The main reason for preparing the list is to provide a strong contrast between these non-speciose forms and those in the speciose genera to be discussed shortly.

In Table 9.5, I present a similar list for forty-two Hawaiian insects, although recognition of a single non-speciose endemic form is considerably more difficult than it is in plants. It will be noted in Table 9.5 that many of the single species listed, especially in the Coleoptera, have been placed by the systematicists into endemic genera. This can mean either one of two things. First, there may have been extensive phyletic evolution in the single specific descendent arising from the ancestral immigrant form. This tempts the systematist to erect a monotypic genus. Secondly, as will be pointed out in the discussion of Table 9.6, there are cases (e.g. Drepanididae or Drosophilidae) where some of the products of a burst of speciation are so bizarre as to lead to the erection of separate genera to encompass each single highly divergent species. Such species, of course, are better interpreted as evidence for the existence of exuberant speciosity after colonization rather than the opposite. Accordingly, a special attempt has been made to exclude such species from the list given in Table 9.5 and to include only those that have formed only a single species following a single immigrant ancestral colonization.

As mentioned in Table 9.1, six species of birds and one mammal, the Hawaiian monk seal, have been placed in the non-speciose category. The recent discovery of an enormous fossil Hawaiian avifauna (Olson & James 1982) has made it necessary to reduce the 'single endemic species' bird list since some groups now represented by single species in the islands were clearly more speciose in the recent past. The category still appears to remain valid, however.

Finally, we come to the exuberant speciating groups for which the Hawaiian biota is famous. As examples, I present in Table 9.6 a list of seventeen plant genera and twenty-one animal genera that are especially speciose. The actual numbers of species should not be taken very seriously. By and large the numbers are based on estimates taken from Zimmerman's 1948 monograph. Modern revisions, many of which are still in progress, will serve to increase the number of species in some genera (e.g.

TABLE 9.5. Non-speciose endemic insects of the Hawaiian Islands

Order	Family	Taxa
Odonata		*Sympetrium blackburni*
		Anax strenuus
Thysanoptera		*Conocephalothrips* tricolor*
Heteroptera		*Nesenicocephalus* hawaiiensis*
		Halobates hawaiiensis
		Speovelia aaa
		Kalania hawaiiensis*
Coleoptera	Anthribidae	*Araeocerus constans*
	Anobiidae	*Tricorynus sharpi*
	Cantharidae	*Caccodes debilis*
	Cerambycidae	2 species in 2 indigenous genera
	Clambidae	*Clambus octobris*
	Colydiidae	*Antilissus* aper*
		3 species in 3 indigenous genera
	Curcilionidae	*Dryotribodes littoralis*
		3 species in 3 endemic genera*
	Discolomidae	*Fallia elongata*
	Dytiscidae	2 species in 2 indigenous genera
	Elateridae	*Anchastus swezeyi*
		Dacnitus currax*
	Erotylidae	*Eidoreus* minutus*
	Hydrophilidae	*Omicrus* brevipes*
	Lucanidae	*Apterocyclus* honoluluensis*
	Melasidae	*Ceratotaxia* tristis*
	Orthoperidae	5 species in 5 indigenous genera
	Propalticidae	*Propalticus oculatus*
	Scaphidiidae	*Scaphisoma perkinsi*
	Staphylinidae	2 species in 2 indigenous genera
Lepidoptera		*Vanessa tameamea*
		Vaga blackburni
		Tinostoma smargditis*

Listed are forty-two endemic Hawaiian species. The ancestor of each apparently arrived by long-distance dispersal, formed a single species and has remained a single species integrated into native Hawaiian ecosystems. Endemic subspecies and varieties as well as species introduced by man are omitted. Asterisks(*) indicate monotypic endemic genera (n = 13). Data from Zimmerman (1948) and unpublished lists of G.A. Samuelson (Coleoptera) and W. Gagné.

Drosophila) or reduce the number in others (e.g. Cyrtandra). The 'over fifteen endemic species in a single genus' criterion was adopted in order to take a conservative view. Unfortunately, however, it makes it less easy to observe the remarkable evolution that has occurred in certain groups, e.g. the Hawaiian tarweeds (Compositae–Madiinae; Carr & Kyhos

TABLE 9.6. Speciose endemic seed plants and insects of the Hawaiian Islands

Family	Genus of seed plant	Approximate number of species	Order	Family	Genus of insect	Approximate number of species
Gramineae	Panicum	30	Homoptera		Nesophosyne*	62
Cyperaceae	Cyperus	15			Nesosydne*	82
Piperaceae	Peperomia	47			Oliarus*	84
Carophyllaceae	Schiedea*	27			Carposina*	40
Rutaceae	Pelea	80	Lepidoptera		Scotorythra*	36
Euphorbiaceae	Euphorbia	16			Scoparia	64
Thymeliaceae	Wikstroemia	29			Hyposmocoma*	350
Labiatae	Phyllostegia*	25	Coleoptera	Carabidae	Mecyclothorax	85
	Stenogyne*	30		Anobiidae	Mirosternus*	70
Gesneriaceae	Cyrtandra	167			Xyletobius*	56
Rubiaceae	Coprosma	18		Cerambycidae	Plagithmysus*	136
	Hedyotis	21		Curcilionidae	Oodemas*	58
Lobeliaceae	Clermontia*	40		Proterhinidae	Proterhinus	159
	Cyanea*	81	Hymenoptera		Eupelmus	54
Compositae	Bidens	45			Hylaeus	55
	Lipochaeta*	28			Sierola	182
	Dubautia*	25			Odynerus	105
			Diptera		Campsicnemus	49
					Lispocephala	38
					Drosophila	359
					Scaptomyza	132
	Total: Genera	17			Total: Genera	21
	Species	724			Species	2256
	Mean No. Species/Genus	43			Mean No. Species/Genus	107

Listed at left are twenty-one selected plant genera having at least fifteen endemic species. (Data from St. John (1973) and *Manual of the Flowering Plants of Hawaii* by W.L. Wagner, D. Herbst & S.H. Sohmer (in prep.). Twenty-one insect genera having thirty-five or more endemic species are listed on the right. The Coleopteran Family Nitidulidae (not included in the list) is represented by thirteen endemic genera and 134 species. It appears to relate to a single immigrant ancestor. Data from Zimmerman *et al.* (1948–81) and G.A. Samuelson (Coleoptera, pers.

1981). Although a single ancestral introduction seems clear, the resulting plants are so extraordinarily different from one another as to have traditionally elicited the establishment of two genera (Argyroxiphium and Wilkesia), each with a number of species in addition to the genus listed in Table 9.6 (Dubautia). In the Hawaiian honeycreepers, there are about twenty-five species included in about ten genera.

In perusing the insect list (right-hand columns of Table 9.6), it may seem that the well-known exuberance of speciation in Drosophila is virtually matched by the moth genus Hyposmocoma. Additionally, there are many other insect groups with profuse speciation. For most of these genera, information does not exist beyond the basic alpha taxonomy. Prior to the initiation of genetic studies on Hawaiian Drosophilidae, nine genera of endemic forms were recognized. Based on evidence from genetics, internal morphology and male genitalia, Hardy & Kaneshiro (1981) have referred flies of four of these genera to Drosophila and, in this paper, the tendency to refer the remaining endemic genera to Scaptomyza has been followed (Table 9.6). The difficulties of deciding about whether or not to erect endemic genera is well illustrated by this case. There has been general concensus that, at most, there have been no more than two ancestral colonizations for the Hawaiian Drosophilidae, one ancestral to Scaptomyza and one to Drosophila. The situation is complicated, however, by the existence of some endemic Hawaiian species showing conditions intermediate between the two, suggesting a single ancestor for both Drosophila and Scaptomyza. As both of these genera are widespread around the world, the single-ancestor hypothesis suggests that one genus (Scaptomyza) may have originated in Hawaii and then colonized the rest of the world (Throckmorton 1966). No one, however, has gone so far as to suggest that the Hawaiian endemic Drosophilidae be reduced to a single genus.

Up to this point, colonization has been considered from the point of view of the archipelago as a whole. In the case of one group, the 'picture-winged' Drosophilas, it has been possible, using the giant polytene chromosomes, to establish putative inter-island ancestries. Summarizing the work on about 103 of these species, Carson (1983) has inferred the existence of forty-five inter-island colonizing events. Almost all of these colonizations have involved the formation of one or more new species so that the inter-island considerations permit a discussion on a smaller scale of the same phenomena that have been invoked for the long-distance colonizations involving the establishment of forms of life derived from populations from outside the archipelago.

These events are most clearly seen when considering the newest island (the 'Big Island' of Hawaii), which is less than 0.4 million years in age (Carson 1984). Of the twenty-six species present, twenty-five are clearly unique to the island; despite some behavioural and fixed chromosomal differences, the questionable species is possibly conspecific with a morphologically similar population on the nearby island of Maui.

With regard to the theme of this chapter, i.e. non-speciose vs. speciose lineages, the facts of the situation on the Big Island are of interest. About half (thirteen) of the picture-winged species found there are single endemic species; nine of these are closely related to known species on Maui. The rest have formed either two species (five cases) or three (one case). Although several of the species pairs could be the result of separate founders, the three-species case (*basisetae*, *pancipuncta* and *prolaticilia*) are chromosomally homosequential and otherwise very similar, suggesting a single, geologically very recent, ancestor.

DISCUSSION

Examples taken from Hawaii suggest that evolutionary processes have produced two kinds of endemic species. First there are those that, following colonization by an ancestral immigrant, are represented by a single species only (non-speciose lineages). Secondly, there are those that have produced two or more and sometimes as many as hundreds of species. In this discussion a genetically-based hypothesis relating to the speciation process in general will be advanced in partial explanation of this difference.

Before presenting this hypothesis, however, I would like to mention the more conventional explanations for the existence of speciose and non-speciose lineages. These are basically ecological and have two aspects. First, if an arriving competent propagule is to produce a speciose lineage, there must be nearby open niches in the ecosystem in which the species finds itself. A propagule landing in an ancient continental rain forest, if lucky enough to become established, is unlikely to spawn a speciose lineage. Secondly, as a subpopulation occupies a new niche there must be sufficient spatial isolation so that the evolving subpopulation is not inundated by gene flow from the main body of the ancestral species.

Without denying the importance of these ecological perspectives, as a geneticist interested in how speciation occurs, I see a number of genetical factors that constrain the process of speciation. First, the population produced by the propagule must be capable of generating a field of genetic variability. This means, as suggested earlier in this paper, that if self-fertilization, novel polyploidy or apomixis characterizes a propagule, the latter will be rendered incapable of generating a population with the necessary field of variability. Thus, the future evolutionary potential of the organism is limited. Accordingly, speciating forms will tend to have diploid, cross-fertilizing genetic systems.

In recent discussions (e.g. Carson 1982b), I have proposed that whereas the processes leading to the production of adaptations are intrapopulational (phyletic), the formation of a species is a process that requires the permanent splitting of a population into daughter populations. The demands of selection for Darwinian fitness within a population elicit a high degree of complexity in the organization of the genome, which becomes an increasingly complex set of heterotic balances of pleiotropic interacting genes, most of which have minor individual effects.

In view of this, I have proposed (Carson 1982b) that, whereas intraspecific phyletic evolution can produce forms, varieties, minor adapted ecotypes and clines, some powerful disorganizing force appears to be necessary to set the stage for the formation of a new species. In other words, an old genetic edifice must be torn down before a new one can be built. It has been suggested that the disorganization phase can be accomplished in one of several ways. Particularly relevant to colonization of oceanic islands is the founder effect, in which the new population may be founded from one or a very few propagules. Also contributing to disorganization would be the formation of hybrid swarms or the reduction of a formerly large population to a small vestige by either a catastrophic or gradual reduction in population size. The latter provides a manner of viewing disorganization from the point of view of vicariance biogeography.

In this view of speciation process, the entity recognized as a new species results from the operation of renewed selection that impinges upon the stochastically-disorganized gene pool. The theory suggests that not all adaptive systems or genetic balances in the genome are disorganized (Carson 1987). Certain ones, especially those recently established in the phylogenetic history of the species, may be especially vulnerable to the disorganization–reorganization cycle. The reorganization process, consisting of the reimposition of natural selection on a particular character set, may occur rapidly in what is sometimes called quantum-phase evolution. It is likely to impinge on only one character set at a time. For example, some exuberant speciating lineages repeatedly revise the system of sexual reproduction through novel adaptations that are accomplished through shifts in the genetic basis of factors contributing to the union of the gametes. The characters involved, for example, may range from those pertaining to sexual selection to the alteration of pollination mechanisms.

Other lineages may result in quantum-phase alterations to adaptations serving response to the ambient environment (soil, temperature, light, desiccation, nutritional sources, etc.).

Applying this idea to the speciation process that may follow a colonization, I suggest that those colonizers that formed only a single species have indeed undergone only *a single* disorganization–reorganization cycle, namely that which occurred at the very first founding of the species by the ancestral immigrant form. The fact that the lineage is species-poor will not affect the profoundness or degree of adaptation achieved by the operation of selection in their populations. Indeed, many of the insects listed in Table 9.5 that belong to monotypic endemic genera are among the most remarkable products of insular evolution.

Unlike the above, the speciose forms such as Drosophila, Hyposmocoma, Cyanaea or Pelea may have undergone repeated disorganization–reorganization cycles which are responsible for their proliferation of recognizably different species. Between phases of disorganization, each of these populations, in turn, may be able to produce novel character sets by ordinary intraspecific phyletic evolution based on mutation, recombination and selection.

In my view, the great exuberance of speciation in the Hawaiian drosophilids has been conditioned by continual disorganization– reorganization cycles that especially affect the sexual phenotypes, particularly characters that serve males in combat and courtship. Thus, the evidence indicates that sexual selection is very important in their populations, tending to build into a complex system that can nevertheless be disorganized at some future founder event. This appears to explain the extraordinary proliferation of secondary sexual characters, particularly in males, of these species.

The geological history of the islands, involving the successive appearance of new volcanoes in the sea in a linear succession seems particularly suited to founder evolution. That not all forms proliferate species actively suggests that their initial ecological situation, their dispersal abilities or the tightness of their adaptations tend to inhibit the onset of disorganization. In fact, as time goes on, one would expect their genomes to become more and more committed by selection to special ecological situations and to the genetic balances that underlie the adaptations. Very little is known of the genetic architecture of many of these species. Future research could well attempt to characterize the genetic system of non-speciators, facilitating a direct comparison with those forms, often closely related to them, that far outstrip them in the formation of species.

ACKNOWLEDGMENTS

I am particularly grateful to my colleagues at The Bishop Museum, Honolulu, who provided valuable help in the preparation of the tables. They are not, however, responsible for any errors of omission or commission on my part. Thus, I am grateful to Drs D. Herbst and W.L. Wagner for help with the plant lists, to Drs Allen Allison, W. Gagné and G.A. Samuelson for preparing lists of animals for my use. The research on Hawaiian Drosophila is supported by grant BSR84-15633 from the National Science Foundation.

REFERENCES

Allard, R.W., Kahler, A.L. & Clegg, M.T. (1975). Isozymes in plant population genetics. *Isozymes IV. Genetics and Evolution* (Ed. by C.L. Markert), pp. 261–72. Academic Press, New York.

Baker, H.G. (1965). Characteristics and modes of origin of weeds. *The Genetics of Colonizing Species* (Ed. by H.G. Baker & G.L. Stebbins), pp. 147–72. Academic Press, New York.

Barker, J.S.F. (1982). Population genetics of *Opuntia* breeding *Drosophila* in Australia. *Ecological Genetics and Evolution* (Ed. by J.S.F. Barker & W.T. Starmer), pp. 209–24. Academic Press, New York.

Carlquist, S. (1970). *Hawaii: A Natural History*. Natural History Press, Garden City, New York.

Carr, G.C. & Kyhos, D.W. (1981). Adaptive radiation in the Hawaiian Silversword alliance (Compositae–Madiinae). I. Cytogenetics of spontaneous hybrids. *Evolution*, **35**, 543–56.

Carson, H.L. (1965). Chromosomal morphism in geographically widespread species of Drosophila. *The Genetics of Colonizing Species* (Ed. by H.G. Baker & G.L. Stebbins), pp. 503–31. Academic Press, New York.

Carson, H.L. (1982a). Evolution of Drosophila on the newer Hawaiian volcanoes. *Heredity*, **48**, 3–25.

Carson, H.L. (1982b). Speciation as a major reorganization of polygenic balances. *Mechanisms of Speciation* (Ed. by C. Barigozzi), pp. 411–433. Alan R. Liss, New York.

Carson, H.L. (1983). Chromosomal sequences and interisland colonizations in Hawaiian Drosophila. *Genetics*, **103**, 465–82.

Carson, H.L. (1984). Speciation and the founder effect on a new oceanic island. *Biogeography of the Tropical Pacific* (Ed. by F.J. Radovsky, P.H. Raven & S.H. Sohmer), pp. 43–54. B.P. Bishop Museum, Honolulu.

Carson, H.L. (1987). Genetic imbalance, realigned selection and the origin of species. *Genetics, Speciation and the Founder Principle* (Ed. by L.V. Giddings, K.Y. Kaneshiro & W.W. Anderson) (in press).

Carson, H.L., Hardy, D.E., Spieth, H.T. & Stone, W.S. (1970). The evolutionary biology of the Hawaiian Drosophilidae. *Essays in Evolution and Genetics in Honor of Theodosius Dobzhansky* (Ed. by M.K. Hecht & W.C. Steere), pp. 437–543. Appleton-Century-Crofts, New York.

Carson, H.L. & Kaneshiro, K.Y. (1976). Drosophila of Hawaii: systematics and ecological genetics. *Annual Review of Ecology and Systematics*, **7**, 311–46.

Carson, H.L. & Yoon, J.S. (1982). Genetics and evolution of Hawaiian Drosophila. *The Genetics and Biology of Drosophila 3b* (Ed. by M. Ashburner, H.L. Carson & J.N. Thompson), pp. 297–344. Academic Press, New York.

Dalrymple, G.C., Silver, E.A. & Jackson, E.D. (1973). Origin of the Hawaiian Islands. *American Scientist*, **61**, 294–308.

Darwin, C. (1873). *The Voyage of H.M.S. Beagle*. D. Appleton Co., New York.

Hardy, D.E. & Kaneshiro, K.Y. (1981). Drosophilidae of Pacific Oceania. *The Genetics and Biology of Drosophila 3a* (Ed. by M. Ashburner, H.L. Carson & J.N. Thompson), pp. 309–347. Academic Press, New York.

Hubbell, T.H. (1968). The Biology of Islands. *Proceedings of the National Academy of Science USA*, **60**, 22–32.

Hunt, J.A. & Carson, H.L. (1983). Evolutionary relationships of four species of Hawaiian Drosophila as measured by DNA reassociation. *Genetics*, **104**, 353–64.

Mueller-Dombois, D., Bridges, K.W. & Carson, H.L. (1981). *Island Ecosystems*. Hutchinson Ross Publ. Co., Stroudsburg, Pa.

Olson, S.L. & James, H.F. (1982). Fossil birds for the Hawaiian Islands: Evidence for Wholesale Extinction by Man Before Western Contact. *Science*, **217**, 633–35.

Parsons, P.A. & Stanley, S.M. (1981). Domesticated and widespread species. *The Genetics and Biology of Drosophila 3a* (Ed. by M. Ashburner, H.L. Carson & J.N. Thompson) pp. 349–393. Academic Press, New York.

Safriel, U.N. & Ritte, U. (1983). Universal correlates of colonizing ability. *The Ecology of Animal Movement* (Ed. by I.R. Swingland & P.J. Greenwood), pp. 213–239. Clarendon Press, Oxford.

Sene, F. de M., Pereira, M.A.Q.R. & Vilela, C.R. (1982). Evolutionary aspects of cactus breeding *Drosophila* species in South America. *Ecological Genetics and Evolution* (Ed. by J.S.F. Barker & W.T. Starmer), pp. 97–106. Academic Press, New York.

Sibley, C.G. & Ahlquist, J.E. (1983). Phylogeny and classification of birds based on the data of DNA–DNA hybridization. *Current Ornithology*, **1**. Plenum Publ. Corp., New York.

St. John, H. (1973). *List and summary of the flowering plants in the Hawaiian Islands*. Pacific Tropical Botanical Garden Memoir No. 1, Lawai, Kauai, Hawaii.

Titus, E.A., Carson, H.L. & Wisotzkey, R.G. (1985). Another new arrival to the Hawaiian Islands: Drosophila bryani Malloch. *Drosophila Information Service 61*, 171.

Throckmorton, L.H. (1966). The relationships of the endemic Hawaiian Drosophilidae. *University of Texas Publication*, **6615**, 335–96.

Tomich, P.Q. (1969). *Mammals in Hawaii*. Publ. No. 57, Bishop Museum Press, Honolulu.

Zimmerman, E.C. (1948–81). *Insects of Hawaii*. Vols 1–14. University of Hawaii Press, Honolulu.

10. COLONIZATION, SUCCESSION AND RESOURCE AVAILABILITY: ECOSYSTEM-LEVEL INTERACTIONS

PETER M. VITOUSEK AND LAWRENCE R. WALKER
Department of Biological Sciences, Stanford University, Stanford, California 94305, USA

INTRODUCTION

When Kilauea Volcano in Hawaii erupts, lava frequently fountains to over 300 m in the air. Wind-borne material covers extensive areas with several metres of tephra (volcanic ash), burying the previous communities and creating a new substratum for primary succession. Even where temperatures are moderate and rainfall heavy, colonization is relatively slow (Smathers & Mueller-Dombois 1974); vegetation cover can remain sparse (<10%) 25 years after an eruption.

Tephra is fairly rich in bases and in phosphorus, but it contains no organic matter or fixed nitrogen at the time it is deposited. The total nitrogen content of soil increases gradually with time in soil development until about 4000 kg ha^{-1} are present in 200-year-old sites (Vitousek *et al.* 1983). Nitrogen availability (estimated by release of inorganic nitrogen from incubated soil) remains low for at least 1000 years, and low foliar nitrogen concentrations suggest that the productivity of early successional vegetation is limited by nitrogen (Balakrishnan & Mueller-Dombois 1983; Vitousek *et al.* 1983). Many of the colonizers are native species with low relative growth rates.

Clearing an established Hawaiian rain forest (while leaving the soil intact) sets entirely different dynamics in motion. Colonization is rapid, and the colonizers are predominantly introduced species with rapid growth rates. Resource availability following forest clearing in Hawaii has not been studied systematically, but considerable information has been obtained from comparable studies of forest clearing in premontane rain forests on tephra soils in Costa Rica. Clearing of Central American rain forests involves felling the vegetation early in the dry season, followed by burning 2–3 months later (Ewel *et al.* 1981). Once the rains begin, revegetation is rapid, often leading to a 3–5 m tall closed canopy within months (Harcombe 1977). Large amounts of both total and available nitrogen are present in volcanic-ash derived Costa Rican soils (Berish 1983; Sollins, Spycher & Glassman 1984). Nitrogen availability further increases to extraordinarily high levels (>100 kg ha^{-1} $month^{-1}$) during the 6 months immediately following clearing and burning, then returns to background levels (Matson *et al.* in press). This brief pulse of available nitrogen is associated with the establishment and growth of nutrient-demanding, often non-mycorrhizal species with high relative growth rates.

Patterns of nitrogen availability during primary succession on tephra in Hawaii are contrasted with those for cleared and burned sites on tephra-derived soils in Costa Rica in Fig. 10.1. The direction, the magnitude, and the time-scale of changes are strikingly different. What controls these differences? What are the consequences for colonizing species? To what extent can colonizing species determine or alter these patterns?

SEMANTICS

Discussions of succession and resource availability generally involve semantic difficulties. We would like to address two potential problems at the beginning: the distinction between primary and secondary succession and that between resource supply and resource demand.

The distinction between primary and secondary succession has long been with us (Clements 1916), and results such as those in Fig. 10.1

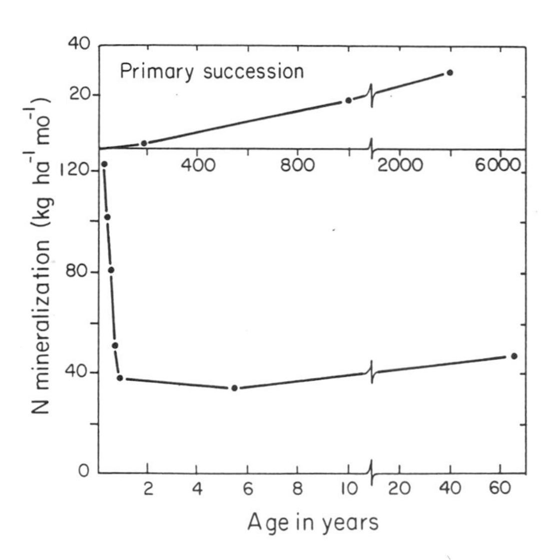

FIG. 10.1. Net nitrogen mineralization during the course of primary (above) and secondary succession on tephra-derived soils. Results for the primary sere on Kilauea Volcano in Hawaii from Vitousek *et al.* (1983); for the secondary sere at Turrialba in Costa Rica from Matson *et al.* (in press).

demonstrate its utility. As originally defined, primary succession was distinguished by the absence of 'reaction' (to the actions of organisms) in the soil at the start of succession. In practice, that meant the absence of soil organic matter that had developed *in situ*. Other definitions have been used (cf. Clements 1928; Lawrence 1979; Miles 1979 and Chapter 1), but we see no reason to prefer them to the original. It must be recognized, however, that many seres are intermediate in character between primary and secondary (Miles 1979). While some are clearly primary (volcanic ash deposits and glacial moraines) and some are clearly secondary (forest harvesting and large-scale windstorms), the dynamics of others can only be obscured by forcing each into a rigid classification. Seres following the most intensive fires, those on long-degraded agricultural land, and perhaps those on river bars fall into the intermediate category. These seres develop in substrates that have some organic matter in the soil at the start, but less than a full complement.

Discussions of resource availability during colonization and succession frequently encounter the difficulty that plants' and microbes' requirements for resources vary systematically with successional time. Accordingly, organisms can develop resource limitations, or begin to compete strongly for resources, even where resource supply is constant. We will consider 'resource availability' to be synonymous with 'resource supply' or 'resource flux density'; all represent the amount of a given resource that enters into available forms in a given period of time.

In practice, it is relatively easy to measure the supply of certain resources and relatively difficult to measure others. Light availability can be determined straightforwardly as the photosynthetically-active photon flux density (PPFD) above a plant canopy. Water is more difficult; inputs are nearly independent of succession (save for situations where a substantial component of horizontal interception is present, cf. Lovett, Reiners & Olson 1982), but soil water holding capacity depends on the texture of soil and on the amount of soil organic matter present. Soil profiles in primary succession generally develop finer soil textures, increased soil organic matter, and therefore increased water holding capacity through time on glacial moraines (Crocker & Major 1955; Viereck 1966), dunes (Olson 1958), volcanic substrates (Tezuka 1961; Vitousek *et al.* 1983), and river floodplains (Walker 1985).

Nitrogen availability is still more difficult. We will define nitrogen availability as the net release of inorganic nitrogen from organically bound forms (net nitrogen mineralization). This definition accounts for the vast majority of nitrogen that moves into available forms in the soil (Rosswall 1976). It does not account for plants with microbial symbioses capable of fixing atmospheric nitrogen, nor does it differentiate ammonium-nitrogen from nitrate-nitrogen, nor does it deal with the dual role of micro-organisms as the major producers of inorganic nitrogen and as its most important consumers (Rosswall 1976). Finally, it neglects the ways in which nitrogen demand and nitrogen supply can interact, feeding back to alter nitrogen supply either positively or negatively—but that will be a major point of this chapter.

The distinction between resource supply and resource demand breaks down when lithophilic (rock- or soil-derived) nutrients are considered. The biological availability of phosphorus and the major cations is regulated by the weathering of primary and secondary minerals and adsorption/desorption reactions in addition to release from organically bound forms. Rates of weathering and/or desorption can be greatly altered by biological activity (Gorham, Vitousek & Reiners 1979; Uehara & Gillman 1981), and can vary systematically during primary succession.

We believe, nonetheless, that it is a useful first step to consider resource supply in isolation from resource demand, and to consider primary succession as wholly distinct from secondary succession. The intermediate situations can be dealt with more clearly once the nature of the extremes is understood.

PRIMARY SUCCESSION

Resource availability

There is a rich body of theory designed to explain patterns of resource availability in primary succession. Work in this area is long-standing (Clements 1916; Cooper 1926), but it has received its most satisfying treatment recently from T.W. Walker and his colleagues (Stevens & Walker 1970; Adams & Walker 1975; Walker & Syers 1976). Their research has emphasized the significance of phosphorus in primary succession, an emphasis that has developed in part because phosphorus, with nitrogen, is in short supply relative to the requirements of organisms in many ecosystems, and in part because phosphorus availability in soils essentially runs down through the course of soil development. Unlike nitrogen, any phosphorus lost from a site cannot be replaced from an inexhaustible atmospheric pool.

Walker & Syers (1976) pointed out that at the beginning of primary succession, most of the phosphorus in the parent material is within primary minerals, especially apatite. Under the influence of water and hydrogen ions, these primary minerals break down to release phosphate in biologically available forms. Much of this P is taken up by organisms, utilized, and eventually returned to the soil as ester-bonded organic phosphorus (P_o). Microbial activity releases biologically available phosphate from P_o. Most of this phosphate is then taken up by organisms or reversibly adsorbed; a small fraction of it is eroded, leached, or irreversibly adsorbed in the soil. Consequently, after many years and many passages through the biota, the total amount of phosphorus in a soil can decline, and the fraction of the remaining phosphorus which is in biologically available forms can decrease to trace amounts. Walker & Syers (1976) suggested that these processes can ultimately lead to a 'terminal steady state' of phosphorus deficiency, low productivity, and a very low capacity for recovery from disturbance; similar observations have been made on geomorphologically old soils in other regions (B.H. Walker *et al.* 1981; J. Walker *et al.* 1981).

These changes in the phosphorus cycle during primary succession are summarized in Fig. 10.2. A similar decline in element availability during primary succession could be anticipated for most of the metallic elements (Gorham, Vitousek & Reiners 1979), although the importance of inorganic adsorption and organic compounds (if any) would certainly differ. Phosphorus is more important in regulating biotic processes and is held more conservatively within soil than most other such elements.

Walker & Syers (1976) concluded that the pattern for phosphorus (Fig. 10.2) has implications for nitrogen availability as well. Most primary substrates start with virtually no nitrogen and with the maximum amount of phosphorus that they will ever have. The relative importance of atmospheric deposition versus weathering in supplying phosphorus and other lithophilic elements to primary seres has been inadequately studied, however, and it is possible that atmospheric deposition of phosphorus may be important during primary succession on sand dunes (Kellman 1985).

Nitrogen-fixing organisms require large amounts of phosphorus, and they should have a substantial competitive advantage early in most primary seres. In fact, primary seres are often dominated by plants with nitrogen-fixing symbioses at some point early in primary succession (Stevens & Walker 1970; Gorham, Vitousek & Reiners 1979). A general pattern for nitrogen availability in primary succession might therefore appear as in Fig. 10.3. This curve suggests a more rapid increase in nitrogen availability than that observed in the Hawaiian tephra sere (Fig. 10.1), reflecting the absence of any symbiotic nitrogen-fixer from that sere.

The patterns summarized by Walker & Syers (1976) were based on

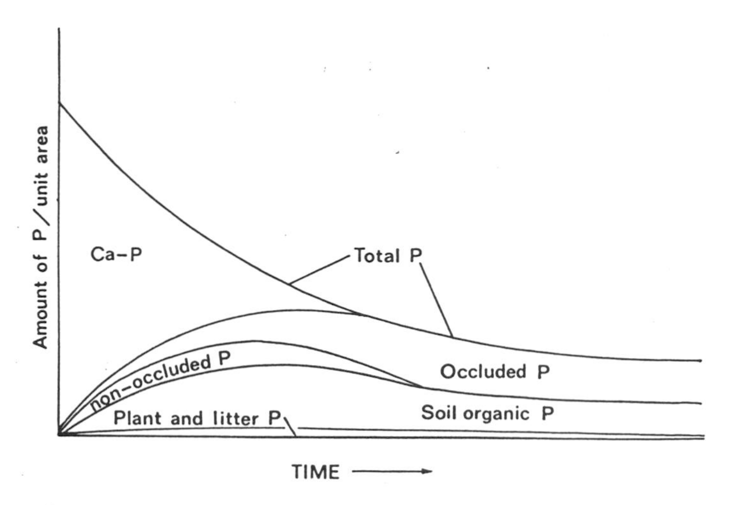

FIG. 10.2. Generalized pattern of change in total phosphorus (top line) and soil phosphorus fractions during soil development. Redrawn from Walker & Syers (1976).

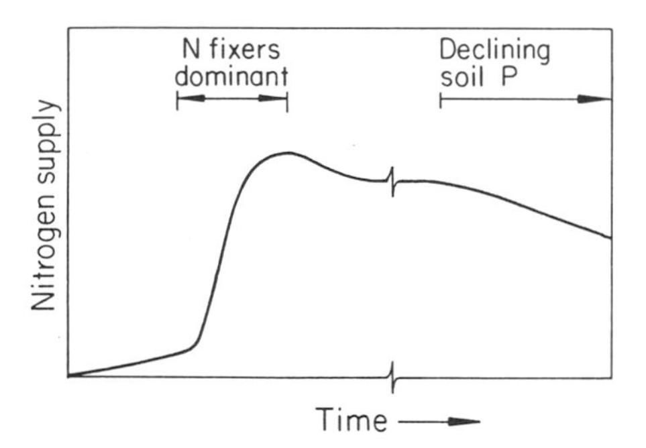

FIG. 10.3. Hypothetical pattern for nitrogen supply (= net nitrogen mineralization) during soil development in a primary sere containing a stage dominated by symbiotic nitrogen-fixers.

extensive chemical fractionation studies of four soil chronosequences in New Zealand. A more recent study in Australia (J. Walker *et al.* 1981) suggested similar patterns of phosphorus availability with primary succession but a rather different mechanism. They reported that most soil phosphorus remained on-site, but that it was transported to a deep B-horizon in the soil with which plants eventually lost contact.

McGill & Cole (1981) extended the Walker & Syers' model by using a more detailed consideration of the chemical forms of organically bound nutrients in the soil. They demonstrated that most of the organic phosphorus and much of the organic sulphur in soil is ester-bonded, while most of the organic nitrogen and some of the sulphur is bonded directly to organic carbon. The ester-bonded forms are subject to cleavage by extracellular enzymes, which can be produced in response to phosphorus or sulphur deficiency. The organic compound containing carbon-bonded elements must be broken down by decomposers in order for these nutrients to be released.

This difference suggests that release of organic phosphorus into biologically available phosphate can be more or less synchronized with microbial, mycorrhizal, and perhaps plant requirements. No such mechanism can synchronize nitrogen supply and demand: in colder areas where decomposition is slow, nitrogen cycling is also delayed. This distinction may explain why phosphorus deficiencies often occur late in primary succession, especially in tropical or subtropical areas, whereas nitrogen is often limiting at all stages of primary succession in temperate and boreal regions (McGill & Cole 1981).

Implications for colonization and succession

The most obvious implications of these patterns is that plants with nitrogen-fixing symbioses should have a strong competitive advantage over non-fixers at some point early in primary succession. Nitrogen-fixing organisms accumulate fixed nitrogen within a site, and it seems reasonable that in doing so they facilitate the growth of non-fixers later in succession. Direct evidence for such facilitation is lacking in primary seres, however, although it can be inferred from some studies of land reclamation (Gadgil 1971; Palaniappan, Marrs & Bradshaw 1979) and secondary succession (Tarrant & Trappe 1971; Binkley, Lousier & Cromack 1984). Moreover, there is contrary evidence which suggests that rapidly growing nitrogen-fixers may inhibit establishment and growth of other species, particularly more slowly growing late successional species (Walker & Chapin, in press). Nonetheless, the nitrogen-fixing stage which commonly occurs in primary succession often adds the majority of the nitrogen present in later successional stages. It must be important in long-term soil development, and the growth of later colonizers is presumably enhanced as a consequence of improved soil nutrition (Reiners 1981; Bolin *et al.* 1983).

A second implication of the patterns of resource availability described above (Figs 10.2, 10.3) is that the earliest colonizers in primary substrates must be tolerant of extreme environmental conditions (Vitousek & White 1981; Finegan 1984). They are likely to have more than adequate lithophilic nutrients and light: PPFD above the canopy is independent of successional age, and cover is generally extremely sparse in early succession. Water input is also largely independent of succession, although fog interception may increase. However, the water-holding capacity of soils is likely to be as low as it will ever be, because physical weathering and pedogenic clay formation make soils finer in texture, and soil organic matter accumulates during primary succession. Finally, the supply of available nitrogen (to a non-fixer) is extremely low. Colonization by plants with inherently low relative growth rates and slow rates of turnover might be anticipated (Grime 1979; Chapin 1980), in contrast with the predictions of Bazzaz (1979) (which were developed for secondary succession).

It is interesting that despite the very low levels of nitrogen early in primary seres, the initial colonizers are often non-fixers (although nitrogen-fixers are frequently the first species to form stands of continuous cover) (Grubb 1986). For example, the first colonizers on river bars on an Alaskan floodplain are herbs and willows (Van Cleve & Viereck 1981); these are succeeded by nitrogen-fixing alder (*Alnus crispa*). This river-bar sere starts out richer in nitrogen than most primary seres (Van Cleve, Viereck & Schlentner 1971; Walker 1985), but similar patterns are observed in other areas. Dispersal ability and stochastic factors may override patterns of nutrient availability, and lead to the initial establishment of non-fixers (Walker, Zasada, & Chapin 1986). Moreover, the low demand for nitrogen by heterotrophs in sites with very little organic matter, together with the high mobility of nitrate, may allow non-fixing species to persist through the utilization of nitrogen inputs from precipitation or flooding (Van Cleve, Viereck & Schlentner 1971; Grubb 1986).

The final implication that we intend to draw is that late-successional species in very old soils must be able to cope with very low levels of available phosphorus (Walker & Syers 1976; B.H. Walker *et al.* 1981). Phosphorus is nearly immobile in soil (Nye & Tinker 1977), so plants in old, low-phosphorus soils must allocate a large proportion of their energy to roots and/or mycorrhizae. Adaptations such as surface root mats, extremely high phosphorus use efficiency, and even tree-climbing roots are observed in old, low-phosphorus tropical soils (Jordan & Herrera 1981).

SECONDARY SUCCESSION

Resource availability

Somewhat surprisingly, there seems to be less solid mechanism-based theory concerning patterns of resource availability in secondary succession than exists for primary succession. A great deal of effort has gone into understanding nutrient losses and ecosystem-level nutrient budgets following forest harvesting and other disturbances (Likens *et al.* 1970; Bormann & Likens 1979; Vitousek & Melillo 1979; Vitousek *et al.* 1982): well over 100 sites have been studied. As discussed below, however, nutrient losses or nutrient budgets more closely reflect the balance between resource supply and demand than any underlying pattern in resource supply alone.

Light availability does not vary with secondary succession, although the amount reaching the soil surface is certainly greatly increased immediately after disturbance. Water-holding capacity in the soil varies much less in secondary succession that in primary succession, so overall water supply also varies little with succession. Clearing a forest greatly reduces transpiration while increasing evaporation from the soil surface to a lesser extent. Consequently, total water yield to streams is generally increased, the soil surface is drier, and the water table is raised in disturbed sites (Swank & Douglass 1974; Bormann & Likens 1979).

A few studies have reported patterns of nitrogen supply in disturbed sites (Matson & Vitousek 1981; Gordon & Van Cleve 1983; Matson & Boone 1984; Vitousek & Matson 1985) and through secondary seres (Lamb 1980; Robertson & Vitousek 1981; Robertson 1984). Fewer have reported patterns for phosphorus or other elements (Black & Marion 1984; M.J. Mazzarino & J.J. Ewel, pers. comm.). Nitrogen supply (net nitrogen mineralization) is generally increased several-fold shortly after disturbance. This pulse of nitrogen supply can continue at a low level for decades in extremely infertile sites (Matson & Boone 1984), for at least 3 years in other temperate forests (Matson & Vitousek 1981), or for only a few months in the Costa Rican site in Fig. 10.1.

In the absence of feedback from colonizing organisms, it seems unlikely that further changes in nutrient supply would be anticipated once the forest canopy closed. Where succession occurs on degraded agricultural land, however, some increase in nitrogen supply is likely as a result of increases in soil nitrogen capital during the course of succession (Nye & Greenland 1960; Zinke, Sabhasri & Kunstadter 1978). Similar increases would be expected following disturbances in which subsoil is brought to the surface, as occurs in the root-throw zone of treefall gaps (Bazzaz 1983).

Following disturbance and during secondary succession, changes in the resource demands of plants are at least as rapid as changes in resource supply. Shortly after a disturbance, the supply of available nitrogen is increased and demand much reduced (Fig. 10.4). The process of vegetation recovery is associated with a decrease in supply and an increase in demand; demand generally outstrips supply within 5–20 years, and remains elevated as long as the successional forest continues to accumulate biomass and nutrients. Finally, demand may come into a rough equilibrium with supply late in secondary succession (Vitousek & Reiners 1975; Bormann & Likens 1979; Silsbee & Larson 1982); this equilibrium may be mediated by small-scale disturbance within the area studied.

The line for total nutrient supply on Fig. 10.4 represents the amount that could be acquired by organisms; the difference between the lines should indicate the potential magnitude of competition for nitrogen as well as potential nutrient losses from a site (Vitousek & Reiners 1975). In practice, large-scale nutrient losses of the magnitude suggested by Fig. 10.4 are rarely observed in disturbed sites (Vitousek & Melillo 1979). The pattern suggested there should be general, and very high losses are indeed observed in some sites (Likens *et al.* 1970; Wiklander 1981; Berish 1983). The major reason for the lack of extreme nutrient losses in many disturbed sites is the dual role of soil micro-organisms. They break down organic compounds and release nutrients in available forms, but they are also major consumers of available nutrients. This microbial uptake of nutrients (immobilization) can prevent or delay losses of nutrients from disturbed sites.

Microbial immobilization is particularly important in sites where nutrients are in short supply prior to disturbance. Under these conditions, plants utilize nutrients efficiently, and the carbon:nutrient ratio of their residue is high (Vitousek 1982; Melillo & Gosz 1983; Shaver & Melillo 1984; Birk & Vitousek 1986). Microbes may then be nutrient- rather

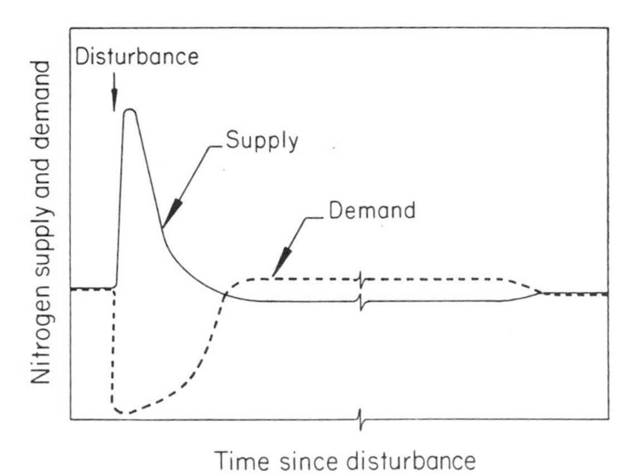

FIG. 10.4. Generalized pattern of change in nitrogen supply in the forest floor and soil (——) and nitrogen demand by plants (----) following disturbance and during secondary forest succession. Redrawn from Vitousek (1984).

than energy-limited, and they are strong competitors for mineralized nutrients (Black 1968). Consequently, nutrient losses can be delayed while micro-organisms utilize much of the pulse of available nutrients in a disturbed site (Vitousek *et al.* 1982). In sites where plants are not limited by a particular nutrient prior to disturbance, plant residue has lower ratios of carbon to that particular nutrient, and microbial immobilization is less (Vitousek 1982).

The importance of microbial uptake in controlling pool sizes of available nitrogen and in preventing large-scale losses of nutrients from disturbed sites was demonstrated in a clearcut area in North Carolina where vegetation regrowth and the amount of organic residue remaining on-site were altered factorially. Removal of organic residue scarcely changed nitrogen supply, but it decreased the immobilization of nitrogen by micro-organisms and led to a large increase in soil nitrate and in nitrogen losses (Vitousek & Matson 1984; 1985).

Implications for colonization and succession

Resource availability is generally at its maximum shortly after disturbance, although soil temperature and moisture cycles on bare ground can inhibit seedling establishment in some sites. Most species which reach open sites grow rapidly (Bazzaz 1979), and dispersal ability and chance elements in establishment seem most likely to control initial composition (Noble & Slatyer 1980; Hils & Vankat 1982). Specific adaptations which tie species life-cycles to high-resource conditions can be observed; for example, a number of early successional species apparently use nitrate levels in the soil as a germination cue (Peterson & Bazzaz 1978; Auchmoody 1979). Despite the dominance of rapidly-growing species, most later successional plants should also be able to make their best growth on newly disturbed sites—if they can get there, and if (once there) they are not suppressed by rapidly growing species (cf. Harcombe 1977).

The rapid growth and high nutrient contents of most early successional pioneers can feed back to affect resource availability at the ecosystem level in a number of ways. The litter of such plants is higher in nitrogen and can be lower in phenolic compounds and lignins than that of later successional plants (Melillo, Aber, & Muratore 1982); it generally decomposes and releases its nutrients more rapidly. Covington (1981) showed that the total amounts of nutrients bound up in the forest floor of a northern hardwoods forest declined for 15 years following clearcutting, despite the fact that productivity and leaf area index reached or exceeded pre-disturbance levels within 4–6 years. High levels of nutrient supply and the chemical composition of the colonizers may thus establish a positive feedback which maintains rapid rates of nutrient cycling and therefore elevated levels of nutrient availability for several years.

Later in succession, plant demand for nutrients generally exceeds the supply in mineralization (Fig. 10.4). Plants can take up slightly more nutrients than are mineralized because they acquire nutrients entering ecosystems through precipitation, rock weathering, and specialized processes such as nitrogen fixation. Nutrient limitation can occur at any stage of succession, but it is likely to be most intense where plant nutrient demand exceeds soil nutrient supply. Where decomposition is slow enough so that both the vegetation and the forest floor are accumulating nutrients simultaneously, the limiting nutrient is very likely to be nitrogen (Bormann & Likens 1979; Miller 1981). This limitation should be most important in low-decomposition sites because nitrogen release is closely coupled to decomposition rather than plant demand (McGill & Cole 1981).

Where potential plant demand for plant nutrients exceeds supply, nutrient use efficiency increases. This increase could come about because of increased retranslocation of nutrients from senescing plant parts (Stachurski & Zimka 1975; Shaver & Melillo 1984), because of increased investment of carbon in nutrient acquisition (Keyes & Grier 1980), because of greater leaf longevity (Small 1972; Chabot & Hicks 1982), or because of greater *in situ* carbon fixation per unit of nutrient in the plant (Birk & Vitousek 1986). Whatever the mechanism, the consequence is higher carbon:nutrient ratios in plant residue (Vitousek 1982). This material can then cause micro-organisms to use nutrients more conservatively, and thereby to decrease nutrient supply to plants. Consequently, a positive feedback towards efficient internal nutrient cycling and low nutrient supply in the soil can be established (Gosz 1981; Vitousek 1982).

In the absence of further large-scale disturbance, an even-aged stand must eventually begin to break up as the first generation trees in the overstorey die (Peet 1981). In special circumstances, this dieback may occur in whole stands (Mueller-Dombois 1983) or in waves (Sprugel 1976; McCauley & Cook 1980); more often it will be a mixture of treefall gaps and small-scale patches (Hartshorn 1978; Reiners & Lang 1979; Denslow 1980). While the large-scale patches clearly enhance nutrient availability in dieback areas (Matson & Boone 1984), it is not clear how strongly small-scale processes such as treefall gaps alter nutrient supply. P.M. Vitousek & J.S. Denslow (in press) found little enhancement of nitrogen supply in the crown-fall zone of treefall gaps in a lowland tropical forest, but background levels of nitrogen supply were so high that any small increase would have been difficult to detect. On the other hand, light levels in gaps were increased fortyfold (Chazdon & Fetcher 1984). Overall, it seems likely that both the mean nutrient supply (on an areal basis) and the variance of that nutrient supply increase later in secondary succession as a consequence of vegetation processes.

ACKNOWLEDGMENTS

We thank P.J. Grubb and P.A. Matson for thoughtful reviews of an earlier draft of this manuscript. The authors were supported by a National Science Foundation grant.

REFERENCES

Adams, J.A. & Walker, T.W. (1975). Some properties of a chronotoposequence of soils from granite in New Zealand, 2. Forms and amounts of phosphorus. *Geoderma*, **13**, 41–52.

Auchmoody, L.R. (1979). Nitrogen fertilization stimulates germination of dormant pin cherry seed. *Canadian Journal of Forest Research*, **9**, 514–516.

Balakrishnan, N. & Mueller-Dombois, D. (1983). Nutrient studies in relation to habitat types and canopy dieback in the montane rainforest ecosystems, Island of Hawaii. *Pacific Science*, **37**, 339–360.

Bazzaz, F.A. (1979). The physiological ecology of plant succession. *Annual Review of Ecology and Systematics*, **10**, 351–371.

Bazzaz, F.A. (1983). Characteristics of populations in relation to disturbance in natural and man-modified ecosystems. *Disturbance and Ecosystems: Components of Response* (Ed. by H.A. Mooney & M. Godron), pp. 259–275. Springer–Verlag, New York.

Berish, C.W. (1983). *Roots, soil, litter, and nutrient changes in simple and diverse tropical successional ecosystems.* Ph.D. thesis, University of Florida.

Binkley, D., Lousier, J.D. & Cromack, K., Jr (1984). Ecosystem effects of Sitka alder in a Douglas-fir plantation. *Forest Science*, **30**, 26–35.

Birk, E.M. & Vitousek, P.M. (1986). Nitrogen availability and nitrogen use efficiency in loblolly pine. *Ecology* **67**, 69–79.

Black, C.A. (1968). *Soil–Plant Relationships.* John Wiley and Sons, New York.

Black, C.H. & Marion, G.M. (1984). Phosphorus availability across a southern California chaparral fire cycle chronosequence. *Bulletin of the Ecological Society of America*, **65**, 271 (Abstract).

Bolin, B., Crutzen, P.J., Vitousek, P.M., Woodmansee, R.G., Goldberg, E.D. & Cook, R.B. (1983). Interactions of biogeochemical cycles. *The Major Biogeochemical Cycles and Their Interactions* (Ed. by B. Bolin & R.B. Cook). SCOPE 21, 1–40, John Wiley & Sons, Chichester.

Bormann, F.H. & Likens, G.E. (1979). *Pattern and Process in a Forested Ecosystem.* Springer–Verlag, New York.

Chabot, B.F. & Hicks, D.J. (1982). The ecology of leaf life spans. *Annual Review of Ecology and Systematics*, **13**, 229–259.

Chapin, F.S., III. (1980). The mineral nutrition of wild plants. *Annual Review of Ecology and Systematics*, **11**, 233–260.

Chazdon, R.L. & Fetcher, N. (1984). Photosynthetic light environments in a lowland rain forest in Costa Rica. *Journal of Ecology*, **72**, 553–564.

Clements, F.E. (1916). *Plant succession: an analysis of the development of vegetation.* Carnegie Institution of Washington Publication, **242**, 1–512.

Clements, F.E. (1928). *Plant Succession and Indicators.* Wilson, New York.

Cooper, W.S. (1926). The fundamentals of vegetation change. *Ecology*, **7**, 391–413.

Covington, W.W. (1981). Changes in forest floor organic matter and nutrient content following clearcutting in northern hardwoods. *Ecology*, **62**, 41–48.

Crocker, R.L. & Major, J. (1955). Soil development in relation to vegetation and surface age at Glacier Bay, Alaska. *Journal of Ecology*, **43**, 427–448.

Denslow, J.S. (1980). Gap partitioning among tropical forest trees in tropical succession. *Biotropica*, **12**, (Suppl.), 47–55.

Ewel, J.J., Berish, C., Brown, B. Price, N. & Raich, J. (1981). Slash and burn impacts on a Costa Rican wet forest site. *Ecology*, **62**, 816–829.

Finegan, G.B. (1984). Forest succession. *Nature*, **312**, 109–114.

Gadgil, R.L. (1971). The nutritional role of *Lupinus arboreus* in coastal sand dune forestry. I. The potential influence of undamaged lupin plants on nitrogen uptake by *Pinus radiata*. *Plant and Soil*, **34**, 357–364.

Gordon, A.M. & Van Cleve, K. (1983). Seasonal patterns of nitrogen mineralization following harvesting in the white spruce forests of interior Alaska. *Resources and Dynamics of Boreal Zone* (Ed. by R.W. Wein, R.R. Pierce & I.R. Methven), pp. 119–130. Association of Canadian Universities for Northern Studies, Sault Sainte Marie, Ontario.

Gorham, E., Vitousek, P.M. & Reiners, W.A. (1979). The regulation of chemical budgets over the course of terrestrial ecosystem succession. *Annual Review of Ecology and Systematics*, **10**, 53–88.

Gosz, J.R. (1981). Nitrogen cycling in coniferous ecosystems. *Terrestrial Nitrogen cycles: Processes, Ecosystem Strategies, and Management Impacts* (Ed. by F.E. Clark & T.H. Rosswall). Ecological Bulletins (Stockholm), **33**, 405–426.

Grime, J.P. (1979). *Plant Strategies and Vegetation Processes.* John Wiley & Sons, Chichester.

Grubb, P.J. (1986). The ecology of establishment. *Ecology and Landscape Design* (Ed. by A.D. Bradshaw, D.A. Goode & E. Thorpe), pp. 83–98. Symposium of the British Ecological Society, 24. Blackwell Scientific Publications, Oxford.

Harcombe, P.A. (1977). The influence of fertilizers on some aspects of succession in a humid tropical forest. *Ecology*, **58**, 1375–1383.

Hartshorn, G.S. (1978). Treefalls and tropical forest dynamics. *Tropical Trees as Living Systems* (Ed. by P.B. Tomlinson & M.H. Zimmerman), pp. 617–638. Cambridge University Press.

Hils, M.H. & Vankat, J.L. (1982). Species removals from a first year old field plant community. *Ecology*, **63**, 705–711.

Jordan, C.F. & Herrera, R. (1981). Tropical rainforests: are nutrients really critical? *American Naturalist*, **117**, 167–180.

Kellman, M. (1985). Nutrient retention by Savanna ecosystems. III. Response to artificial loading. *Journal of Ecology*, **73**, 963–972.

Keyes, M.R. & C.C. Grier (1980). Above- and below-ground net production in 40 year old Douglas-fir stands on high and low productivity sites. *Canadian Journal of Forest Research*, **11**, 599–605.

Lamb, D. (1980). Soil nitrogen mineralisation in a secondary rainforest succession. *Oecologia*, **47**, 257–263.

Lawrence, D.B. (1979). Primary versus secondary succession at Glacier Bay National Monument, southeastern Alaska. *Transactions and Proceedings Series, United States Department of the Interior* (Ed. by R.E. Linn), pp. 213–224. No. 5, 2 volumes, Washington, D.C.

Likens, G.E., Bormann, F.H., Johnson, N.M., Fisher, D.W. & Pierce, R.S. (1970). Effects of forest cutting and herbicide treatment on nutrient budgets in the Hubbard Brook ecosystem in New Hampshire. *Ecological Monographs*, **40**, 23–47.

Lovett, G.M., Reiners, W.A. & Olson, R.K. (1982). Cloud droplet deposition in subalpine balsam fir forests; hydrological and chemical inputs. *Science*, **218**, 1303–1305.

Matson, P.A. & Boone, R.D. (1984). Natural disturbance and nitrogen mineralization: wave-form dieback of mountain hemlock in the Oregon Cascades. *Ecology*, **65**, 1511–1516.

Matson, P.A. & Vitousek, P.M. (1981). Nitrification potentials following clearcutting in the Hoosier National Forest, Indiana. *Forest Science*, **27**, 781–791.

Matson P.A., Vitousek, P.M., Ewel, J.J., Mazzarine, M.J. & Robertson, G.P. Nitrogen transformations following forest felling and burning on a volcanic soil. *Ecology* (in press).

McCauley, K.J. & Cook, S.A. (1980). *Phellinus weirii* infestation of two mountain hemlock forests in the Oregon Cascades. *Forest Science*, **26**, 23–29.

McGill, W.B. & Cole, C.V. (1981). Comparative aspects of cycling of organic C, N, S and P through soil organic matter. *Geoderma*, **26**, 267–286.

Melillo, J.M. & Gosz, J.R. (1983). Interactions of the biogeochemical cycles in forest ecosystems. *The Major Biogeochemical Cycles and Their Interactions* (Ed. by B. Bolin & R.B. Cook). SCOPE 21, 177–222. John Wiley & Sons, Chichester.

Melillo, J.M., Aber, J.D. & Muratore, J.F. (1982). Nitrogen and lignin control of hardwood leaf litter decomposition dynamics. *Ecology*, **63**, 621–626.

Miles, J. (1979). *Vegetation Dynamics*. Chapman & Hall, London.

Miller, H.G. (1981). Forest fertilization: some guiding concepts. *Forestry*, **54**, 157–167.

Mueller-Dombois, D. (1983). Canopy dieback and successional processes in Pacific forests. *Pacific Science*, **37**, 317–326.

Noble, I.R. & Slatyer, R.O. (1980). The use of vital attributes to predict successional changes in plant communities subject to recurrent disturbances. *Vegetatio*, **43**, 5–21.

Nye, P.H. & Greenland, D.J. (1960). *The soil under shifting cultivation*. Technical Bulletin 51, Commonwealth Bureau of Soils, Harpenden, England.

Nye, P.H. & Tinker, P.B. (1977). *Solute Movement in the Soil–Root System*. University of California Press, Berkeley, California.

Olson, J.S. (1958). Rates of succession and soil changes on southern Lake Michigan sand dunes. *Botanical Gazette*, **119**, 125–170.

Palaniappan, V.M., Marrs, R.H. & Bradshaw, A.D. (1979). The effect of *Lupinus arboreus* on the nitrogen status of china clay wastes. *Journal of Applied Ecology*, **16**, 825–837.

Peet, R.K. (1981). Changes in biomass and production during secondary forest succession. *Forest Succession: Concept and Application* (Ed. by D. West, H.H. Shugart & D.B. Botkin), pp. 324–338. Springer–Verlag, New York.

Peterson, D.L. & Bazzaz, F.A. (1978). Life cycle characteristics of *Aster pilosus* in early successional habitats. *Ecology*, **59**, 1005–1013.

Reiners, W.A. (1981). Nitrogen cycling in relation to ecosystem succession. *Terrestrial Nitrogen Cycles: Processes, Ecosystem Strategies, and Management Impacts* (Ed. by F.E. Clark & T. Rosswall), Ecological Bulletin (Stockholm), **33**, 507–528.

Reiners, W.A. & Lang, G.E. (1979). Vegetational patterns and processes in the balsam fir zone, White Mountains, New Hampshire. *Ecology*, **60**, 403–417.

Robertson, G.P. (1984). Nitrification and nitrogen mineralization in a lowland rainforest

succession in Costa Rica, Central America. *Oecologia (Berlin)*, **61**, 99–104.
Robertson, G.P. & Vitousek, P.M. (1981). Nitrification potentials in primary and secondary succession. *Ecology*, **62**, 376–386.
Rosswall, T. (1976). The internal nitrogen cycle between microorganisms, vegetation, and soil. *Nitrogen, Phosphorus, and Sulphur Global Cycles* (Ed. by B.H. Svensson & R. Soderlund). SCOPE 7. Ecological Bulletin (Stockholm), **22**, 157–167.
Shaver, G.R. & Melillo, J.M. (1984). Nutrient budgets of marsh plants: efficiency concepts and relationship to availability. *Ecology*, **65**, 1491–1510.
Silsbee, D.G. & Larson, G.L. (1982). Water quality of streams in the Great Smoky Mountains National Park. *Hydrobiologia*, **89**, 97–115.
Small, E. (1972). Photosynthetic rates in relation to nitrogen recycling as an adaptation to nutrient deficiency in peat bog plants. *Canadian Journal of Botany*, **50**, 2227–2233.
Smathers, G.A. & Mueller-Dombois, D. (1974). *Invasion and Recovery of Vegetation after a Volcanic Eruption in Hawaii.* Natural Park Service Science Monograph No. 5.
Sollins, P., Spycher, G. & Glassman, C.A. (1984). Net nitrogen mineralization from light- and heavy-fraction forest soil organic matter. *Soil Biology and Biochemistry*, **16**, 31–37.
Sprugel, D.G. (1976). Dynamic structure of wave-regenerated *Abies balsamea* forests in the northeastern United States. *Journal of Ecology*, **64**, 889–912.
Stachurski, A. & Zimka, J.R. (1975). Methods of studying forest ecosystems: leaf area, leaf production, and withdrawal of nutrients from leaves of trees. *Ekologia Polska*, **23**, 627–648.
Stevens, P.R. & Walker, T.W. (1970). The chronosequence concept and soil formation. *Quarterly Review of Biology*, **45**, 333–350.
Swank, W.T. & Douglass, J.E. (1974). Streamflow greatly reduced by converting deciduous hardwood forests to pine. *Science*, **185**, 857–859.
Tarrant, R.F. & Trappe, J.M. (1971). The role of *Alnus* in improving the forest environment. *Plant and Soil, Special Volume*, 335–348.
Tezuka, Y. (1961). Development of vegetation in relation to soil formation in the volcanic island of Oshima, Izu, Japan. *Japan Journal of Botany*, **17**, 371–402.
Uehara, G. & Gillman, G. (1981). *The Mineralogy, Chemistry, and Physics of Tropical Soils with Variable Charge Clays.* Westview Press, Boulder, Colorado, USA.
Van Cleve, K., Viereck, L.A. & Schlentner, R.L. (1971). Accumulation of nitrogen in even-aged alder ecosystems near Fairbanks, Alaska. *Arctic and Alpine Research*, **3**, 101–114.
Van Cleve, K. & Viereck, L.A. (1981). Forest succession in relation to nutrient cycling in the boreal forests of Alaska. *Forest Succession: Concepts and Applications* (Ed. by D.C. West, H.H. Shugart & D.B. Botkin), pp. 185–211. Springer–Verlag, New York.
Viereck, L.A. (1966). Plant succession and soil development on gravel outwash of the Muldrow Glacier, Alaska. *Ecological Monographs*, **36**, 181–191.
Vitousek, P.M. (1982). Nutrient cycling and nutrient use efficiency. *American Naturalist*, **119**, 553–572.
Vitousek, P.M. (1984). A general theory of forest nutrient dynamics. *State and Change of Forest Ecosystems: Indicators in Current Research* (Ed. by G.I. Agren) Swedish University for Agricultural Sciences, Department of Ecology and Environmental Research Report, **13**, 121–135.
Vitousek, P.M. & Denslow, J.S. (1986). Nitrogen and phosphorus availability in treefall gaps of a lowland tropical rainforest. *Journal of Ecology* (in press).
Vitousek, P.M. & Matson, P.A. (1984). Mechanisms of nitrogen retention in forested ecosystems: a field experiment. *Science*, **225**, 51–52.
Vitousek, P.M. & Matson, P.A. (1985). Disturbance, nitrogen availability, and nitrogen losses in an intensively managed loblolly pine plantation. *Ecology*, **66**, 1360–1376.

Vitousek, P.M. & Melillo, J.M. (1979). Nitrate losses from disturbed forests: patterns and mechanisms. *Forest Science*, **25**, 605–619.

Vitousek, P.M. & Reiners, W.A. (1975). Ecosystem succession and nutrient retention: a hypothesis. *BioScience*, **25**, 376–381.

Vitousek, P.M. & White, P.S. (1981). Process studies in succession. *Forest Succession: Concepts and Applications* (Ed. by D. West. H.H. Shugart & D.B. Botkin), pp. 267–276. Springer–Verlag, New York.

Vitousek, P.M., Gosz, J.R., Grier, C.C., Melillo, J.M. & Reiners, W.A. (1982). A comparative analysis of potential nitrification and nitrate mobility in forest ecosystems. *Ecological Monographs*, **52**, 155–177.

Vitousek, P.M., Van Cleve, K., Balakrishnan, N. & Mueller-Dombois, D. (1983). Soil development and nitrogen turnover in montane rainforest soils on Hawaii. *Biotropica*, **15**, 268–274.

Walker, L.R. (1985). *The processes controlling primary succession on an Alaskan floodplain*. Ph.D. dissertation, University of Alaska.

Walker, B.H., Ludwig, D., Holling, C.S. & Peterman, R.M. (1981). Stability of semi-arid savanna grazing systems. *Journal of Ecology*, **64**, 473–498.

Walker, J., Thompson, C.H., Fergus, I.F. & Tunstall, B.R. (1981). Plant succession and soil development in coastal sand dunes of subtropical eastern Australia. *Forest Succession: Concepts and Applications* (Ed. by D.C. West, H.H. Shugart & D.B. Botkin), pp. 107–131. Springer–Verlag, New York.

Walker, L.R., Zasada, J.C. & Chapin, F.S., III. (1986). The role of life history processes in primary succession on an Alaskan floodplain. *Ecology* **67**, 1243–1253.

Walker, L.R. & Chapin, F.S., III. (1986). Physiological controls over seedling growth in primary succession on an Alaskan floodplain. *Ecology* (in press).

Walker, T.W. & Syers, J.K. (1976). The fate of phosphorus during pedogenesis. *Geoderma*, **15**, 1–19.

Wiklander, G. (1981). Rapporteur's comment on clearcutting. Terrestrial Nitrogen Cycles: Processes, Ecosystem Strategies, and Management Impacts (Ed. by F.E. Clark & T. Rosswall), *Ecological Bulletin (Stockholm)*, **33**, 642–647.

Zinke, P.J., Sabhasri, S. & Kunstadter, P. (1978). Soil fertility aspects of the Lua' forest fallow system of shifting cultivation. *Farmers in the Forest* (Ed. by P. Kunstadter, E.C. Chapman & S. Sabhasri), pp. 134–159. University of Hawaii Press.

11. ARE THERE ASSEMBLY RULES FOR SUCCESSIONAL COMMUNITIES?

J.H. LAWTON
Department of Biology, University of York, Heslington, York, YO1 5DD

INTRODUCTION

'Assembly' is putting together a number of parts to make something; 'rules' imply that the process is not random. Probably the first, and certainly the best known use of the term 'assembly rule' for ecological communities was by Diamond (1975). Diamond highlighted interspecific competition for resources as the major factor underlying rules of assembly in sets of New Guinea birds; but it is easy to imagine other forms of interspecific or species–environment interactions as part of the rules, particularly in successional systems (McIntosh 1980). Odum (1969) had no doubt that succession was governed by rules. He wrote (p. 262): Succession 'is an orderly process of community development that is reasonably directional and, therefore, predictable. It results from modification of the physical environment by the community; that is, succession is community-controlled.'

If there are assembly rules for successional communities, what are they, and how well do we understand them? This chapter tries to provide answers, starting with assembly rules for plant populations during succession, before moving up one or more trophic levels to consider animals. I then add complexity by looking at the effects of plant succession on animals, and the impact of animals (herbivores) on plant succession. There follow brief remarks about plant pathogens and more extensive speculations on the role of mutualists. Finally, individual 'pair-wise' interactions are subsumed into community-wide webs of interactions. I have basically taken the view that assembly rules arise from species interactions. Table 11.1 summarizes the 'pair-wise' processes dealt with, however briefly, in the text. I think it is probably impossible, given the relatively primitive state of current knowledge, to rank these processes according to their importance in community assembly.

I have restricted the scope of the review almost entirely to terrestrial ecosystems, and to successions based on green plants. Succession in freshwater and marine habitats, and heterotrophic successions in dung, dead wood, detritus and corpses are wilfully ignored.

TABLE 11.1. Summary of the 'pair-wise' interactions dealt with in the chapter that might, or are known to, contribute to assembly rules for successional communities. The blank cells are fairly easily filled in, although they are not explicitly discussed here; for example, consider the role of carnivores in regulating herbivore populations, with potentially important effects on succession via the herbivores. The entire web of community interactions, both direct and indirect, may potentially influence succession

	Plants	Herbivores	Carnivores	Mutualists	Pathogens
Plants	Effects of plants on plants: Interspecific competition Facilitation Inhibition Tolerance Allelopathy	Effects of herbivores on plants: Herbivory modifies succession		Effects of mutualists on plants: Mycorrhiza N-fixing organisms Pollinators Seed dispersers Ant–extrafloral nectary interactions	Effects of pathogens on plants: Diseases may modify succession
Herbivores Carnivores	Effects of plants on animals: Animal populations wax and wane along successional gradients created by plants. Animals respond 'passively' to changing habitats created by plant succession	Effect of herbivores on herbivores: Interspecific competition in changing environments may drive some species turnover	Effect of carnivores on herbivores: Not discussed but by regulating herbivore populations, the potential effect of carnivores is clearly important Effect of carnivores on carnivores: Interspecific competition in changing environments may drive some species turnover	Effect of mutualists on herbivores: Ants attending herbivorous insects	

ASSEMBLY RULES FOR VEGETATION

The quintessence of terrestrial succession is a change in the species composition and structure of vegetation. Studies on these phenomena have a long pedigree in ecology, inspiring classical work by F.E. Clements, H.C. Cowles, H.A. Gleason, Sir Arthur Tansley, and A.S. Watt to name but a few. Jackson (1981) provides an introduction to this early literature, drawing attention to approximately 100 papers published in *Journal of Ecology*, *Journal of Animal Ecology*, *Ecology* and *American Naturalist* between 1919 and 1959 that deal in one way or another with the role of competition between plant species as a driving force in succession. Jackson notes (p. 893): 'Much of what is considered original to modern niche theory of competition, except the mathematics, was well formulated and understood by many plant ecologists, especially in England as early as 1914.' This body of work makes plain that interspecific competition, Diamond's *raison d'être* for assembly rules, is a major ingredient in the assembly of successional plant communities. (See McIntosh 1980, and Grime, Chapter 20, for further discussions).

Obviously, however, interspecific competition is not the only cause of vegetation change, a fact that has again been known for a long time (McIntosh 1980). As early as 1916, for example, Clements recognized the importance of chance colonization, as well as habitat alteration and interspecific competition in plant succession (Clements 1916; MacMahon 1981). Contemporary understanding of the processes determining patterns of succession in vegetation (i.e. processes involving only the plants) can be gathered under four headings (there are other ways of classifying the problem, involving greater subdivision of processes, but these four will do) (Connell & Slatyer 1977; Golley 1977; McIntosh 1980; Schowalter 1981; Horn 1981; Hils & Vankat 1982; Crawley 1983; Finegan 1984; Krebs 1985). Each model specifies a particular set of rules (or lack of them) for community assembly during succession:

The facilitation model. Species occurring early in succession modify the habitat, making it less suitable for themselves and more suitable for later colonists.

The inhibition model. Whichever plant species reaches a site first, holds it against all subsequent invaders until it dies.

The tolerance model. Slower growing, more tolerant (competitively superior) plant species invade and mature in the presence of earlier, faster growing, but less tolerant species, and eventually exclude them.

The random colonization model. Succession involves no more than chance survival of different species at the time succession is initiated, and subsequent random colonization by new species; species then grow and mature at different rates. There is no facilitation, nor are interspecific interactions important.

Except insofar as species differ in dispersal abilities, the fourth and last possibility comes nearest to a 'no assembly rule' model of succession. All the other possibilities constitute assembly rules. Uncertainty therefore lies not with formulating what the rules might be, but with their relative contributions in different habitats and biomes, or along primary and secondary successions. There is also disagreement about the importance of particular means of facilitation or inhibition, for example on the role of allelopathy (see Golley 1977, and Harper 1977 for discussions). Several recent reviews come to markedly different conclusions about the relative importance of the four models (Connell & Slatyer 1977; MacMahon 1981; Finegan 1984; Krebs 1985), with no clear consensus about how succession works (McIntosh 1980; Crawley 1983). Hils & Vankat (1982) modestly conclude that more than one mechanism probably operates at the same time, even in the same abandoned old-field. In other words, as is so often the case in ecology, we must seek a multiplicity of explanations. There is not one universal truth.

For plant succession, possible assembly rules, embodied in the four models, are well defined; relative roles are much harder, and will take much longer, to work out.

CHANGES IN ANIMAL SPECIES ALONG SUCCESSIONAL GRADIENTS

Interspecific competition or habitat selection?

It is hardly surprising that as plant communities wax and wane, so do associated animals, influenced by changing microclimates, food supplies, hiding places, resting sites and so on (e.g. Lowrie 1948; Lack & Lack 1951; Yapp 1962; Shelford 1963; Chevin 1966; Karr 1968; Lack 1971; Healey 1972; Glowaciński & Järvinen 1975; MacMahon 1981; Fuller 1982; Price 1984).

Diamond (1975) clearly recognizes that particular habitat preferences by different species result in bird communities that are not random samples drawn from a pool of all potential colonists. Such individual habitat preferences constitute simple assembly rules, applicable to successional systems or indeed to any type of community. However, as we have already noted, Diamond argues that individual habitat selection alone is insufficient to account for the structure of New Guinea bird communities. The missing ingredient, he suggests, is interspecific competition, excluding species from otherwise suitable habitats.

This important question has triggered a large, confusing and at times acrimonious debate among a small group of animal ecologists (see Connor & Simberloff 1979, 1984; Diamond & Gilpin 1982; Gilpin & Diamond 1982, 1984). Two points are worth making. Ecology as a science has undoubtedly been guilty of too easily embracing competition between animal species as a major force structuring communities, without giving critical thought to how competitive explanations might be distinguished from other explanations. Although we now seem to have cleared this particular hurdle, it turns out that the evolutionary and ecological effects of interspecific competition can be bafflingly (and frustratingly!) difficult to distinguish from random processes (Colwell & Winkler 1984), often requiring field studies of great sophistication (e.g. Schluter & Grant 1984; Schluter, Price & Grant 1985).

The available evidence suggests that the importance of interspecific competition as an assembly rule for animal communities depends very much on the kinds of organisms involved, particularly their size and trophic level, and the nature of their habitat (e.g. Connell 1983; Lawton & Hassell 1984; Schoener 1983; Strong Lawton & Southwood 1984). Unfortunately, none of this information is specifically directed at successional systems; indeed the dearth of studies clearly demonstrating that competitive exclusion is the reason why one animal species replaces another along a terrestrial succession is quite remarkable.

Of course, there are hints of such effects. A number of studies on small mammals show that habitat choice in one or more species is influenced by competitors (see Connell 1983 and Schoener 1983 for reviews), although it is not clear, at least to me, whether the habitats involved are really seral stages. Yapp (1962) suggests that the occasional early appearance in heathland succession of birds more typical of woodland may be due to chance absences of typical heathland bird species, but the data are anecdotal. Spiders provide a possible counter-example. Although spider species show clear and well documented habitat preferences along successional gradients (e.g. Lowrie 1948; van der Aart in Price 1984), there is, as yet, little good evidence for significant interspecific competition among spiders (Wise 1984).

We urgently need studies on animal species in a variety of taxa, specifically designed to test whether interspecific competition is, or is not, a significant cause of species replacements along successional gradients.

Birds on Hawaiian islands

The nearest I can come to a detailed study of competitive exclusion by animals in succession is work by Moulton & Pimm (1983, 1986) on the introduced avifaunas of six Hawaiian islands. (Exotic birds constitute almost the entire land-avifaunas of these islands below 600m.) Build-up of these exotic bird communities from 1860 onwards is well documented, and yields moderately good evidence for interspecific competition. For example, extinction rates increase as the number of introduced species on each island increases; extinctions are more likely when species have similar sized bills (and hence presumably share similar foods); and some species appear to be kept rare by competitors, even though they are not actually exterminated.

A number of important differences exist between the introduced avifaunas of Hawaiian islands, and a real succession. Rates of arrival and 'discovery' of suitable habitats by birds are certainly different in the two situations, and a significant part of the build-up in avian species richness in real successions can be attributed to changes in the complexity of vegetation (see below) rather than to periodic introductions by man into heavily modified habitats. But if Moulton and Pimm's results have any generality, they suggest that at least part of the turnover of bird species along a succession could be due to interspecific competition, a conclusion that marches in accord with Lack's (1971) views, but for which there is currently no good experimental evidence.

Habitat 'architecture' and animal species richness

Not only do animal species come and go along successional gradients, but as Odum (1969) noted, increased habitat stratification (or vegetation 'architecture') generated by plant succession promotes enhanced animal species diversity. That is, generally more animal species invade the system than disappear as succession proceeds. Again, the phenomenon has a long pedigree, with examples from many taxa (e.g. Lowrie 1948; Duffey 1966; Karr 1968; Healey 1972; Shugart & Hett 1973; Shugart & James 1973; Southwood 1977; Southwood, Brown & Reader 1979; MacMahon 1981; Fuller 1982; Lawton 1983). It constitutes another, simple assembly rule.

Habitat stratification alone, however, is unlikely to be the sole explanation for a general increase in animal species richness during succession, particularly for herbivores. Changes in host-plant diversity must also play a part. The effects of plant species richness, habitat stratification and resource diversity on insect species richness during succession are documented by Southwood, Brown & Reader (1979). (See also Lawton 1983, and Brown & Southwood, Chapter 15 for further discussion.)

Other changes in vegetation

During succession, populations of particular plant species change in density, spatial distribution, patch size, and 'purity', i.e. the extent to which they grow intermingled with other species, (e.g. Yarranton & Morrison 1974). All these properties of vegetation markedly influence the distribution and abundance of phytophagous insects (e.g. Kareiva 1982; Strong, Lawton & Southwood 1984) and must contribute to the turnover of animal species in succession, directly so for phytophages, indirectly for their enemies. With two notable exceptions (Southwood 1977; Southwood Brown & Reader 1983) the problem is ignored in the literature on animal succession; it deserves more attention.

IMPACT OF OTHER ORGANISMS OF PLANT SUCCESSION

Herbivores

Assembly rules for succession become more interesting, and potentially more complicated, when animals are no longer viewed as passive agents responding to changes in vegetation. Herbivores eat plants, and in so doing introduce new rules and alter successional processes (Edwards & Gillman, Chapter 14). Yet one recent review on succession (Finegan 1984) effectively ignores herbivores.

The impact of vertebrate herbivores on the successional dynamics of plant populations has been known for a long time (e.g. Tansley 1935, 1939); more modern reviews are in Crawley (1983), Harper (1977), Jackson (1981), MacMahon (1981), McIntosh (1980) and Whittaker (1953). In marked contrast, the effects of insects on succession have only recently received the attention they deserve although Tansley (1939) guessed that they were important (p. 144), and Costello (1944) provided data on harvester ants. There is now little room to doubt that insect herbivores influence both the rate and the direction of succession, at all seral stages (e.g. Brown 1982, 1985; Brunsting 1982; Brunsting & Heil 1985; Kulman 1971; McBrien, Harmsen & Crowder 1983; Schowalter, Hargrove & Crossley 1986; Tenow 1983). Edwards & Gillman (Chapter 14) and Brown & Southwood (Chapter 15) review the evidence.

Pathogens

Chestnut blight and Dutch elm disease are powerful reminders that plant population dynamics can be profoundly altered by pathogens. Build-up of plant diseases is one of the reasons why farmers are forced to rotate crops. 'It seems probable that an entirely similar disease accumulation occurs in soil under more natural vegetation ... (and) may need to be taken into account in interpreting successions' (Harper 1977, p. 110). It would be interesting to know how much apparent 'facilitation' (p. 227) is attributable to pathogens. Burdon & Chilvers (1974) suggest that species diversity in maturing eucalyptus associations may be maintained, in part, by pathogenic fungi, whilst poor seedling performance in the vicinity of parent trees may be due to an accumulation of soil pathogens (Connell & Slatyer 1977). The implications for patterns of succession are obvious, and suggest that tracing assembly rules for vegetation requires the active co-operation not only of botanists and zoologists, but also of plant pathologists.

Mutualists

We know even less about the role of mutualists in the assembly of successional communities. Obvious mutualisms of interest to students of succession are:

- lichens;
- plants and mycorrhizal fungi;
- plants and nitrogen-fixing micro-organisms;
- plants and pollinators;
- plants and seed dispersers;
- plants and ant-visitors to extrafloral nectaries;
- herbivorous insects and attendant ants.

For most of these mutualisms, I can find little consistent evidence to support Odum's (1969) generalization that 'internal symbiosis' (i.e. mutualism) is 'undeveloped' in early seral stages, but well developed in mature communities. Indeed, often the reverse is true.

Lichens are among the classical early colonists of bare rock in primary successions, and hence do not appear to be consistent with Odum's views. Data from mycorrhizal fungi are also equivocal (see Cromack (1981) for a review of the biology of mycorrhizas during succession). In some seres, mycorrhizas establish very early, with levels of infection showing no consistent change as succession proceeds (Fig. 11.1; Nicolson 1960) (see also Daft & Nicolson 1973). Tropical successions may conform more closely to Odum's model (Janos 1980), with mycorrhizas effectively absent from pioneer plant communities; facultatively associated with intermediate seres; and forming obligatory associations with mature trees.

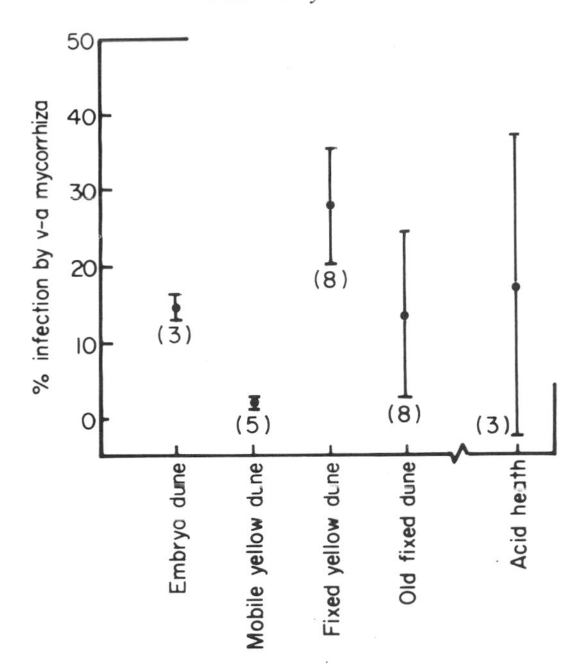

FIG. 11.1. Mean percentage (and 95% CI) of roots of grasses infected with vesicular–arbuscular mycorrhizas in sand dune successions. Data from Nicolson (1960), Tables 1 and 2. Sample sizes in parentheses. The dune successions are at Gibraltar Point, Lincolnshire; Southport, Lancashire and St. Cryus, Kincardineshire. The acid heathland, representative of some late stages of dune succession, is from an inland site at Charnwood Forest. No significant trend exists in the data on percentage infection as succession proceeds, except the low levels of infection in mobile yellow dunes (stage 2), which Nicolson believes to be genuine, and a result of the harsh habitat created by moving sand.

Mycorrhizas infected most species of canopy trees in later stages of succession in Zambia, from open fire degraded chipya woodland to closed evergreen mateshi forest via miombo woodland, although the balance of species changed from predominantly endomycorrhizas to predominantly ectoomycorrhizas, or mixtures of both types (Högberg & Piearce 1986). The mycorrhizal status of seres preceding the growth of trees in these areas is apparently unknown. Even less is known about how mycorrhizal infection influences the course and rate of succession. Janos (1980) develops a good circumstantial case for a significant effect on tropical successions (see also p. 239), but experimental data are lacking, and would appear to be very difficult to obtain.

A rather clearer picture emerges from a limited number of studies on plants and symbiotic nitrogen-fixing micro-organisms. Such plants are often conspicuous early colonists of primary successions; for example, legumes on china-clay works (Marrs *et al.* 1983) and alders on glacial moraines (Crocker & Major 1955). This vital mutualism contributes to a build-up in soil nitrogen and undoubtedly facilitates colonization by plant species characteristic of later successional stages, and hence influences rates, if not patterns of succession (Cromack 1981; Marrs *et al.* 1983; Tilman 1982; Vitousek & Walker, Chapter 10). Such facilitation may be less important in secondary successions, on soils with a greater accumulated capital of organic nitrogen (see Finegan 1984). Data from secondary successions show that chemosynthetic bacteria involved in the nitrogen cycle, namely *Nitrosomonas* (converting ammonia to nitrite) and *Nitrobacter* (converting nitrite to nitrate), are conspicuously more abundant in early seral stages (Rice & Pancholy 1972; Fig. 11.2), and may be virtually absent from climax communities. It seems not to be known how other components of the nitrogen cycle, particularly nitrogen-fixing micro-organisms, change in abundance during secondary succession.

I suspect that a number of other mutualisms may turn out to be very important early in community development, but they have been too poorly studied in this context to be sure. Limited data on mutualistic associations between ants and plants with extrafloral nectaries point to greater ant activity and a higher density of plant species with extrafloral nectaries in early and mid-successional communities, at forest edges and in clearings (Bentley 1976, 1977). The differential protection against herbivory offered by ants to plant species bearing extrafloral nectaries could clearly influence succession.

So too, could ant protection of insect herbivores, particularly Homoptera and butterfly caterpillars in the family Lycaenidae. Ant attendance often (but not always) leads to enhanced herbivore survival, and ultimately to a detrimental impact on selected host-plant species (Strong, Lawton & Southwood 1984). Pierce (1984) shows that a high proportion of ant-attended lycaenid caterpillars exploit legumes as host-plants, perhaps suggesting that this fascinating mutualism is again more frequent in the early stages of succession. (A comprehensive study of the biology and host-plant utilization of all lycaenids for which data are available may soon clarify the frequency of this particular mutualism on plants characteristic of different stages of succession, e.g. annuals and perennials: N.E. Pierce, in prep.)

Finally, the impact of seed dispersers on rates of succession deserves comment. (Seed dispersers undoubtedly benefit plants; the reciprocal benefits are sometimes less certain. Hence, the processes may either be mutualistic or commensal.) Both Finegan (1984) and MacMahon (1981) provide brief reviews. As Edwards & Gillman (Chapter 14) point out,

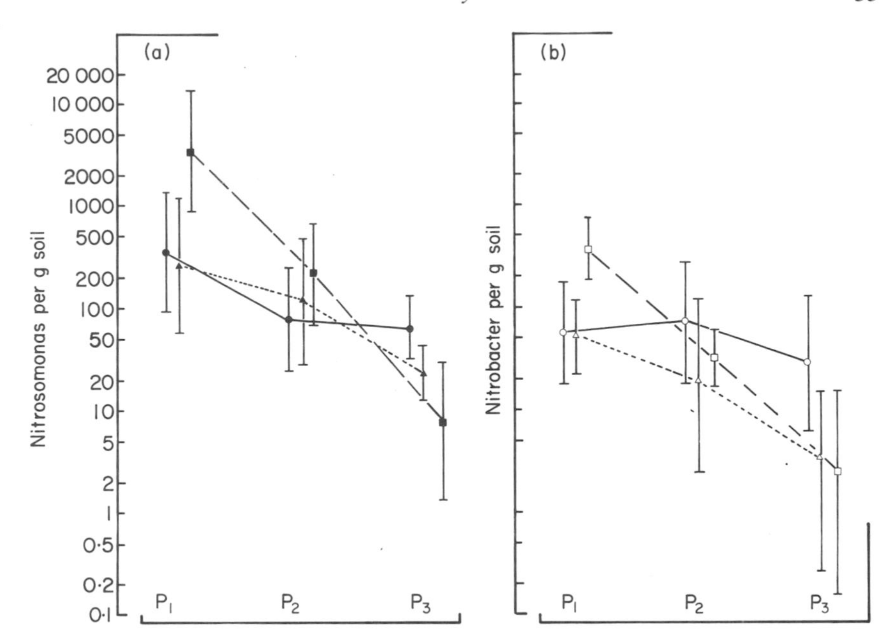

FIG. 11.2. Numbers of micro-organisms g^{-1} soil in three secondary successional sequences (from data in Rice & Pancholy 1972). (a) *Nitrosomonas* (b) *Nitrobacter*. Plotted are means (±95% CI) of six samples, taken at 2-month intervals between April of one year and February of the next, each sample itself being the mean of four determinations from soil 0–15 cm deep. Data were transformed to $\log_{10}(n + 1)$ before calculating, and back-transformed prior to plotting. Successions are to climax tall grass prairie (circles and solid lines), post oak–blackjack oak (triangles and dotted lines) and oak–pine (squares and broken lines), all commencing in abandoned agricultural fields. P_{1-3} are sampling stations as specified by authors: P_1, 1–2 year old abandoned fields: P_2, 6-year old prairie succession, 8-year old post-oak succession, 25 year old oak–pine succession: P_3, climax communities (undated). With one exception, numbers of micro-organisms from this stage in the nitrogen cycle fall during succession.

seed dispersal by animals over short (e.g. by ants) or long (e.g. by birds) distances probably plays a vital part in determining both patterns and rates of succession.

Drawing these arguments together, it is clear that mutualisms are among the most interesting, but also one of the most poorly studied processes contributing to assembly rules for successional communities. Studies to document the frequency and intensity of various mutualistic associations in different seral stages would be valuable. Even more enlightening would be experiments to 'uncouple' or disrupt particular mutualisms, and then follow what happens to succession.

WEBS OF INTERACTIONS

Studies on interspecific competition, predation, herbivory and mutualisms ultimately mesh to yield community-wide webs of interactions. A rapidly developing theoretical literature points to some important, and hitherto unexpected, assembly rules that are properties of the entire interactive web (Pimm 1982); once again, pertinent field data from successional systems, useful for testing these ideas, are few.

Attempts to show that the broad trophic structure of developing communities (the proportions of herbivores, detritivores, predators and so on) is somehow constrained by species interactions, and converges to relatively fixed proportions during community development irrespective of the species involved (Heatwole & Levins 1972), are at best equivocal (Simberloff 1976). As with competitive interactions (p. 229) it has proved extremely difficult to distinguish community-wide patterns generated by trophic interactions, from random effects (Cole 1980; Diamond & Gilpin 1982; Van Valen 1982). However, an approximately constant ratio for number of predator species/number of prey species emerges as a clear property of model food-webs, allowed to 'grow' under a process akin to succession (Mithen 1984; Mithen & Lawton 1986; Drake 1985) (Figs. 11.3 and 11.4), and is consistent with some field data (e.g. Jeffries & Lawton 1985). Comparable information on predator: prey and other 'trophic-ratios' (e.g. Brown & Southwood 1983) from successional communities would be valuable.

A number of recent studies have modelled community development by drawing species from a pool of potential colonists (e.g. Pimm 1982; Post 1983; Post & Pimm 1983; Drake 1983; 1985). These studies, and the comparable one by Mithen & Lawton summarized in the previous paragraph, model species interactions using Lotka-Volterra equations (see legend to Fig. 11.3), with all their attendant simplifications and uncertainties. Their results, however, broadly match what we might reasonably expect to happen during succession. For example:

(a) Species accumulate in the model systems with time, but at steadily decreasing rates. It gets harder and harder for new species to invade the model communities (Post 1983; Post & Pimm 1983; Drake 1985).

(b) 'Turnover' of species slows down. Species persist in the community for longer, later in the life of the model (Pimm 1982; Drake 1985).

These assembly rules are generated by the whole web of interactions, with community development constrained by levels of connectance,

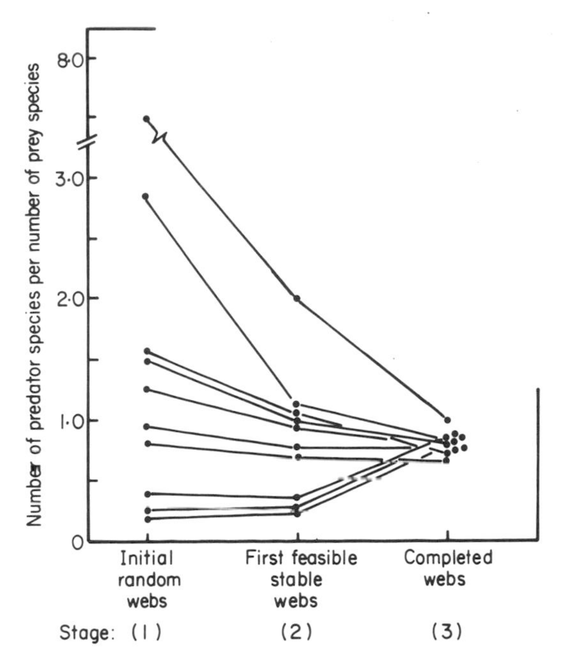

FIG. 11.3. Ratio: Number of predator species/number of prey species in simple two trophic-level Lotka-Volterra models simulating community assembly from a pool of potential colonists (Mithen & Lawton 1986). The ratio varies widely in initial, random draws from the pool (stage 1). The ratio is less variable in the first feasible (i.e. all populations have positive equilibria), stable webs (stage 2). Repeated invasions, extinctions, and species establishment ultimately results in final communities (stage 3) with very tightly constrained ratios. The exact value of this ratio depends on the parameter values used in the model (i.e. upon the 'biology' of the model species making up the web). The model is:

$$\frac{dN_i}{dt} = N_i\left\{b_i - a_iN_i - \sum_{j=1}^{P} \alpha_{ij}P_j\right\}$$

$$\frac{dP_j}{dt} = P_j\left\{-b_j - a_jP_j + \sum_{i=1}^{n} \alpha_{ji}N_i\right\}$$

where N_i = density of ith prey species ($i = 1 --- n$); P_j = density of jth predator species ($j = 1 --- p$); a_i, a_j = intraspecific competition coefficient; b_i = birth rate of ith prey species; b_j = death rate of jth predator species; α_{ij} = effect of jth predator on the ith prey; α_{ji} = effect of ith prey on jth predator. Parameter values were chosen according to the logic in Pimm & Lawton (1977) and Pimm (1982). In this run they were selected at random from the following ranges: b_i (0, 0·1), b_j : (−0·1, 0), a_i (−1·0, 0), a_j (−1·0, 10), $|\alpha_{ij}|/|\alpha_{ji}|$ 10, with α_{ij} min = α_{ji} min = 0, and the constraint that connectance (proportion of α_{ij}s > 0) = 0·4 in the species pool from which colonists were drawn.

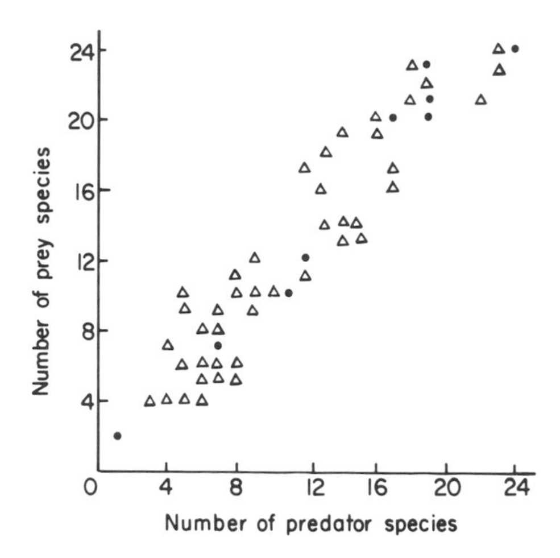

FIG. 11.4. Plot of number of predator species: number of prey species in Lotka-Volterra food-web models, at the end of the community assembly process (stage 3 of Fig. 11.3). Despite a wide range of species occupying the final communities, the ratio of predator to prey species is constrained within narrow limits, in a manner reminiscent of that seen in some field data (e.g. Jeffries & Lawton 1985). The slope of the line (i.e. the predator: prey ratio) depends upon the parameter values in the model (Mithen & Lawton 1986). (△ single point; ● two coincident points).

'apparent competition' and other indirect species interactions, degrees of omnivory, food-chain length and so on (see Holt 1977; Pimm 1982; Pimm & Lawton 1983; Dethier & Duggins 1984; Bender, Case & Gilpin 1984). It is not yet clear how generally these same constraints operate in real successional communities. It *is* known that the rate of species loss from real communities decreases steadily as succession proceeds (Shugart & Hett 1973; Głowaciński & Järvinen 1975). Unfortunately, these data measure extinction rates in real time, not generation times, and since plants from later successional communities live much longer than primary colonists, I would expect species extinction rates (be they for plants, or the animals that depend on plants) to be slower later in succession. Model species, in contrast, have fixed, arbitrary generation times, making comparisons with extinction rates of real organisms based on generation times (the only sensible comparison) very difficult.

There are also other reasons for interpreting these models very carefully; some at least (e.g. Post & Pimm 1983) fail to mimic the decline in plant species richness observed in many late successional communities, in studies ranging from Clements (1905) to Southwood, Brown & Reader (1979). Clearly, scope exists to make such models more realistic.

One assembly rule to emerge from these models is startling. Drake (1985) shows that initial random differences in the order in which species colonize the model community can 'lock' the subsequent development of the system into quite different pathways. This means that model 'climax communities' have alternative, relatively *deterministic* end-points dependent upon *random* initial conditions. Moreover, he can reproduce the same 'priority effects' in laboratory microcosms, subject to colonization from a pool of species under his experimental control. Drake is sensibly cautious to point out that his results cannot be applied directly to community assembly in the field. Nevertheless, a growing number of field and laboratory studies show, or are consistent with, similar priority effects (e.g. Talling 1951; Cole 1983; Kneidel 1983; Morin 1984). In the successional literature the idea is hinted at by Whittaker (1953, p. 55) and explicitly developed by Horn (1981) who postulates that initial differences in the abundance of tree species may generate alternative climax forest communities 'dependent upon accidents of history'. Janos (1980) has also suggested that initial differences in soil fungi, particularly whether or not mycorrhizas are present in nutrient-poor tropical soils, may create priority effects that alter the whole course of succession, again culminating in alternative climax communities.

If such priority effects turn out to be common in the field, they will make it doubly difficult to discover whether assembly rules exist from studies based solely on the end-points of community development. Priority effects mean that although there are rules, the way in which they work out in particular circumstances depends upon who got there first, i.e. upon a mixture of initial chance and subsequent determinism.

Unfortunately, the problem is even harder than this, because in order to demonstrate priority effects we must be able to eliminate other explanations for alternative end-points to succession; for example, local differences in habitat management, soil or climate (Tansley 1939; see Sousa & Connell 1985 for a recent discussion). We must also be able to identify the habitat patches (or 'communities') that have developed in different directions, a problem that has proved to be notoriously intractable (see, for example, Whittaker's 1953 exposition on the 'climax pattern hypothesis').

Distinguishing between the effects of chance and determinism on community assembly undoubtedly constitutes one of the most important and difficult problems in contemporary ecology.

ARE THERE ASSEMBLY RULES FOR SUCCESSIONAL COMMUNITIES?

The broad answer to this question is clearly 'yes'; the details are considerably more fuzzy. Assembly rules for successional communities certainly involve more than interspecific competition between plant species, as the range of possible 'pairwise' interactions in Table 11.1 makes plain. Beyond that it is not possible to say which of the many possible species interactions discussed in this chapter generally contribute most to determining patterns and rates of succession; probably all are important some of the time. It is remarkable that a subject with so long a pedigree as community assembly during succession should be so poorly understood, and so lacking in basic data.

ACKNOWLEDGMENTS

A.H. Fitter and B. Hopkins provided help with literature on mycorrhizal fungi and nitrogen-fixing organisms; N.E. Pierce discussed lycaenid–ant mutualisms with me. J.A. Drake kindly allowed me to quote from his unpublished manuscripts. S.L. Pimm provided valuable discussions on 'growing' food-webs, and on effects of community-wide webs of interactions. The following colleagues made helpful and constructive comments on the manuscript: V.K. Brown, M.J. Chadwick, M.J. Crawley, R. Law, M.B. Usher and M.H. Williamson.

REFERENCES

Bender, E.A., Case, T.J. & Gilpin, M.E. (1984). Perturbation experiments in community ecology: theory and practice. *Ecology*, **65**, 1–13.

Bentley, B.L. (1976). Plants bearing extrafloral nectaries and the associated ant community: interhabitat differences in the reduction of herbivore damage. *Ecology*, **57**, 815–20.

Bentley, B.L. (1977). Extrafloral nectaries and protection by pugnacious bodyguards. *Annual Review of Ecology and Systematics*, **8**, 407–27.

Brown, V.K. (1982). The phytophagous insect community and its impact on early successional habitats. *Proceedings 5th International Symposium, Insect–Plant Relationships*, Wageningen, pp. 205–13. Pudoc.

Brown, V.K. (1985). Insect herbivores and plant succession. *Oikos*, **44**, 17–22.

Brown V.K. & Southwood, T.R.E. (1983). Trophic diversity, niche breadth and generation times of exopterygote insects in a secondary succession. *Oecologia*, **56**, 220–25.

Brunsting, A.M.H. (1982). The influence of the dynamics of a population of herbivorous beetles on the development of vegetation patterns in a heathland system. *Proceedings 5th International Symposium, Insect–Plant Relationships*, Wageningen, pp. 215–23. Pudoc.

Brunsting, A.M.H. & Heil, G.W. (1985). The role of nutrients in the interactions between a herbivorous beetle and some competing plant species in heathland. *Oikos*, **44**, 23–6.

Burdon, J.J. & Chilvers, G.A. (1974). Fungal and insect parasites contributing to niche differentiation in mixed species stands of eucalypt saplings. *Australian Journal of Botany*, **22**, 103–14.

Chevin, H. (1966). Végétation et peuplement entomologique des terrains sablonneux de la côte ouest du Cotentin. *Memoires de la Société Nationale de Sciences Naturelles et Mathématiques de Cherbourg*, **52**, 8–138.

Clements, F.E. (1905). *Research Methods in Ecology*. Lincoln, Nabraska. University Publishing Co.

Clements, F.E. (1916). Plant succession: an analysis of the development of vegetation. *Carnegie Institute of Washington Publication*, **242**, 1–512.

Cole, B.J. (1980). Trophic structure of a grassland community. *Nature*, **288**, 76–7.

Cole, B.J. (1983). Assembly of mangrove ant communities: patterns of geographical distribution. *Journal of Animal Ecology*, **52**, 339–47.

Colwell, R.K. & Winkler, D.W. (1984). A null model for null models in biogeography. *Ecological Communities. Conceptual Issues and the Evidence* (Ed. by D.R. Strong Jr., D. Simberloff, L.G. Abele & A.B. Thistle), pp. 344–59. Princeton University Press, Princeton, N.J.

Connell, J.H. (1983). On the prevalance and relative importance of interspecific competition: evidence from field experiments. *American Naturalist*, **122**, 661–96.

Connell, J.H. & Slatyer, R.O. (1977). Mechanisms of succession in natural communities and their role in community stability and organisation. *American Naturalist*, **111**, 1119–44.

Connor, E.F. & Simberloff, D. (1979). The assembly of species communities: chance or competition? *Ecology*, **60**, 1132–40.

Connor, E.F. & Simberloff, D. (1984). Neutral models of species' co-occurrence patterns. *Ecological Communities. Conceptual Issues and the Evidence.* (Ed. by D.R. Strong Jr., D. Simberloff, L.G. Abele & A.B. Thistle), pp. 316–31, 341–3. Princeton University Press, Princeton, N.J.

Costello, D.F. (1944). Natural revegetation of abandoned plowed land in the mixed prairie association of northeastern Colorado. *Ecology*, **25**, 312–26.

Crawley, M.J. (1983). *Herbivory. The Dynamics of Animal–Plant Interactions.* Blackwell Scientific Publications, Oxford.

Crocker, R.L. & Major, J. (1955). Soil development in relation to vegetation and surface age at Glacier Bay, Alaska. *Journal of Ecology*, **43**, 427–48.

Cromack, K. Jr. (1981). Below-ground processes in forest succession. *Forest Succession. Concepts and Application* (Ed. by D.C. West, H.H. Shugart & D.B. Botkin), pp. 361–73. Springer-Verlag, New York.

Daft, M.J. & Nicolson, T.H. (1973). Arbuscular mycorrhizas in plants colonizing coal wastes in Scotland. *New Phytologist*, **73**, 1129–38.

Dethier, M.N. & Duggins, D.O. (1984). An 'indirect commensalism' between marine herbivores and the importance of competitive hierarchies. *American Naturalist*, **124**, 205–19.

Diamond, J.M. (1975). Assembly of species communities. *Ecology and Evolution of Communities* (Ed. by M.L. Cody & J.M. Diamond), pp. 342–444. Belknap Press, Cambridge, Mass.

Diamond, J.M. & Gilpin, M.E. (1982). Examination of the 'null' model of Connor and Simberloff for species co-occurrences on islands. *Oecologia*, **52**, 64–74.

Drake, J.A. (1983). Invasibility in Lotka-Volterra interaction webs. *Current Trends in Food Web Theory. Report on a Food Web Workshop* (Ed. by D.L. DeAngelis, W.M. Post & G. Sugihara), pp. 83–9. Oak Ridge National Laboratory, Tennessee.

Drake, J.A. (1985). *Some theoretical and empirical explorations of structure in food webs.* Ph.D. thesis, Purdue University, West Lafayette, Indiana.

Duffey, E. (1966). Spider ecology and habitat structure. *Senckenbergiana Biologica*, **47**, 45–9.

Finegan, B. (1984). Forest succession. *Nature*, **311**, 109–14.
Fuller, R.J. (1982). *Bird Habitats in Britain*. Poyser, Calton, Staff.
Gilpin, M.E. & Diamond, J.M. (1982). Factors contributing to non-randomness in species co-occurrences on islands. *Oecologia*, **52**, 75–84.
Gilpin, M.E. & Diamond, J.M. (1984). Are species co-occurrences on islands non-random, and are null hypotheses useful in community ecology? *Ecological Communities. Conceptual Issues and the Evidence* (Ed. by D.R. Strong Jr., D. Simberloff, L.G. Abele & A.B. Thistle), pp. 297–315, 332–41. Princeton University Press, Princeton, N.J.
Głowaciński, Z. & Järvinen, O. (1975). Rate of secondary succession in forest bird communities. *Ornis Scandinavica*, **6**, 33–40.
Golley, F.E. (Ed) **(1977).** Ecological succession. *Benchmark Papers in Ecology* **5**. Dowden, Hutchinson & Ross, Stroudsburg, Penn.
Harper, J.L. (1977). *Population Biology of Plants*. Academic Press, London.
Healey, I.N. (1972). The habitat, the community and the niche. *Biology of Nutrition* (Ed. by R.N. Fiennes), pp. 307–49. Pergamon Press, Oxford.
Heatwole, H. & Levins, R. (1972). Trophic structure stability and faunal change during colonisation. *Ecology*, **53**, 531–4.
Hils, M.H. & Vankat, J.L. (1982). Species removals from a first-year old-field plant community. *Ecology*, **63**, 705–11.
Högberg, P. & Piearce, G.D. (1986). Mycorrhizas in Zambian trees in relation to host taxonomy, vegetation type and successional patterns. *Journal of Ecology*, **74**, 775–786.
Holt, R.D. (1977). Predation, apparent competition, and the structure of prey communities. *Theoretical Population Biology*, **12**, 197–229.
Horn, H.S. (1981). Succession. *Theoretical Ecology. Principles and Applications* (Ed. by R.M. May), pp. 253–71. Blackwell Scientific Publications, Oxford.
Jackson, J.B.C. (1981). Interspecific competition and species' distributions: the ghost of theories and data past. *American Zoologist*, **21**, 889–901.
Janos, D.P. (1980). Mycorrhizae influence tropical succession. *Biotropica*, **12** (Supplement), 56–64.
Jeffries, M.L. & Lawton, J.H. (1985). Predator–prey ratios in communities of freshwater invertebrates: the role of enemy free space. *Freshwater Biology*, **15**, 105–112.
Kareiva, P. (1982). Influence of vegetation texture on herbivore populations: resource concentration and herbivore movement. *Variable Plants and Herbivores in Natural and Managed Ecosystems* (Ed. by R.F. Denno & M.S. McClure), pp. 259–89. Academic Press, New York.
Karr, J.R. (1968). Habitat and avian diversity on strip-mined land in east-central Illinois. *Condor*, **70**, 348–57.
Kneidel, K.A. (1983). Fugitive species and priority during colonization in carrion-breeding Diptera communities. *Ecological Entomology*, **8**, 163–9.
Krebs, C.J. (1985). *Ecology. The Experimental Analysis of Distribution and Abundance*, 3rd edn. Harper & Row, New York.
Kulman, H.M. (1971). Effects of insect defoliation on growth and mortality of trees. *Annual Review of Entomology*, **16**, 289–324.
Lack, D. (1971). *Ecological Isolation in Birds*. Blackwell Scientific Publications, Oxford.
Lack, D. & Lack, E. (1951). Further changes in bird-life caused by afforestation. *Journal of Animal Ecology*, **20**, 173–9.
Lawton, J.H. (1983). Plant architecture and the diversity of phytophagous insects. *Annual Review of Entomology*, **28**, 23–39.
Lawton, J.H. & Hassell, M.P. (1984). Interspecific competition in insects. *Ecological Entomology* (Ed. by C.B. Huffaker & R.L. Rabb), pp. 451–95. John Wiley, New York.
Lowrie, D.C. (1948). The ecological succession of spiders of the Chicago area dunes. *Ecology*, **29**, 334–51.

MacMahon, J.A. (1981). Successional processes: comparisons among biomes with special reference to probable roles of and influences on animals. *Forest Succession: Concepts and Application* (Ed. by D.C. West, H.H. Shugart & D.B. Botkin), pp. 277–304. Springer-Verlag, New York.

Marrs, R.H. Roberts, R.D., Skeffington, R.A. & Bradshaw, A.D. (1983). Nitrogen and the development of ecosystems. *Nitrogen As An Ecological Factor* (Ed. by J.A. Lee, S. McNeill & I.H. Rorison), pp. 113–36. Symposium of the British Ecological Society 22. Blackwell Scientific Publications, Oxford.

McBrien, H., Harmsen, R. & Crowder, A. (1983). A case of insect grazing affecting plant succession. *Ecology*, **64**, 1035–9.

McIntosh, R.P. (1980). The relationship between succession and the recovery process in ecosystems. *The Recovery Process in Damaged Ecosystems* (Ed. by J. Cairns Jr.), pp. 11–62. Ann Arbor Science, Ann Arbor, Michigan.

Mithen, S.J. (1984). *Growing foodwebs: a simulation model exploring constant pedator prey ratios in ecological communities*. M.Sc. thesis, University of York.

Mithen, S.J. & Lawton, J.H. (1986). Food-web models that generate constant predator–prey ratios. *Oecologia*, **69**, 542–550.

Morin, P.J. (1984). Odonate guild composition: experiments with colonisation history and fish predation. *Ecology*, **65**, 1866–73.

Moulton, M.P. & Pimm, S.L. (1983). The introduced Hawaiian avifauna: biogeographic evidence for competition. *American Naturalist*, **121**, 669–90.

Moulton, M.P. & Pimm, S.L. (1986). The extent of competition in shaping an introduced avifauna. *Community Ecology* (Ed. by J.M. Diamond & T.J. Case), pp. 80–97. Harper & Row, New York.

Nicolson, T.H. (1960). Mycorrhiza in the Graminae. II. Development in different habitats, particularly sand dunes. *Transactions of the British Mycological Society*, **43**, 132–45.

Odum, E.P. (1969). The strategy of ecosystem development. *Science*, **164**, 262–70.

Pierce, N.E. (1984). Amplified species diversity: a case study of an Australian lycaenid butterfly and its attendant ants. *Symposium of the Royal Entomological Society*, **11**, 197–200.

Pimm, S.L. (1982). *Food Webs*. Chapman & Hall, London.

Pimm, S.L. & Lawton, J.H. (1977). The number of trophic levels in ecological communities. *Nature*, **268**, 329–31.

Pimm, S.L. & Lawton, J.H. (1983). The causes of foodweb structure: dynamics, energy flow, and natural history. *Current Trends in Food Web Theory. Report on a Food Web Workshop* (Ed. by D.L. DeAngelis, W.M. Post & G. Sugihara), pp. 45–9. Oak Ridge National Laboratory, Tennessee.

Post, W.M. (1983). Dynamic patterns of randomly assembled food webs. *Current Trends in Food Web Theory. Report on a Food Web Workshop* (Ed. by D.L. DeAngelis, W.M. Post & G. Sugihara), pp. 69–76. Oak Ridge National Laboratory, Tennessee.

Post, W.M. & Pimm, S.L. (1983). Community assembly and food web stability. *Mathematical Biosciences*, **64**, 169–92.

Price, P.W. (1984). The concept of the ecosystem. *Ecological Entomology* (Ed. by C.B. Hufaker & R.L. Rabb), pp. 19–50. John Wiley, New York.

Rice, E.L. & Pancholy, S.K. (1972). Inhibition of nitrification by climax ecosystems. *American Journal of Botany*, **59**, 1033–40.

Schluter, D. & Grant, P.R. (1984). Determinants of morphological patterns in communities of Darwin's finches. *American Naturalist*, **123**, 175–96.

Schoener, T.W. (1983). Field experiments on interspecific competition. *American Naturalist*, **122**, 240–85.

Schluter, D., Price, T.D. & Grant, P.R. (1985). Ecological character displacement in Darwin's finches. *Science*, **227**, 1056–9.

Schowalter, T.D. (1981). Insect herbivore relationship to the state of the host plant: biotic regulation of ecosystem nutrient cycling through ecological succession. *Oikos*, **37**, 126–30.

Schowalter, T.D., Hargrove, W.W. & Crossley, D.A. Jr. (1986). Herbivory in forested ecosystems. *Annual Review of Entomology*, **31**, 177–196.

Shelford, V.E. (1963). *The Ecology of North America.* University of Illinois Press, Urbana, Illinois.

Shugart, H.H. & Hett, J.M. (1973). Succession: similarities in species turnover rates. *Science*, **180**, 1379–81.

Shugart, H.H. Jr. & James, D. (1973). Ecological succession of breeding bird populations in northwestern Arkansas. *Auk*, **90**, 62–77.

Simberloff, D. (1976). Trophic structure determination and equilibrium in an arthropod community. *Ecology*, **57**, 395–8.

Sousa, W.P. & Connell, J.H. (1985). Further comments on the evidence for multiple stable points in natural communities. *American Naturalist*, **125**, 612–5.

Southwood, T.R.E. (1977). Habitat, the templet for ecological strategies? *Journal of Animal Ecology*, **46**, 337–65.

Southwood, T.R.E., Brown, V.K. & Reader, P.M. (1979). The relationship of plant and insect diversities in succession. *Biological Journal of the Linnean Society*, **12**, 327–48.

Southwood, T.R.E., Brown, V.K. & Reader, P.M. (1983). Continuity of vegetation in space and time: a comparison of insects' habitat templet in different successional stages. *Researches on Population Ecology Supplement*, **3**, 61–74.

Strong, D.R., Lawton, J.H. & Southwood, T.R.E. (1984). *Insects on Plants. Community Patterns and Mechanisms.* Blackwell Scientific Publications, Oxford.

Talling, J.F. (1951). The element of chance in pond populations. *The Naturalist 1951* (Oct.–Dec.), 157–70.

Tansley, A.G. (1935). The use and abuse of vegetation terms and concepts. *Ecology*, **16**, 284–307.

Tansley, A.G. (1939). *The British Isles and their Vegetation.* Cambridge University Press, Cambridge. (Reprinted 1965.)

Tenow, O. (1983). Topoclimatic limitations to the outbreaks of *Epirrita* (= *Oporinia*) *autumnata* (Bkh.) (Lepidoptera: Geometridae) near the forest limit of the mountain birch in Fennoscandia. *Tree-Line Ecology* (Ed. by P. Morisset & S. Payette), pp. 159–64. Centre d'études Nordiques, Université Laval, Quebec.

Tilman, D. (1982). *Resource Competition and Community Structure.* Princeton University Press, Princeton, N.J.

Van Valen, L.M. (1982). A pitfall in random sampling. *Nature*, **295**, 171.

Whittaker, R.H. (1953). A consideration of climax theory: the climax as population and pattern. *Ecological Monographs*, **23**, 41–78.

Wise, D.H. (1984). The role of competition in spider communities: insights from field experiments with a model organism. *Ecological Communities. Conceptual Issues and the Evidence* (Ed. by D.R. Strong Jr., D. Simberloff, L.G. Abele & A.B. Thistle), pp. 42–53. Princeton University Press, Princeton, N.J.

Yapp, W.B. (1962). *Birds and Woods.* Oxford University Press, London.

Yarranton, G.A. & Morrison, R.G. (1974). Spatial dynamics of a primary succession: nucleation. *Journal of Ecology*, **62**, 417–28.

12. EXPERIMENTAL STUDIES ON THE EVOLUTION OF NICHE IN SUCCESSIONAL PLANT POPULATIONS

F.A. BAZZAZ

Department of Organismic and Evolutionary Biology, Harvard University, Cambridge, Massachusetts 02138, USA

INTRODUCTION

The question of how species coexist in a community has been central in ecology. There has been disagreement over the relative roles of biotic interactions versus physical environment and disturbance patterns in explaining community structure (reviews in Diamond & Case 1986). MacArthur (1972) and others have emphasized the role of coevolutionary interactions in reducing competition and facilitating coexistence by what is called niche differentiation. Communities are seen to reach an equilibrium. Recently, however, there has been an increased emphasis on the role of non-equilibrium processes, e.g. disturbance, in species coexistence and community organization.

There is some evidence for each of the theories of community organization and it is very likely that there are ecological situations where the effects of any one of these forces may be overriding (Grime 1977; Schoener 1982). In unstable, repeatedly disturbed habitats, competition is interrupted, the identity of neighbours changes, and coevolutionary niche differentiation is minimized or altogether prevented. In stable environments, where encounters between species may be prolonged and consistent, coevolutionary niche differentiation may occur and therefore may contribute to coexistence.

This chapter explores niche relations in successional plant populations. It is largely based on work in my laboratory that involved a long range experimental program on this subject. It is not intended to review the literature on coexistence and niche relations in general. The recent book *Community Ecology*, edited by J. Diamond & T.D. Case (1986), addresses these issues in detail. I begin with a discussion of the concepts of niche, discuss niche separation and niche differentiation, identify a series of predictions about niche and succession, present experimental tests for these predictions, and finally discuss the evolution of niche in successional plant communities.

THE CONCEPT OF NICHE

Despite its central role in community ecology and its evolutionary implications (Whittaker, Levin & Root 1973; Whittaker & Levin 1976), the concept of niche remains ambiguous and, in the extreme, may be of questionable utility (see a critique in Hurlbert 1977). Its major impact on ecology was to shift the orientation and emphasis of community ecology from a largely descriptive and classificatory one toward a population-based, dynamic view of community organization. But, unlike several other ecological concepts, the niche has not undergone a progressive evolution in meaning, clarity, and utility. Rather, the concept has been defined differently by various authors, and the corollary concepts of niche differentiation and niche separation have been used interchangably and in a confusing manner. There are four main ways in which the concept has been used:

(a) The habitat in which a species makes its living (Grinnell 1928).
(b) The role of the species in the biological environment; what it does and how it makes its living (Elton 1927).,
(c) The hyperspace–hypervolume defined by a number of physical and biological axes relative to the species and in which a species can exist indefinitely: the *n*-dimensional niche (Hutchinson 1958).
(d) The way a species' population is specialized within a community; the species position in space and time and its functional relationship to other species in that community (Whittaker 1972).

In practice, most studies of the niche concerned animals and emphasized the response of species to one resource axis, especially that of size of food items. However, some studies have used more than one axis and the concept has increasingly become associated with the range of resources related to food acquisition (Pianka 1981). Whittaker (1972) proposed that plants differ in resource use, time of activity, vertical location (including rooting depth), relation to horizontal pattern, etc., and suggested that they are axes of the niche. He (Whittaker 1975) used some of these variables to study niche relations of plants of the Sonoran semi-desert. Like Whittaker, Cody (1986) considered life-form diversity as a fundamental niche axis and concluded that differences in strategies of light interception and nutrient and water uptake among species promote their coexistence in some desert and Mediterranean-type communities. Grubb (1977) recognized four components of a plant's niche, viz., the habitat niche, the life-form niche, the phenological niche, and the regeneration niche. He suggested that coexistence could be promoted by diversification of any or all of them, but emphasized the role of the regeneration niche.

Severe limitations may be imposed on niche differentiation in plants for the following reasons. Firstly, plant resources are continuous and are usually not presented in discrete packages (Harper 1977). Secondly, autotrophy constrains possible divergence (Harper 1968). Thirdly, plants are thought to compete for space (Connell 1980). Furthermore, predation, herbivory, and pollination are strong interactions in many plant communities; therefore, it is more likely that niche differentiation may occur in plants along these parameters rather than along physical factors.

If the niche concept is to become clearly and directly relevant to the central issues of community organization (including species diversity, competitive exclusion, character displacement, maximal tolerable overlap, and species packing) the concept (i) must be clearly understood and unambiguously defined, (ii) should rest on the response of organisms to both physical and biological gradients, and (iii) ought to be applicable to both sessile and mobile organisms.

While a thorough review of the literature is not attempted here, an examination of theoretical and experimental work on the niche concepts suggests that (i) there is a large number of resources and controllers to which an organism responds, (ii) these resources and controllers are both physical and biological in nature, (iii) an organism's response is rarely, if ever, equable across different resource gradients, (iv) commonly, more than one resource determines the coexistence of organisms in a community, (v) it is highly unlikely that these resource axes are simple or are orthogonal, and (vi) the response may be influenced by the identity of neighbours. A workable definition of the niche must consider all these aspects and integrate them such that they are related to the evolution of various traits and strategies in a community context. A *niche* may therefore be defined as *the pattern of response of an individual, a population or a species to the physical and biological gradients of its environment.*

While data on performance of plants on single environmental gradients are quite informative, the quantification of an organism's niche remains incomplete without simultaneously considering performance on several gradients. Accordingly, performance on single gradients should be referred to as *response breadth* rather than *niche breadth*.

NICHE SEPARATION VERSUS NICHE DIFFERENTIATION

It has been assumed that in order for species to coexist they must differ in resource use. Theoretical analyses have produced a range of minimum values of differences required for coexistence (e.g. May 1974; Newman 1982). It is further assumed that coevolutionary interactions among coexisting species have resulted in changes in niche breadths or their dislocation over resource gradients to generate lower levels of niche overlap.

Ecologists have used, sometimes interchangably, the terms niche difference, niche separation, and niche differentiation with regard to the relative locations of response of species on resource gradients. This uncritical usage has created further confusion in understanding community organization. *Niche differentiation* should refer to *differential resource use which results from long-term, consistent competitive interactions between or among species in a community where the species populations act as selective agents on each other; it results from coevolutionary resource-use displacement. Niche separation* should refer to *the differences in resource-use patterns which may or may not involve niche differentiation. They may have evolved independently as a result of competitive displacement with species in a different community*. Niche differentiation is therefore a subset of niche separation.

Trifolium repens clones from specific biological neighbourhoods dominated by different grass species performed best when grown in competition with the grass species from whose neighbourhood they originated (Turkington & Harper 1979). This suggests that, within the pasture, strains of clover have been selected by competitive interactions with the associated grasses (Harper 1983). However, in many other communities coexistence is promoted by fitting together preadapted species (Grant 1975; Connell 1980). For example, the two winter annuals, *Erigeron annuus* and *Lactuca scariola*, that co-occur in old-fields in the mid-western US have overlapping flowering seasons and share pollinators. However, the two species show marked differences in the daily time of flower opening and, therefore, the timing of flower visitation. This results in large niche separation between the two species in use of pollinators. The origins and life histories preclude the possibility that competition for pollinators between the two species resulted in niche differentiation. *Lactuca* is an introduced species from Europe while *Erigeron* is native to North America (details in Parrish & Bazzaz 1978).

NICHE RELATIONS IN SUCCESSIONAL HABITATS

Disturbances generate early successional habitats and select for plants with short life span, fast growth, and high reproductive output. These plants possess several other life-history features (Bazzaz 1983) and the attendant physiological attributes (Bazzaz 1979) appropriate for these habitats. Because of large fluctuation in resource availability in these habitats, Odum (1969) predicted that these plants should have broad responses on resource gradients (broad niches) and therefore show much resource use (niche) overlap.

A number of other predictions could be made on the basis of the above. Firstly, communities made up of broad-niched species are likely to have low species diversity as each species occupies a large portion of total available resources (niche space). Low diversity has been shown for a number of early successional sequences, e.g. Whittaker (1975), Bazzaz (1975) and May (1981). Secondly, because niche overlap is high among plants of early succession, the species are expected to experience a high degree of competition and consequently should, on the average, show much reduction in biomass in pairwise competition. Thirdly, because of the high degree of overlap in resource use and intense competition, the plants in pairwise combinations should be able to convert less of the available resources into plant biomass. Fourthly, because of the broad niches and high degree of niche overlap, these plants should exhibit a higher equivalency of neighbours, i.e. it makes little difference to species A if it is competing with species B, species C or species D. Late-successional plants are predicted to behave in the opposite manner. Finally, because early-successional environments vary greatly within one growing season early-successional plants must deal with such variation, their response breadths will vary ontogenetically.

Over the last decade my laboratory has been engaged in an experimental programme to test these predictions. We have measured environmental variability for a number of physical parameters. We have quantified response breadth and overlap for important species of herbaceous early, mid, and late-successional communities on both continuous (e.g. soil moisture, nutrients, light, temperature) and discontinuous (pollinator) gradients. We also compared the responses of early-successional trees with those of late-successional trees. The responses considered were seed germination, seedling growth, survivorship, and mature plant vegetative biomass and reproductive biomass.

LEVELS OF ENVIRONMENTAL VARIABILITY

It has been generally assumed and repeatedly stated that early-successional habitats are more variable and unpredictable than are late-successional habitats. This notion is implicit in the incorrectly alleged correspondence of successional gradients with the $r-K$ continuum of life-history strategies (Grubb, Chapter 4). However, there has been little detailed and convincing measurement of many environmental parameters along successional gradients. While it is obvious that early-successional environments are open and sunny, at least early in the growing season, it is not clear how this openness would change as succession proceeds to older communities in different seres. The light environment clearly changes greatly as succession proceeds from a forest gap or an open field (early-successional) to a forest with well-developed canopy (late-successional). But the extent of that change with succession will be difficult to assess if late-successional communities (e.g. grassland) are of low stature and of less structural complexity than a forest. Data are still very limited for this and other aspects of the physical environment, e.g. soil moisture, nutrient content, air humidity, all of which may vary spatially and temporally. Information on the variation of the biological environment in any successional site is virtually non-existent. We know very little about variability in predators, herbivores, pollinators, pathogens, mycorrhiza, etc., in relation to succession. If we are to understand the evolution of life-history design of species along successional gradients, we must have more data and a better understanding of the levels of variation and unpredictability (which are likely to be positively correlated) of the important environmental factors that affect plants and their animal and microbe associates. Furthermore, the degree of concordance among these factors must also be understood, because resources rarely act singly or independently (discussed below). The capacity of some plants to accumulate a resource and store it until other necessary resources become available is also not understood. The luxury uptake and hoarding of nutrients in excess of current demand in some plants may be a mechanism for both depriving a neighbour of these nutrients, and for rapid growth when a limiting resource (e.g. sunlight) becomes abundant. The response to timing of nutrient availability varies for different species (Benner & Bazzaz 1985) and this may influence the competitive interactions among plants of the early-successional community.

We have compared light and soil moisture variability over a 3-year period during the growing season in three types of communities: an early-successional annual-dominated community, a late-successional perennial grassland, and a mature mesic deciduous forest S.A. Rice & F.A. Bazzaz, unpublished). In the mid-western US, the latter two represent possible end-points of succession. The sites in which measurements were taken were within a few hundred metres of each other and were on level terrain and similar soils of postglacial origin. We measured photosynthetically active radiation (PAR), leaf area index, plant height profiles, and soil water potential. We also used cloned phytometers to assess the level of variability in the three habitat types, detected by differential performance of the clones. The sampling procedure for both light and moisture allowed estimation of fine-level and coarse-level heterogeneity and temporal variability. Variability in moisture conditions is determined by spatial and temporal patterns of rainfall, soil type and topography. Furthermore, the pattern of distribution of individuals and their identity modify soil moisture heterogeneity by differential interception of precipitation, stem flow, absorption, and transpiration. The variability in the light environment is also governed by weather patterns, and within the canopy largely by the three-dimensional distribution of leaf area.

In general, measurements of light and soil water potential showed that early successional fields were more spatially and temporally variable than the grasslands, which were in turn more variable than the forest floor (Fig. 12.1). Furthermore, the field had a higher fine-resolution spatial heterogeneity in soil water potential than did the prairie and the forest. Height profiles interfaced with leaf area profiles gave frequency distributions of the light conditions as experienced by individual plants in the community. Plants in the early-successional fields experienced a greater variance in light conditions relative to individuals in the grassland. Phytometers placed in the field had a much greater variance in biomass and reproduction than did those placed in the grassland.

RESPONSES OF SPECIES ON ENVIRONMENTAL GRADIENTS

Our experimental work examined the response to several resource gradients of a number of species from an early successional annual community, a mid-successional perennial community, an early-successional forest community, a late-successional herbaceous community (tall grass prairie) and a late-successional forest community (mesic deciduous forest). The species and the communities chosen for study are widely distributed in the mid-western and eastern US and the conclusions reached from this work are very likely applicable to a large number of seres. The annual community is a mixture of native and introduced species, but the later-successional communities are dominated (with only a few exceptions) by native species. The late-successional species have been present together for a long time (but see Davis 1981; Webb 1984), thus the probability of coevolutionary niche differentiation is higher in these communities than in the early-successional communities.

The experiments involved germinating seeds or growing a number of individuals on a range of environmental gradients that included both shortages and excesses of resources. The response of the plants to these

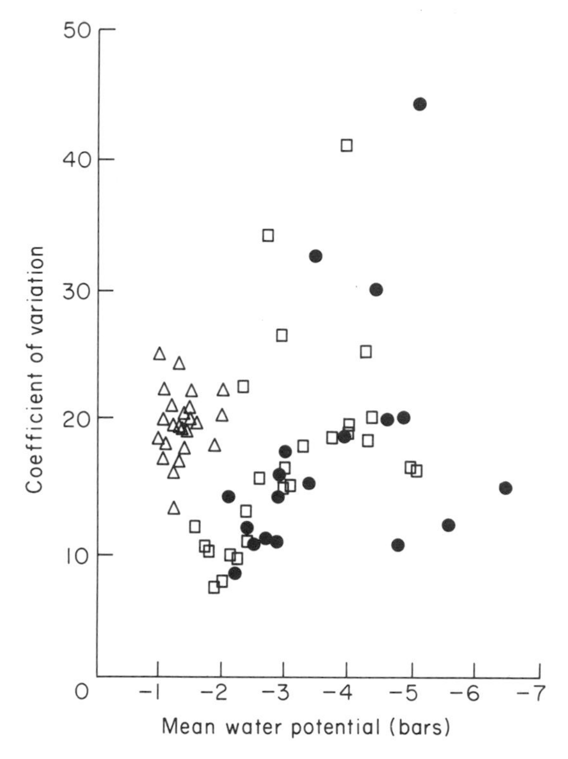

FIG. 12.1. Coefficient of variation in soil water potential of an early-successional field (●), a late-successional grassland (□), and a late-successional deciduous forest (△) during a growing season.

gradients were measured in terms of germination, survivorship, and growth as vegetative and reproductive biomass. Response breadths were calculated using Levins' B (1968) formula $B = 1/(P_i^2)S$, where P_i is the proportional response in resource state i and S is the number of resource states. Response overlap was estimated as the degree of similarity between species: $PS_{ij} = 1 - 1/2\Sigma|P_{ih} - P_{jh}|$ as used by Schoener (1970) where PS_{ij} is the proportional similarity of species i and j, P_{ih} is the proportion of species i in the hth resource state, P_{jh} is the proportion of species j in the same resource state.

The data on response breadths and overlaps were compared at two levels: between communities and within communities (among species). On a between-community scale (Table 12.1) we consistently found that the early-successional communities have species with broader and more overlapping responses than do late-successional communities, irrespective of their stature. In cases where mid-successional communities were studied the response breadths were between those of early and late-successional plants. These results confirm the initial general prediction based on measurements of the levels of environmental variability in these habitats.

On a within-community level, however, each species differed in its response on different environmental gradients and different species differed in their response on the same gradient. An example is given in Table 12.2.

TABLE 12.1. Response breadths (Levins' *B*) and proportional similarities of plants from early and late succession on several niche axes

	Early succession	Late succession
(a) Mean niche breadth of all species in experimental assemblage		
Herbaceous		
Underground space	0·71	0·29
Pollinators	0·20	0·16
Nutrients	0·77	0·70
Trees		
Nutrients	0·91	0·82
Moisture	0·89	0·65
(b) Mean proportional similarity of all species in experimental assemblage		
Herbaceous		
Underground space	0·68	0·43
Pollinators	0·31	0·19
Nutrients	0·85	0·83
Trees		
Nutrients	0·94	0·83
Moisture	0·82	0·70

TABLE 12.2 Response breadths calculated as Levins' *B* of the early-successional annuals on four environmental gradients

	Moisture	Light	Temperature	Nutrients
Abutilon theophrasti	0·88	0·71	0·62	0·60
Amaranthus retroflexus	0·49	0·83	0·78	0·69
Ambrosia artemisiifolia	0·76	0·91	0·95	0·72
Chenopodium album	0·87	0·88	0·54	0·81
Polygonum pensylvanicum	0·99	0·90	0·91	0·82
Setaria faberii	0·99	0·86	0·92	0·76

RESPONSE ON SOIL MOISTURE GRADIENTS

We compared the response of two prominent members of the annual community, *Polygonum pensylvanicum* and *Abutilon theophrasti*, on soil moisture gradients in pure and mixed stands (Pickett & Bazzaz 1976). In pure stands (exploring axes of fundamental niche) both species had broad germination, height, biomass, seed production, and seed weight responses. The overlap in their responses was especially high for the first three characters (proportional similarities PS = 0·93, 0·88, and 0·82 respectively). In competition (exploring axes of realized niche), the response centre for *Polygonum* was markedly displaced toward the wetter portion of the gradient, while *Abutilon* was displaced toward the drier end. In fact, *Abutilon* reproduced mainly in drier portions of the gradient where *Polygonum* canopy height was reduced (Fig. 12.2). The results clearly show that competitive divergence can occur in early-successional communities.

In another study, all six dominant species of the annual community were grown individually and all together on a moisture gradient (Pickett & Bazzaz 1978a). The species had similar broad vegetative and reproductive responses on the gradient. They also had a high degree of response overlap (mean = (0·798 ± 0·055.) The responses of some species shifted greatly or were reduced along the gradient in competition while those of others did not (Fig. 12.3). For example, *Polygonum* response breadth was reduced from 0·73 to 0·43 while that of *Ambrosia* dropped only from 0·94 to 0·92.

Populations of *Polygonum* and *Abutilon* responded differently in the presence of other competitors than they did when they competed only with each other. Instead of shifting toward the opposite portions of the gradient as before, they both shifted toward the wetter end of the gradient. This indicates the presence of 'refugia' for these species to which they would withdraw in the face of competition. These refugia may be shifted, reduced, or remain unchanged depending on the identity of the species and its competitors. Coexistence among these species is undoubtedly promoted by species having different refugia independent of their response similarities along the gradient when grown individually.

Seed germination response to soil moisture in the annual community was broad (mean B = 0·862 ± 0·093), but was narrower than the growth and reproduction response on the same gradient. The species were quite similar to each other in germination (PS = 0·862 ± 0·057) (Pickett & Bazzaz 1978b). In contrast, seed germination response on a temperature gradient differed for these species (some were broad and others were

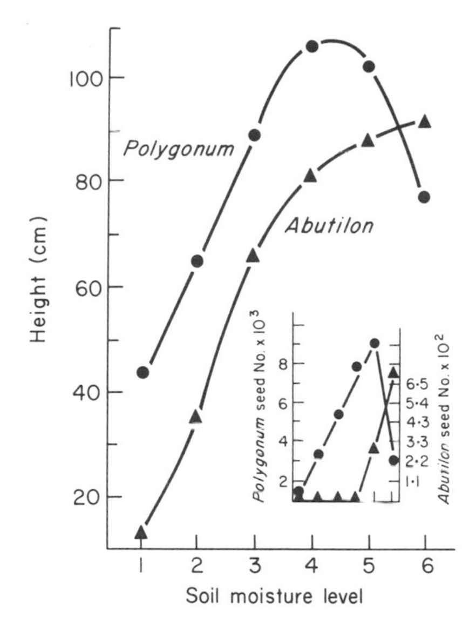

FIG. 12.2. Mean height of *Polygonum pensylvanicum* and *Abutilon theophrasti* grown together on a soil moisture gradient. Inset shows seed production. Reproduction in *Abutilon* occurs only in states where its canopy reaches or exceeds that of *Polygonum*.

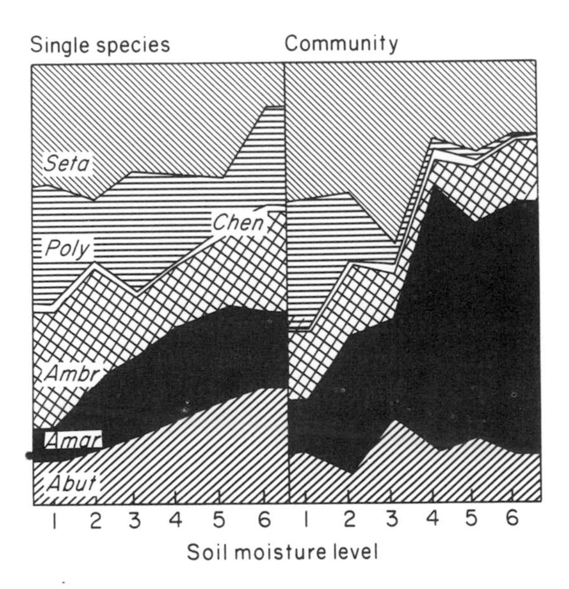

FIG. 12.3. Proportional contributions of six annuals of the early-successional community grown separately and in competition on a soil moisture gradient. The species from top to bottom are: *Setaria faberii*, *Polygonum pensylvanicum*, *Chenopodium album*, *Ambrosia artemisiifolia*, *Amaranthus retroflexus*, and *Abutilon theophrasti*.

narrow) and, in combination with stratification requirements, may influence population structure in this community (Bazzaz 1984).

Members of the annual community also exhibit other differences in response. The species have different root habits and they place their roots in different locations in the soil profile (Wieland & Bazzaz 1975). There were also differences among the species in the timing of root growth (Fig. 12.4) but the similarity among them was quite high (mean PS = 0·88). Compared with the root distribution of members of a late-successional community (a prairie), response breadth and overlap were significantly higher in the annual community (Parrish & Bazzaz 1976) (Table 12·1).

RESPONSE ON NUTRIENT GRADIENT

The response of plants from early, mid and late-successional communities were investigated on a nutrient gradient (Parrish & Bazzaz 1982a). The species were compared with regard to germination, survival, and growth. All species germinated over a wide range of nutrient concentrations. Their germination response breadths were similar within each community but were slightly broader in the early than in the late-successional community (mean values of B were 0·94 and 0·89, respectively). Mean overlap in response was also high (PS = 0·84, 0·84, 0·81 for early, mid, and late-successional communities). The small differences in response breadth and proportional similarity for germination were not statistically significant however.

The early-successional species survived over a wider range on the nutrient gradient than did the mid and late-successional species (Fig. 12.5). Species of the early successional community had the broadest biomass response along the gradient, whether the plants of each community were grown in pure or in mixed stands. The reduction in response breadth for the late-successional community was much more pronounced in mixed than in pure stands (Table 12.3).

There were significant differences in the breadth of growth response among the three assemblages in both pure and mixed stands. Species from the early-successional communities had the broadest and, especially in mixed stands, species of late-successional communities had the narrowest response (Table 12.3). Proportional similarities were high in the three assemblages. Considered over all nutrient concentrations, the average individual biomass in mixture was 1·04 times that in pure for the early-successional species and 1·66 for the late-successional species. Thus, individuals of the early-successional species experienced more competitive reduction in mixed stands than did species of late succession. We

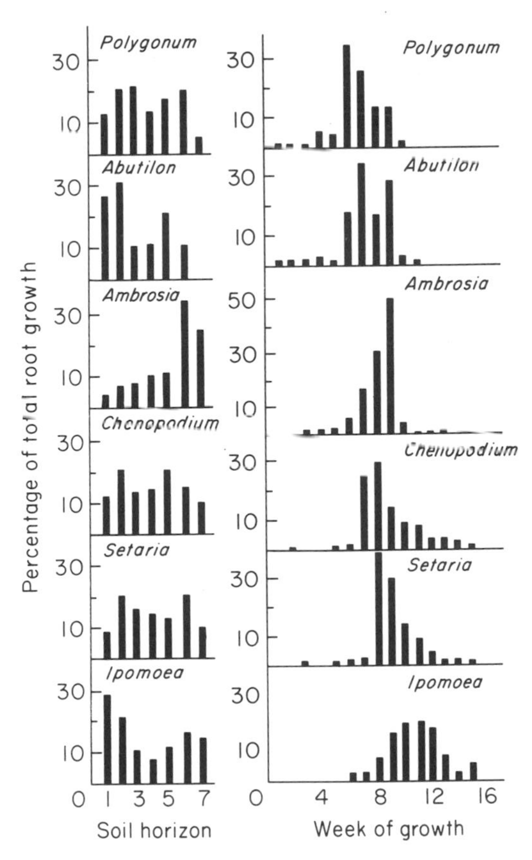

FIG. 12.4. Differences among the early-successional annuals in the location of roots at different depths in soil and the times of root growth during the growing season.

concluded that past competition resulting in niche divergence was more important in the organization of late-successional communities.

Contrary to the results on the moisture gradient discussed previously, the germination response was broader than the growth response on the nutrient gradient. When germination and growth responses were combined in the calculation of response breath, using the method of Colwell & Futuyma (1971), differences among the communities became large: B = 0·93 for germination and 0·61 for growth for the early-successional community; 0·72 for germination and 0·42 for growth for mid-successional community; and 0·75 for germination and 0·51 for growth for late-successional.

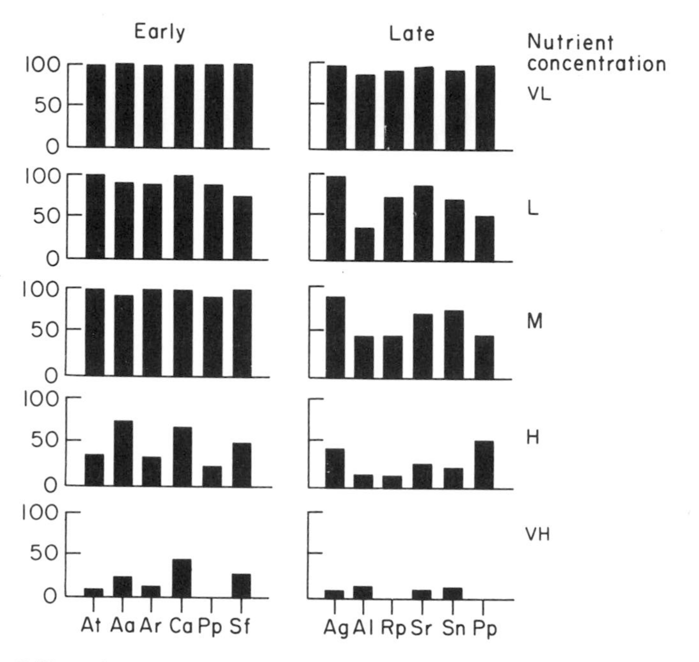

FIG. 12.5. Differential survivorship of early- and late-successional plants on a nutrient gradient. Species from left to right are *Abutilon theophrasti*, *Ambrosia artemisiifolia*, *Amaranthus retroflexus*, *Chenopodium album*, *Polygonum pensylvanicum*, *Setaria faberii*, *Andropogon gerardii*, *Aster laevis*, *Ratibida pinnata*, *Solidago rigida*, *Sorghastrum nutans*, *Petalostemum purpureum*. VL is very low nutrient concentration, VH is very high nutrient concentration.

RESPONSE BREADTH OF WOODY PLANTS

Trees of the eastern deciduous forest and elsewhere can also be viewed as early, mid or late-successional. Their position along successional gradients is thought to be largely determined by their response to the light environment and the attendant physiological adaptations. To test the applicability of predictions about niche relations to woody plants we chose three early-successional species (*Gleditisia triacanthos*, *Crataegus mollis* and *Pinus taeda*) and three late-successional species (*Quercus rubra*, *Tilia americana*, and *Acer saccharum*) and measured the responses of their seedlings to light, moisture, and nutrient gradients (Parrish & Bazzaz 1982b). Again, mean response breadth and proportional similarity on the moisture and nutrient gradients were higher for the early-successional than for the late-successional tree species. Surprisingly, on the light gradient, however, all species had a broad response (Table 12.4). Further-

TABLE 12.3. Mean response breadth (B) and proportional similarity of species in pure and mixed stands calculated on the basis of plant biomass for plants grown on a nutrient gradient

	Mean B		*Mean PS*	
	Pure	Mixed	Pure	Mixed
Early successional	0·77	0·66	0·89	0·77
Late successional	0·70	0·50	0·83	0·81

TABLE 12.4. Response breadth of early- and late-successional tree seedlings on three resource gradients

Gradient	Moisture	Nutrients	Light
Early-successional			
Gleditsia	0·77	0·91	0·95
Pinus	0·97	0·91	0·97
Crataegus	0·92	—	—
Mean	0·89	0·91	0·96
Late-successional			
Tilia	0·86	0·78	0·95
Quercus	0·54	0·82	0·97
Acer	0·56	0·82	0·99
Mean	0·65	0·81	0·97

more, as in the herbaceous communities discussed earlier, not all species in any group consistently had the highest or lowest values and the response was dependent on the gradient and the species considered.

RESPONSE BREADTH ON BIOLOGICAL RESOURCES

Plant response to biotic factors is more likely to result in niche differentiation of species than is response to physical environment. As noted previously, it may be argued that the non-discrete nature of plant resources, the similarity in resource requirements by plants, and competition for space may preclude niche differentiation. Pollinators are presented to the plants as discrete entities and, since pollinators are more-or-less specific, pollination has been shown to be important in community organization. There is good evidence that competition for pollinators may result in reduced seed set (review in Waser 1983).

We examined pollination niche relations of plants from three successional communities (Parrish & Bazzaz 1979). We considered seasonal and daily time of flowering, the species of pollinator, daily time of visits to flowers, and calculated a two-dimensional niche parameter (species of pollinator × daily time of visit). Seasonal flowering was clumped in all communities and, in two communities, occurred at three distinct seasonal times: spring, summer, and autumn. In the annual community the differences among species in flowering time reflect differences in life history of the individual species (e.g. *Erigeron annuus* and *Erigeron canadensis*) as related to time and spread of germination (which are controlled by rainfall and temperature), photosynthetic response to changing temperatures, and other factors of the physical environment (details in Bazzaz 1984). However, in the late-successional community (the prairie) there was no ready explanation of flowering time in terms of responses to the physical environment. The species are all perennials and they do not obviously differ in their vegetative life cycles. The three aggregations in seasonal flowering in this community are probably formed through adaptation to the biotic environment. Response breadths were higher in the early-successional community than in the late-successional community (Table 12.5). The prairie species had small seasonal flowering response breadth, but a high seasonal response overlap. In fact, in this community there was convergence in flowering time of species attracting similar pollinators. Response overlaps for pollinator species used were significantly lower than obtained for computer-generated random assemblages, suggesting some degree of niche separation among the plants. The mean similarities between plants for pollinator species used and of the two-dimensional variable (species of pollinator × daily time) were lower for the prairie community than for the annual community. These results suggest that coevolutionary niche differentiation on pollination axes may have occurred in the late-successional community.

NICHE SHIFTS DURING ONTOGENY

As sessile organisms, plants experience the environment as it varies throughout their entire life cycles. Seeds of many early-successional plants are often long-lived and are able to await conditions favourable for germination and growth. Such seeds are expected to have a relatively narrow germination response, especially on the moisture gradient. In contrast, recruitment of the initial colonists in open, early-successional habitats occurs usually when the soil surface is bare and the seedlings may experience extreme variation in temperature (Raynal & Bazzaz 1975) and soil moisture (Regehr & Bazzaz 1979). Later in the season, as the

TABLE 12.5. Mean response breadth on several pollination axes for several species of an early- and a late-successional community. 2-D = two-dimensional species of pollinator × time of visit

	Seasonal flowering time	Daily flowering time	Pollinator species	2-D
Annual	0·213	0·556	0·198	0·153
Late-successional	0·163	0·516	0·163	0·096

canopy develops, there is usually a decrease in environmental variability caused in part by the vegetation itself (Bazzaz & Mezga 1973). Mature plants may be buffered against changes in soil moisture because of rooting patterns and against changes in temperature and nutrient availability by their large above- and below-ground mass: they have a wide access to resources. We have found that seedlings and mature individuals of the early-successional species *Ambrosia artemisiifolia* (Bazzaz 1974) and *Aster pilosus* (Peterson & Bazzaz 1978) do differ in photosynthetic response to light, temperature and moisture with age as predicted.

A detailed examination of plants of the early successional community as germinating seeds, seedlings and adults on four environmental gradients revealed significant differences in response breadth among the three life stages (Parrish & Bazzaz 1985) (Fig. 12.6) These species responses differed on various gradients but generally seem to be consistent with the nature of variability of the environmental parameter considered. Furthermore, plant responses were more similar to those of other species of the same age than to different age individuals of the same species, especially on the nutrient and soil moisture gradient (Table 12.6).

Perennials may exhibit ontogenetic niche shifts as well. The shift may not be as pronounced as in the annuals. We observed differences in the response of individuals to environmental gradients in herbaceous perennials grown under the same conditions from seed or from rhizome. In *Polygonum virginianum*, a species which is found in the understorey and on the edges of deciduous forests in the US, individuals derived from seeds were more responsive to resource deficiencies than were individuals started from rhizomes (Zangerl & Bazzaz 1983). Further analysis (Lee *et al.* 1986) using this species and its early-successional annual congener, *P. pensylvanicum*, showed that variation in response of conspecific individuals of different ages differed as much in response to environmental gradients as did same-aged individuals of separate species. In general, individuals of the two species were more similar to one another

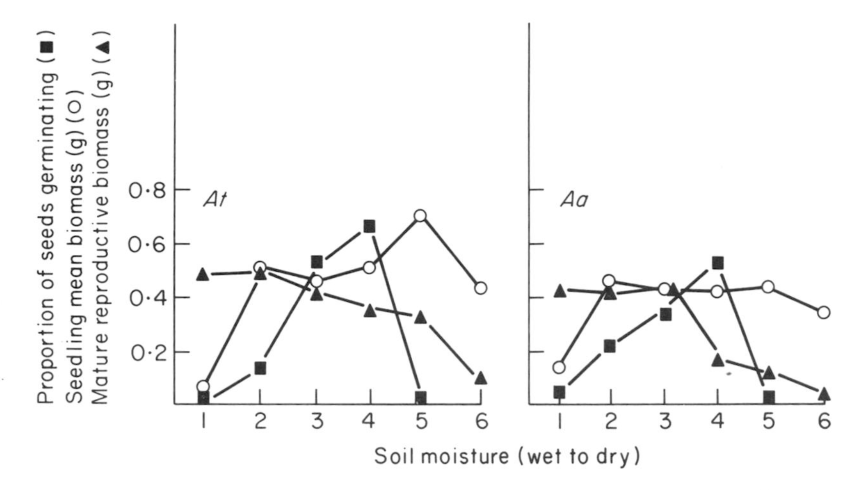

FIG. 12.6. Response shifts with ontogeny as seed germination, seedling growth and reproductive biomass of mature individuals of *Abutilon theophrasti* and *Ambrosia artemisiifolia* on a soil moisture gradient.

TABLE 12.6. Number of pairwise comparisons in which species are most similar to (a) other species of the same age, or (b) other ages of the same species

Gradient	(a)	(b)	*P*
Nutrient	27	9	<0.002
Moisture	33	3	<0.001
Temperature	21	15	>0.20
Light	4	6	>0.50

when both were started from seed than when individuals of *P. virginianum* were started from rhizome.

RESPONSE ON MULTIDIMENSIONAL GRADIENTS

The plant's environment is always multidimensional. A number of resources and controllers act upon an individual and its response must encompass all these factors. A single factor may be limiting and thus override the effects of other factors, but only in certain locations and at certain times. The plants must respond to the simultaneous influences of these factors and their changing interactions. Therefore, while single-factor results may shed some light on species strategies, multifactor data are necessary for an understanding of life-history evolution. Unfortunately, data on multifactor responses are nearly non-existent for plants. As mentioned previously, we demonstrated that the behaviour of species in a two-dimensional resource space may not be predictable on the basis of each species response to separate resources. The degree of similarity among species with regard to placement of roots in the soil and the timing of nutrient uptake by these roots was greater than predicted by considering these activities separately.

We compared the behaviour of two species of *Polygonum* on single gradients of soil, moisture, nutrients, and light, and on the two-factor crossed gradient of nutrients and light (Zangerl & Bazzaz 1983). The species behaved somewhat differently from each other on the gradients. *Polygonum pensylvanicum* (early-successional annual) was more productive and behaved more opportunistically to resource addition. In contrast, *Polygonum virginianum* (late-successional perennial) was more responsive to deficiency in resources. In the single gradient experiment, there were significant interactions between levels of soil moisture and competition, and between light and competition in both species. However, in the double-gradient experiments, there were no significant interactions for *P. virginianum* but all interactions were significant for *P. pensylvanicum*. Furthermore, the contrasts between the two species were more pronounced in double-gradient experiments. Proportional similarities in response of the two species in the double-gradient experiment were higher than in the single-gradient experiment.

Tilman (1986) considers that soil nutrients and light form a natural complex gradient. Each plant has a certain region along this gradient in which it is a superior competitor. He feels that such gradients may have been a major axis of differentiation and evolution of early plants and may help explain the life-history patterns of current plants.

Experiments that simultaneously consider several resource gradients are necessary for a clear understanding of niche relations in plants. In order to satisfactorily determine the multi-gradient responses, the following points must be considered. Firstly, the relevant resources and controllers of the plant's niche must be correctly identified. Light, water, nutrients, CO_2, and temperature are important to all plants and must be included. But it is conceivable that specific nutrients or even ratios of these to each other (e.g. Tilman 1982) are also important. Furthermore, the importance of biological resources may be different for different plant species. Specializations in pollination, dispersal, herbivory, mycorrhizal associations and pathogens are being discovered in many ecosystems. Secondly, little is known about factor interaction and the degree to which plants can compensate for excesses or shortages of one resource by the acquisition of other resources. For example, how much reduction in light could a plant tolerate if it were to receive an additional quantity of phosphorus? Thirdly, the experimental design of multifactor experiments may not be straightforward. The appropriate ranges for each resource may not be easy to determine prior to experimental work. Furthermore, while there are satisfactory methods (review in Smith 1982, a critical evaluation in Petraitis 1981) for analysing response data on single gradients (e.g. Levins' *B*, Colwell & Futuyma's 1971 method of calculating weighted niche breadths, partitioning of niche breadth into between- and within-phenotype (Roughgarden 1979) and modification of the Finlay & Wilkinson method for magnitude of response (Garbutt & Zangerl 1983), multivariate methods are still evolving (Dueser & Shugart 1982; Carnes & Slade 1982; Van Horne & Ford 1982), and some of these methods may not be applicable to plants.

Non-orthogonality, response interaction, and resource compensation have important evolutionary implications. Responses of plants cannot be maximized for individual factors in isolation of other factors because of genetic correlation (e.g. Antonovics 1976). Rather, evolution will focus on integrated plant responses to resource interactions. It may be possible by rearrangement of genes on chromosomes (by crossing over, translocation, transposable elements, etc.) even to increase the levels of correlation between certain axes of the niche—a process of 'biological eigenvectorization!'

COMPETITIVE INTERACTIONS AND NICHE BREADTH

Since there may be real limitations to niche differentiation among plants, it is possible that competition may not be of primary importance in plant community organization. Nevertheless, there are many studies that demonstrate the importance of interference in plant communities. We have experimentally shown interference in successional processes in some old-field ecosystems (e.g. Raynal & Bazzaz 1975). Furthermore, we have demonstrated in several communities that differences in response breadth and magnitude along gradients among species could be modified by competitive interactions (Fig. 12.3). These species, therefore, show clear niche separation on these gradients and can coexist because of these differences. Flexibility in timing of reproduction, exhibited by several species in the early-successional community, may contribute to coexistence as well. For example, when grown together, *Abutilon theophrasti* reproduces successfully only when water deficit limits the stature of *Polygonum pensylvanicum* canopy and where the *Polygonum* canopy disintegrates after reproduction. Reproduction of *Chenopodium album* in the field occurs only after the disintegration of the canopy of associated annuals. Suppressed individuals remain in the understorey of *Ambrosia trifida* stands after all other species have been eliminated by competition from *Ambrosia* (Abul-Fatih & Bazzaz 1979). Again flowering occurs after the disintegration of the *Ambrosia* canopy.

Based on our data on response breadth and overlap (Table 12.1) we predicted that competitive interaction among species of early-successional communities would be intense because a number of factors prevent directional selection to reduce competition. In contrast, past competition and directional selection in late-successional communities could have resulted in niche differentiation and reduced competition. If resources are presented to plants in limited supply we predict that the species with lower response overlap (late-successional) would experience less reduction in biomass in pairwise heterospecific competition than should highly overlapping species of early-successional communities. Furthermore, in mixed stands late-successional species should be able to convert relatively more of the available resources into plant matter than do species of the early-successional community.

We tested these predictions by growing six species each of early, mid, and late-successional communities in containers singly, in conspecific and heterospecific pairs, and in pure and mixed stands (Parrish & Bazzaz 1982c). The mass of individuals grown in heterospecific pairs (relative biomass) divided by the mass of individuals of the same species grown in conspecific pairs was significantly higher for the late-successional than for the early-successional community. In most of the pairwise combinations of the early-successional species, individuals of both competitors had the same biomass in mixtures and in pure stands. In contrast, in the late-successional and the mid-successional communities all mixed pairs had a winner and a loser. Biomass in heterospecific pairs divided by biomass of the same species grown singly showed that early-successional species suffered most reduction in total mass ($\bar{X} = 0{\cdot}50$) and the late-successional species the least ($\bar{X} = 0{\cdot}76$). The results show that there was less competitive reduction in performance in the more niche-separated late-successional community. There was some evidence that mixtures of the late-successional community are able to convert relatively more of the available resources into plant matter. In mixed stands the late-successional species had a higher relative yield total and a slightly greater mixed/pure stand ratio than did the early-successional assemblage.

Therefore, this study showed that members of the late-successional community suffered less reduction in the presence of heterospecific neighbours than did species of early succession. This is strong evidence that niche separation results in reduced competition within communities and that selection to reduce competition could have been more important in the evolution of late-successional species than of early-successional species. Furthermore, the coefficient of variation of the ratio of mean individual biomass in mixed stand to mean individual biomass in pure stand for all species was much lower in the annual community than for the prairie community (0·23 vs. 0·85). That is, there was less difference in the performance among species in the annual group. This similarity suggests that early-successional species are more equivalent as neighbours. This may reduce the possibility of niche differentiation in the annual community, as will be discussed below.

EVOLUTION OF NICHE BREADTH IN SUCCESSIONAL PLANTS

Coexistence in plants may occur primarily because of sufficient niche separation (including niche differentiation) among them. However, Aarssen (1983) argues that selection for balanced competitive ability may provide an alternative evolutionary mechanism for coexistence which does not require that every species occupy a different niche. Also, the potential to generate appropriate genetic variants in response to changing relative competitive ability in each species permits coexistence (Turkington & Aarssen 1984).

While it is practically impossible to prove the existence of coevolutionary niche differentiation in plants, our work discussed above strongly suggests that some niche differentiation has occurred in late-successional communities and is the basis for some of the clear niche separation found in these communities. In contrast, it is less likely that differentiation has occurred in early-successional communities and that many of the observed differences in response among some species on some gradients is more likely caused by difference of evolutionary history and preadaption to conditions in other communities from which these species are derived.

In the early-successional community, selection by the physical environment is strong and may override most biological interactions. The advantage of broad niches for species colonizing early-successional habitats are obvious. The progeny of these individuals must disperse widely in space and in time (by seed longevity) in order to exploit uncontested resources in disturbed, unoccupied, or sparsely occupied locations. The unpredictability of the environment of these patches would favour species with broad niches. In the early-successional community we found that mean similarity in response of species in mixed stands along the moisture gradients was not significantly different from that of a computer-generated random community (0·66 ± 0·18 vs. 0·68 ± 0·10). Thus, the species are not differentially packed on this gradient. In fact, considered individually without competition with the other members of the community, the similarity in response breadths of the annual community (0·798 ± 0·055) was higher than for the random community. Therefore, selection by the physical environment has increased the similarity of the niches of these species above random expectations (Pickett & Bazzaz 1978b). There has been niche convergence rather than divergence.

Another manifestation of the strength of physical selection and response convergence in the early-successional community is found in ontogenetic response shifts described earlier. Individuals of different species, but of the same age (life state) have more similar responses on the gradients than do individuals of the same species of different ages. Multiple patches occupied by seedlings of different species may be more similar in environment than is the same patch at different times in the growing season.

The response of a species or a population on a resource gradient (niche breadth) is the sum of the response of the individuals of that population on the gradient (Roughgarden 1972; Pianka 1983). The manner in which the niches of individuals contribute to the population niche has not been studied. In the extreme a population may be made up of either (i) similar, broad-niched (plastic) individuals or (ii) narrow-niched individuals each specializing on a different portion of the resource gradients. It is likely that populations will be composed of various combinations of generalists and specialists.

The structure of a population niche may be influenced by many factors including the nature of the breeding system, efficiency of gene flow, the level of patchiness in resource availability, the nature of competitors, pathogens and herbivores, etc. Studies of morphological traits as well as enzyme loci have generally shown the presence of large amounts of genetic variation within and between plant populations (review in Hamrick, Linhart, & Mitton 1979). Species with large ranges, outcrossing, wind pollination, high fecundity, long generation times, and from later-successional habitats seem to have more genetic variation than do other types of species. Furthermore, in plants plasticity plays a major role in their response to the environment (review in Bradshaw 1974, Sultan 1987). This seems to be especially true for early-successional plants. It remains to be seen whether differences in partitioning of response breadth among individuals between and within successional groups of plants exist and how they are related. Nevertheless, it may be predicted that early-successional plants would have a lower genotypic diversity and broader individual niche breadth than do late-successional plants.

The response of individual genotypes to each other and to their environment is not a trivial matter for the evolution of niche breadth in interacting plant populations. In nature, interactions do not take place among species, but among individuals. The genotype identity of competing neighbours may be highly variable in time and space. Therefore, if any niche differentiation occurs, it has to be on a genotype-to-genotype basis. Models of competition among genotypes can maintain genetic polymorphism (Antonovics 1978). Furthermore, differences in response may involve separate characters such that a genotype's morphological and physiological traits will behave differently on different resource states. Thus, evolutionary interactions, their strengths, consequences, etc. will depend on the identity of interacting genotypes, the specific character (or characters) involved and the nature of the gradient.

Since interactions are usually more common among genotypes of the same species (especially in communities of low species diversity such as early-successional ones), it may be argued that any niche differentiation could occur within species rather than between species. But within species in a given location, niche differentiation may be severely limited by gene flow. It may not matter to which species a genotype belongs: the ecological response rather than the specific identity determines the influence of a genotype on its neighbours. The notion of equivalency of neighbours (Parrish & Bazzaz 1982c; Goldberg & Werner 1983) could therefore be extended from species to genotype.

Though these questions are very crucial to the understanding of the central issues of the evolution of diversity and to the maintenance of variation in plant populations, they remain unresolved and, at this time, no specific predictions can be confidently made about the genetic structures of niches for plant populations. Investigation of the response of individual genotypes on various resource gradients would go a long way to clarify this situation.

ACKNOWLEDGMENT

I thank J.A.D. Parrish for substantial contributions to the research discussed in this paper.

REFERENCES

Aarssen, L.W. (1983). Ecological combining ability and competitive combining ability in plants: towards a general evolutionary theory of coexistence in systems of competition. *American Naturalist*, **122**, 707–731.

Abul-Fatih, H.A. & Bazzaz, F.A. (1979). The biology of *Ambrosia trifida* L. I. Influence of species removal on the organization of the plant community. *New Phytologist*, **83**, 813–816.

Antonovics, J. (1976). The nature of limits to natural selection. *Annals of Missouri Botanical Garden*, **63**, 224–247.

Antonovics, J. (1978). The population genetics of mixtures. *Plant Relations in Pastures* (Ed. by J.R. Wilson), pp. 233–252. CSIRO, East Melbourne, Australia.

Bazzaz, F.A. (1974). Ecophysiology of *Ambrosia artemisiifolia*: a successional dominant. *Ecology*, **55**, 112–119.

Bazzaz, F.A. (1975). Plant species diversity in old-field successional ecosystems in southern Illinois. *Ecology*, **56**, 485–488.

Bazzaz, F.A. (1979). Physiological ecology of plant succession. *Annual Review of Ecology and Systematics*, **10**, 351–371.

Bazzaz, F.A. (1983). Characteristics of populations in relation to disturbance in natural and man-modified ecosystems. *Disturbance and Ecosystems: Components of Response* (Ed. by H.A. Mooney & M. Godron), pp. 259–275. Springer-Verlag, Berlin.

Bazzaz, F.A. (1984). Demographic consequences of plant physiological traits: some case studies. *Perspectives in Plant Population Ecology* (Ed. by R. Dirzo & J. Sarukhan), pp. 324–346. Sinauer Publishers, Sunderland, Massachusetts.

Bazzaz, F.A. & Mezga, D. (1973). Primary productivity and microenvironment in an Ambrosia-dominated old field. *American Midland Naturalist*, **90**, 70–78.

Benner, B.L. & Bazzaz, F.A. (1985). Response of the annual *Abutilon theophrasti* medic. (Malvaceae) to timing of nutrient availability. *American Journal of Botany*, **72**, 320–323.

Bradshaw, A.D. (1974). Environment and phenotypic plasticity. *Brookhaven Symposium in Biology*, **25**, 75–94.

Carnes, A. & Slade, N.A. (1982). Some comments on niche analysis in canonical space. *Ecology*, **63**, 888–893.

Cody, M.L. (1986). Structural niches in plant communities. *Community Ecology* (Ed. by J. Diamond & T.J. Case), pp. 381–405. Harper and Row, New York.

Colwell, R.K. & Futuyma, D.J. (1971). On the measurement of niche breadth and overlap. *Ecology*, **52**, 567–576.

Connell, J.H. (1980). Diversity and the coevolution of competitors, or the ghost of competition past. *Oikos*, **35**, 131–138.

Davis, M.B. (1981). Quaternary history and stability of forest communities. *Forest Succession: Concepts and Applications* (Ed. by D.C. West, H.H. Shugart & D.B. Botkin), pp. 134–153. Springer-Verlag, New York.

Diamond J. & Case, T.J. (Eds) **(1986).** *Community Ecology*. Harper and Row, New York.

Dueser, R.D. & Shugart, H.H. (1982). Reply to comments by Van Horne & Ford and by Carnes & Slade. *Ecology*, **63**, 1174–1175.

Ellenberg, H. (1963). Vegetation Mitteleuropas mit den Alpen. *Einfuhrung in die Phytologie*, Vol. 4 (Ed. by J. Walter), pp. 1–943. Ulmer, Stuttgart.

Elton, C.S. (1927). *Animal Ecology*. Sidgwick and Jackson, London.

Garbutt, K. & Zangerl, A.R. (1983). Application of genotype-environment interaction analysis to niche quantification. *Ecology*, **64**, 1292–1296.

Goldberg, D.E. & Werner, P.A. (1983). Equivalence of competitors in plant communities: A null hypothesis and field experimental approach. *American Journal of Botany*, **70**, 1098–1104.

Grant, P. (1975). The classical case of character displacement. *Evolutionary Biology*, **8**, 237–337.

Grime, J.P. (1977). Evidence for the existence of three primary strategies in plants and its relevance to ecological and evolutionary theory. *American Naturalist*, **11**, 1169–1194.

Grinnell, J. (1928). Presence and absence of animals. *University of California Chronicle*, **30**, 429–450.

Grubb, P.J. (1977). The maintenance of species richness in plant communities: the importance of the regeneration niche. *Biological Reviews*, **52**, 107–145.

Hamrick, J.L., Linhart, Y.B. & Mitton, J.B. (1979). Relationships between life history characteristics and electrophoretically detectable genetic variation in plants. *Annual Review of Ecology and Systematics*, **10**, 173–200.

Harper, J.L. (1968). The regulation of numbers and mass in plant populations. *Population Biology and Evolution* (Ed. by R.C. Lewontin), pp. 139–158. Columbia University Press. New York.

Harper, J.L. (1977). *Population Biology of Plants*. Academic Press, London.

Harper, J.L. (1983). A Darwinian plant ecology. *Evolution from Molecules to Man* (Ed. by D.S. Bendall), pp. 323–345. Cambridge University Press. Cambridge.

Hurlbert, S.H. (1977). A gentle depilation of the niche: Dicean resource sets in resource hyperspace. *Evolutionary Theory*, **5**, 177–184.

Hutchinson, G.E. (1958). Concluding remarks. *Cold Spring Harbor Symposium on Quantitative Biology*, **22**, 415–427.

Lee, H.S., Zangerl, A.R., Garbutt, K. & Bazzaz, F.A. (1986). Within and between species variation in response to environmental gradients in *Polygonum pensylvanicum* and *Polygonum virginianum*. *Oecologia* (in press).

Levins, R. (1968). *Evolution in Changing Environments*. Princeton University Press, Princeton, N.J.

MacArthur, R.H. (1972). *Geographical Ecology*. Harper and Row, New York.

May, R.M. (1974). On the theory of niche overlap. *Theoretical Population Biology*, **5**, 297–332.

May, R.M. (1981). Patterns in multi-species communities. *Theoretical Ecology: Principles and Applications* (Ed. by R.M. May), pp. 197–227. Blackwell Scientific Publications, Oxford.

Newman, E.I. (1982). Niche separation and species diversity in terrestrial vegetation. *The Plant Community as a Working Mechanism* (Ed. by E.I. Newman), pp. 61–78. Special Publications No. 1 of the British Ecological Society. Blackwell Scientific Publications, Oxford.

Odum, E.P. (1969). The strategy of ecosystem development. *Science*, **164**, 262–270.

Parrish, J.A.D. & Bazzaz, F.A. (1976). Underground niche separation in successional plants. *Ecology*, **57**, 1281–1288.

Parrish, J.A.D. & Bazzaz, F.A. (1978). Pollination niche separation in a winter annual community. *Oecologia*, **35**, 133–140.

Parrish, J.A.D. & Bazzaz, F.A. (1979). Difference in pollination niche relationships in early and late successional plant communities. *Ecology*, **60**, 597–610.

Parrish, J.A.D. & Bazzaz, F.A. (1982a). Responses of plants from three successional communities to a nutrient gradient. *Journal of Ecology*, **70**, 233–248.

Parrish, J.A.D. & Bazzaz, F.A. (1982b). Niche responses of early and late successional tree seedlings on three resource gradients. *Bulletin of the Torrey Botanical Club*, **109**, 451–456.

Parrish, J.A.D. & Bazzaz, F.A. (1982c). Competitive interactions in plant communities of different successional ages. *Ecology*, **63**, 314–320.

Parrish, J.A.D. & Bazzaz, F.A. (1985). Ontogenetic niche shifts in old-field annuals. *Ecology*, **66**, 1296–1302.

Peterson D.L. & Bazzaz, F.A. (1978). Life cycle characteristics of *Aster pilosus* in early successional habitats. *Ecology*, **59**, 1005–1013.

Petraitis, P.S. (1981). Algebraic and graphical relationships among niche breadth measures. *Ecology*, **62**, 545–548.

Pianka, E.R. (1981). Competition and niche theory. *Theoretical Ecology: Principles and Applications* (Ed. by R.M. May), pp. 167–196. Blackwell Scientific Publications, Oxford.

Pianka, E.R. (1983). *Evolutionary Ecology*, 3rd edn. Harper and Row, New York.

Pickett, S.T.A. & Bazzaz, F.A. (1976). Divergence of two co-occurring successional annuals on a soil moisture gradient. *Ecology*, **57**, 169–176.

Pickett, S.T.A. & Bazzaz, F.A. (1978a). Organization of an assemblage of early successional species on a soil moisture gradient. *Ecology*, **59**, 1248–1255.

Pickett, S.T.A. & Bazzaz, F.A. (1978b). Germination of co-occurring annual species on a soil moisture gradient. *Bulletin of the Torrey Botanical Club*, **105**, 312–316.

Raynal, D.J. & Bazzaz, F.A. (1975). Interference of winter annuals with *Ambrosia artemisii-folia* in early successional fields. *Ecology*, **56**, 35–49.

Regehr, D.L. & Bazzaz, F.A. (1979). The population dynamics of *Erigeron canadensis*, a successional winter annual. *Journal of Ecology*, **67**, 923–933.

Roughgarden, J. (1972). Evolution of niche width. *American Naturalist*, **106**, 683–718.

Roughgarden, J. (1979). *Theory of Population Genetics and Evolutionary Ecology: An Introduction.* Macmillan Publishing Co., Inc. New York.

Schoener, T.W. (1970). Non-synchronous spatial overlap of lizards in patchy habitats. *Ecology*, **51**, 408–418.

Schoener, T.W. (1982). The controversy over interspecific competition. *American Scientist*, **70**, 586–595.

Smith, E.P. (1982). Niche breadth, resource availability and interference. *Ecology*, **63**, 1675–1681.

Sultan, S. (1987). Evolutionary implications of phenotypic plasticity in plants. *Evolutionary Biology* (in press).

Tilman, D. (1982). *Resource Competition and Community Structure.* Princeton University Press, Princeton, N.J.

Tilman, D. (1986). Evolution and differentiation in terrestrial plant communities: the importance of the soil resource: light gradients. *Community Ecology* (Ed. by J. Diamond & T.J. Case), pp. 359–380. Harper and Row, New York.

Turkington, R.A. & Aarssen, L.W. (1984). Local-scale differentiation as a result of competitive interactions. *Perspectives in Plant Population Ecology* (Ed. by Dirzo & J. Sarukhan), pp. 107–127. Sinauer Publishers, Sunderland, Massachusetts.

Turkington, R.A. & Harper, J.L. (1979). The growth and distribution and neighbor relationship in *Trifolium repens* in a permanent pasture. IV. Fine scale biotic differentiation. *Journal of Ecology*, **64**, 245–254.

Van Horne, B. & Ford, R.G. (1982). Niche breadth calculation based on discriminant analysis. *Ecology*, **63**, 1172–1174.

Waser, N.M. (1983). Competition for pollination and floral character differences among sympatric plant species: a review of evidence. *Handbook of Experimental Pollination Ecology* (Ed. by C.E. Jones & R.J. Little), pp. 277–293. Van Nostrand, New York.

Webb, T.W. III (1984). Criteria for inferring climatic equilibrium in the temporal and spatial variation of plant taxa. *Proceedings of the 6th International Palynological Conference*, Calgary (Abstracts).

Whittaker, R.H. (1972). Evolution and measurements of species diversity. *Taxon*, **21**, 217–251.

Whittaker, R.H. (1975). *Communities and Ecosystems*, 2nd edn. Macmillan, New York.

Whittaker, R.H. & Levin, S.I. (Eds) **(1976).** *Niche: Theory and Application.* Dowden, Hutchinson and Ross, Stroudsburg, PA.

Whittaker, R.H., Levin, S.I. & Root, R. (1973). Niche habitat and ecotope. *American Naturalist*, **107**, 321–338.

Wieland, N.K. & Bazzaz, F.A. (1975). Physiological ecology of three co-dominant successional annuals. *Ecology*, **56**, 681–688.

Zangerl, A.R. & Bazzaz, F.A. (1983). Responses of an early and a late successional species of *Polygonum* to variations in resource availability. *Oecologia*, **56**, 397–404.

13. GENETIC CHANGE DURING SUCCESSION IN PLANTS

ALAN J. GRAY
Institute of Terrestrial Ecology, Furzebrook Research Station, Furzebrook, Wareham, Dorset, BH20 5AS

INTRODUCTION

Natural populations of plants frequently display high levels of intraspecific variation. At one extreme this may be measured as electrophoretically-detectable isoenzyme variation coded by single Mendelian loci but of unknown ecological significance. At another it may be in continuously varying traits such as life-history attributes, under complex genetic control, often with low heritabilities and negatively correlated with similar traits, but with obvious implications for individual fitness. Whether simple or complex, the presence of a large heritable component to the variation, and the pattern of its distribution within and between families and populations, has prompted considerable interest in the processes which generate and maintain it. There is, for example, no shortage of theoretical models attempting to explain why so much genetic diversity exists in natural populations (see e.g. review by Ennos 1983).

In this chapter I examine the idea that successional change in plant communities may be a factor maintaining genetic diversity in some species populations. Doomed to local extinction, and able to survive only by dispersal to similar stages elsewhere, is the temporal niche of species in successions ever extended by genetic change? Is such change ever detectable as a significant shift in gene or genotype frequencies, or in the variance of heritable traits, between the early, colonist, generation(s) and those which succeed them? Is the consistent pattern of differences between early and late-successional species revealed by large-scale comparative surveys (Bazzaz, Chapter 12 and references therein) and, most importantly, the selective process it implies, reflected at the single-species population level (e.g. are colonist generations generally shorter-lived, earlier reproducing, smaller seeded, etc, than successive ones?).

First, however, it is necessary to consider the conditions under which successional change may act as a selective force, and the species which may respond to it.

PATTERNS OF DIFFERENTIATION: SPATIAL AND TEMPORAL ELEMENTS

In almost all empirical studies in this field population differentiation is observed, and measured, at one point in time. Inferences are then made about the processes which may have led to the observed pattern of differences. Rarely, as in the seminal study of the corm-forming herb *Liatris cylindracea* (Schaal 1975; Schaal & Levin 1976), it has been possible to determine the age of individuals (but see Werner 1978) and to compare gene and genotype frequencies in different age-classes. In this type of study (which has been called 'cross-sectional' by Lande & Arnold 1983) attributing between-generation changes to the effects of selection is not without problems (see, for example, in relation to the *Liatris* work, the comments of Clegg, Kahler & Allard 1978 and Brown 1979). Nonetheless, such studies provide powerful ways of generating hypotheses about selection which can be tested by experiment.

Ideally the population should be observed before, during, and after selection and the relative fitness of all individuals measured in some way: a 'longitudinal' study in the terminology of Lande & Arnold. This involves combining demographic observations over time with measures of genetic variation and, except in artificial populations, has only infrequently been attempted (but see later references).

The problems of ascribing population differentiation to selection by successional change can be highlighted with a simple illustration. Suppose we observe a successional gradient occupied by a high-amplitude species, as in Fig. 13.1, in which the early, middle and late stages are characterized by populations with a high frequency of three different phenotypes. (We will assume that the differences in phenotypes have a genetic basis and the populations are thus genetically differentiated.) This pattern might be observed, for example, along the zones of a salt marsh or sand dune, on a post-fire heathland or in a woodland clearing. There are at least four ways in which such differentiation may have occurred. In the first (model 1 in Fig. 13.1) differential selection of phenotypes has occurred at each stage from a more or less homogeneous seed population. In this model the habitats and populations over which differentiation occurs are temporally related and the possibility exists that some aspect of successional change has been a force contributing to micro-evolutionary change in these populations.

In the second model, differences in phenotype frequency have arisen from differences in individual rates of development among the founder population. Analagous to the initial floristic composition model of vegetation succession (Egler 1954), such differentiation might stem from genetic

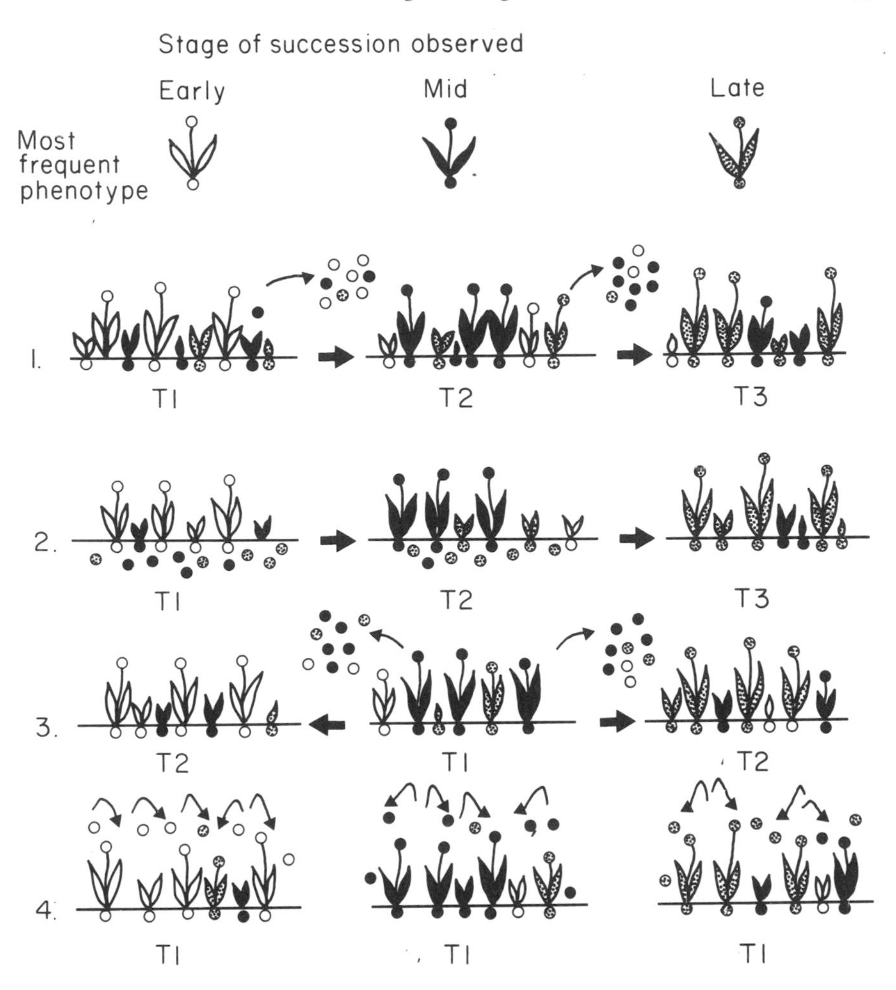

FIG. 13.1. Four models of population differentiation along a successional gradient (for explanation see text; T1 = Time 1). Arrows indicate a temporal link between populations.

differences in phenology or from germination polymorphisms of the type known to exist in some populations (see later examples). Models 3 and 4 also represent cases in which the populations do not represent successive generations. They can occur not only where habitats are unrelated temporally (e.g. on salt marshes where the upper zones have not developed from a vegetation type at present characterizing the lower zones) but also where vegetation succession has occurred, but has not involved the species of interest. Thus, in model 3, central populations disperse offspring to nearby areas, above and below them in the successional gradient, where environments which are marginal for the species select different phenotypes (a commonly envisaged situation, e.g. Antonovics 1976), and in model 4 discrete populations occupy areas which happen to be on a successional gradient.

These simple pictorial models represent cases which are not mutually exclusive. Elements of all four could be expected to operate simultaneously over any set of adjacent successional stages. However, only in the first model has the perceived pattern of spatial heterogeneity arisen, at least partly, by the process assumed to underlie the questions posed at the outset. Although succession may have generated the requisite environmental heterogeneity for selection in models 3 and 4, it is not itself an agent of selection. In the absence of other evidence, genetic drift and similar non-selective processes are equally possible causes of the observed non-random distribution of phenotypes. Such processes will, of course, play a part in model 1. Indeed, changes in population size and density in plants carry inescapable structural effects, notably in genetic neighbourhoods (Crawford 1984; Loveless & Hamrick 1984) which may confound, and be inseparable from, the effects of selection.

The models also emphasize that the species able to change genetically during succession are likely to occur in populations which are reasonably persistent, yet within which there is a turnover of generations. Thus, individuals may be short-lived. In addition the genetic system, and particularly the breeding system, would need to be capable of generating, by recombination, the variability on which selection might act (and be sufficiently protean, for example, to generate new early-successional phenotypes from late-successional populations). As a corollary, the rates of change in surrounding vegetation, soil properties, and so on, which characterize successions must be sufficiently rapid in relation to the species' temporal niche to act as a selective force. This suggests that genetic tracking of successional change is most likely to be displayed by annual or short-lived perennial species which are outbreeding and which invade the early (generally more rapid) stages of either primary or secondary successions and persist for periods exceeding average individual longevity.

It is not surprising, therefore, that in the few cross-sectional studies available of long-lived species, the pattern of variability is dominated by spatial heterogeneity. In *Pinus contorta* stands studied by Knowles & Grant (1985) populations of recently recruited saplings did not have a significantly different genetic composition (using isozyme analysis) from that of the pioneer population. Spatial heterogeneity contributed more to variation in genetic composition than did temporal heterogeneity, as had been found by Linhart *et al.* (1981) in *Pinus ponderosa*. Nor perhaps is it surprising that where succession is rapid in relation to the longevity of the species the patterns of genetic differentiation appear unrelated to current patterns of environmental heterogeneity, as, for example, in *Salix repens* on sand dunes (Fowler, Zasada & Harper 1983).

THE DEMOGRAPHIC GENETICS OF *AGROSTIS CURTISII*: A POST-FIRE SUCCESSION

Agrostis curtisii (formerly *A. setacea*) is a perennial grass of well-drained acid soils, principally heathland, in southern Britain and Western Europe. It is diploid ($2n = 14$) and self-incompatible (Gray & Bates 1979) and is often one of the first invaders of recently-burned heathland. When, in August 1976, a severe fire destroyed the surface vegetation on part of Hartland Moor, Dorset, the opportunity was taken to monitor demographic and genetic changes in a colonizing population. The first appearance (birth), size, flowering performance, and death of all individuals was recorded within an area of 40 m^2 on sloping ground where all recruitment was from seed. Individuals were sampled for electrophoresis by removing 3–4 green leaves. To minimize the effect of defoliation only those plants with 15 or more leaves were analysed.

The population reached a peak (563 individuals) 4 years after the fire (Fig. 13.2) and has gradually decreased as ericaceous species, mainly *Erica cinerea* and *Calluna vulgaris*, have become dominant. Recruitment, erratic in the first 4 years, has also gradually declined following a second peak in 1980, the year after most of the pioneer cohort (the 1976–77

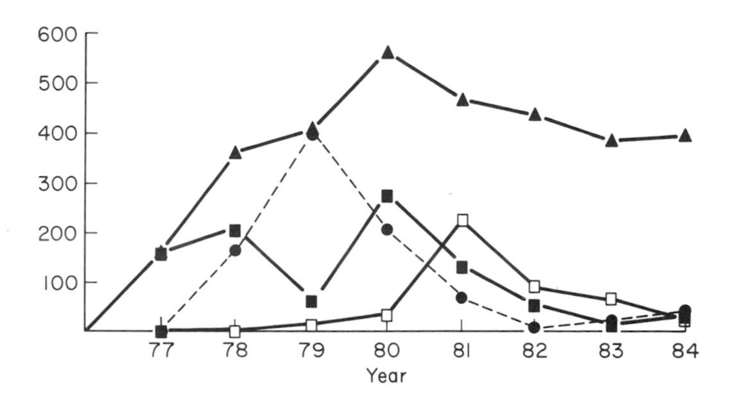

FIG. 13.2. Changes in a population of *Agrostis curtisii* colonizing an area of Hartland Moor burnt in August 1976: (▲) numbers of plants; (■) numbers of new recruits; (□) numbers of deaths; (●) numbers of inflorescences divided by 10.

cohort) flowered, producing more than 4000 flowering tillers: an estimated 10^6 seeds (Gray & Bates 1979). The sharp rise in the death rate the following year was largely made up of deaths among the cohorts from the two previous years. Survivorship curves for individual cohorts (Fig. 13.3a) show that very few deaths have occurred among plants which established in the first 2 years after the fire. Only 3 of 161 plants in the 1976–77 cohort had died by 1984. In contrast, individuals recruited after 1979 have a very reduced chance of surviving, e.g. the 1979–80 cohort was reduced from 174 to 11 plants in 4 years. This pattern of survivorship produces a population with a very uneven age structure (Fig. 13.3b), dominated by individuals from the two oldest year-classes. These individuals are also the most fecund. Of the total of 9215 inflorescences produced up to 1984, 8877 (97·3%) have been from the 1976–77 cohort. In fact, one individual (400 inflorescences) recruited in 1976–77 exceeded more than twofold the entire flowering tiller production of the 363 individuals recruited from 1977 onwards. No plants recruited after the third year have flowered. Furthermore, only 6 of these plants have reached the 15-leaf stage, enabling them to be sampled for electrophoresis.

These results indicate that recruitment to the sexual population has effectively ceased, or will be a rare event, in this colonizing *A. curtisii* population. The pioneer generations continue to dominate. There has effectively been no turnover of generations and the attempt to make a longitudinal study of genetic change has been thwarted, both by this and by the fact that later recruits were too small to sample. Interestingly, among those individuals which were sampled (263 plants) there is a marked spatial genetic heterogeneity. The subpopulations occupying the upper and lower halves of the slope displayed a significant heterogeneity of allele frequencies at eight out of nine isoenzyme loci for which the plants were routinely assayed (Gray, Stephens & Ambrosen 1985). However generated, this population structuring occurred during the colonizing stages when the site was largely open ground with scattered seedlings of *A. curtisii* and other pioneer species.

Successional change is, therefore, unlikely to have influenced the pattern of genetic differentiation. However, succession has been important in its effects on longevity and fecundity. Pioneer plants are long-lived (9+ years) and several produced more than 200 flowering tillers at 3 years old. In contrast, plants which established 3 years after the fire tend to be short-lived (2–3 years) and have produced no seed to date. This difference, believed to be caused by phenotypic plasticity, may reflect the early capture of resources by pioneer plants in the uncrowded post-fire conditions compared with the difficult conditions of establishment below an already closed canopy of dwarf shrubs faced by the later arrivals. The

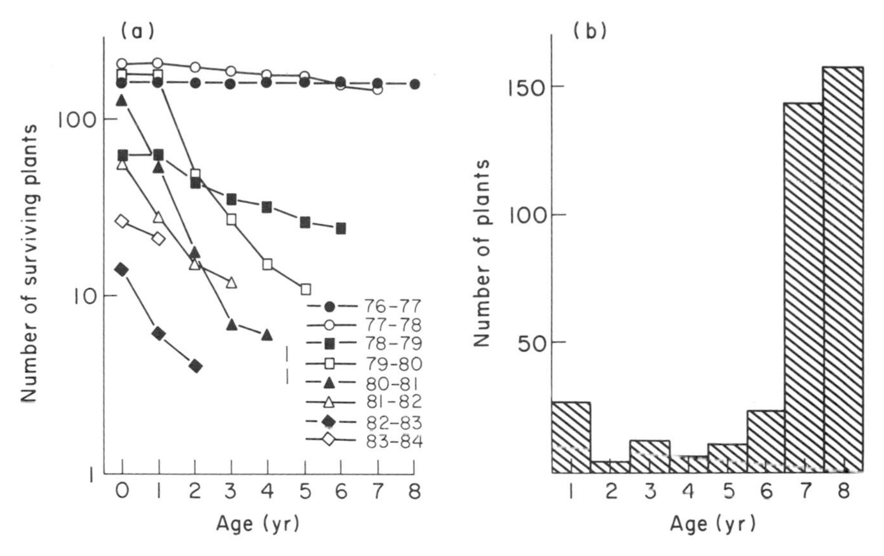

FIG. 13.3. (a) Survivorship curves for each year cohort of *Agrostis curtisii* colonizing Hartland Moor since it was burnt in August 1976. (b) Age structure in 1984 of the population of *Agrostis curtisii* on Hartland Moor.

effects of such plasticity on micro-evolutionary changes are, as yet, unknown. Extreme plasticity of size in a population of the annual plant *Stephanomeria exigua* ssp. *coronaria* studied by Gottlieb (1977) had little evolutionary significance, large and small individuals having similar enzyme genotype frequencies and growth rates in uniform environments.

Clearly the *A. curtisii* population should be measured over a longer period of time, but it seems likely that here genetic change will consist of the gradual depletion of the variability present at the initial establishment phase (I return to this idea later). The salutory lesson of this study, and of others which have looked in detail at the demography of colonizing plants (see for example Gross & Werner (1982) on four biennial species in successions) is that convenient categories such as 'short-lived perennial', and 'biennial', may mask great variation in regenerative properties and in the potential for plants to track the successional changes going on around them.

THE RAYLESS GENES IN *ASTER TRIPOLIUM*: A LONGER-TERM STUDY

Aster tripolium (Compositae) occurs in saline inland and coastal marshes throughout western and central Europe, extending southwards to North Africa and eastwards into Central Asia and Japan. Within British salt marshes it has a high ecological amplitude, occurring from pioneer to mature zones. It occurs mainly as a rayed plant in which the capitulum consists of a central disc of yellow hermaphrodite tubular florets surrounded by a ring of pale-blue female ray florets. A form lacking the ray florets, which has been called var. *discoideus*, has been known from Britain since at least 1838 (Gray 1971). Today it is particularly common in south-east England where almost entirely rayless populations may be found in the lower zones of many marshes. The differential ecological distribution of rayed and rayless plants in eastern England, with the frequency of rayed individuals increasing towards the higher elevations, has also been reported from European salt marshes (Gray 1971; Sterk & Wijnands 1970).

Crossing experiments (Gray 1971) have indicated that the variation is genetically controlled. Unlike the ray/rayless variation in *Senecio vulgaris* (Trow 1912) and *S. squalidus* (Ingram & Taylor 1982), where heterozygotes display short-rayed capitula, variation in the *A. tripolium* ray floret character is meristic, with heterozygotes possessing a variable, but intermediate, number of rays. The pattern of recovery of parental types in the F_2 generation is consistent with the hypothesis that two genes are responsible for ray number, with rayless and fully-rayed being double homozygotes. However, these genes were found to vary considerably in their penetrance; for example, in heterozygotes both the mean number and variance of rays might differ on the same plant scored at different times in the flowering season (Gray 1971).

Populations were scored for the ray character in 1965 on a series of east coast salt marshes, grouped on the basis of elevation, tidal relations, and associated species into low, mid, and high marsh types. Eight populations, ranging in size from 53 to 1302 plants, were scored again in 1984. The results are equivocal (Table 13.1). In low marsh areas where the vegetation had changed considerably in 19 years (mainly by the invasion of *Halimione portulacoides*, *Puccinellia maritima* and *Suaeda maritima* into formerly almost pure *Aster tripolium* and *Spartina anglica* communities) the frequency of rayless phenotypes had not significantly changed. Nor had it changed in the high marsh populations of largely radiate plants where successional change, as indicated by change in the spectrum and density of species, had been very slow (these areas were mainly dominated by *Scirpus maritimus* and *Elymus pycnanthus* surrounding brackish pools). The two mid-marsh populations showed an increase in rayless phenotypes.

The ecological distribution of the rayless genes suggests that they are marking a polymorphism which is balanced via, presumably, the action of

TABLE 13.1. Frequency of rayless *Aster tripolium* in eight populations 19 years apart

Marsh type	Popn.	Frequency rayless phenotypes 1965	1984	Vege. change	% Annuals	$\bar{x}$ Disc fruit wt (mg)
Low	Gib Spa	0·96	0·94	+++	0	1·41
	Scolt Cob	1·0	0·93	+++	0	1·56
	Scolt Bch	0·96	0·91	++	0	1·38
Mid	Gib Ast	0·64	0·81	++	8	1·15
	Scolt Mis	0·48	0·70	+	0	0·64
High	Gib Lim	0·17	0·28	+	12	0·89
	Gib Agr	0	0	.	55	0·53
	Cley	0	0·06	.	60	—

disruptive selection on pleiotropically associated or linked fitness traits. If so, an increase in radiate phenotypes would be expected as succession proceeded in those low marsh populations where the ray genes were present. This has not happened, even though two of the low marshes in Table 13.1 would now be classified as mid marshes from their vegetation. In fact, the change in rayless phenotype frequencies on the mid marshes has been in the unpredicted direction (suggesting perhaps a slowly transient polymorphism?).

It is clearly important to know how many generations have passed in the 19 years since the populations were first scored. This leads to a further interesting aspect of the *A. tripolium* study: namely that most of the life-history and other attributes vary along the successional gradient in a manner completely opposite to that predicted from comparative surveys of early and late-successional plants (Bazzaz 1979, Chapter 12; Finegan 1984). Comparative cultivation of a wide range of salt marsh populations reveals that *A. tripolium* plants from early-successional stages are generally longer-lived, have a longer pre-reproductive phase and have fewer but heavier fruits (Gray 1971, 1974; Gray, Parsell & Scott 1979). Data from the populations shown in Table 13.1 are typical of many populations, including those from west coast marshes where rayless plants are absent. High marsh populations frequently contain a proportion of plants which (in cultivation and in the field) are annual or pauciennial (Bøcher, Larsen & Rhan 1955), whereas plants from low marshes may not flower for the first time until they are 3 years old and then flower repeatedly each year after that (the longevity of these plants is not known). Some individuals scored on low marshes in 1984 may have been present in 1965, whereas up to nineteen generations may have elapsed on high marshes. This contrast in age structure and generation turnover rate will have profound consequences for the ability of the respective populations to become genetically differentiated during succession.

It is possible to relate the pattern of variation in *A. tripolium* along the salt marsh gradient with respect to life-history and seed attributes to variation in several environmental factors affecting establishment and survival (Gray *et al.* 1979, Gray 1985, and Huiskes, van Soelen & Markusse 1985 who describe similar patterns in Dutch populations of the species). In trying to understand why these patterns are contrary to those expected for the different successional stages it is important to realize that *A. tripolium* populations on the low marshes are often very much denser than on high marshes. The species enters the succession in *Spartina* swards under conditions where mortality is more likely to be density-dependent than in the later stages where the plant occurs on the edges of brackish pools and other scattered and more ephemeral sites within the marsh. For this particular species, increasing successional maturity does not involve a shift towards a more crowded, less disturbed environment, with higher density-dependent mortality. Since these features are felt to be important in shaping life-history attributes (e.g. Gadgil & Solbrig 1972) it is perhaps not surprising that *A. tripolium*, and some other species described below, fail to conform to theoretical predictions based purely on relative successional maturity.

GENETIC CHANGE AND SUCCESSION: FURTHER EVIDENCE FROM NATURAL POPULATIONS

The prediction that plants occupying the early stages of successions will tend to allocate relatively more of their resources to sexual reproduction, have a shorter pre-reproductive period, be shorter-lived, produce more but lighter seeds, and so on (Gadgil & Solbrig 1972; Finegan 1984) is generally supported by evidence from comparative surveys (Bazzaz 1979) and from studies of closely related species, e.g. species of *Polygonum* (Zangerl & Bazzaz 1983).

At the level of individual species the evidence is more equivocal. Populations which are genetically differentiated with respect to life-history traits are known in *Taraxacum officinale* (Solbrig & Simpson 1974) and *Poa annua* (Law, Bradshaw & Putwain 1977) where degree of disturbance was a major variable distinguishing the population types. In *Andropogon scoparius* (Roos & Quinn 1977) differences in phenology and reproductive allocation were found among populations from old-fields of different ages; although they were largely attributable to phenotypic plasticity. Variation among five populations of *Danthonia spicata* of different successional age was also found to be dominated by plasticity (Scheiner & Goodnight 1984). Although genetic variation existed, for example in light-saturated photosynthetic rate (Scheiner, Gurevitch & Teeri 1984), there were no significant differences between populations—contrasting with the pattern derived from a comparative survey of the photosynthetic responses of species occupying different successional habitats (Bazzaz & Carlson 1982). Reproductive allocation was also environmentally rather than genetically determined in the annual *Polygonum cascadense* where it varied with the openness of the habitat (Hickman 1975) and in *Fragaria virginia* from habitats of different successional maturity (Holler & Abrahamson 1977). Six perennial herbs at three sites studied by Stewart & Thompson (1982) did not show a straightforward decline in reproductive effort with successional age, a fact they attribute to variation in levels of competition.

Even where genetic differentiation occurs, as in *Aster tripolium* (above), the role of successional change, as emphasized at the beginning, remains to be demonstrated. For example, populations of *Salicornia europea* from different salt marsh areas, the substantial genetic differentiation between which has been demonstrated by perturbation and reciprocal transplant experiments to have clear adaptive value (Davy & Smith 1985), are thought to be affectively separate (Fig. 13.1, model 4) or nearly so. Whatever the selection pressures were historically, current successional changes, although providing the template, are not driving the differentiation.

Significant differences in genotype frequencies (at isoenzyme loci) between three populations of *Erigeron annuus* from adjacent fields abandoned at different times in the past provide strong circumstantial evidence of successional age being a prime selective agent (Hancock & Wilson 1976). However, I know of only two relevant longitudinal studies which unequivocally demonstrate genetic changes in the same natural population with time. In *Verbascum thapsus* in North Carolina significant genetic variation (in relative biomass allocation to leaves and stalks and in leaf shape) between first and second cohorts was thought to result from differential selection acting on the parental generation (analogous to Fig. 13.1, model 2) (Reinartz 1984). In the annual species *Echium plantagineum* in Australia, significant directional changes in allele frequencies at two isoenzyme loci and in heterozygosity occurred during the season in a single population, but possible selective forces are not discussed (Burdon,

Marshall & Brown 1983). Successional change, other than the increase in plant density, would clearly have to be rapid to have an effect.

There is, therefore, very little empirical evidence that successional change is maintaining genetic variation in natural populations, despite broad comparisons of different species.

BIOTYPE OR GENOTYPE DEPLETION AS A FEATURE OF SUCCESSION

Puccinellia maritima is a perennial grass which invades the early stages of primary salt marsh successions in Britain and Europe and has a high ecological amplitude, persisting into the upper marsh (Gray & Scott 1977, 1980). Comparisons were made between pioneer populations, where individuals were scattered on open mudflats, and mature populations, where *P. maritima* was the dominant plant in dense swards at higher elevations, plants from ten populations sampled both as tillers and seed being grown in spaced pots in a large, completely randomized clone trial (Gray *et al.* 1979, Gray 1985); (for methods of measuring traits see Gray & Scott 1980).

For most traits pioneer populations were more variable than the mature populations immediately above them. There was a greater range of vegetative biotypes (a contrast between individuals with comparatively few, tall, long-leaved tillers and those with many small, short-leaved tillers) of plant sizes (as measured by above-ground dry weight) and of reproductive allocation (the proportion of flowering to vegetative tillers) on the pioneer marshes. This was interpreted not as close tracking of higher environmental heterogeneity but resulting from a relaxing of the selection associated with interference from neighbouring plants. This hypothesis assumes that there is an increase in density-dependent mortality as the sward closes and that the pioneer individuals represent (in respect of their 'competitive ability') an 'unselected' population, establishing from tiller fragments and seedlings, and being derived from genetic recombination following sexual reproduction in this largely outbreeding species. The hypothesis is supported by four lines of evidence.

First, the possibility that reduced variation in mature populations arises from sampling error, specifically the repeated sampling of large clones, was discounted by the use of isoenzyme markers. Although grazed marshes contained fewer (larger) clones per unit area than ungrazed marshes (an average of 2·3 and 4·5 genets m^{-2} respectively) the mature swards were not dominated by a small number of very large individuals (cf. Harberd 1961, 1962 and 1967 for populations of *Festuca ovina*, *F. rubra*

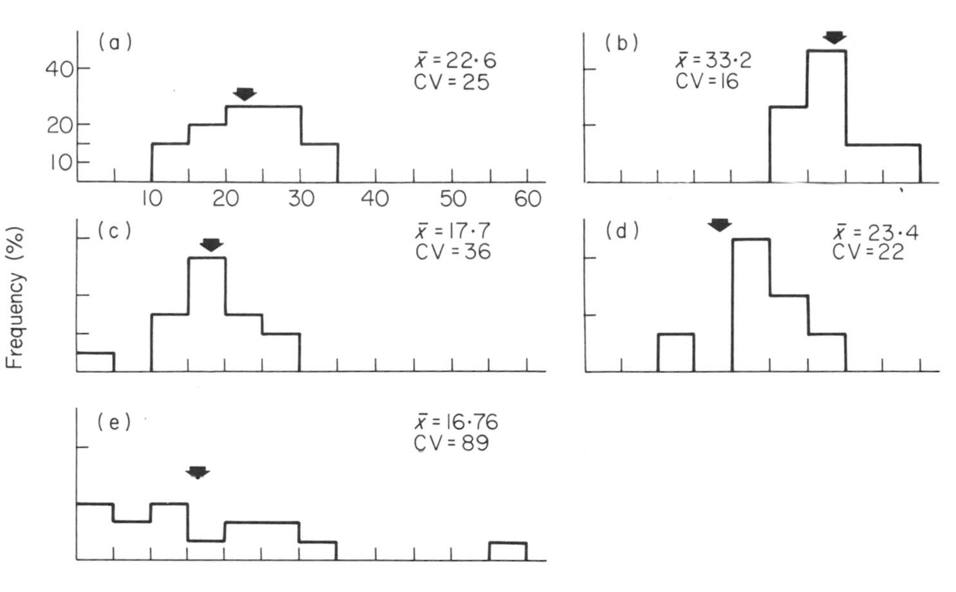

FIG. 13.4. Variation in reproductive allocation (% of above-ground dry weight of plant consisting of flowering tillers) in *Puccinellia maritima* plants grown in a spaced trial and originally collected as tillers from (a) an ungrazed pioneer marsh, (b) the mature marsh above (a), (c) an adjacent grazed pioneer marsh, (d) the mature marsh above (c), and (e) plants originally collected as seed from the ungrazed mature marsh. Arrows indicate means ($\bar{x}$), CV = coefficient of variation.

and *Holcus mollis* respectively). Second, a comparison of mature populations alone provides strong genotype–environment correlations. For example, plants from mature grazed populations were significantly more prostrate, had smaller leaves, flowering tillers and panicles, and flowered later than those from mature ungrazed populations. Furthermore, clipping experiments (R. Scott & A.J. Gray, unpublished data) indicated that plants with grazed-marsh vegetative biotype had significantly increased yields and survivorship under conditions of repeated defoliation. Third, the populations of plants grown from seed collected on the mature marsh displayed a greater range of variation for many traits than that shown by the population grown from tiller samples. The adult ('selected') population represented a narrow segment of the variation present in the seed ('unselected') population. These last two points are illustrated with respect to reproductive allocation (the percentage of the total above-ground dry weight made up of flowering tillers) in Fig. 13.4, which shows variation for this character in populations from both pioneer marshes and from field-collected seed to be higher than in the mature populations from adjacent ungrazed and heavily sheep-grazed marshes (see Gray *et al.* 1979, Gray 1985). Plants from the ungrazed population allocate significantly more resources to flowering tillers (*c.* 33% of above-ground dry weight) than do those from the grazed population (*c.* 23%).

A fourth line of evidence that selection during succession in these populations is related to the effects of increasing plant density stems from an overall comparison of pioneer and mature populations. The average effect of being in a pioneer population has been estimated for the ten populations, six from Morecambe Bay and four from North Norfolk, and is given in Table 13.2. (which gives methods of estimating this effect and its variance and significance.) For most traits there is a significant negative effect of being pioneer. Plants from mature populations are, on average, significantly larger, have longer leaves, produce more vegetative and flowering tillers, have longer inflorescences and panicles, and produce more seed than their counterparts from pioneer populations. Only flowering time and the weight of vegetative tillers are not significantly different in the two population types. The increase in average size and vigour of plants in mature marshes agrees with general predictions but the fact that proportionally more of the extra production is allocated to sexual reproduction as additional inflorescences, and ultimately almost a thousand extra seed per plant, is perhaps surprising. It contrasts with the preferential allocation of resources to sexual reproduction in less crowded, more disturbed environments with higher density independent mortality which is a feature of variation in some other grass species, e.g. *Bromus* species (Wu & Jain 1979), *Poa annua* (Law *et al.* 1977; Warwick & Briggs 1978; McNeilly 1981), *Typha latifolia* (Grace & Wetzel 1981) and *Spartina patens* (Silander & Antonovics 1979). However, if, as we suggest, the pioneer environment is unselected (and not *r*-selected) the comparison is probably not valid (Parry 1981). Moreover, it is unclear whether higher seed production in a spaced trial in a single year in a perennial iteroparous grass, without an assessment of the cost to vegetative maintenance and survival, signifies greater fitness in the field.

The picture which emerges from this study is one in which successional change affects the survival of only a segment of the variation present in the colonizing population. A different segment of that variation is selected in grazed and ungrazed marshes, but the depletion of biotypes is a consistent feature of the change from early to late-successional stage. A similar pattern is seen in populations of *Plantago maritima* along a Baltic shore gradient with, for example, seed size showing an increased mean and reduced variance in the higher zone (unlike *Aster tripolium*, in the predicted direction for seed size; Fig. 13.5, L. Jerling, pers. comm.).

TABLE 13.2. Effect of 'pioneer' on *Puccinellia maritima* populations. Population means were compared using the method given in Snedecor & Cochran (1967, p. 268). The average effect of being pioneer is estimated by computing the difference between the mean of the pioneer population means and that of the mature population means by adding together all the population means when each has been multiplied by an appropriate coefficient (λ), according to which class (pioneer or mature) it belongs. The effect is in the same units of measurement (cm, g) as the trait value. An unbiased estimate of the error variance of the effect is $V = \Sigma\lambda_i^2 S_i^2/n_i$ where λ_1 is the comparison coefficient, S_i^2 the variance and n_i the number of individuals, each in the ith population. The n_i are all equal to n and the degrees of freedom are $(n - 1)(\Sigma V_i)^2/\Sigma V_i^2$ where $V_i = \lambda_i^2 S_i^2/n_i$ (Snedecor & Cochran 1967, p. 324). The significance of the effect can be tested by Students' t, with this number of degrees of freedom.

Trait	Pioneer vs. mature pops	
	Effect	Significance (t)
Plant height (cm)	−0·73	***
Plant width (cm)	−1·90	***
Leaf length (cm)	−0·97	***
Tiller no. (7 m)	−3·25	**
Infl. length (cm)	−4·53	***
Panicle length (cm)	−1·48	***
No. inf./plant	−4·78	***
Flow. tiller weight (g)	−0·90	***
Veg. tiller weight (g)	−0·29	NS
Weight flow till (%)	−9·16	***
No. caryopses	−951	***
Date ear emerg. (d)	0·04	NS
Total a/g dry weight (g)	−1·19	**

A dramatic reduction in genetic diversity during the establishment phase is a feature of Charles' classic work on sown swards of grass and clover cultivars (Charles 1961, 1964, 1966). High early mortality in these swards, up to 90% of seedlings being eliminated in a year (Charles 1961), may be random but is followed by a later phase of selective mortality (Charles 1966) (but see Hayward 1970; Hayward, Gottlieb & McAdam 1978). McNeilly & Roose's (1984) study of *Lolium perenne* swards suggests that individuals continue to be eliminated over considerable time periods, although at a slower rate (genotype numbers declining from around 40 m^{-2} in 10-year old swards to only 5 m^{-2} in a 40-year old pasture). Recently, Aarssen & Turkington (1985) report a decline in variance for many characters in *Lolium perenne* and *Trifolium repens* with increasing pasture age and suggest this may be attributable to 'continuous elimination of the less-fit genotypes by competition and grazing'. However, not all traits show a decline in variation in these two species

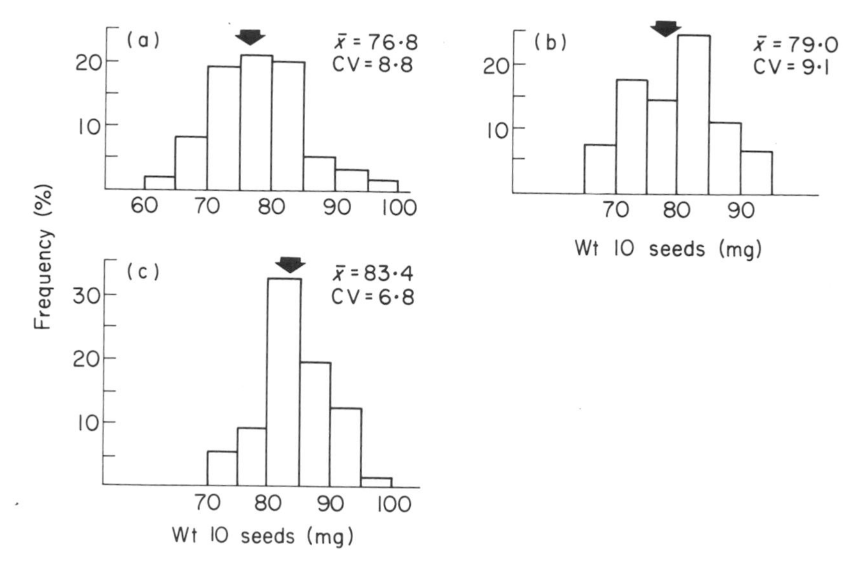

FIG. 13.5. Variation in seed size of *Plantago maritima* populations grown in a spaced trial and collected as seed from the (a) lower, (b) middle, and (c) upper shore of a Baltic shore meadow (data of L. Jerling). Arrows indicate means ($\bar{x}$), CV = coefficient of variation.

and only one (out of fifteen) showed such a decline in *Holcus lanatus* from the same pastures. It should also be noted that the decline in the number of genotypes in a sown population can, where there is little or no seedling recruitment, generate a decline in diversity in the absence of selection (Soane & Watkinson 1979).

The high levels of genetic diversity reported in pasture populations of several species (e.g. Turkington & Harper 1979; Burdon 1980; Aarssen & Turkington 1985) indicate that, even with low recruitment levels, genetic variation is generated and maintained. This is generally believed to occur under the influence of both abiotic and biotic small-scale heterogeneity which may actually increase with pasture age.

The *Agrostis curtisii* study described earlier suggests that 'natural' populations of perennial plants may have recruitment patterns not too far removed from those of sown pastures. Bursts of colonization following fire, treefall or a good seed year, may punctuate prolonged periods with little or no establishment. There is good empirical evidence that seedling recruitment in established vegetation is relatively rare (Harper 1977). Certainly it is never possible to assume that populations will have stable age distributions. The age structure of both short-lived species such as *A. curtisii* and long-lived ones such as *Pinus ponderosa* (Linhart *et al.* 1981) indicate that rates of recruitment and mortality may vary with time in populations in a manner which has little connection with successional changes.

CONCLUDING REMARKS

Despite high levels of genetic diversity in natural plant populations, despite clear patterns of differences between early- and late-successional species and even congeners, and despite the fact that 'most plants occupy stages in successions' (Harper 1977, p. 771) there is a surprising lack of empirical evidence that successional change is a major factor maintaining genetic diversity in species populations. This finding accords with theoretical predictions that spatial heterogeneity in the environment will be more important in maintaining genetic polymorphisms than temporal heterogeneity, and that the conditions under which temporal variation in selection might operate effectively to maintain diversity are rather restricted (Hedrick 1976; Hedrick, Ginevan & Ewing 1976; Ennos 1983).

However, the attempt to measure the effects of selection in field populations, intrinsically difficult even under ideal conditions, is dogged in the case of successional change by several, almost intractable, problems. These include the fact that successionally-related environments are often also spatially-related, as along primary successional gradients. The untangling of spatial and temporal influences can only be achieved by longitudinal studies (Lande & Arnold 1983) which measure genetic change between generations born into successionally-different habitats. Results from one such study, on a post-fire population of *Agrostis curtisii*, add to the increasing number of demographic analyses which have detected unstable age distributions in natural populations. Even short-lived outbreeding perennials which invade the early stages of succession may display regeneration cycles which effectively prevent them from genetic tracking of successional change. Finally, the selective effects of succession *per se* may be inseparable in practice from the genetic consequences of changes in plant distribution and density imposed, for example, by changing pollination distances and neighbourhood relationships (Antonovics & Levin 1980).

In spite of these difficulties, indirect evidence of differences between populations of the same species in different successional habitats is available from several studies. This evidence has two main features. First, it does not always agree with the pattern predicted by comparative surveys: individuals in early populations are not always shorter-lived or smaller-seeded and so on. This may be because, as suggested here for *Aster tripolium*, different species 'perceive' successional change very differently and not always as a progressive increase in crowding, freedom from disturbance and density-dependent mortality. Second, the between-population differences often arise by phenotypic plasticity. Indeed, plasticity may produce the highly skewed size and fecundity distributions which complicate longitudinal studies. As a corollary, successional changes are all-pervading and, at the species population level may involve cyclical but unpredictable episodes of local colonization followed by prolonged periods of persistence. Therefore, it is not unreasonable to suppose that succession is a potent force retaining plasticity for many traits in plants. For example, the ability to vary reproductive allocation or to adjust photosynthetic efficiency at different light intensities is likely to be of clear adaptive value to plants in habitats varying unpredictably in resource levels.

Finally, one feature of what may be a rather restricted group of species has been noted. This is the biotype, and in some instances genotype, depletion which characterizes initial changes in populations of perennial species invading (or being sown into) formerly open habitats. The decline in variation with successional age may be little more than a trivial consequence of the decline in numbers, but sometimes, as in *Puccinellia maritima*, the changes are directional and can be related to increasing density and/or to environmental management. This process may be viewed as the relatively rapid removal of the less fit genotypes generated by outbreeding and as a prelude to a longer period of disruptive selection as small-scale environmental heterogeneity increases. The generality of this process remains to be tested.

ACKNOWLEDGMENTS

I am grateful to Richard Ennos and Lenn Jerling for their comments on an earlier draft, to Lenn Jerling for the use of the unpublished data in Figure 13.5., and to Ralph Clarke for advice on the analysis of the data in Table 13.2.

REFERENCES

Aarssen, L.W. & Turkington, R. (1985). Within-species diversity in natural populations of *Holcus lanatus*, *Lolium perenne* and *Trifolium repens* from four different-aged pastures. *Journal of Ecology*, **73**, 869–886.

Antonovics, J. (1976). The nature of limits to natural selection. *Annals of the Missouri Botanic Garden*, **63**, 224–247.

Antonovics, J. & Levin, D.A. (1980). The ecological and genetic consequences of density-dependent regulation in plants. *Annual Review of Ecology and Systematics*, **11**, 411–452.

Bazzaz, F.A. (1979). The physiological ecology of plant succession. *Annual Review of Ecology and Systematics*, **10**, 351–371.
Bazzaz, F.A. & Carlson, R.W. (1982). Photosynthetic acclimation to variability in the light environment of early and late successional plants. *Oecologia*, **54**, 313–316.
Bøcher, T.W., Larsen, K. & Rhan, K. (1955). Experimental and cytological studies on plant species. II. *Trifolium arvense* and some other pauciennial herbs. *Dansk Biologisk Skrifter*, **8**, 3–31.
Brown, A.H.D. (1979). Enzyme polymorphism in plant populations. *Theoretical Population Biology*, **15**, 1–42.
Burdon, J.J. (1980). Intra-specific diversity in a natural population of *Trifolium repens*. *Journal of Ecology*, **68**, 717–735.
Burdon, J.J., Marshall, D.R. & Brown, A.H.D. (1983). Demographic and genetic changes in populations of *Echium plantagineum*. *Journal of Ecology*, **71**, 667–679.
Charles, A.H. (1961). Differential survival of cultivars of *Lolium*, *Dactylis* and *Phleum*. *Journal of the British Grassland Society*, **16**, 69–75.
Charles, A.H. (1964). Differential survival of plant types in swards. *Journal of the British Grassland Society*, **19**, 198–204.
Charles, A.H. (1966). Variation in grass and clover populations in response to agronomic selection pressures. *Proceedings of the Xth International Grassland Congress, Helsinki, Finland*, pp. 625–629.
Clegg, M.T., Kahler, A.L. & Allard, R.W. (1978). Genetic demography of plant populations. *Genetics and Ecology: The Interface* (Ed. by P.F. Bussard), pp. 173–188. Springer, Berlin.
Crawford, T.J. (1984). What is a population? *Evolutionary Ecology* (Ed. by B. Shorrocks), pp. 135–173. Blackwell Scientific Publications, Oxford.
Davy, A.J. & Smith, H. (1985). Population differentiation in life-history characteristics of salt-marsh annuals. *Vegetatio*, **61**, 117–125.
Egler, F.E. (1954). Vegetation science concepts. 1. Initial floristic composition—a factor in old-field vegetation development. *Vegetatio*, **4**, 412–417.
Ennos, R.A. (1983). Maintenance of genetic variation in plant populations. *Evolutionary Biology*, **16**, 129–155.
Finegan, B. (1984). Forest succession. *Nature*, **312**, 109–114.
Fowler, N., Zasada, J. & Harper, J.L. (1983). Genetic components of morphological variation in *Salix repens*. *New Phytologist*, **95**, 121–131.
Gadgil, M.D. & Solbrig, O.T. (1972). The concept of *r*- and *k*-selection: evidence from wild flowers and some theoretical considerations. *American Naturalist*, **106**, 14–31.
Gottlieb, L.D. (1977). Genotypic similarity of large and small individuals in a natural population of the annual plant *Stephanomeria exigua* ssp. *coronaria* (Compositae). *Journal of Ecology*, **65**, 127–134.
Grace, J.B. & Wetzel, R.G. (1981). Phenotypic and genotypic components of growth and reproduction in *Typha latifolia*: experimental studies in marshes of differing successional maturity. *Ecology*, **62**, 789–801.
Gray, A.J. (1971). *Variation in* Aster tripolium *L., with particular reference to some British populations*. Ph.D. thesis, University of Keele.
Gray, A.J. (1974). The genecology of salt marsh plants. *Hydrobiological Bulletin (Amsterdam)*, **8**, 152–165.
Gray, A.J. (1985). Adaptation in perennial coastal plants—with particular reference to heritable variation in *Puccinellia maritima* and *Ammophila arenaria*. *Vegetatio*, **61**, 179–188.
Gray, A.J. & Bates, H.E. (1979). The breeding system of *Agrostis setacea*. *Annual Report of the Institute of Terrestrial Ecology* (1979), 85–86.
Gray, A.J., Parsell, R.J. & Scott, R. (1979). The genetic structure of plant populations in

relation to the development of salt marshes. *Ecological Processes in Coastal Environments* (Ed. by R.L. Jefferies & A.J. Davy), pp. 43–64. Blackwell Scientific Publications, Oxford.

Gray, A.J. & Scott, R. (1977). *Puccinellia maritima (Huds.) Parl.* Biological Flora of the British Isles. *Journal of Ecology*, **65**, 699–716.

Gray, A.J. & Scott, R. (1980). A genecological study of *Puccinellia maritima (Huds.) Parl.* 1, Variation estimated from single-plant samples from British populations. *New Phytologist*, **85**, 89–107.

Gray, A.J., Stephens, D. & Ambrosen, H.E. (1985). Demographic genetics of the perennial heathland grass *Agrostis curtisii. Annual Report of the Institute of Terrestrial Ecology* (1984), 96–100.

Gross, K.L. & Werner, P.A. (1982). Colonizing abilities of 'biennial' plant species in relation to ground cover: implications for their distributions in a successional sere. *Ecology*, **63**, 921–931.

Hancock, J.F. & Wilson, R.E. (1976). Biotype selection in *Erigeron annuus* during old-field succession. *Bulletin of the Torrey Botanical Club*, **103**, 122–125.

Harberd, D. (1961). Observations on population structure and longevity in *Festuca rubra L. New Phytologist*, **60**, 184–206.

Harberd. D. (1962). Some observations on natural clones in *Festuca ovina. New Phytologist*, **60**, 85–100.

Harberd, D. (1967). Observations on natural clones of *Holcus mollis. New Phytologist*, **66**, 401–408.

Harper, J.L. (1977). *The Population Biology of Plants.* Academic Press, London.

Hayward, M.D. (1970). Selection and survival in *Lolium perenne. Heredity*, **25**, 441–447.

Hayward, M.D., Gottlieb, L.D. & McAdam, N.J. (1978). Survival of allozyme variants in swards of *Lolium perenne L. Zeitschrift für Pflanzenzüchtung*, **81**, 228–234.

Hedrick, P.W. (1976). Genetic variation in a heterogeneous environment. II. Temporal heterogeneity and directional selection. *Genetics*, **84**, 145–157.

Hedrick, P.W. Ginevan, M.E. & Ewing, E.P. (1976). Genetic polymorphism in heterogeneous environments. *Annual Review of Ecology and Systematics*, **7**, 1–32.

Hickman, J.C. (1975). Environmental unpredictability and plastic energy allocation strategies in the annual *Polygonum cascadense* (Polygonaceae). *Journal of Ecology*, **63**, 689–701.

Holler, L.C. & Abrahamson, W.G. (1977). Seed and vegetative reproduction in relation to density in *Fragaria virginiana* (Rosaceae). *American Journal of Botany*, **64**, 1003–1007.

Huiskes, A.H.L., van Soelen, J. & Markusse, M.M. (1985). Field studies on the variability of populations of *Aster tripolium L.* in relation to salt marsh zonation. *Vegetatio*, **61**, 163–169.

Ingram, R. & Taylor, L. (1982). The genetic control of a non-radiate condition in *Senecio squalidus L.* and some observations on the role of ray florets in the Compositae. *New Phytologist*, **91**, 749–756.

Knowles, P. & Grant, M.C. (1985). Genetic variation of lodgepole pine over time and microgeographical space. *Canadian Journal of Botany*, **63**, 722–727.

Lande, R. & Arnold, S.J. (1983). The measurement of selection on correlated characters. *Evolution*, **37**, 1210–1226.

Law, R., Bradshaw, A.D. & Putwain, P.D. (1977). Life-history variation in *Poa annua. Evolution*, **31**, 233–246.

Loveless, M.D. & Hamrick, J.L. (1984). Ecological determinants of genetic structure in plant populations. *Annual Review of Ecology & Systematics*, **15**, 65–95.

Linhart, Y.B., Mitton, J.B., Sturgeon, K.B. & Davis, M.L. (1981). Genetic variation in space and time in a population of ponderosa pine. *Heredity*, **46**, 407–426.

McNeilly, T. (1981). Ecotypic differentiation in *Poa annua*: interpopulation differences in response to competition and cutting. *New Phytologist*, **88**, 539–547.

McNeilly, T. & Roose, M.L. (1984). The distribution of perennial rye grass genotypes in swards. *New Phytologist*, **98**, 503–513.

Parry, G.D. (1981). The meanings of *r*- and *k*-selection. *Oecologia*, **48**, 260–264.

Reinartz, J.A. (1984). Life history variation of common mullein (*Verbascum thapsus*). II. Plant size, biomass partitioning and morphology. *Journal of Ecology*, **72**, 913–925.

Roos, F.H. & Quinn, J.A. (1977). Phenology and reproductive allocation in *Andropogon scoparius* (Graminae) populations in communities of different successional stages. *American Journal of Botany*, **64**, 535–540.

Schaal, B.A. (1975). Population structure and local differentiation in *Liatris cyclindracea*. *American Naturalist*, **109**, 511–528.

Schaal, B.A. & Levin, D.A. (1976). The demographic genetics of *Liatris cyclindracea Mich.* (Compositae). *American Naturalist*, **110**, 191–206.

Scheiner, S.M. & Goodnight, C.J. (1984). A comparison of phenotypic plasticity and genetic variation in populations of the grass *Danthonia spicata*. *Evolution*, **38**, 845–855.

Scheiner, S.M., Gurevitch, J. & Teeri, J.A. (1984). A genetic analysis of the photosynthetic properties of populations of *Danthonia spicata* that have different growth responses to light level. *Oecologia*, **64**, 74–77.

Silander, J.A. & Antonovics, J. (1979). The genetic basis of the ecological amplitude of *Spartina patens*. I. Morphometric and physiological traits. *Evolution*, **33**, 1114–1127.

Snedecor, G.W. & Cochran, W.G. (1967). *Statistical Methods*, 6th edn. Iowa State University Press, Ames, Iowa.

Soane, I.D. & Watkinson, A.R. (1979). Clonal variation in populations of *Ranunculus repens*. *New Phytologist*, **82**, 557–573.

Solbrig, O.T. & Simpson, B.B. (1974). Components of regulation of a population of dandelions in Michigan. *Journal of Ecology*, **62**, 473–486.

Sterk, A.A. & Wijnands, D.O. (1970). On the variation in the flower heads of *Aster tripolium L.* in The Netherlands. *Acta Botanica Neerlandica*, **19**, 436–444.

Stewart, A.J.A. & Thompson, K. (1982). Reproductive strategies of six herbaceous perennial species in relation to a successional sequence. *Oecologia*, **52**, 269–272.

Trow, A.H. (1912). On the inheritance of certain characters in the common groundsel, *Senecio vulgaris L.*, and its segregates. *Journal of Genetics*, **6**, 1–12.

Turkington, R. & Harper, J.L. (1979). The growth, distribution and neighbour relationships of *Trifolium repens* in a permanent pasture. IV. Fine-scale biotic differentiation. *Journal of Ecology*, **67**, 245–254.

Warwick, S.I. & Briggs, D. (1978). The genecology of lawn weeds. II. Evidence for disruptive selection in *Poa annua L.* in a mosaic environment of bowling green lawns and flower beds. *New Phytologist*, **81**, 725–737.

Werner, P.A. (1978). On the determination of age in *Liatris aspera* using cross-sections of corms: implications for past demographic studies. *American Naturalist*, **112**, 1113–1120.

Wu, K.K. & Jain, S.K. (1979). Population regulation in *Bromus rubens* and *B. mollis*: life cycle components and competition. *Oecologia*, **39**, 337–357.

Zangerl, A.R. & Bazzaz, F.A. (1983). Plasticity and genotypic variation in photosynthetic behaviour of an early and a late successional species of *Polygonum*. *Oecologia*, **57**, 270–273.

14. HERBIVORES AND PLANT SUCCESSION

P.J. EDWARDS AND M.P. GILLMAN
Department of Biology, The University, Southampton, SO9 5NH

INTRODUCTION

Most models of succession attribute only minor importance to herbivory as a process bringing about change in plant species composition. In the classical theory of Clements (1916), succession is a sequence of plant communities culminating in a climatically determined climax community in which the vegetation is in equilibrium with the whole of its environment. In Clements' model the most important process bringing about change is site modification by the plants themselves. Under favourable local conditions, a subclimax may persist which represents a phase below that of climatic climax. Grazing is viewed as only one of many possible causes for the appearance of a subclimax. Although Tansley (1939) recognized the profound effects that animals may have on vegetation, including small mammals such as mice and voles, he tended to regard grazing as an external factor—a kind of treatment—which may arrest or deflect a natural succession. Perhaps this was because he was chiefly concerned with the effects of cattle, sheep and rabbits (e.g. Tansley & Adamson 1925) in populations influenced or managed by man. For the practical purpose of understanding the causes of differences in vegetation this approach is adequate, since the presence or absence of one or a few species of large herbivore can have a major and obvious effect on vegetation. In contrast, although most plant communities support a large diversity of invertebrate herbivores the effects of a single species on the plant community may often be rather small. The combined effects on the vegetation of all the smaller herbivores may be considerable, but we rarely have the opportunity to detect them since we can never find a 'control' community from which herbivory is absent.

This chapter is about herbivory in the broadest sense. It considers all animals which feed on living plant material, whether foliage, ovules, pollen, fruits or seeds. Part of our purpose is to emphasize that rather than being an external factor which may, often through the agency of man, deflect or interrupt a succession, herbivory is a process which influences the organization of almost every plant community (Crawley 1983). Although no ecologist doubts the effects that large herbivores such as cattle, deer or rabbits may have on vegetation, most descriptions of vegetation processes assume, at least implicitly, that invertebrate herbivores are unimportant. This assumption has rarely been tested, and rests chiefly on the general impression that in most communities small herbivores do not do enough damage (e.g. Hairston, Smith & Slobodkin 1960). Because of what we regard as undue neglect of the role of invertebrate herbivores in plant communities, we shall give them particular emphasis in this article.

ROLE OF HERBIVORY IN PLANT COMMUNITIES

Ecologists commonly measure the effects of herbivores upon vegetation in terms of plant *species* composition. However, we should recognize that herbivory represents a process of selection which acts at the level of the individual plant. The selection may be direct, as, for example, when a mollusc chooses to consume a particular seedling, or indirect as when defoliation affects the competitive balance between plants. Under a particular regime of herbivory some plant genotypes may be favoured more than others; sometimes a change in the regime may alter very rapidly the genetic composition of the plant population. If none of the individuals within a plant population proves tolerant then the plant species will disappear from the community. It follows that in any community the plant species present and the genetic composition of their populations are to some extent products of past herbivory. In stable conditions the relative abundance of plant species and the genetic make-up of their populations reflects an approximate equilibrium between all the various selective pressures, including herbivory, upon the plants.

In practice, it is difficult to define precisely what we mean by the role of herbivory within a community. Harper (1982) emphasizes the need to distinguish between proximal explanations in ecology (i.e. in terms of the present properties of organisms) and ultimate explanations in terms of the evolutionary forces which have acted upon their populations in the past. Since the process of selection through herbivory is continuous and often very rapid this distinction is crucial yet may be difficult to establish; we may be in danger of looking for an ecological answer to what is essentially an evolutionary problem. For example, if we perform an experiment in which all herbivores are removed from a community, various adjustments may occur which alter the relative abundance of plant species, perhaps eliminating some while allowing new species to become established. Does the experiment demonstrate the role of herbivory in that community? Since the system we began with was the product of herbivory, the available species, and the genetic variation within them, were themselves determined by herbivory. In particular, we can never restore all of the genotypes which have been eliminated by herbivory. The most that we can expect from our experiment is to demonstrate the proximal influence of herbivory in determining the relative abundance of the available species.

It is a common misconception that the impact of a herbivore upon the plant species composition of a community can be quantified in terms such as energy flow or the proportion of net primary production consumed. Within a stable plant community the role of herbivores in regulating the abundance of a plant species may be quite unrelated to the quantity of plant material they consume. As Harper (1977) says, 'It is perhaps a valid generalization that if in natural vegetation a predator or pathogen is important in the population dynamics of its prey, it will rarely be seen to be so. At the stage at which we are likely to observe the system, the feedbacks within it will be operating at a fine scale of adjustment which is rarely obvious'. It follows that the best way to investigate the role of herbivory is to perturb the system; for example, by the removal or introduction of a herbivore, or through control of herbivore activity (Harper 1969).

Compelling evidence of the potential role of insects in restricting the distribution of plants comes from successful attempts at biological control of introduced species such as *Hypericum perforatum* in California (Huffaker & Kennett 1959), and *Opuntia stricta* in Australia (Holloway 1964). However, these examples are to some extent artificial, since they concern an introduced plant species and an introduced insect, both of which are largely free from predators and pathogens which control their populations in their natural environment. We can learn more about the proximal role of herbivory from experiments in which all herbivores are removed from a plant community. The few experiments of this kind in which insects were the main herbivores suggest that they may have a substantial influence upon plant species composition. For example, McBrien, Harmsen & Crowder (1983) found considerable changes in the vegetation of an old-field community treated with insecticide for 5 years compared with the control site. The experiment suggested that grazing of *Solidago canadensis*, chiefly by three species of *Trirhabda*, reduced its cover from 40–70% to <1%, and allowed a number of earlier successional species, notably perennial grasses and *Fragaria virginiana* to increase. Brown (1982, 1985) also found significant changes when early successional vegetation was treated with insecticide. In the insecticide plots there was a higher rate of plant species accumulation through the year and a higher cover of vegetation was achieved. There was also a change in the relative abundance of plant species, with a higher proportion of grasses in the insecticide plots and a much higher survival of plant seedlings. Experiments of this kind are valuable in revealing the role of invertebrate herbivores, though possible phytotoxic effects of the insecticide mean that results must be viewed with caution (Shure 1971).

INFLUENCE OF HERBIVORY AT VARIOUS STAGES IN THE REGENERATION CYCLE

The emphasis in much ecological work upon measuring the quantity of plant material consumed by herbivores means that attention has largely been directed towards mature plants, since these make up the bulk of vegetation. However, the effect of herbivory on an individual plant may be unrelated to the quantity of plant material consumed because of the unequal importance to the plant of the various tissues which may be selected (Chew 1978). In particular, an emphasis upon quantity neglects the possibility that for many plant species the greatest effect of herbivores upon their populations occurs by seed predation or the destruction of seedlings. At these stages the total amount of plant material removed will probably be small, but the ecological consequences may be profound. It also ignores the influences of herbivores on vegetation through activities other than feeding (e.g. through dispersal of seed, and creation of gaps in which seeds may become established), which may be equally or more important.

A useful framework for understanding the effects of herbivory on plant communities has been provided by Grubb (1977) who argued that we can only understand the diversity and relative abundance of plant species in communities if we consider their requirements at all stages of the regeneration cycle. Grubb divides the regeneration cycle into the following stages: production of viable seed, dispersal of seed, germination, establishment, and onward growth. He places particular emphasis on the 'regeneration niche' of plants since there appear to be almost limitless possibilities for differences between species in their requirements for regeneration. The important point for our argument is that herbivory may exert significant selection at any stages in regeneration, and not only in the mature phase.

Production of viable seed

Any effect of grazing upon the performance of a plant is likely to affect indirectly the production of seed. In this way large herbivores in particular can have a major impact upon the production of viable seed. In the same way insects can also affect seed production. For example, Crawley

(1985) investigated the effects of reducing insect herbivore populations on oak, *Quercus robur*, using an insecticide treatment which reduced insect grazing from 8–12% of leaf area in controls to about 5%. He could find no effect of the treatment upon twig growth or wood increment, but there was a substantial effect upon acorn production which was 2·5–4.5 times higher than that of controls.

Perhaps the greatest potential for insects to influence plant population dynamics is through their direct effects upon the supply of viable seed, by predation both before and after dispersal. In a study of *Haplopappus squarrosus* (Louda 1982), an insecticide treatment reduced seed predation from 94% to 41%; as a result recruitment of seedlings to the next generation was 23 times as high as in control plots. More often the evidence for the importance of seed loss for plant populations is less conclusive. For example, Tevis (1958) showed that a colony of harvester ants in the Sonoran desert gathered an average of 7000 seeds per day, and although the ants did not deplete the total seed supply they were highly selective in the species gathered. The ants may have been a factor leading to the dominance of *Plantago insularis*, since they did not gather the seed of this species although it represented 86% of total seed production. Bohart & Koerber (1972) review the extent of seed loss by insects and point to much circumstantial evidence that seed insects influence the species composition of plant communities. For example, during one recent period in New England the beetle *Conophthorus conifera* was the principal cause of white pine (*Pinus strobus*) trees producing small seed crops for 10 years in succession (Fowells 1965). It seems probable that few white pine seedlings became established in that period, and proportionately more trees of other species are present as a consequence. Similarly, the almost total failure of the acorn crop in many parts of Britain during the early 1980s because of the Knopper gall wasp *Andricus quercus-calicis* probably reduced the proportion of oaks (*Quercus robur*) that became established in deciduous woodland (Crawley 1984).

Dispersal

Although the fact that animals are vectors for the seeds of many species is well understood, and the importance of seed dispersal in the maintenance of plant populations is widely accepted (e.g. Harper 1977; Pijl 1972), the consequences of dispersal for the species composition of vegetation have rarely been demonstrated. Perhaps they can be seen most clearly in early-successional scrub vegetation which often contains a high proportion of bird-dispersed species. For example, the scrub which develops in grassland in the New Forest in southern England is chiefly composed of berry-bearing species such as *Frangula alnus*, *Prunus spinosa*, *Ilex aquifolium*, *Crataegus monogyna* and *Rubus fruticosus*. Since seed deposition by birds is usually beneath perches (Sorensen 1981), such scrub begins to develop in a patchy fashion around isolated trees. In Australia the berry-bearing shrub *Pittosporum undulatum* is spreading outside its former range and invading wet and dry eucalypt forests in Central Victoria (Gleadow & Ashton 1981, Gleadow 1982). The invading *Pittosporum* tends to be clumped around established trees, apparently because of preferential deposition of seeds in these sites by the introduced European blackbird (*Turdus merula*) which appears to be the main vector for dispersal. *Ilex aquifolium* and *Cotoneaster pannosa*, also berried species, are spreading into the eucalypt forests for the same reason. In Northern Europe the jay (*Garrulus glareolus*) seems to be a key species for the dynamics of some forests (Bossema 1979), and was probably one of the species that dispersed beech (*Fagus sylvatica*) northwards during the Holocene (Nilsson 1985). Similarly, the blue jay (*Cyanocitta cristata*) was probably responsible for the spread of *Fagus grandifolia* in North America (Davis, Chapter 18).

Large herbivores transport seeds both internally and externally. Shmida & Ellner (1983) suggested that external seed transport on grazing livestock may contribute to the maintenance of annual plant populations in Mediterranean chaparral. Welch (1985), in a study of the seed in dung on heather moorland in Scotland, found that the mean numbers of seedlings which could be germinated from individual faecal deposits were eighty-six for cattle, ten for red deer, fifteen for sheep, seven for hare and two for rabbits. From one cow pat he obtained 662 seedlings of twenty-four different species. The proportions of different plant species in the dung of these animals varied and broadly reflected known differences in the diet. Welch showed that the contribution of seed in the dung of cattle significantly increased the cover of grasses in the vegetation.

Among invertebrates, ants are probably the most important agents of dispersal, and many seeds bear elaiosomes which apparently make them attractive to ants (Thompson 1981). In some habitats ant-dispersed plants form a very large proportion of the species present (Beattie & Culver 1982); for example, Australian heathlands are said to contain more than 1500 ant-dispersed species (Berg 1981).

Germination

Many activities of herbivores produce gaps in vegetation which provide conditions favourable for seed germination, and the size and dynamics of these gaps may influence the kinds of plant species which persist in the community (Fenner 1985, and Chapter 5). For example, in pastures there is usually a well-defined regime of gaps produced by the death of the sward beneath dung and urine; these gaps provide the opportunity for particular species to germinate, such as ragwort (*Senecio jacobaea*) and thistles (*Cirsium* spp.) (Edwards & Hollis 1982). The excreta of large herbivores also alter soil pH and nutrient conditions and may favour the germination of certain species, such as those stimulated by high nitrate concentrations (Steinbauer & Grigsby 1957). Hoof-prints represent another way in which herbivores can create gaps, and thus conditions suitable for germination.

At a larger scale burrowing animals such as rabbits, pocket gophers, and badgers create distinctive pockets of vegetation in the regions of their burrows (Chew 1978). Very often particular plant species are associated with these areas of high disturbance, such as the guilds of 'fugitive' species including *Mirabilis hirsuta*, *Solidugo rigida* and *Apocynum sibiricum* found near the burrows of American badgers, *Taxidea taxus* (Platt 1975).

As well as dispersing seeds, herbivores often provide suitable conditions for their survival and germination by burying them in favourable conditions. Watt (1919) showed that the viability of acorns of *Quercus robur* depended on the maintenance of their moisture content; those seeds which had been trampled into the ground by grazing animals were usually more successful than those which remained on the soil surface. Similarly beech (*Fagus sylvatica*) seeds on the surface of the litter have a lower chance of germinating than those which have been buried (Watt 1923), and Jensen (1985) suggested that few if any seeds would germinate without the caching activity of rodents. The seeds carried by ants to their nests may also find a more favourable environment in which the chances of predation from rodents are reduced (Heithaus 1981), and where nutrient levels are higher (Culver & Beattie 1980).

Establishment

It is clear that large herbivores can have an overwhelming effect upon vegetation through browsing of seedlings. In the New Forest in southern England, for example, which is grazed by cattle, ponies and deer, very few beech and oak seedlings survive for longer than a year or two, and in many places regeneration has almost ceased. Peterken & Tubbs (1965) showed that unenclosed woodlands in the New Forest consist of three generations of trees which became established in periods when browsing pressure was low.

Smaller herbivores can also exert a significant influence at this stage, since even limited browsing of a small plant can cause its death. Pigott (1985) investigated the damage to a range of deciduous tree seedlings by bank voles (*Clethrionomys glareolus*). He found that small saplings of *Betula* spp. were rarely damaged, and that while many seedlings of *Quercus* spp. and *Fagus sylvatica* were damaged or destroyed, saplings of these species suffered little damage from voles. Lime (*Tilia cordata*) suffered more damage and higher mortality than any of these species, and Pigott suggested that voles may be an important factor in the failure of lime regeneration in English woodland, because of the normally low density of seedlings which become established.

Insects are often responsible for high mortality of tree seedlings, especially those close to mature trees of the same species which provide the source of animals. For example, Lemen (1981) found that seedlings of *Ulmus parviflora* directly beneath adult trees were often completely defoliated by the elm leaf beetle (*Pyrrhalta luteola*). These seedlings carried an average of 852 times as many egg masses as those further away. In tropical rain forest, insects can be important seedling predators (Janzen 1970; Connell 1971, 1979), though most seedling deaths are probably caused by fungal pathogens (Augspurger 1984).

Further development

The effects of large herbivores upon plant communities occur chiefly through defoliation of established plants, and in particular through removal of aerial meristems. Under conditions of heavy browsing or grazing the plant species which succeed are those in some way resistant to these pressures, either because they are protected physically or chemically, or because their growth form makes their meristems or foliage inaccessible to herbivores. Furthermore, grazing-resistant plants may exhibit a remarkable capacity for compensatory growth in response to defoliation (McNaughton 1979). In response to an increase in grazing pressure, there may be rapid selection for resistant ecotypes (e.g. *Puccinellia maritima* grazed by sheep; Gray & Scott 1977), and many species may be eliminated from a cummunity (e.g. Allen, Payton & Knowlton 1984).

Although their effects may be less conspicuous, smaller herbivores can also influence the composition of vegetation through altering the competitive balance between species (Reid & Harmsen 1974). A clear demonstration comes from Bentley & Whittaker's study (1979) of the effects of the beetle *Gastrophysa viridula* upon competition between two species of *Rumex*. *Rumex crispus* proved to be more vulnerable to grazing than *R. obtusifolius* when the two species were growing together. Indeed, it was suggested that the absence of *R. crispus* from some apparently suitable sites was because of large populations of *Gastrophysa* on nearby *R. obtusifolius*. Exclusion effects of this kind can only be revealed by experiment, but there is no reason to suppose that this was an isolated example.

A major outbreak of an insect species can significantly alter the development of vegetation by promoting the growth of suppressed plants. For example, heavy defoliation of aspen forest in Minnesota by the forest tent caterpillar (*Malacosoma distria*) led to reduced growth of aspen (*Populus tremuloides*), but greatly enhanced growth of the understorey balsam fir (*Abies balsamea*) (Duncan & Hodson 1958). Similarly, red maples (*Acer rubrum*) growing beneath oaks defoliated by gypsy moth (*Porthetria dispar*) made much more growth than those beneath intact trees (Collins 1961).

ROLE OF HERBIVORY IN SUCCESSION

Connell & Slatyer (1977) examine three possible mechanisms for species change in plant communities, which they describe as the facilitation, tolerance, and inhibition models. According to the facilitation model the established plants modify their environment in such a way that they bring about change in species composition. Since the vegetation provides the habitat of animals, in a general sense facilitation must occur whenever herbivores are responsible for successional change. For example, it seems likely that the presence of pioneer woody species often increases the rate of colonization of a site by other woody species whose seeds are dispersed by animals and birds, since the pioneers make the site more attractive to the dispersal vectors (Finegan 1984). In some cases the established vegetation also provides protection for the invading species from grazing animals, as when yew (*Taxus baccata*) develops beneath juniper (*Juniperus communis*) (Watt 1926).

In the tolerance model the later species are those able to tolerate the environmental conditions which develop through succession. Connell and Slatyer use the example of shade tolerance, but point out that tolerance to other environmental factors such as allelochemicals and grazing may be equally or more important in particular cases. This chapter has argued that the composition of a plant community nearly always reflects its history of herbivory; as the animal community changes during the course of succession, plant species which are intolerant of the changing herbivore regime will be eliminated. There is evidence (e.g. Cates & Orians 1975; Janzen 1976), though much of it open to criticism (e.g. Maiorana 1978), that plants of later successional stages have generally higher levels of secondary chemicals and are more resistant to attack from insects.

In the inhibition model we see a major way in which herbivory may affect the course of succession by allowing new species to invade established vegetation. We have seen how activities of herbivores, e.g. grazing, dunging, urination and trampling, create gaps in which plants can become established. Thus, all three models of succession described by Connell and Slatyer suggest possible ways in which herbivory may bring about change in plant communities (Crawley 1983). To illustrate these processes we will focus on three examples of communities of plants and herbivores which have been chosen because they represent different successional stages with different kinds of herbivore.

Early successional weed community grazed by molluscs

Our first example is an early successional weed community in its first year after ploughing in the Southampton University experimental grounds. Gastropods are the most important grazers in this community, and at the experimental site a total of thirteen species were recorded including *Arion ater*, *Deroceras reticulatum*, and *Helix aspersa*. The density of these three species together was up to twenty-five individuals per square metre. Herbivory accounted for about 40% of all seedling mortality, of which at least half could be attributed to gastropods (Gillman 1986). Some simple experiments were performed to find whether the predation of seedlings by gastropods is selective. Individuals of *Helix aspersa* and *Deroceras reticulatum* were allowed to graze in trays of seedlings which had germinated in soil taken from the study site. The null hypothesis was that seedlings would be taken in proportion to their abundance, but the results show that with both gastropods some plant species were grazed far more than expected (those plants below the line in Fig. 14.1), and vice versa. Many factors were probably important in determining the selection of seedlings including palatability (Grime, MacPherson & Dearman 1968) and relative abundance (Cottam 1985). These experiments suggest that the growth form of the seedling may also be important, with erect seedlings such as *Lamium purpureum* and *Chenopodium album* proving more vulnerable than those with a shorter, rosette growth form such as *Crepis capillaris*.

A second experiment was performed to investigate the consequence of selective grazing for the species composition of the vegetation. Equal numbers of seeds of four common species in the successional plots (*Hypochoeris radicata*, *Lamium purpureum*, *Senecio jacobaea* and *Taraxacum officinale*) were sown in trays in the greenhouse. There were three grazing treatments, each using one individual of *Helix aspersa* per tray: (i) grazing of young seedlings for 5 days starting 3 weeks after germination; (ii) grazing of older plants for 5 days starting 7 weeks after germination;

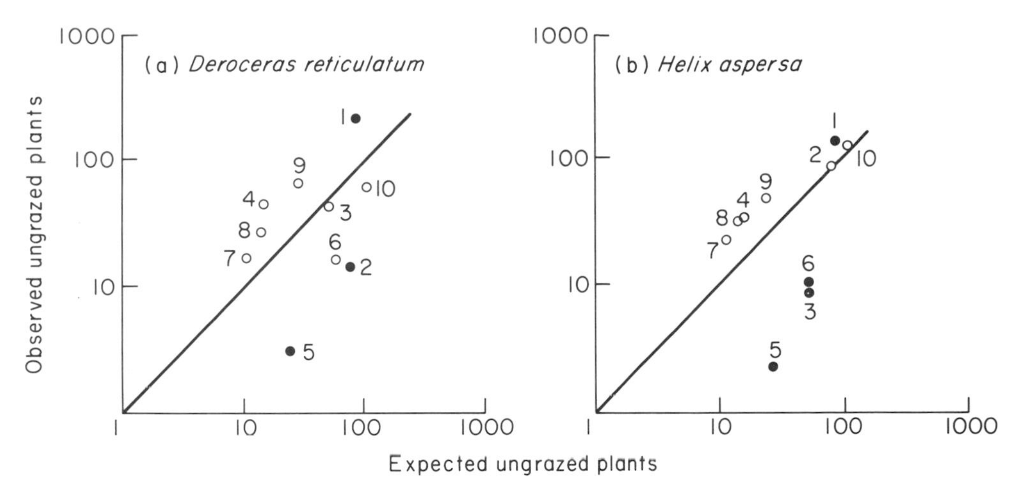

FIG. 14.1. The numbers of seedlings left ungrazed by (a) *Deroceras reticulatum* and (b) *Helix aspersa*, plotted against the expected numbers based on the total numbers of seedlings which germinated in soil taken from recently ploughed experimental plots. The plant species are: 1, *Aphanes arvensis*; 2, *Cerastium fontanum* ssp. *glabrescens*; 3, *Chenopodium album*; 4; *Crepis capillaris*; 5, *Lamium purpureum*; 6, *Plantago lanceolata*; 7, *Sonchus asper*; 8, *Senecio vulgaris*; 9, *Veronica arvenis*; 10, *Viola arvensis*. The straight line represents the null hypothesis of no selection by gastropods. ● = difference significant at $P < 0{\cdot}05$; ○ = difference not significant.

(iii) grazing throughout weeks 3–7. There was also an ungrazed control. After week 7 the plants were left to grow without grazing for 3 weeks and were then harvested. The results (Table 14.1) show that the relative abundance of plant species was considerably affected by the grazing treatments. *Hypochoeris radicata* was eliminated from trays with early grazing and much reduced in trays with late grazing. The proportion of *Lamium purpureum* was also substantially reduced by grazing, while the proportion of *Senecio vulgaris* was higher in grazed treatments. The early grazing treatment had a much greater effect on the final species composition than the later treatment; however, the greatest differences from the controls were in trays which were grazed throughout. The conclusion from these experiments is that terrestrial gastropods can potentially exert a very considerable effect upon the species composition of early-successional weed communities, especially through selective destruction of seedlings.

Grazing by large herbivores in mid-successional communities

Our second example concerns the influence of large herbivores—cattle, ponies and roe and fallow deer—upon a mid-successional community in

TABLE 14.1. Effects of grazing by *Helix aspersa* upon (a) plant survival (percentage of number germinating) and (b) plant species composition (percentage of above-ground dry weight) in trays with equal numbers of four plant species

	Ungrazed		Early grazing		Late grazing		Grazing throughout	
	A	B	A	B	A	B	A	B
Hypochoeris radicata	100	19	1	0	48	5	0	0
Lamium purpureum	100	22	20	6	58	9	3	0
Senecio vulgaris	98	51	87	84	99	72	93	98
Taraxacum officinale	99	8	63	10	91	14	30	2

the New Forest, England. Low-lying and poorly drained areas inaccessible to large herbivores are often occupied by nearly pure stands of the tussock-forming grass *Molinia caerulea*. In places *Molinia* may be invaded by various shrubs, particularly *Myrica gale* and *Salix atrocinerea*, leading to the development of carr woodland; however, in the absence of disturbance the process is very slow and pure stands of *Molinia* can persist indefinitely. It is a good example of a long-lived perennial which can persist and reproduce vegetatively, so that for long periods the reproductive phases of the life cycle are unimportant in maintaining the presence of the species at a particular site; only herbivores which affect the performance of the mature plant are likely to influence its abundance once it has become established.

In more accessible areas, and particularly on the flood plains of streams in the New Forest, there are small, rounded mounds of silt and clay similar in size and spacing to the tussocks of *Molinia* (Edwards 1985). There is little doubt that most of these are former tussocks, and in some places we can see all stages in a sequence of change from large tussocks growing in pure stands to small, grassy mounds containing no *Molinia*. The change is apparently due to increased grazing in recent years. Table 14.2 shows the mean percentage cover of plant species in tussocks or mounds at three pairs of sites chosen because the first of each pair was less grazed than the second: in one case (A) the less grazed site was protected by a fence; in the other two cases the less grazed areas were pony latrine areas. The results suggest that even relatively light grazing is sufficient to reduce the vigour of *Molinia* and allow other species like *Sphagnum* spp., *Hydrocotyle vulgare* and grasses, to invade the tussock. Further successional change may follow: the mounds are better aerated than the surrounding soil and, in the absence of a dense cover of grass foliage and litter, provide

TABLE 14.2. Percentage cover of plant species (+ indicates the presence of a species with cover of <1%) in *Molinia caerulea* tussocks or grassy mounds at three pairs (A, B, C) of adjacent sites with low and high grazing pressure from large herbivores

Plant species	A		B		C	
	Low	High	Low	High	Low	High
Achillea ptarmica	–	+	–	+	–	–
Agrostis canina	–	3	–	35	–	–
Agrostis capillaris	–	+	–	–	–	10
Anagallis tenella	–	+	–	–	–	–
Angelica sylvestris	+	–	+	–	–	–
Betula pubescens	–	+	–	+	–	–
Calluna vulgaris	–	–	–	–	5	–
Carex panicea	–	10	–	2	–	5
Cirsium dissectum	3	+	+	–	–	3
Danthonia decumbens	–	–	–	–	–	3
Erica tetralix	–	5	–	–	10	–
Festuca ovina	–	5	–	–	–	–
Festuca rubra	–	–	–	–	–	25
Holcus lanatus	–	–	+	–	–	–
Hydrocotyle vulgaris	3	18	–	5	–	–
Juncus acutiflorus	+	–	+	–	–	–
Lotus uliginosus	–	–	–	3	–	+
Molinia caerulea	95	8	95	30	75	35
Nardus stricta	–	2	–	–	–	–
Pinus sylvestris	–	–	–	–	–	+
Potentilla erecta	–	5	+	3	–	+
Salix repens	–	5	–	–	5	–
Scutellaria minor	–	–	–	2	–	–
Serratula tinctoria	–	–	–	–	–	+
Sphagnum auriculatum ssp. *auriculatum*	–	40	–	20	5	15

a suitable seed bed for tree species. Seedlings of *Salix atrocinerea*, *Quercus robur* and *Betula pubescens* establish in the mounds, though they usually fail to grow into trees because of browsing. However, *Pinus sylvestris* is avoided by ponies, and in several places areas formerly dominated by *Molinia* have developed into pine woodland.

Insect herbivory in woodland

For our third example we consider the consequences of herbivory in mixed woodland in which the most important herbivores are a wide diversity of phytophagous insects. To simplify the discussion we will ignore other herbivores such as small mammals and gastropods, although they may also be important in particular circumstances. It must be said that to a large extent we do not know the effects of insects in maintaining the structure and composition of woodland. However, we speculate that the processes of selection through grazing are different from those in the previous two examples for the following reasons.

(a) There is usually a very large number of herbivore species.

(b) Among these species there is a high level of feeding specialization; most phytophagous insects are oligophagous or monophagous. Although the important defoliators are often polyphagous species such as gypsy moth (*Porthetria dispar*) and winter moth (*Operophtera brumata*), there is evidence that many polyphagous species consist of monophagous or oligophagous local populations (Fox & Morrow 1981; Strong, Lawton & Southwood 1984).

(c) There is very wide variation in population sizes of many insect species from year to year; for example Strong, Lawton & Southwood (1984) give typical fluctuations in phytophagous insects between generations of 10 to 100-fold.

(d) The generation times of insect herbivores are very short compared with those of the dominant plant species.

Given these differences, the processes of selection through herbivory are also likely to be different. Where there are successive outbreaks of different herbivore species, it seems likely that the direction of selection must change from year to year. For example, let us assume that in an oak wood each mature tree is replaced by another every 200 years or so, and that each year a small number of future trees become established. It is probable that in different years the selective pressures from herbivory faced during the critical phases of seed production, germination and establishment will be different. This may be one reason for the apparently high genetic diversity that we find among many forest trees; in the oak (*Quercus robur*), for example, individual trees growing together at one site typically vary widely in characters likely to affect insect performance, such as leaf phenology (e.g. Rackham 1975), phenolic content and palatability (P.J. Edwards, unpubl.). In conifers, such as *Pinus ponderosa* and *Pseudotsuga menzesii*, the profiles of resin terpenes are often extremely variable within populations (Hanover 1975).

Because of the longevity of trees relative to the life cycle of insects, over many generations an insect population may become increasingly adapted to the defensive characteristics of a single host-tree (Edmunds & Alstad 1978). 'Host tracking' of this kind occurs in relatively immobile species, such as the black scale of the ponderosa pine (*Nuculaspis californica*; Edmunds & Alstad 1978) and the beech scale (*Cryptococcus fagisuga*;

Wainhouse & Howell 1983), in which there is little gene flow between populations on different trees. Variability between trees in their defensive characteristics means that insects differ in their ability to colonize individual trees, leading to a discontinuous distribution of infested trees.

Another possible difference in the processes of selection in woodland, given the high level of feeding specialization among insects, is that density-dependent effects may reduce the likelihood of over-exploitation of one host species. An exception to this may be that plant species which share the same herbivore species but differ in their resistance to it—as in the case of two *Rumex* species grazed by *Gastrophysa viridula* (Bentley & Whittaker 1979)—are unable to coexist.

We conclude that the composition of each of the three plant communities we have considered—the plant species present, their relative abundance, their spatial organization, and the genetic make-up of their populations—reflect their history of herbivory. In the various cases different phases of the regeneration cycle may have been affected, and the consequences for plant populations and the composition of vegetation have differed. In Table 14.3 we compare the three plant–herbivore communities in terms of the selective processes acting upon the plants. In the weed community and the mid-successional community the main herbivore species are few in number and highly polyphagous; palatability and growth form of plants appear to be important in determining patterns of food selection. In the examples chosen, herbivory appears to have had strongly directional effects, leading to the elimination of certain species and encouraging others. In the woodland commuity there are a large number of insect herbivore species, many of which have a very restricted host-plant range. For these species the possible food plants have been determined by evolution, but characteristics of the plant community such as species diversity, and the density and distribution of host-plants may influence the extent to which individual plants are attacked. Differences in food quality between individuals may also be important. Although a woodland community may achieve equilibrium with its herbivore community in the sense that there are no directional changes in plant species composition, we suggest that there are likely to be fluctuations in the relative abundance of plant species and in the genetic composition of their populations which reflect variations in the insect community.

The importance of herbivory relative to other factors in bringing about succession is uncertain. It seems probable that in the early stages of most primary successions herbivory is unimportant since it may take a relatively long time before vegetation is sufficiently developed to provide an adequate habitat for animals. If the succession occurs in a remote or very

TABLE 14.3. Comparison of three plant–herbivore communities to show some of the factors affecting plant selection by herbivores, and the phase of the plant reproductive cycle most affected

	Example 1 Early successional weed community Gastropods	Example 2 Mid-successional grassland Large herbivores	Example 3 Late-successional deciduous woodland Insects
Number of herbivore species	13	4	100s
Variation in herbivore populations	±Stable	Increasing	Very variable
Herbivore food range	Highly polyphagous	Highly polyphagous	Mainly oligophagous or monophagous
Plant factors affecting level of herbivory	Palatability Growth form Relative frequency	Palatability Growth form —	— — Density and distribution of host-plants
Phase of plant reproductive cycle most affected	Establishment	All phases	Seed production, establishment

extensive area then the rate of colonization by some animal species may be delayed because of their slow rates of dispersal. In most primary successions facilitation—for example, by building up soil nitrogen or through ameliorating the microclimate close to the soil surface—is probably most important in determining plant succession (Vitousek & Walker, Chapter 10). However, as the herbivore community increases and becomes more diverse, so herbivory may become an important factor in primary succession. In contrast, herbivory may be important from the beginning of some secondary successions. As the areas affected are often rather small, there is likely to be adjacent vegetation which provides both a source of animals and a suitable habitat from which they may forage. Often the disturbance which begins the succession does not totally eliminate all herbivore species or destroy the essential features of their habitat. For example, in our studies of early weed successions, we found that the slug *Arion ater* was at its highest density in newly cleared land, and its eggs almost certainly survived the process of ploughing.

Before we can understand the importance of herbivory as a factor bringing about change in plant communities we need to know much more about patterns of change in herbivore populations during the course of succession; with a few notable exceptions (e.g. Cameron & Morgan-Huws 1975; Brown & Southwood, Chapter 15) very little systematic information is available on this topic. We also need more experimental studies of successional communities in which herbivore populations are manipulated (e.g. Cantlon 1969; Watt 1981; Brown 1982), and especially in which the resultant effects on the plant community are recorded at the level of the individual (e.g. Mills 1983; Silander 1983; Gillman 1986). In the absence of such experiments, the role of herbivory in bringing about change in plant communities must remain largely a matter for speculation.

REFERENCES

Allen, R.B., Payton, I.J. & Knowlton, J.E. (1984). Effects of ungulates on structure and species composition in the Urewera forests as shown by exclosures. *New Zealand Journal of Ecology*, **7**, 119–130.

Augspurger, C.K. (1984). Seedling survival of tropical tree species; interactions of dispersal distance, light-gaps and pathogens. *Ecology*, **65**, 1706–1712.

Beattie, A.J. & Culver, D.C. (1982). Inhumation: how ants and other invertebrates help seeds. *Nature*, **297**, 627.

Bentley, S. & Whittaker, J.B. (1979). Effects of grazing by a chrysomelid beetle, *Gastrophysa viridula*, on competition between *Rumex obtusifolius* and *Rumex crispus*. *Journal of Ecology*, **67**, 79–90.

Berg, R.Y. (1981). The role of ants in seed dispersal in Australian lowland heathland.

Heathlands and related shrublands of the World B. Analytical Studies (Ed. by R.L. Sprecht), pp. 51–59. Elsevier, Amsterdam.

Bohart, G.E. & Koerber, T.W. (1972). Insects and seed production. *Seed Biology*, Vol. III (Ed. by T.T. Kozlowski), pp. 1–53. Academic Press, New York.

Bossema, I. (1979). Jays and oaks: an eco-ethological study of a symbiosis. *Behaviour*, **70**, 1–117.

Brown, V.K. (1982). The phytophagous insect community and its impact on early successional habitats. *Proceedings of the 5th International Symposium on Insect–Plant Relationships* (Ed. by J.H. Visser & A.K. Minks), pp. 205–213. Pudoc, Wageningen.

Brown, V.K. (1985). Insect herbivores and plant succession. *Oikos*, **44**, 17–22.

Cameron, R.A.D. & Morgan-Huws, D.I. (1975). Snail faunas in the early stages of a chalk grassland succession. *Biological Journal of the Linnean Society*, **7**, 215–229.

Cantlon, J.E. (1969). The stability of natural populations and their sensitivity to technology. *Brookhaven Symposium in Biology*, **22**, 197–205.

Cates, R.G. & Orians, G.H. (1975). Successional status and the palatability of plants to generalised herbivores. *Ecology*, **56**, 410–418.

Chew, R.M. (1978). The impact of small mammals on ecosystem structure and function. *Populations of Small Mammals under Natural Conditions* (Ed. by D.P. Snyder), pp. 167–180. *Pymatuning Symposia in Ecology*. Special Publications Series, University of Pittsburgh, Pennsylvania.

Clements, F.E. (1916). *Plant succession: an analysis of the development of vegetation.* Carnegie Institute, Washington, Publication No 242.

Collins, S. (1961). Benefits to understorey from canopy defoliation by gypsy moth larvae. *Ecology*, **42**, 836–838.

Connell, J.H. (1971). On the role of natural enemies in preventing competitive exclusion in some marine animals and in rain forest trees. *Proceedings of the Advanced Study Institute on Dynamics of Numbers in Populations*, Oosterbeek, 1970 (Ed. by P.J. den Boer & G.R. Gradwell), pp. 298–312. Centre for Agricultural Publishing and Documentation, Wageningen.

Connell, J.H. (1979). Tropical rain forests and coral reefs as open non-equilibrium systems. *Population Dynamics* (Ed. by R.M. Anderson, B.D. Turner & L.R. Taylor), pp. 141–163. Symposium of the British Ecological Society 25. Blackwell Scientific Publications, Oxford.

Connell, J.H. & Slatyer, R.O. (1977). Mechanisms of succession in natural communities and their role in community stability and organisation. *American Naturalist*, **111**, 1119–1144.

Cottam, D.A. (1985). Frequency-dependent grazing by slugs and grasshoppers. *Journal of Ecology*, **73**, 925–933.

Crawley, M.J. (1983). *Herbivory. The Dynamics of Animal–Plant Interactions*. Blackwell Scientific Publications, Oxford.

Crawley, M.J. (1984). The gall from Gaul. *International Dendrological Society Yearbook* 1983, pp. 135–137.

Crawley, M.J. (1985). Reduction of oak fecundity by low-density herbivore populations. *Nature*, **314**, 163–164.

Culver, D.C. & Beattie, A.J. (1980). The fate of *Viola* seeds dispersed by ants. *American Journal of Botany*, **67**, 710–714.

Duncan, D.P. & Hodson, A.C. (1958). Influence of the forest tent caterpillar upon the aspen forests of Minnesota. *Forest Science*, **4**, 71–93.

Edmunds, G.F. & Alstad, D.N. (1978). Coevolution in insect herbivores and conifers. *Science*, **199**, 941–945.

Edwards, P.J. (1985). Some effects of grazing on the vegetation of streamside lawns in the New Forest. *Proceedings of the Hampshire Field Club*, **41**, 45–50.

Edwards, P.J. & Hollis, S. (1982). The distribution of excreta on New Forest grassland used by cattle, ponies and deer. *Journal of Applied Ecology*, **19**, 953–964.

Fenner, M. (1985). *Seed Ecology*. Chapman and Hall, London.

Finegan, B. (1984). Forest succession. *Nature*, **312**, 109–114.

Fowells, H.A. (1965). *Silvics of forest trees in the United States*. United States Department of Agriculture Forest Services **71**, 329.

Fox, L.R. & Morrow, P.A. (1981). Specialization: species property or local phenomenon? *Science*, **211**, 887–893.

Gillman, M.P. (1986). *The effect of grazing by gastropods on vegetation dynamics in early secondary succession*. Ph.D thesis, University of Southampton.

Gleadow, R.M. (1982). Invasion by *Pittosporum undulatum* of the forests of Central Victoria. 2. Germination and establishment. *Australian Journal of Botany*, **30**, 185–198.

Gleadow, R.M. & Ashton, D.H. (1981). Invasion by *Pittosporum undulatum* of the forests of Central Victoria. 1. Invasion patterns and plant morphology. *Australian Journal of Botany*, **29**, 705–720.

Gray, A.J. & Scott, R. (1977). *Puccinellia maritima* (Huds) Parl. Biological Flora of the British Isles. *Journal of Ecology*, **65**, 699–716.

Grime, J.P. Macpherson-Stewart, S.F. & Dearman, R.S. (1968). An investigation of leaf palatability using the snail *Cepaea nemoralis* L. *Journal of Ecology*, **56**, 405–420.

Grubb, P.J. (1977). The maintenance of species richness in plant communities: the importance of the regeneration niche. *Biological Reviews*, **52**, 107–145.

Hairston, N.G., Smith, F.F. & Slobodkin L.B. (1960). Community structure, population control and competition. *American Naturalist*, **94**, 421–425.

Hanover, J.W. (1975). Physiology of tree resistance to insects. *Annual Review of Entomology*, **20**, 75–95.

Harper, J.L. (1969). The role of predation in vegetational diversity. *Diversity and Stability in Ecological Systems*. Brookhaven Symposia in Biology, **22**, 48–62.

Harper, J.L. (1977). *Population Biology of Plants*. Academic Press, London.

Harper, J.L. (1982). After description. *The Plant Community as a Working Mechanism* (Ed. by E.T. Newman), pp. 11–25. Special Publications No. 1 of the British Ecological Society. Blackwell Scientific Publications, Oxford.

Heithaus, E.R. (1981). Seed predation by rodents on three ant-dispersed plants. *Ecology*, **62**, 136–145.

Holloway, J.K. (1964). Projects in biological control of weeds. *Biological Control of Pests and Weeds* (Ed. by P. DeBach), pp. 650–670. Chapman and Hall, London.

Huffaker, C.B. & Kennett, C.E. (1959). A ten year study of vegetational changes associated with biological control of Klamath weed. *Journal of Range Management*, **12**, 69–82.

Janzen, D.H. (1970). Herbivores and the number of tree species in tropical forests. *American Naturalist*, **104**, 501–528.

Janzen, D.H. (1976). *Ecology of Plants in the Tropics*. Studies in Biology No. 133, Arnold, London.

Jensen, T.S. (1985). Seed–seed predator interactions of European beech, *Fagus silvatica* and forest rodents, *Clethrionomys glareolus* and *Apodemus flavicollis*. *Oikos*, **44**, 149–156.

Lemen, C. (1981). Elm trees and elm leaf beetles—patterns of herbivory. *Oikos*, **36**, 65–67.

Louda, S.M. (1982). Limitations of the recruitment of the shrub *Haplopappus squarrosus* (Asteraceae) by flower- and seed-feeding insects. *Journal of Ecology*, **70**, 43–53.

McBrien, H., Harmsen, R. & Crowder, A. (1983). A case of insect grazing affecting plant succession. *Ecology*, **64**, 1035–1039.

McNaughton, S.J. (1979). Grazing as an optimization process: grass–ungulate relationships in the Serengeti. *American Naturalist*, **113**, 691–703.

Maiorana, V.C. (1978). What kinds of plants do herbivores really prefer? *American Naturalist*, **112**, 631–635.

Mills, J.N. (1983). Herbivory and seedling establishment in post-fire southern Californian chaparral. *Oecologia (Berlin)*, **60**, 267–270.

Nilsson, S.G. (1985). Ecological and evolutionary interactions between reproduction of beech *Fagus sylvtica* and seed eating animals. *Oikos*, **44**, 157–164.

Peterken, G.F. & Tubbs, C.R. (1965). Woodland regeneration in the New Forest, Hampshire, since 1650. *Journal of Applied Ecology*, **2**, 159–170.

Pigott, C.D. (1985). Selective damage to tree seedlings by bank voles (*Clethrionomys glareolus*). *Oecologia (Berlin)*, **67**, 367–371.

Pijl, L. van der. (1972). *Principles of Dispersal in Higher Plants*, Springer-Verlag, New York.

Platt, W.J. (1975). The colonization and formation of equilibrium plant species associations on badger disturbances in tall-grass prairie. *Ecological Monographs*, **45**, 285–305.

Rackham. O. (1975). *Hayley Wood, its history and ecology*. Cambridgeshire and Isle of Ely Naturalist's Trust.

Reid, D.G. & Harmsen, R. (1974). *Trirhabda borealis* Blake (Coleoptera: Chrysomelidae): a major phytophagous species on *Solidago canadensis* L. (Asteraceae) in south-eastern Ontario. *Proceedings of the Entomological Society of Ontario*, **105**, 44–47.

Shmida, A. & Ellner, S. (1983). Seed dispersal on pastoral grazers in open Mediterranean chaparral, Israel. *Israel Journal of Botany*, **32**, 147–159.

Shure, D.J. (1971). Insecticide effects on early succession in an old-field ecosystem. *Ecology*, **52**, 271–279.

Silander, J.A. Jr (1983). Demographic variation in the Australian desert Cassia under grazing pressure. *Oecologia (Berlin)*, **60**, 227–233.

Sorensen, A.E. (1981). Interactions between birds and fruit in a temperate woodland. *Oecologia (Berlin)*, **50**, 242–249.

Steinbauer, G.P. & Grigsby, B. (1957). Interaction of temperature, light and moistening agent in the germination of weed seeds. *Weeds*, **5**, 157.

Strong, D.R., Lawton, J.H. & Southwood, T.R.E. (1984). *Insects on Plants*. Blackwell Scientific Publications, Oxford.

Tansley, A.G. (1939). *The British Isles and their Vegetation*. Cambridge University Press, Cambridge.

Tansley, A.G. & Adamson, R.S. (1925). Studies of the vegetation of the English Chalk. III. The chalk grasslands of the Hampshire–Sussex border. *Journal of Ecology*, **13**, 177–223.

Tevis, L. (1958). Interrelations between the harvester ant *Veromessor pergandei* (Mayr) and some desert ephemerals. *Ecology*, **39**, 695–704.

Thompson, J.N. (1981). Elaiosomes and fleshy fruits: phenology and selection pressures for ant-dispersed seeds. *American Naturalist*, **117**, 104–108.

Wainhouse, D. & Howell, R.S. (1983). Intraspecific variation in beech scale populations and in susceptibility of their host *Fagus sylvatica*. *Ecological Entomology*, **8**, 351–359.

Watt, A.S. (1919). On the causes of failure of natural regeneration in British oakwoods. *Journal of Ecology*, **7**, 173–203.

Watt, A.S. (1923). On the ecology of British beechwoods with special reference to their regeneration. *Journal of Ecology*, **11**, 1–48.

Watt, A.S. (1926). Yew communities of the South Downs. *Journal of Ecology*, **14**, 282–316.

Watt, A.S. (1981). A comparison of grazed and ungrazed grassland A in East Anglian Breck. *Journal of Ecology*, **69**, 499–508.

Welch, D. (1985). Studies in the grazing of heather moorland in north-east Scotland. IV. Seed dispersal and plant establishment in dung. *Journal of Applied Ecology*, **22**, 461–472.

15. SECONDARY SUCCESSION: PATTERNS AND STRATEGIES

VALERIE K. BROWN[1] AND T.R.E. SOUTHWOOD[2]
[1]*Imperial College at Silwood Park, Ascot, Berkshire, SL5 7PY and*
[2]*Department of Zoology, University of Oxford, South Parks Road, Oxford*

INTRODUCTION

Changes in ecological patterns and strategies have featured prominently in models of succession (e.g. Margalef 1968; Odum 1969; Van Hulst 1978; Peet & Christensen 1980). Indeed, Odum recognized twenty-four ecosystem attributes that may change during succession, including community energetics, community structure and the characteristics of the organisms themselves. Many of these structural and functional attributes have since been assessed over a portion of a successional gradient and there has been some contention over Odum's predictions (e.g. Drury & Nisbet 1973). However, most studies have described temporal changes in the plant community, especially in terms of species composition (e.g. Bazzaz 1975; Mellinger & McNaughton 1975; Nicholson & Monk 1975; Tramer 1975; White 1979; Denslow 1980), while far fewer have even mentioned the associated organisms or pathogens (e.g. Murdoch, Evans & Peterson 1972; Witkowski 1973; Nagel 1979; Schimpf, Henderson & MacMahon 1980; Itamies 1983). We therefore believe that Ricklef's (1973) statement that 'the complexity of community organization along a successional gradient has never been measured' still applies. Obviously, to attempt to fill this gap in ecological knowledge is a bewildering prospect, requiring unlimited resources and ecologists with high longevity. But we believe that relatively long-term, holistic field studies can make a substantial contribution, enable theoretical predictions and generalizations to be tested and thereby advance ecological theory.

Several authors have suggested that the spatial and temporal patterns of organisms and the evolutionary strategies should be considered in terms of the properties of the habitat (e.g. Grime 1977; Southwood 1977; Greenslade 1983; Begon 1985; Sibly & Calow 1985). The habitat templet represents a type of ecological periodic table in which the main axes describe features of habitats important for plants and animals; properties such as life-history strategies can be expected to vary in a predictable fashion across this templet. One of us (TRES) proposed a scheme in which the two cardinal axes of the templet are the durational stability of the habitat and the constancy of favourableness (or adversity). The concept of adversity or stress, as developed by Whittaker (1975) and Grime (1977, 1979), is largely a reflection of climatic and soil conditions. During the course of a succession the habitat changes in a sequential fashion along the axis of durational stability.

Our three aims in this paper are to describe and quantify the following features of a secondary succession:

(a) Changes in the habitat in terms of temporal and spatial dimensions and the relation of the former to the vertical axis of the habitat templet.

(b) Successional patterns in the plants, their associated fauna and pathogens.

(c) Adaptive strategies of the organisms characteristic of different stages in succession and an assessment of whether they follow a consistent pattern.

As the backbone of this paper we shall use results from an intensive field study of the characters and attributes of the macro-organisms (mainly green plants and insects) of a secondary succession on sandy soil at Silwood Park in southern Britain; where appropriate we shall also refer to comparable work from elsewhere. The study at Silwood Park, Berkshire, began in 1977 and is based on a range of experimental sites of different ages which have been allowed to recolonize and develop naturally. Each year since 1977 new sites have been created by harrowing the ground; these sites will be referred to as 'Young Fields'. Permanent grassland, established 7 years before the start of the study, provides an 'Old Field' site, while a later stage in the succession is represented by a predominantly birch woodland (*Betula pendula* and *B. pubescens*) with an estimated age of 60 years ('Woodland'). Details of the experimental sites and sampling programme are given in Southwood, Brown & Reader (1979).

THE CHANGING HABITAT TEMPLET

The plant community constitutes a key feature of the habitat of both individual plants and animals; indeed, so far as many animals are concerned, a particular plant species may be an absolute prerequisite of their habitat. Of course, the plants themselves have adaptive strategies and, later, we will explore the extent to which these have consistent trends and parallel those amongst the animals. In this section we consider changes in the above-ground vegetation in both temporal and spatial dimensions.

Temporal dimensions

If, during a secondary succession, a site moves vertically up the axis representing durational stability in the habitat templet (Fig. 15.1) then the rate of change, measured in various ways, should decrease with successional age. Differences in the plant species diversity with time (β-diversity) can be expressed by means of a dendrogram, based on values for Sorensen's index of similarity (Fig. 15.2a). In the Woodland these differences are small and reflect only seasonal variation, while the differences in the Young Field are large and reflect an underlying change in the composition of the plant community. The changes in species composition of the Young and Old Field plant communities can be demonstrated more clearly by ordination techniques (Fig. 15.2b). The analysis was based on a 5-year (1977–81) data set for 'Young Field' sites which were created each year and for the single 'Old Field' site. In the first 5 years there is a relatively rapid change in the position of the sites on the first two axes (the development of a single site is shown by consistent symbols) until relative stability is reached between years 7 and 11. Further details of the ordination of the plants (and phytophagous Coleoptera) are given in Brown & Hyman (in press). This reduced rate of change in plant species composition is probably a reflection of the increasing longevity of individual plants in the older sites. There have been relatively few studies of the longevity of individual plants in natural communities

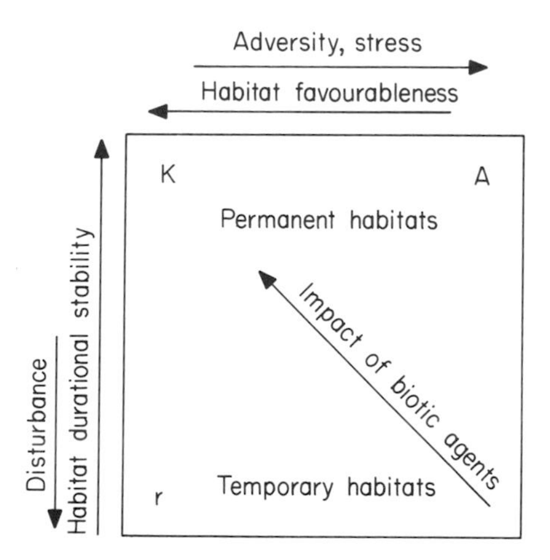

FIG. 15.1. The habitat templet (after Southwood 1977; Greenslade 1983): A, adversity selection; K, *K*-selection; *r*, *r*-selection.

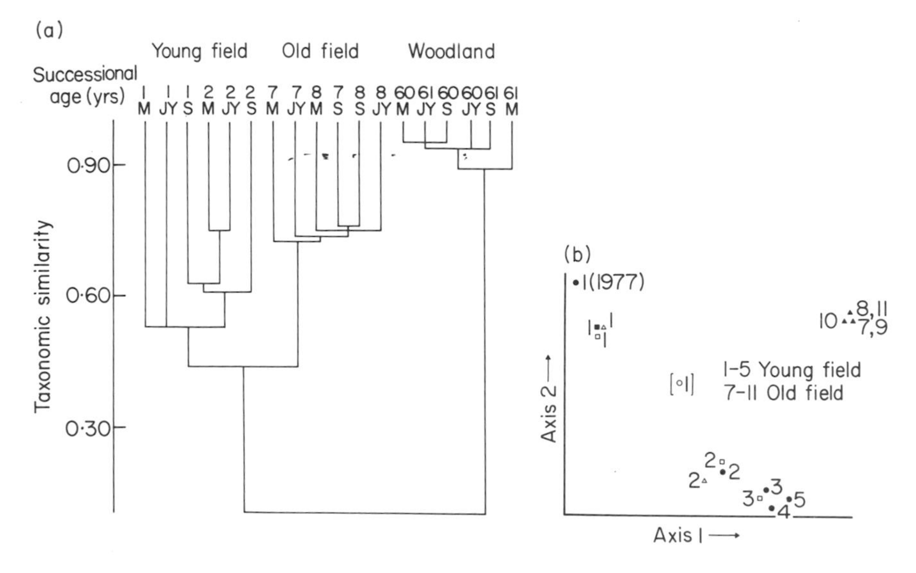

FIG. 15.2. (a) Dendrogram based on values for Sorensen's index of similarity to show plant taxonomic similarity in three sites of different successional age over a 2-year period (after Southwood, Brown & Reader 1979). M, May; JY, July; S, September. (b) Reciprocal averaging ordination of early and mid-successional plant communities (after Brown & Hyman, in press). The site in parenthesis had a different land history (see Stinson & Brown 1983).

(Tamm 1972; Watt 1960). From studies on fixed quadrats we estimated the probability of an insect, or other orgnism, finding its host-plant in the same position after various periods of time (Southwood, Brown & Reader 1983). It was found that the expectancy of a host-plant being present in the same site increases with successional age (Fig. 15.3a). In terms of the habitat templet (Southwood 1977), the durational stability of the early successional habitat may not be as low as these results suggest, because durational stability is calculated as the ratio of habitat life to the organism's generation time and there is a tendency for insects associated with very early successional host-plants to have shorter generation times (Brown & Southwood 1983; Brown 1985, and in press).

Spatial dimensions

The spatial attributes of a habitat are evaluated here in four ways: (i) the amount of biomass, including measures of litter, present in the site; (ii) the structure of the plant community in terms of the vertical distribution

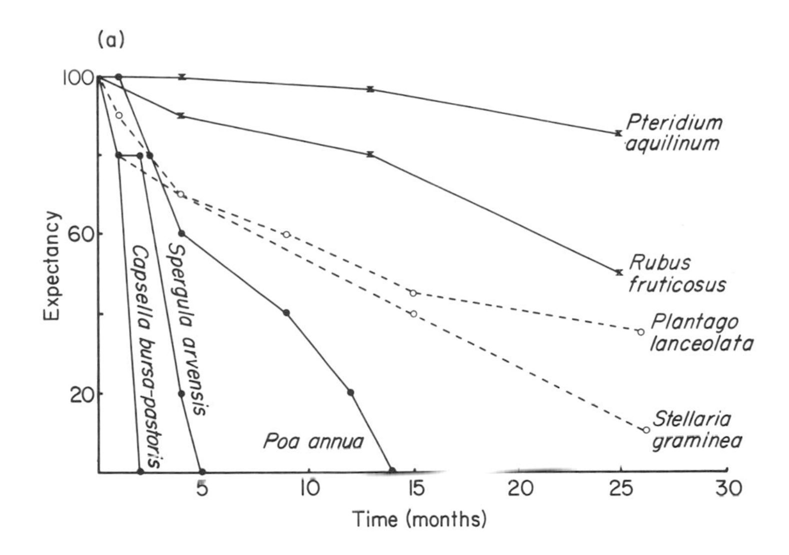

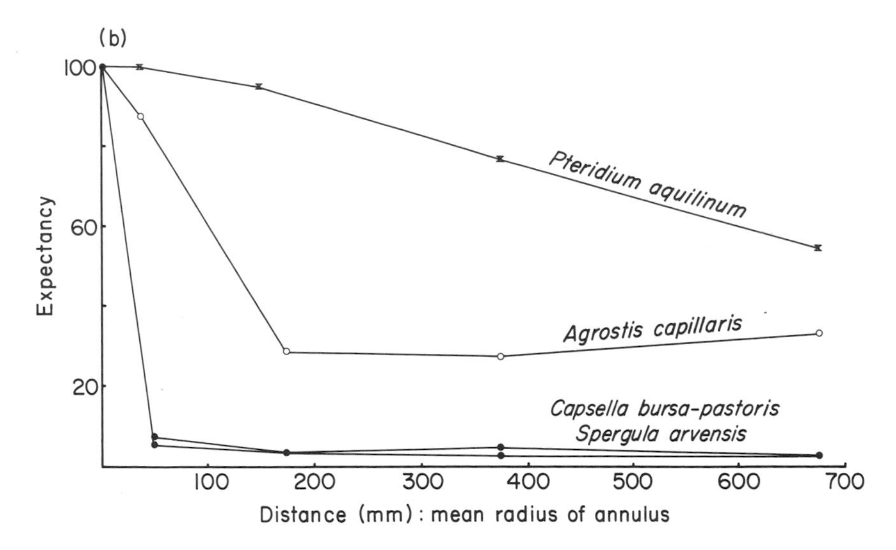

FIG. 15.3. (a) The changing expectancy with time of a host-plant being present in the same position: (●) abundant Young Field species (survivorship curves); (○) Old Field species; (X) Woodland species (depletion curves). (b) The expectancy over distance of an abundant host-plant being encountered by an insect moving from a similar plant at distance zero: (●) Young Field species; (○) Old Field species; (X) Woodland species. (After Southwood, Brown & Reader 1983.)

of biomass (its structural complexity) and its architectural complexity (*sensu* Lawton 1978), the types of plant structure; (iii) the form of the individual components of the plant (i.e. the shape, surface area and nature of the leaf surface) which defines the micro-habitat for many insects and pathogens; and (iv) the distribution of different plant species in space. Studies describing patterns in biomass along a successional gradient are relatively few, although Törmälä & Raatikainen (1976) provide data for the seasonal dynamics of an Old Field in Finland that illustrate the relative proportions, in terms of biomass, for above- and below-ground components of the vegetation. Our work only provides information on above-ground vegetation but for three successional stages and shows the steady increase in the standing crop and the depth of the litter layer. It has been claimed (Margalef 1968; Odum 1969; Whittaker 1970 (p. 69)) that primary productivity increases through a succession; such arguments are often based on 'development of the soil', as well as on 'community structure and increasing utilization by the community of environmental resources'. In our sites, although total biomass (living and dead) increased with successional age (from 275 g m^{-2} y^{-1} in the Young Field to 5450 g m^{-2} y^{-1} in the Woodland), the net annual primary productivity showed no trend (being 970 g m^{-2} y^{-1} in the Young Field in its second season and 980 g m^{-2} y^{-1} in the Woodland); the lowest estimate (750 g m^{-2} y^{-1}) was for the grassland of the Old Field. There are, of course, no major changes in soil depth in these stages of the secondary succession, but our results suggest that, contrary to the views of Margalef (1968) and others, increased 'community structure' does not lead to higher primary productivity. This similarity in primary productivity (annual dry matter production) for areas with the same climate accords with the findings of Phillipson (1973) and the model of Lieth (1972).

In quantifying the plant structure of our successional communities we have recognized both structural and architectural attributes (e.g. Southwood, Brown & Reader 1979; Stinson & Brown 1983). The importance of structural attributes to birds is well established (e.g. MacArthur & MacArthur 1961; Karr 1968; Recher 1969; James 1971). For smaller organisms, such as insects, structural attributes reflect the size of the target; larger plants offer a greater potential surface for colonization and can support larger populations, so that the chances of extinction through stochastic events are reduced (May 1973). Structural diversity, as shown in Fig. 15.4, is a measure of the distribution of plant structures (of any type or species) in the vertical plane and increases along the successional gradient.

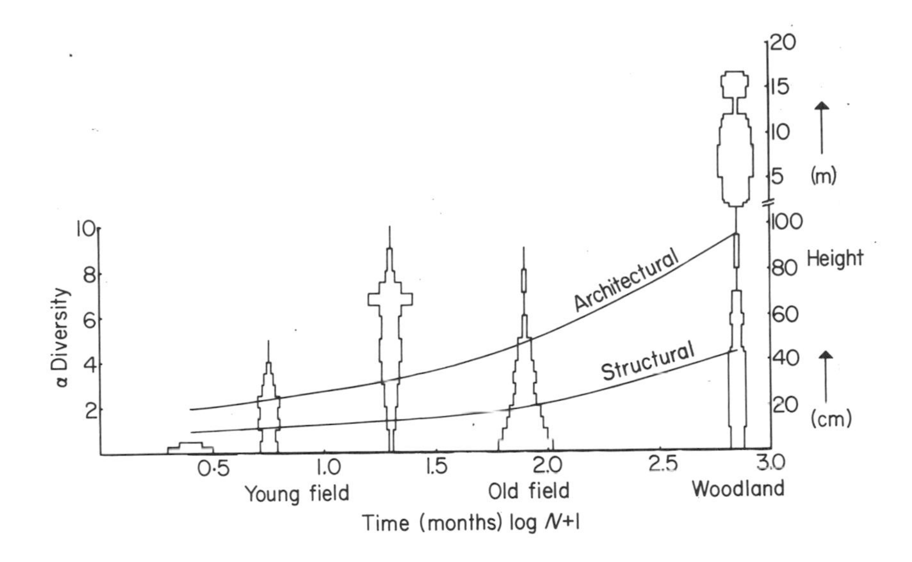

FIG. 15.4. Spatial and architectural diversity (William's α) of the vegetation along a successional gradient, based on the number of height categories and plant structures respectively (see Southwood, Brown & Reader 1979; Stinson & Brown 1983). The height profile for the early Young Field represents early May, the remainder mid-July.

The height profiles, based on mid-season (July) emphasize this trend. Superimposed on this is a seasonal cycle, as described by Southwood, Brown & Reader (1979) and Stinson & Brown (1983), although this is far less marked in the Woodland where the trees make the community structurally diverse even in spring. Architectural attributes refer to the availability and distribution of plant parts or structures in space above ground level. The relevance of plant architecture to insects has long been recognized, but was emphasized by Lawton (1978) who has subsequently attempted to describe plant architecture in terms of fractal surfaces (Morse *et al.* 1985). Architectural complexity relates directly to resource diversity and availability. A more complex plant community will present a greater diversity of structures for specialization by feeding guilds. It will also provide more non-trophic resources such as resting, overwintering and oviposition sites and related to this is the idea of enemy-free space (Askew 1961; Lawton & Strong 1981; Price *et al.* 1980).

We quantified architectural diversity based on the recognition of thirty-one different plant structures. Although these were defined mainly on botanical grounds, cognisance was taken of the way organisms might exploit the plant structures, e.g. to an insect the upper and lower surfaces of a leaf provide different habitats. Architectural diversity *per se* increased along the successional gradient (Fig. 15.4). However, in explaining these changes in the plant community it must be remembered that structural and architectural measures often covary with plant chemistry and climatic factors.

The form of the individual components of the plant, which is the proximate environment of many organisms, has received relatively little attention (e.g. Lawton & Price 1979; Neuvonen & Niemelä 1983). Variation in leaf form of a successional plant community may be worth pursuing, but this would need to be considered in relation to leaf persistence, a subject which we deal with later in this chapter.

The dispersion of plant species (i.e. their distribution in the horizontal plane) was quantified on our sites, by the use of fixed quadrats, so as to determine the expectancy of a host-plant being encountered by an organism moving from a plant of a similar species (see Southwood, Brown & Reader 1983). The expectancy, over distance, of an abundant host-plant being encountered by an insect moving from distance zero increases with successional age (Fig. 15.3b). The extent to which the expectancy falls is related to the development and size of patches of the host-plant, especially when these are due to modular growth.

PATTERNS IN THE PLANTS OF SUCCESSIONAL COMMUNITIES

Growth form and diversity

By monitoring vegetation change in single sites for an 8-year period and categorizing plants into annual, monocarpic perennial (biennial) and perennial herbs, grasses, and shrubs and trees it can be seen that, in this particular sere, there is a very rapid transition from an annual herb-dominated plant community to a perennial herb and grass community before shrub and tree establishment begins (Fig. 15.5). It is convenient to recognize four types of plant community: ruderal, typically the first year of succession when the annuals dominate; early-successional, the second to fifth year where annual and biennial herbs are declining but perennials and grasses are establishing; mid-successional, from the fifth to fifteenth year when grasses and perennials dominate although tree and shrub establishment is beginning; late-successional, when the latter are dominant. This sequence is typical of many secondary successions in temperate

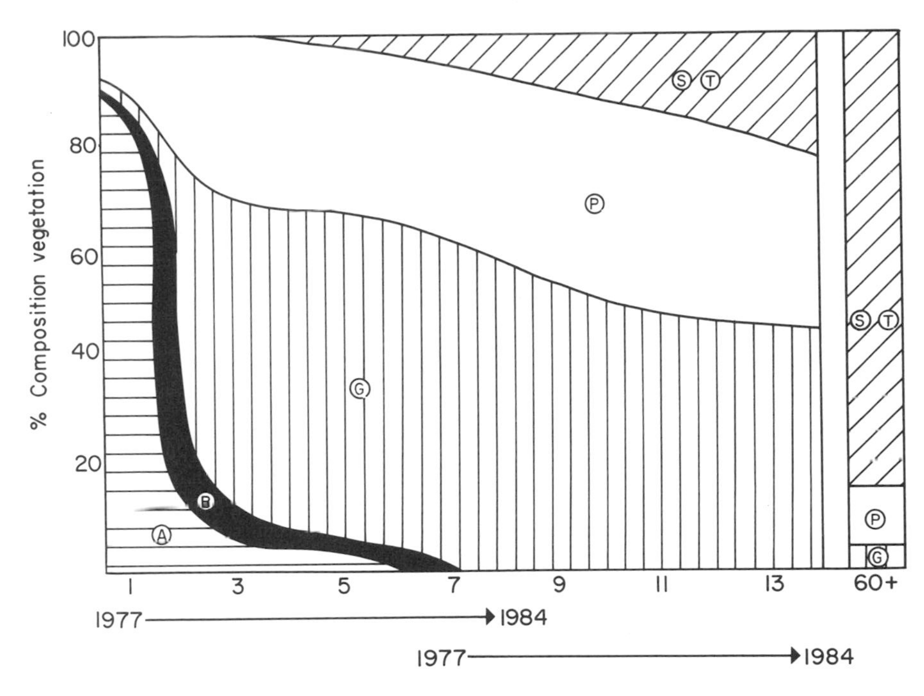

FIG. 15.5. Model of vegetation change in the same sites in terms of plant growth form: A, annual herbs and grasses; B, monocarpic perennials (biennials); P, perennials; G, perennial grasses; S & T, shrubs and trees.

regions, though the relative timing of these events will vary from succession to succession (e.g. Törmälä 1982). These successional plant communities will be referred to later, since they have an important bearing on the associated organisms.

The species diversity (William's α) of green plants (mosses, ferns and flowering plants) is shown in Fig. 15.6, the highest and most seasonally variable levels being seen in the early-successional plant community which consists of a few persistent annual and biennial species in addition to the establishing perennial grasses and herbs. A similar pattern of diversity was found by Tramer (1975) in a study of the first 4 years of succession on an Old Field in Ohio. However, the diversity of the macrofungi, as seen by their fruiting bodies, rises slowly. It is also interesting that the diversity distances (β-diversity) between the groups of samples appear to correspond proportionally to the successional age differences on a logarithmic scale. This suggests that in this particular sere the turnover rate is linear with regard to time expressed as log successional age.

Recruitment and seedling survival

The potential for the recruitment of plant species into the sites was established by a study of the seed bank to be reported elsewhere (T.R.E. Southwood, V.K. Brown, P.M. Reader & E. Mason, unpublished). The first point on the diversity curve in Fig. 15.6 was derived from the species composition of the seedling community germinating from the seed bank in the Young Field.

A study of the germination and survival of seedlings in the field provides a clear indication of the role of seed germination in the floral changes of the various seral stages. By following the fate of individual seedlings in the Young Field we showed that, of the ruderals which germinate on bare ground, 64–100% (according to species) survive until flowering or, if longer-lived, until the end of the season, whereas only 1·8–24% of those germinating in autumn, when the vegetation cover was higher, survived. However, even with these relatively low rates of survival, new species arise in the plant community and so germination as well as

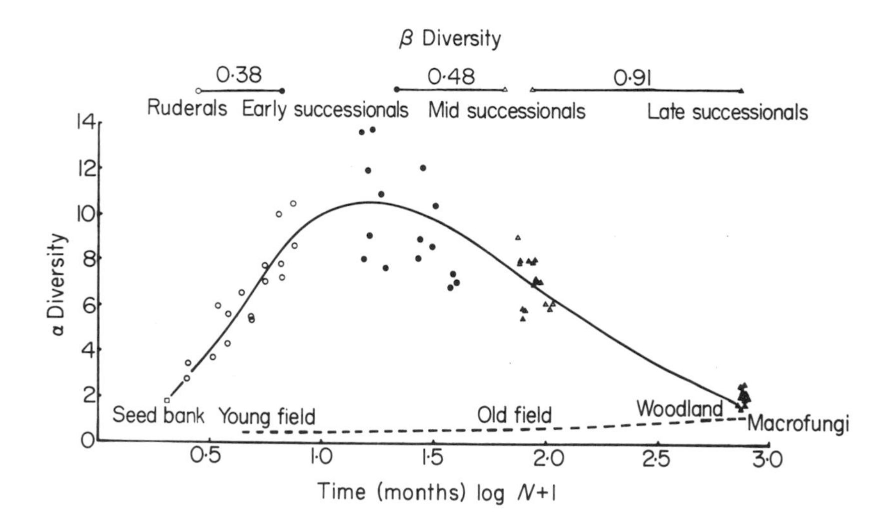

FIG. 15.6. Taxonomic α diversity of green plants (solid line) and macrofungi (dotted line) in sites of different successional age. At top of figure the distances between the different communities in terms of β-diversity (values for Is). (After Southwood, Brown & Reader 1979, but with additional data.)

modular growth make a major contribution to changes in the plant community. In the Old Field very few seedlings germinated (only eleven recorded in the field) and none survived until the following season. Such a lack of seedling success may explain, at least in part, the uniformity of the Old Field vegetation (Fig. 15.2). By contrast, in the Woodland many seedlings germinated, but failed to survive once shading by the birch and field-layer forbs (mainly *Pteridium aquilinum*) began (Fig. 15.7).

The germination and survival of the dominant tree in the Woodland, birch, seemed to be linked to germination conditions and propagule size, rather than to its relative abundance in the seed bank. Large numbers of birch seeds germinated in the Woodland and were noticeably associated with the gaps in the herb canopy but, with their limited reserves, few survived (Southwood, Brown, Reader & Mason, unpublished).

Allocation to reproduction

Several studies have investigated the proportion of resources allocated to reproductive structures by comparing the biomass of the flowering and fruiting structures with the total for all parts of the plant (e.g. Harper &

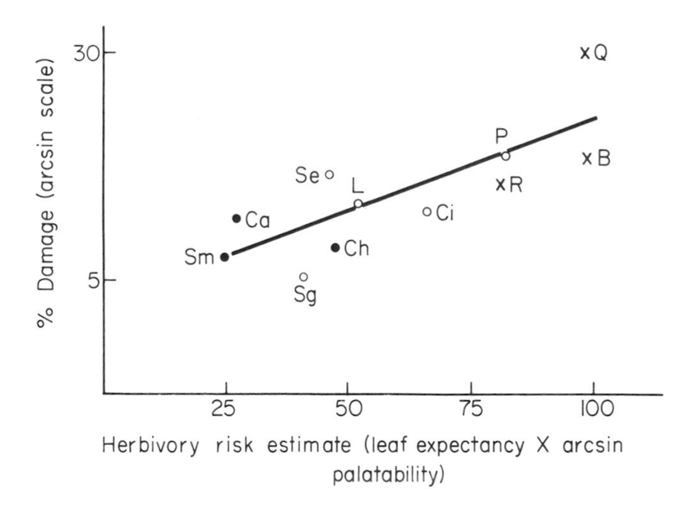

FIG. 15.7. Relationship between level of damage and a measure of the risk of herbivory (equivalent to exposure time × vulnerability) in eleven plant species characteristic of different stages in succession. $y = 11{\cdot}1 + 0{\cdot}16x$, $r = 0{\cdot}78$ ($P < 0{\cdot}01$.. (●) Early-successional species (Ca, *Capsella bursa-pastoris*; Ch, *Chenopodium album*; Sm, *Stellaria media*). (○) Mid-successional species (Ci, *Cirsium arvense*; L, *Lotus corniculatus*; P, *Plantago major*; Se, *Senecio jacobaea*). (X) Late-successional species (B, *Betula pendula*; Q, *Quercus robur*; R, *Rubus fruticosus*).

Ogden 1970; Abrahamson & Gadgil 1973; Hickman 1975). In general it has been found that annual plants, especially those of early-successional stages or open habitats, allocate more of their assimilates to reproduction than do perennial species in climax vegetation. Our data, based on the allocation of biomass to reproductive and non-reproductive structures, for some of the more abundant species of each habitat exhibited the same trend. It was especially noteworthy that the one species that occurred abundantly in all three sites, *Holcus lanatus*, showed a similar pattern to that observed between species. The relative frequencies of flowering structures and vegetative parts was determined by quadrat sampling in the three summer months: these results indicate that the plant community as a whole reflects the trends noted for the commoner species (Table 15.1).

Allocation to defence

The life expectancy of leaves of plants characteristic of different stages in succession show a clear trend with successional age (Southwood, Brown, Reader & Mason, unpublished). The life expectancy of leaves is shortest in the ruderal plant species and greatest in climax species. One might predict that the longer a leaf persists, the more it is likely to be attacked by a herbivore or pathogen and thus the greater the evolutionary advantage to the plant of defending it, making it unpalatable. The palatability of a range of plants from different successional stages was assessed by bioassay using five species of generalist herbivore (Reader & Southwood 1981). The product of life expectancy of a leaf and its palability can be considered as a measure of the risk of herbivore damage (exposure time × vulnerability) and is considered in more detail in Southwood, Brown & Reader (in press). Here we have related this measure of the risk of herbivory to the level of herbivore damage as recorded in the field (Fig. 15.7). There is a significant positive relationship between percentage damage and the risk of herbivory ($r = 0{\cdot}78$, $P < 0{\cdot}005$). The amount of herbivory increases with the successional age of the habitat. The short-lived but palatable plants of early succession are less apparent (*sensu* Feeny 1976) and are able to escape phytophagous insects in space and time, whilst the persistent plants of the later successional stages, although less palatable, are more likely to be discovered, colonized and fed upon by insects. The apparent conflict with Coley's results (Coley 1980, 1983) in terms of the habitat templet can be explained, since the habitats of the species she studied differ more in their adversity than in their durational stability.

TABLE 15.1. Allocation of resources to reproduction in the plant community in sites of different successional ages

Site	Ratio dry weights reproductive/non-reproductive structures: Most abundant herb	Grass (*Holcus*)	Index reproductive structures/vegetation density
Young Field (0–2 y)	0·76 (*Spergula*)	0·146	263
Old Field (6–8 y)	0·082 (*Lotus*)	0·112	52
Woodland (*c.* 60 y)	0·001 (*Rubus*)	0·017	—

In addition to assessing defoliation, we also measured the level of attack, in the field, by plant pathogens. This was highest in early succession and lower in the Old Field and Woodland; when infestations occurred in ruderal plants leaves were often extensively damaged. This may be related to their paucity of structural and biochemical defences, whilst their strategy of escape in space and time is less effective in avoiding pathogen infestations than for insects.

PATTERNS IN THE ANIMALS OF SUCCESSIONAL COMMUNITIES

Macro-invertebrates

Much of our attention has been focused on the insects (e.g. Southwood, Brown & Reader 1979; Brown & Southwood 1983; Brown 1982a, b, 1984, 1985, in press; Brown & Hyman in press), although all other taxa dwelling above ground have been considered. The macro-invertebrates have been described in terms of density per unit area and where possible guild structure and species diversity. The density (mean number m^{-2} for four sampling occasions per season) of macro-invertebrates excluding Collembola increases markedly with successional age (Fig. 15.8) with over 600 individuals m^{-2} in late succession. The Collembola are very numerous in our successional sere, but are so variable, both between samples and within a site, that they are to be treated separately elsewhere. Of four major guilds (Root 1973; Moran & Southwood 1982), the plant suckers are the most abundant throughout the succession with numbers in excess

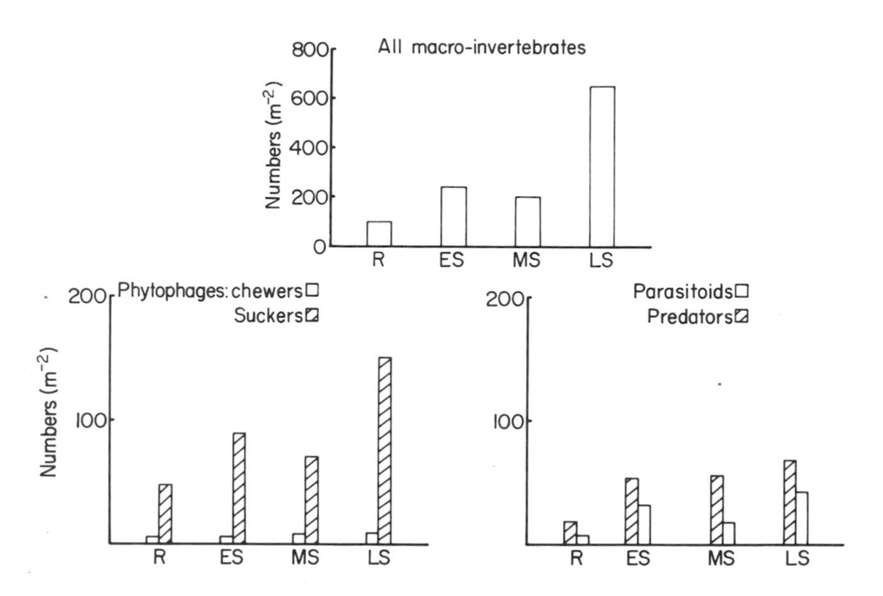

FIG. 15.8. Density of all macro-invertebrates, phytophages (external chewers and suckers), parasitoids and predators associated with four successional plant communities (R, ruderal; ES, early-successional; MS, mid-successional; LS, late-successional).

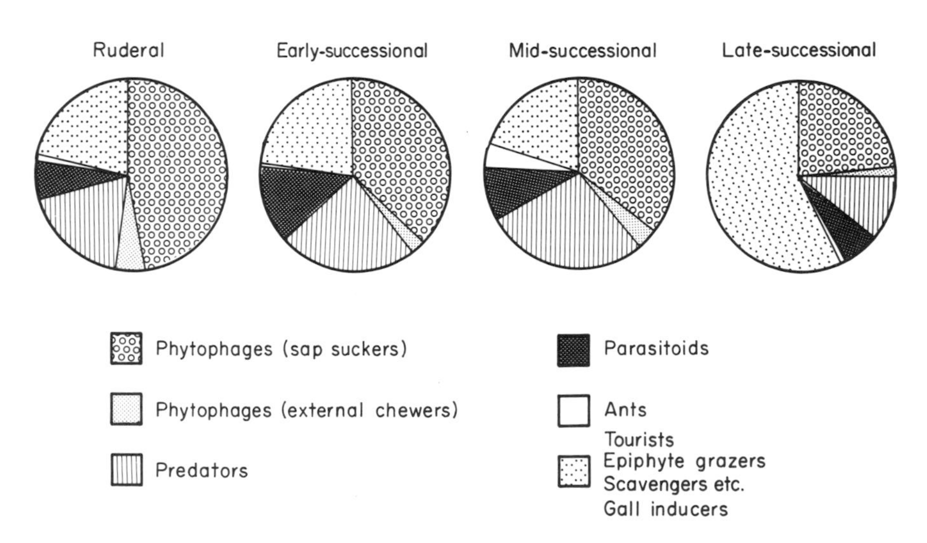

FIG. 15.9. Guild structure of macro-invertebrates from four stages in a secondary succession.

of 150 m^{-2} in late succession. The numbers of predators and parasitoids also show a general increase with successional age. Based on an analysis of some 34 500 individuals, we have compared the relative contribution of these four major guilds but also include information on the ants and another group including the tourists, epiphyte grazers, scavengers and gall-inducers which have not yet been completely identified (Fig. 15.9). The numerically dominant gall-inducers, the Cecidomyiidae, constitute between 0·9 and 4·0% of the fauna. There are clear trends in guild structure associated with the four successional communities; the most obvious being the marked increase in the tourists, epiphyte grazers, and scavengers. This trend has already been observed in the exopterygote insects (Brown & Southwood 1983). In the earlier stages the pattern tends to be more consistent, although the increase in predators, parasitoids and ants with increasing successional age is clear, as is the decrease in plant suckers. Our results, based on number of individuals, do not therefore suggest a consistent guild structure along a successional gradient. Continuing work on identification in some other groups will permit analysis at the species level.

Species diversity (William's α) was calculated for all phytophagous species and predators (excluding a few Dipteran species) (Fig. 15.10). These show the same general pattern; diversity rises rapidly in the early successional community of the Young Field and conforms to the trends already described by Southwood, Brown & Reader (1979) on a more restricted data set. The number of leaf miner species shows a general increase along the successional gradient (Godfray 1985). Unfortunately, we do not yet have any comprehensive data on gall-inducers. We have already suggested that, although the plant species diversity falls in late succession, the insect diversity is maintained by the high level of structural diversity of the late-successional vegetation.

Birds and small mammals

The bird communities of different habitats, as revealed from an analysis of the BTO Register of Ornithological Sites, have been described by Fuller (1982). Species richness in open, early-successional habitats is generally less than in the succeeding scrub and woodland and the frequency-rank curves are, as with insects (Southwood, Brown & Reader 1979), very steep in the early stages, becoming shallower in later seral stages. Our areas were small in relation to bird territories, but regular observations at dawn over a 2-year period were made of the utilization of each site by birds, recording the length of time they were present. Details

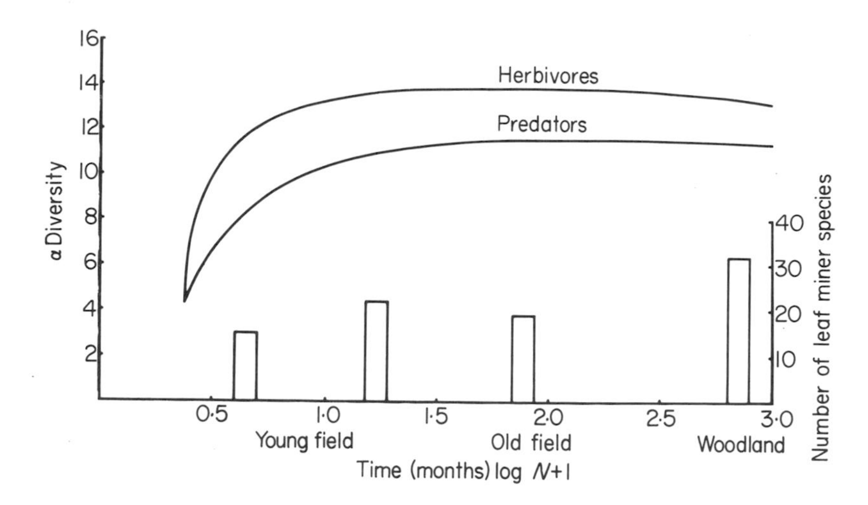

FIG. 15.10. Species diversity of herbivores and predators (excluding Diptera) along a secondary successional gradient. Leaf miners are expressed as number of species found on herbs at each successional stage (Godfray 1985).

of the recording and analysis of results are given in Southwood *et al.* (in press). These observations showed increasing species richness with succession and, in accord with Fuller's (1982) findings, that smaller birds were most frequent in the Woodland (Southwood, *et al.*, in press). However, if the biomass of the different species is weighted by the amount of time spent on each site, measures of the average bird biomass present on the site can be obtained (Table 15.2). It will be noted that, notwithstanding the trend in average weights, the total bird biomass utilizing the Old Field is very low.

Small mammals have only rarely featured in studies of succession (e.g. the Old Field succession described by Hirth 1959). In this sere a number of small mammal species were found by regular trapping (mainly *Microtus agrestis*, *Apodemus sylvaticus* and *Sorex araneus* and *S. minutus* with the occasional record of other species) (J.S. Churchfield, unpublished). The populations of small mammals are relatively high throughout the succession and even the first year site was utilized by the field vole, *M. agrestis*. The highest population levels occur in the Old Field where four species are common; earlier in succession the grass-feeding *M. agrestis* is dominant, while in the Woodland and Old Field the seed-feeding *A. sylvaticus* is common. (Larger mammals, of which rabbits were the most abundant, were excluded from the experimental sites.) Variations in the

TABLE 15.2. The utilization of the different sites by vertebrates (weights in g)

	Young Field	Old Field	Woodland
Mean weight of bird	457	135	19
Mean biomass of birds present	328	21·5	111
Mean biomass of small mammals present	75	359	275
Mean total vertebrate biomass	403	381	386
% Total vertebrate biomass: herbivores	93	51	44

biomass of small mammals present on each site—calculated using the average mature weights from Southern (1964)—and in the biomass of birds appear to compensate each other (Table 15.2). Thus the total vertebrate biomass is surprisingly constant at around 400 g in each of our sites, i.e. about 1 g m^{-2}.

Birds and small mammals fill rather similar roles in the community as grazers, seed feeders and insectivores. The proportion of the total biomass of these groups for a site contributed by herbivorous species falls through the successional stages; conversely, the density of insects and the biomass (and numbers) of insectivores rises (Fig. 15.11). It may be argued that the higher vertebrate herbivore biomass is a reflection of the simpler (more efficient) ecosystem and that the intervention of linkages involving macro-invertebrates in later successional stages reduces the carrying capacity of the system for vertebrates (Southwood *et al.*, in press).

STRATEGIES THROUGH SUCCESSION

In the foregoing sections we have shown how there are certain overall trends in both community patterns and the attributes of the associated organisms as one moves along a successional gradient. Figure 15.12 summarizes some of these trends.

The Young Field represents a stage on which there have been comparatively few studies, yet it provides the natural homologue of arable ecosystems. The organisms that exploit this stage have the characteristics of opportunists or exploiters; they represent a *r*-strategy (MacArthur & Wilson 1967; Pianka 1970; Southwood 1977). Both plants and insects have relatively short generation times and high investments in reproduction; plants show rapid leaf turnover and a low allocation to leaf defence. In the insects flight ability is well developed, enabling rapid colonization, and many of the phytophages are generalists. In more than one sense the

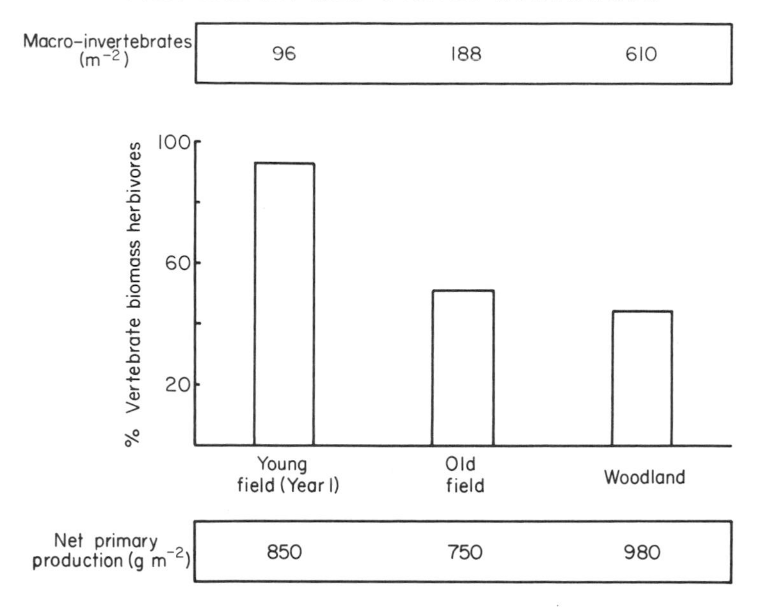

FIG. 15.11. Summary of site utilization by vertebrate herbivores and macro-invertebrates, when viewed against the constant net annual primary production.

community is structurally simple: the structural and architectural complexity of the vegetation is low, many animals (particularly vertebrates) feed directly on the plants (i.e. there are simple food chains) and the range of size and form is low in several insect groups and in birds. The plants are not in a competitive equilibrium for there is bare ground in this habitat and we would suggest that local populations are not at equilibrium; they are likely to boom and burst. For example, this is the only site where we have evidence of plant pathogens 'booming'. In this habitat, with its low durational stability, species are not 'tightly packed' and the low morphological diversity is a reflection of this. Another indication of 'vacant niches' is the occurrence in the Young Field, but not elsewhere in our study, of plants (six species) that are not native to Britain.

The organisms of the Old Field exhibit very different traits from those of the Young Field, and we believe it unfortunate that habitats comparable to our Old Field have so often been regarded as 'early-successional sites'. Bare ground has virtually disappeared and the low germination and even lower survival of seeds is evidence of the increasing competition between the plant species. Predaceous insects have become an important and diverse component in the fauna and nearly half the vertebrate biomass consists of secondary consumers (insectivores). Both food

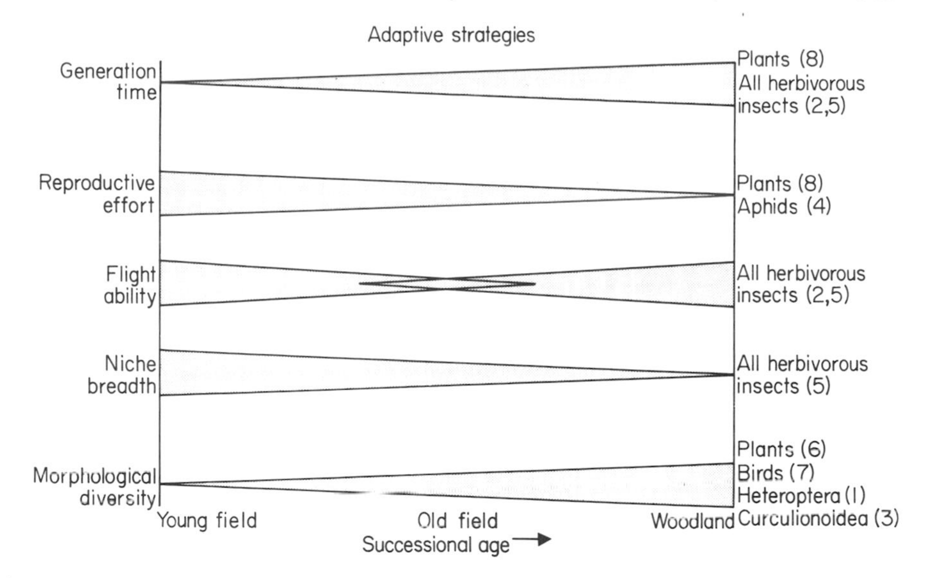

FIG. 15.12. Adaptive strategies of plants and animals along a successional gradient. Numbers in parentheses indicate references: (1) Brown 1982b; (2) Brown 1985, in press; (3) Brown & Hyman in press; (4) Brown & Llewellyn 1985; (5) Brown & Southwood 1983; (6) Southwood, Brown & Reader 1979, (7) Southwood, Brown, Reader & Green, in press; (8) Southwood, Brown, Reader & Mason (unpublished).

chains and the structural organization of the habitat are more complex; the tussock form of the grasses and deepening litter layer provide spaces that can be exploited by small non-flying vertebrates (i.e. small mammals). Among the insects though, a higher proportion of species are apterous or brachypterous, a feature which may be attributed to more uniformly distributed resources than in early or late-succession (Waloff 1983; Brown 1985, in press), where plant patch size is larger (Southwood, Brown & Reader 1983).

In the Woodland the structural diversity of the habitat, provided by the vegetation, has increased many fold (Southwood 1978). The great number of small and special structures provide a wide variety of niches for specialists: morphological diversity in insects and birds is high, as is taxonomic and trophic diversity for almost every group studied (from macrofungi to birds). The plants and their leaves are long-lived, and more resources are allocated to survival than to reproduction. Similar trends are seen in the insects which appear to use flight more for distribution on the host-plant than for dispersal (Waloff 1983). In general, organisms display many features of the *K*-strategy.

Notwithstanding these general trends that conform to the predictions of the habitat templet, there are many exceptions, e.g. the ruderal plant, *Spergula arvensis*, is particularly unpalatable, and weevils appear to be less specialized in their feeding habits in later succession. Predictions about organism size in relation to *r*- and *K*-strategies appear confounded by our observations on birds (Southwood *et al.*, in press) and Heteropteran bugs (Brown 1982b). We believe that the trends we have demonstrated are important, not only for the organization of facts, but more significantly for the insight they provide on the functioning of natural ecosystems. However, one must not become mesmerized by the patterns; the exceptions are not simply troublesome anomalies, but pointers to a need for further understanding.

ACKNOWLEDGMENTS

We are grateful to many colleagues who have helped with the field work associated with this project, especially Dr P. Reader, Ms M. Reese, Ms E. Mason, and to Dr J.S. Churchfield for providing the data on small mammals.

REFERENCES

Abrahamson, W.G. & Gadgil, M. (1973). Growth form and reproductive effort in Goldenrods (Solidago, Compositae). *American Naturalist*, **107**, 651–661.

Askew, R.R. (1961). On the biology of the inhabitants of oak galls in Britain. *Transactions of the Society for British Entomology*, **14**, 237–268.

Bazzaz, F.A. (1975). Plant species diversity in old-field successional ecosystems in southern Illinois. *Ecology*, **56**, 485–488.

Begon, M. (1985). A general theory of life-history variation. *Population Dynamics* (Ed. by R.M. Anderson, B.D. Turner & L.R. Taylor, pp. 91–7. Symposium of the British Ecological Society 25. Blackwell Scientific Publications, Oxford.

Brown, V.K. (1982a). The phytophagous insect community and its impact on early successional habitats. *Proceedings of the 5th International Symposium on Insect–Plant Relationships, Wageningen, 1982* (Ed. by J.H. Visser & A.K. Minks), pp. 205–213. Pudoc, Wageningen.

Brown, V.K. (1982b). Size and shape as ecological discriminants in successional communities of Heteroptera. *Biological Journal of the Linnean Society of London*, **18**, 279–290.

Brown, V.K. (1984). The ecology of secondary succession: insect–plant relationships. *Bioscience*, **34**, 710–716.

Brown, V.K. (1985). Insect herbivores and plant succession. *Oikos*, **44**, 17–22.

Brown, V.K. (in press). Life cycle strategies and plant succession. *The Evolution of Insect Life Cycles* (Ed. by F. Taylor & R. Karban), Springer Verlag.

Brown, V.K. & Southwood, T.R.E. (1983). Trophic diversity, niche breadth and generation times of exopterygote insects in a secondary succession. *Oecologia*, **56**, 220–225.

Brown, V.K. & Hyman, P.S. (in press). Successional communities of plants and phytophagous Coleoptera. *Journal of Ecology*, **74**.

Coley, P.D. (1980). Effects of leaf age and plant life history patterns on herbivory. *Nature*, **284**, 545–546.

Coley, P.D. (1983). Herbivory and defensive characteristics of tree species in a lowland tropical forest. *Ecological Monographs*, **53**, 209–33.

Denslow, J.S. (1980). Patterns of plant species diversity during succession under different disturbance regimes. *Oecologia*, **46**, 18–21.

Drury, W.H. & Nisbet, I.C.T. (1973). Succession. *Journal of the Arnold Arboretum*, **54**, 331–368.

Feeny, P.P. (1976). Plant apparency and chemical defense. *Biochemical Interactions Between Plants and Insects* (Ed. by J. Wallace & R. Mansell), pp. 1–40. Recent Advances in Phytochemistry 10, Plenum Press, New York, London.

Fuller, R.J. (1982). *Bird Habitats in Britain*. T. & A.D. Poyser, Calton.

Godfray, H.C.J. (1985). The absolute abundance of leaf miners on plants of different successional stages. *Oikos* **45**, 17–25.

Greenslade, P.J.M. (1983). Adversity selection and habitat templet. *American Naturalist*, **122**, 352–365.

Grime, J.P. (1977). Evidence for the existence of three primary strategies in plants and its relevance to ecological and evolutionary theory. *American Naturalist*, **111**, 1169–94.

Grime, J.P. (1979). *Plant Strategies and Vegetation Processes*. John Wiley, New York.

Harper, J.L. & Ogden, J. (1970). The reproductive strategy of higher plants. I. The concept of strategy with special reference to *Senecio vulgaris* L. *Journal of Ecology*, **58**, 681–98.

Hickman, J.C. (1975). Environmental unpredictability and plastic energy allocation strategies in the annual *Polygonum cascadensa* (Polygonaceae). *Journal of Ecology*, **63**, 689–701.

Hirth, H.F. (1959). Small mammals in Old Field successions. *Ecology*, **40**, 417—425.

Itamies, J. (1983). Factors contributing to the succession of plants and Lepidoptera on the islands off Rauma, S.W. Finland. *Acta Universitatis Oulerensis, Series A. Scientiae Perum Naturalium No. 142, Biologica*, **18**, 1–52.

James, F.C. (1971). Ordinations of habitat relationships among breeding birds. *Wilson Bulletin*, **83**, 215–236.

Karr, J.R. (1968). Habitat and avian diversity on strip-mined land in East Central Illinois. *Condor*, **70**, 348–357.

Lawton, J.H. (1978). Host-plant influence on insect diversity: the effects of space and time. *Diversity of Insect Faunas* (Ed. by L.A. Mound & N. Waloff), pp. 105–125. Symposium of the Royal Entomological Society of London 9. Blackwell Scientific Publications, Oxford.

Lawton, J.H. & Price, P.W. (1979). Species richness of parasites on hosts: agromyzid flies on the British Umbelliferae. *Journal of Animal Ecology*, **48**, 619–637.

Lawton, J.H. & Strong, D.R.Jr (1981). Community patterns and competition in folivorous insects. *American Naturalist*, **118**, 317–338.

Lieth, H. (1972). Modelling the primary productivity of the world. *Nature and Resources*, **8**, 5–10.

MacArthur, R.H. & MacArthur, J.W. (1961). On bird species diversity. *Ecology*, **42**, 594–598.

MacArthur, R.H. & Wilson, E.O. (1967). *The Theory of Island Biogeography*. Princeton University Press, Princeton, New Jersey.

Margalef, R. (1968). *Perspectives in Ecological Theory*. University of Chicago Press, Chicago.

May, R.M. (1973). *Stability and Complexity in Model Ecosystems*. Princeton University Press, Princeton.

Mellinger, M.V. & McNaughton, S.J. (1975). Structure and function of successional vascular plant communities in central New York. *Ecological Monographs*, **45**, 161–182.

Moran, V.C. & Southwood, T.R.E. (1982). The guild composition of arthropod communities in trees. *Journal of Animal Ecology*, **51**, 289–306.

Morse, D.R., Lawton, J.H., Dodson, M.M. & Williamson, M.H. (1985). Fractal dimension of vegetation and the distribution of arthropod body lengths. *Nature*, **314**, 731–732.

Murdoch, W.W., Evans, F.C. & Peterson, C.H. (1972). Diversity and pattern in plants and insects. *Ecology*, **53**, 819–28.

Nagel, H.G. (1979). Analysis of invertebrate diversity in a mixed prairie ecosystem. *Journal of the Kansas Entomological Society*, **52**, 777–786.

Neuvonen, S. & Niemelä, P. (1983). Species richness and faunal similarity of arboreal insect herbivores. *Oikos*, **40**, 452–459.

Nicholson, S.A. & Monk, C.D. (1975). Changes in several community characteristics associated with forest formation in secondary succession. *American Midland Naturalist*, **93**, 302–310.

Odum, E.P. (1969). The strategy of ecosystem development. *Science (New York)*, **164**, 262–270.

Peet, R.K. & Christensen, N.L. (1980). Succession: a population process. *Vegetatio*, **43**, 131–140.

Phillipson, J. (1973). The biological efficiency of protein production by grazing and other land-based systems. *The Biological Efficiency of Protein Production* (Ed. by J.G.W. Jones), pp. 217–235. Cambridge University Press, Cambridge.

Pianka, E.R. (1970). On *r*- and K-selection. *American Naturalist*, **104**, 592–7.

Price, P.W., Bouton, C.E., Gross, P., McPheron, B.A., Thompson, J.N. & Weis, A.E. (1980). Interactions among three trophic levels: the influence of plants on interactions between insect herbivores and natural enemies. *Annual Review of Ecology and Systematics*, **11**, 41–65.

Reader, P.M. & Southwood, T.R.E. (1981). The relationship between palatability to invertebrates and the successional status of a plant. *Oecologia*, **51**, 271–275.

Recher, H.F. (1969). Bird species diversity and habitat diversity in Australia and North America. *American Naturalist*, **103**, 75–80.

Ricklefs, R.E. (1973). *Ecology*, Nelson, London.

Root, R.B. (1973). Organisation of a plant-arthropod association in simple and diverse habitats: the fauna of collards (*Brassica oleracea*). *Ecological Monographs*, **43**, 95–124.

Schimpf, D.J., Henderson, J.A. & MacMahon, J.A. (1980). Some aspects of succession in the spruce-fir forest zone of northern Utah. *The Great Basin Naturalist*, **40**, 1–26.

Sibly, R. & Calow, P. (1985). Classification of habitats by selection pressures: a synthesis of life-cycle and *r*/*K* theory. *Population Dynamics* (Ed. by R.M. Anderson, B.D. Turner & L.R. Taylor), pp. 75–90. Symposium of the British Ecological Society 25. Blackwell Scientific Publications, Oxford.

Southern, H.N. (1964). *The Handbook of British Mammals*. Blackwell Scientific Publications, Oxford.

Southwood, T.R.E. (1977). Habitat, the templet for ecological strategies. *Journal of Animal Ecology*, **46**, 337–365.

Southwood, T.R.E. (1978). The components of diversity. *Diversity of Insect Faunas* (Ed. by L.A. Mound & N. Waloff), pp. 19–40. Symposium of the Royal Entomological Society of London 9. Blackwell Scientific Publications, Oxford.

Southwood, T.R.E., Brown, V.K. & Reader, P.M. (1979). The relationships of plant and insect diversities in succession. *Biological Journal of the Linnean Society of London*, **12**, 327–348.

Southwood, T.R.E., Brown, V.K. & Reader, P.M. (1983). Continuity of vegetation in space and time: a comparison of insects' habitat templet in different successional stages. *Researches on Population Ecology*, Supplement 3, 61–74.

Southwood, T.R.E., Brown, V.K. & Reader, P.M. (in press). Leaf palatability, life expectancy and herbivore damage. *Oecologia.*

Southwood, T.R.E., Brown, V.K., Reader, P.M. & Green, E.E. (in press). Comparative utilization of different stages of a secondary succession by birds. *Bird Study.*

Stinson, C.S.A. & Brown, V.K. (1983). Seasonal changes in the architecture of natural plant communities and its relevance to insect herbivores. *Oecologia*, **56**, 67–69.

Tamm, C.O. (1972). Survial and flowering of some perennial herbs. II and III. *Oikos*, **23**, 23–28, 159–166.

Törmälä, T. (1982). Structure and dynamics of reserved field ecosystem in central Finland. *Biological Research Reports from the University of Jyvaskyla, Finland*, **8**, 1–58.

Törmälä, T. & Raatikainen, M. (1976). Primary production and seasonal dynamics of the flora and fauna of the field stratum in a reserved field in Middle Finland. *Journal of the Scientific Agricultural Society of Finland*, **48**, 363–385.

Tramer, E.J. (1975). The regulation of plant species diversity on an early successional old field. *Ecology*, **56**, 905–914.

Van Hulst, R. (1978). On the dynamics of vegetation: patterns of environmental and vegetational change. *Vegetatio*, **38**, 65–75.

Waloff, N. (1983). Absence of wing polymorphism in the arboreal, phytopagous species of some taxa of temperate Hemiptera: an hypothesis. *Ecological Entomology*, **8**, 229–232.

Watt, A.S. (1960). Population changes in acidophilous grass heath in Breckland, 1936–57. *Journal of Ecology*, **48**, 605–629.

White, P.S. (1979). Pattern, process and natural disturbance in vegetation. *The Botanical Review*, **45**, 229–299.

Whittaker, R.H. (1970). *Communities and Ecosystems*. Macmillan, New York.

Whittaker, R.H. (1975). The design and stability of some plant communities. *Unifying Concepts in Ecology* (Ed. by W.H. van Dobben & R.H. Lowe-McConnell), pp. 169–181. Junk, The Hague.

Witkowski, Z. (1973). Species diversity, stability and succession. Studies on weevils (Curculionidae, Coleoptera) and their host plants during the succession of mowed meadows under the influence of draining. *Bulletin de l'academie polonaise des Sciences*, **21**, 223–228.

16. CHANGE AND PERSISTENCE IN SOME MARINE COMMUNITIES

JOSEPH H. CONNELL
Department of Biological Sciences, University of California, Santa Barbara, California 93106, USA

INTRODUCTION

The species composition in a local site may change over time, for several different reasons. First, it may change due to gradual shifts in local physical conditions. An example of this is the seasonal succession of short-lived species in the plankton of lakes or shallow seas. Longer-term variations in the physical environment may cause changes in longer-lived species over large areas, e.g. changes in marine intertidal populations along the English Channel over several decades (Southward 1976; Southward & Crisp 1954) or in the temperate forests of eastern North America over several millenia (Davis 1981).

Secondly, in periods without such gradual shifts in physical conditions, the species composition of a local patch may change for several reasons. (a) Disturbances may kill all organisms in a local patch, and the species recolonizing it may differ from the previous residents. (b) In the interval between such disturbances, some or all of the residents may die and be replaced. This gradual replacement process may result in: (i) persistence of the same species composition, (ii) a predictable, progressive succession of species, or (iii) an unpredictable, non-successional pattern of change.

Even though the species composition is changing on the small scale of local patches, it may remain constant at larger spatial scales. If the scale of observation is expanded to an area encompassing many patches disturbed at different times in the past, the average species composition of this larger area may remain constant, even though each patch within it is changing (Sousa 1984a; Shugart 1984).

These ideas have seldom been tested against reality, in part because few natural communities exist in which all species are short-lived enough to permit observations of change or constancy as well as experimental tests of hypotheses, within the working lifetime of a scientist. However, in some aquatic communities such observations and experiments are possible. The purpose of this paper is to summarize some of the evidence of community change and persistence in natural communities of marine sessile organisms on natural hard substrata. (Assemblages on artificial substrata have recently been reviewed in Cairns 1982.) In contrast to terrestrial systems, the organisms covering the substratum may be either sessile animals or plants. Change or persistence within a single patch will be considered first, then among many patches.

LOCAL WITHIN-PATCH DYNAMICS

Persistence with little change

Populations on some patches have been observed to persist over many generations. For example, populations of four species of intertidal barnacles in three different geographical locations have been observed over many years; most persisted for one or more complete replacements of all individuals (Connell 1985). Coexisting with these barnacle populations were six mobile species of grazing gastropods, predatory snails and starfish; five of the six grazing gastropods also persisted for one or more complete turnovers. Two populations of the alga *Eisenia arborea* also persisted over one turnover (J. Kastendiek, in Connell 1985). Other examples of apparently persistent intertidal species such as mussels (Paine 1966, 1976, Paine, Castillo & Cancino 1985), large red algae (Sousa 1979a) and sea grasses (Turner 1983) are long-lived and, to my knowledge, have not been studied long enough to demonstrate that their populations are self-replacing. In fact, long-term records of two mussel beds created by the temporary removal of starfish predators indicate that neither is maintaining itself (Paine, Castillo & Cancino 1985). Both are shrinking and new recruits are being eliminated by resident predators.

Species may persist despite disturbances. When disturbance clears a patch, the first species colonizing it may be the same one that was present beforehand. For example, Foster (1982) cleared patches in a bed of the alga *Iridaea flaccida* which covered 92% of the substratum. Within 2 months the first visible settlers were a mixture of diatoms and algal cells forming red patches. The cells grew into *Iridaea flaccida* blades which persisted over the next 2 years. Other examples of species persistence involve vegetative growth. If very small patches are opened up in stands of colonial invertebrates such as sponges, bryozoans, corals or ascidians, the patches may be recolonized by vegetative growth of the same species into the opening (see review in Connell & Keough 1985).

In the subtidal, anecdotal observations of persistence of populations have been made, e.g. kelp beds that have persisted at the same sites for periods longer than the maximum life of individuals (North 1971; Dayton *et al.* 1984; Dayton 1985). Sebens (1985) suggests that at local sites on vertical rock walls in the subtidal, any of three species of invertebrates could persist in the absence of heavy predation by sea urchins. In the presence of sea urchins, one coralline alga could persist in a patch since it is apparently the most resistant to heavy grazing. While the period of observation (6 years) was probably shorter than the time of a complete turnover of these long-lived species, the indirect evidence supports Sebens' (1985) suggestions.

The intertidal barnacle populations described above replaced themselves and persisted over many generations because their juveniles were better able to become established and to survive and grow in the presence of adults of the same species, than could other space-occupying organisms, such as algae or mussels. Experimental or natural reductions of grazers (see reviews by Lubchenco & Gaines 1981; Hawkins & Hartnoll 1983) or predators (Menge 1976) on barnacle-covered shores resulted in invasion and dominance of algae or mussels, indicating that the barnacles formerly had resisted the attacks by grazers and predators better than did other potential dominants. Long-term persistence of these sessile barnacle populations is thus dependent upon the persistence of associated mobile grazers and predators (Dungan 1986).

Successional changes

Ecological succession usually refers to a predictable, orderly or progressive change in species composition within a local patch after it has been opened up by a disturbance. Such successions have been described quantitatively several times in naturally-occurring open patches in marine rocky intertidal habitats (see reviews by Connell 1972; Connell & Slatyer 1977; Sousa 1979a; Paine & Levin 1981). The first species to appear in the newly-opened patch are short-lived plants or animals (e.g. microorganisms, diatoms, certain algae, small species of barnacles). Other species such as larger algae, barnacles, mussels or sea grasses appear later (Hawkins, Southward & Barrett 1983).

In contrast to these demonstrations of succession in intertidal habitats, there are apparently few quantitative accounts of succession in natural clearings in subtidal sites, although several anecdotal ones have been published. An example of a quantitative study is that of Paine & Vadas (1969). They found that when sea urchins were removed from subtidal boulders, the algal composition changed from a dominance by coralline algae and sheet-forming green algae to one of large brown kelps. After 2 years, the early dominant kelp *Nereocystis* had been replaced by another species of *Laminaria*. An example involving physical disturbance is that of Harris *et al.* (1984), who studied a subtidal reef in southern California. They found that on large patches of rock newly bared by a storm, a diatom layer appeared first, followed by filamentous algae and young kelp.

There have been a great number of publications on disturbances on subtidal coral reefs in tropical latitudes, but little detailed quantitative measurement of the sequence of colonization afterwards. Storms cause varying degrees of damage on coral reefs, reducing living cover of corals or algae (see review by Pearson 1981). Within a few weeks after hurricanes in Jamaica (Woodley *et al.* 1981) and Queensland (Connell unpublished), openings were colonized by filamentous and sheet-forming algae. This cover disappeared in a few more weeks, presumably as a result of heavy grazing by herbivorous invertebrates and fish. In Queensland, corals then either recruited from larvae or dispersed fragments, or in some cases regrew from damaged colonies that had survived the storms.

Succession from algae to corals has been observed in large openings created by herds of the coral-eating starfish, *Acanthaster planci* (see review by Pearson 1981) and on submerged lava flows (Grigg & Maragos 1974). Contrasts in life-history attributes of different coral species suggest that succession among the corals themselves might be expected. A few species colonize open patches quickly and have fast growth rates. Examples are *Pocillopora damicornis* in Queensland (Stephenson & Stephenson 1933) and Hawaii (Grigg & Maragos 1974), *Stylophora pistillata* in Israel (Loya 1976) and *Agaricia agaricites* and *Leptoseris cucullata* in the Caribbean (Bak & Engel 1979; Rylaarsdam 1983; Hughes & Jackson 1985). In contrast, other species have lower rates of recruitment and growth, greater competitive ability and greater resistance to storm damage, so would be expected to replace the early colonists. However, despite these inferences, I am unaware of any direct evidence of a successional sequence among corals.

Any of several mechanisms may determine the sequence of species in these marine successions. A species may appear because its propagules happen to be present and available to settle in the patch. This probably explains why certain species invade throughout the year whereas others appear only at particular seasons (Sousa 1979a). It also may account for the variability in succession among patches. For example, Lubchenco (1983) excluded herbivores from twenty cleared intertidal patches and recorded the invasion of algae. In two patches, only *Fucus vesiculosus* invaded; in five both *Fucus* and various ephemeral algae invaded and in the remainder only the ephemerals appeared. Variations in propagule availability is a reasonable explanation for this pattern.

In some algal species, propagules disperse only a very short distance. Thus, the position of an open patch relative to that of adult plants may determine which species invade. Sousa (1984b) found that the recruitment of several species of algae that invaded open patches late in the succession sequence was highly correlated with the abundance of conspecific adults within 1 m of the edge of the patch.

Some species may invade later in succession because their propagules attach more readily to other organisms than to bare rock. For example, young mussels attach more readily to filaments or in crevices than to bare rock (see review by Suchanek 1985) and some sea grass seeds are modified to attach to certain algae (Turner 1983). Their invasion is apparently 'facilitated' (Connell & Slatyer 1977) by the prior occupation of the site. Other later-appearing species are not so adapted, and apparently only invade small bare patches. Sousa (1979a) and Lubchenco (1983) demonstrated experimentally that thick covers of earlier colonizing green algae inhibited the invasion of red or brown algae.

Various mechanisms could cause the replacement of species in a successional sequence. Early species may be eliminated in competition with later ones, or may be killed more readily by harsh physical conditions or by grazers and predators. The growth of the more resistant later species may then pre-empt the space and prevent recolonization by the early species.

Evidence for these mechanisms comes from laboratory and field experiments, mainly on intertidal species. Early-successional algal species were preferred over later appearing species in laboratory tests of feeding by two intertidal species of invertebrate grazers in California (Sousa 1979a), and by one species in New England (Lubchenco & Gaines 1981). Subtidal sea urchins preferred an early-successional alga to a later one in Washington (Vadas 1977) and California (Littler & Littler 1980). In studies of the intertidal species (Sousa 1979a; Lubchenco 1983) the early species was replaced more rapidly when these herbivores were present than when they were excluded in field experiments.

Early species are also replaced because they are less resistant to extremes of physical environment than are late-successional species. Sousa (1979a, 1980) demonstrated that later colonizing red algae resisted the effects of desiccation and of storms (which overturn the boulder substrates in this study) better than did the earlier colonizing species.

In all these cases, the later species gradually took over the empty space released when the early species died. Thus, replacement is by gradual pre-emption of space rather than by direct interference. The latter may occur, however, when grazers or predators are removed by disturbances.

For example, when sea urchin grazing is intense, the substrate is usually occupied mainly by red coralline crusts; these areas have been termed 'barren grounds' (see reviews by Lawrence 1975; Lubchenco & Gaines 1981). When the urchins were killed by storms or removed experimentally, erect algae invaded and overgrew the corallines (op. cit.; Harris *et al.* 1984; Ebeling, Laur & Rowley 1985). In these instances, it is possible that the corallines were eliminated by direct interference; field experiments testing this hypothesis would be useful. Direct competitive elimination of sessile animals has been demonstrated in field experiments by Menge (1976). He showed that, when predators were excluded from stands of intertidal barnacles, they were overgrown and killed by invading mussels.

Non-successional changes

In contrast to the previous category, mixed-species assemblages could change in species composition without an orderly succession. Several possible mechanisms could be responsible for non-successional changes within a single patch. Variations in physical conditions, in abundance of grazers or predators, or in numbers of propagules available for colonization, are all possibilities. Even without these variations, erratic changes could take place if the sessile species in the assemblage cannot be consistently ranked in their abilities to colonize, compete or resist harsh physical conditions and attacks by natural enemies. If so, each replacement could be any member of the set of species capable of colonizing and surviving in that habitat. Thus, the species composition could change or remain constant, depending upon which species colonizes the space left after a resident dies.

Examples of assemblages in which the species were not consistently ranked in competitive ability come from observations of encounters between individuals or colonies on hard substrata. (I am not aware of any studies indicating that species were equivalent in their ability to colonize or to resist disturbing forces.) When two individuals or colonies grow until they meet, the border of one may die back while that of the other advances. Alternatively, one may grow over the other, or occasionally undercut it. Lastly, the two may remain opposed in a 'standoff', sometimes for long periods (Connell 1976; Kay & Keough 1981; Russ 1982). The evidence of such encounters is of two sorts. The first is of static observations of overlap, in which the loser is scored as the colony whose edge is below that of its neighbour, or which has a dead edge next to a living one. Such evidence may be misleading, since the apparently overgrown loser may actually be undercutting the 'winner'; also, this static method cannot detect standoffs. Better evidence comes from observations of the progress of an encounter over an interval of time, when wins, losses and standoffs can be scored with confidence.

Connell & Keough (1985) reviewed many studies of organisms on hard substrates from subtidal habitats. A tally of interspecific interactions among 219 pairs of species of sessile invertebrates as well as algae, showed that in half of the cases, the members were statistically equivalent in competitive ability. Either there were many standoffs, or each species won some of the encounters, or both, so that neither was judged to be consistently superior in competitive ability. In these cases, competitive ability was a characteristic of individuals, not of species. Outcomes of encounters were determined by the relative sizes of colonies or by the positions (i.e. elevation or orientation) of their opposing edges.

In some assemblages the species can be ranked in competitive ability; in each pair one wins consistently over the others. In such cases of asymmetric competition the species could either be ranked in a transitive hierarchy (A over B, B over C, A over C) or in an intransitive network (A over B, B over C, C over A), as suggested by Jackson & Buss (1975). However, it is also possible that in some pairs one species does not win consistently over the other; they are symmetrical in competitive abilities (Kay & Keough 1981). Such assemblages cannot be completely arranged in either a hierarchy or network. Connell & Keough (1985), reviewing nine studies of competitive encounters, found only one in which all species in the assemblage could be ranked consistently; it formed an intransitive network. Networks did not occur in any of the other studies.

Assemblages in which several species are equivalent or symmetrical in competitive ability should show no predictable succession within a patch. For example, during 21 years of observation (Connell unpublished) of species composition of corals on two permanently marked square metres on the reef crest at Heron Island, all but four of the original 163 colonies had died. During this period, a total of 705 new coral genets recruited onto the site and there were 266 genets present at the end. (A genet is defined as a colony derived from a single zygote (Hughes & Jackson 1985).) The species density changed very little, ranging between 18 and 23 species, whereas the live coral coverage ranged from 58% to 18%. The relative abundance of each species was compared at start and end of the 21-year period, excluding the four colonies that survived throughout, so that there was a complete turnover. In a succession, these relative abundances would have been negatively correlated, as some species declined or disappeared while others increased or invaded, with few persisting at the same abundance. However, they were not negatively correlated, either in number of colonies or in coverage (Fig. 16.1). This indicates that, amidst all this turnover, there was no succession.

Another example in which species changed without a predictable succession comes from a study of invertebrates colonizing empty tests of the large intertidal barnacle, *Tetraclita stalactifera*, in Panama (Reimer 1976). A comparison of the assemblages at 1 month and 14 months after the barnacles died showed a high turnover but no orderly succession. As in the coral example above, the abundances at start and end were not significantly negatively correlated as would be expected in a succession.

BETWEEN-PATCH DYNAMICS AND COMMUNITY CONSTANCY

Within each patch on a local area the species composition may be changing, as described above. However, on a larger spatial scale, the species composition may remain constant, averaging out the different assemblages on patches with changing populations. Even if early-successional species are always replaced within a patch by later colonists, the former will persist over a larger area if enough new patches are opened by disturbances at a regular rate within the dispersal distance of their propagules. This subject was recently reviewed by Sousa (1984a, p. 380) who posed the following question: 'Do within-patch, nonequilibrium conditions average to an equilibrium pattern when one considers the mean dynamics of an area containing many such patches?' He pointed out that 'Recent simulation studies ... indicate that the likelihood of a large-scale steady state is a function of the total landscape area and the size of the individual disturbances. The larger the area affected by a single disturbance, the more extensive the landscape must be to average out its effects'.

A few examples from marine populations relate to this question. Paine (1979) studied the persistence of the annual alga *Postelsia palmaeformis* which tends to be eliminated in competition with mussels. The alga persisted at seven sites that had regular rates of formation of open patches. Sousa (1979b) studied succession in marine algae on boulders. Three species of red algae colonized at an intermediate time after disturbance of a boulder, but eventually were replaced there by a different long-lived persistent red alga. However, by observing many boulders over a year and a half, Sousa (1979b, Table 4) found that although each of the three species decreased or went extinct on some boulders, they invaded or increased on others, so continued to persist at about the same abundance in the boulder field as a whole. The coexistence of four species of ascidians having very different life-history patterns on vertical rock walls

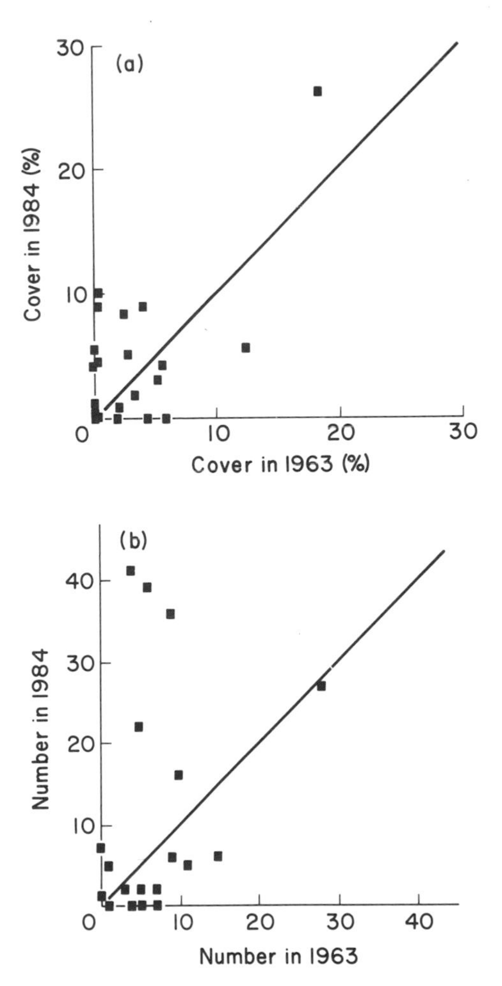

FIG. 16.1. Abundances of coral species on a permanently marked quadrat (2 m^2) on the protected south reef crest at Heron Island, Queensland, in 1963 and 1984. Of the 163 colonies in 1963, four survived in 1984; these four were excluded from this figure and the analysis, which therefore represents a complete turnover of all colonies. Each point refers to a single species; the line indicates no change. (a) Percentage of total coral cover at each date; $r^2 = 0{\cdot}43$, $P = 0{\cdot}001$. (b) Number of colonies at each date; $r^2 = 0{\cdot}15$, $P > 0{\cdot}05$. (If the point for the commonest species in 1963 is excluded, neither correlation is significant. The two lower points on the abscissa in (a) and (b) each include two species, and the lowest point on the ordinate in (b) includes four species.)

at 20 m depth in Sweden may be another example of constancy on larger scales (Svane 1982).

EVOLUTIONARY ORIGINS OF 'SUCCESSIONAL' SPECIES

Species that appear early or late in succession possess life-history traits which, it has been suggested, evolved in response to the environmental conditions existing in open patches at successive times after these are created. The term 'fugitive' was first applied to early colonists which 'inevitably succumb to competition with any other species capable of entering the same niche' (Hutchinson 1951, p. 575).

An alternative hypothesis is that these species evolved, not in different temporary stages in a successional sequence, but in habitats with average environments that resemble the different successional stages. If so, early colonists would already be adapted to living in habitats that always have an abundance of open space, either because the environmental regime prevents the build-up of populations of persistent large organisms or because frequent disturbances continually create relatively large new openings. Late-successional species are adapted to living in fully occupied habitats where disturbance rates are low and open patches are almost always small.

This hypothesis has recently been developed for land plants that are early invaders of abandoned fields in eastern North America (Marks 1983). Before man cleared the forests in this region such plants occurred in places where trees could not grow, e.g. on shallow soils, on bluffs and cliffs, on river bars or in bogs rather than in the relatively small light gaps in forests. Marks (1983) suggests that most early-successional plants in that region evolved in these permanently open sites rather than in the temporary openings we find them in today.

The same reasoning may apply in some marine habitats. For example, filamentous or sheet-like green algae were usually the first colonizers on large overturned boulders in Sousa's (1979a) study, but were always replaced by later colonizing red algae. However, some of these green algae also live in habitats kept open by low salinity in freshwater seeps or brackish estuaries (Abbott & Hollenberg 1976), or on smaller boulders with high rates of overturning. Rather than having evolved as 'fugitives' in a successional sequence, it is as likely that they evolved in continually open or disturbed environments. Their persistence in the boulder field is ensured by these permanently open habitats, which are always available for colonization, rather than by the temporary openings occurring on large boulders.

Another example comes from the subtidal. When Reed & Foster (1984) removed the canopy from a kelp forest in a protected cove, species of annual brown algae (*Desmarestia* and *Nereocystis*) invaded. These species were rare in the cove, but very common in forests in more disturbed conditions on nearby wave-beaten shores. Thus, some of the species colonizing early in the experimental large openings may be responding to conditions similar to those of continually disturbed habitats.

The terms 'fugitive' or 'early-successional' vs. 'climax' or 'late-successional' as descriptions of evolved life-history traits may not apply in all habitats. In the marine examples above, the terms high-, intermediate- and low-disturbance species might more accurately describe the conditions in which they evolved. At this point these two evolutionary hypotheses seem to be equally likely alternatives.

CONCLUSIONS

Changes in populations of sessile organisms on natural marine hard substrata are discussed in this chapter. In the intervals between disturbances the species composition of an assemblage on a single patch may either change in a progressive, orderly sequence (a 'succession'), in a non-successional sequence, or not at all. A progressive successional sequence implies that species can be ranked in a hierarchy on their abilities to colonize, compete or resist deleterious environmental conditions. Some successional sequences have been documented quantitatively in the intertidal zone, but seldom in subtidal zones.

Populations that persist with little change over more than a complete turnover exist in the marine intertidal and probably also in the subtidal. Their persistence probably depends upon that of associated mobile grazers and predators which remove superior competitors.

In some patches the species composition changes, but not in an orderly progressive successional sequence. In some coral assemblages many of the species cannot be ranked in a hierarchy or circular network of competitive abilities. Being in effect equivalent, there is no orderly progression in their replacement sequence.

On a scale large enough to include many disturbed patches, all three categories should persist; two cases have been described for intertidal algae.

Two evolutionary hypotheses to account for 'fugitive' species are compared. The traditional one suggests that such species evolved as early invaders of open patches which eventually succumb to competition from later invaders. An alternative hypothesis is that they evolved in continually disturbed sites in which no later species ever invade.

ACKNOWLEDGMENTS

I thank R. Ambrose, J. Bence, M. Carr, C. D'Antonio, T. Dean, J. Dixon, R. Doyle, M. Dungan, M. Foster, S. Holbrook, T. Hughes, J. Kastendiek, D. Lohse, W. Murdoch, S. Pennings, P. Raimondi, D. Reed, P. Ross, R. Schmitt, S. Schroeter and W. Sousa for critical comments on the manuscript; the British Ecological Society and the National Science Foundation (BSR83-07119 and OCE84-08610) provided financial assistance.

REFERENCES

Abbott, I.A. & Hollenberg, G.J. (1976). *Marine Algae of California.* Stanford University Press, Stanford, California.

Bak, R.P.M. & Engel, M.S. (1979). Distribution, abundance and survival of juvenile hermatypic corals (Scleractinia) and the importance of life history strategies in the parent coral community. *Marine Biology*, **54**, 341–352.

Cairns, J. (Ed.) **(1982).** *Artificial Substrates.* Ann Arbor Science Publishers, Inc., Ann Arbor, Michigan.

Connell, J.H. (1972). Community interactions on marine rocky intertidal shores. *Annual Review of Ecology and Systematics*, **3**, 169–192.

Connell, J.H. (1976). Competitive interactions and the species diversity of corals. *Coelenterate Ecology and Behavior* (Ed. by G.O. Mackie), pp. 51–58. Plenum, New York.

Connell, J.H. (1985). Variation and persistence of rocky shore populations. *The Ecology of Rocky Coasts* (Ed. by P.G. Moore & R. Seed), pp. 57–69. Hodder and Stoughton, Sevenoaks.

Connell, J.H. & Slatyer, R.O. (1977). Mechanisms of succession in natural communities and their role in community stability and organization. *American Naturalist*, **111**, 1119–1144.

Connell, J.H. & Keough, M.J. (1985). Disturbance and patch dynamics of subtidal marine animals on hard substrata. *The Ecology of Natural Disturbance and Patch Dynamics* (Ed. by S.T.A. Pickett & P.S. White), pp. 125–151. Academic Press. New York.

Davis, M.B. (1981). Quarternary history and the stability of forest communities. *Forest Succession: Concepts and Application* (Ed. by D.C. West, H.H. Shugart & D.B. Botkin), pp. 132–153. Springer-Verlag, New York.

Dayton, P.K. (1985). Ecology of kelp communities. *Annual Review of Ecology and Systematics*, **16**, 215–245.

Dayton, P.K., Currie, V., Gerrodette, T., Keller, B.D., Rosenthal, R. & Ven Tresca, D. (1984). Patch dynamics and stability of some California kelp communities. *Ecological Monographs,* **54**, 253–289.

Dungan, M.L. (1986). Three-way interactions: barnacles, limpets, and algae in a Sonoran desert rocky intertidal zone. *American Naturalist.* **127**, 292–316.

Ebeling, A.W., Laur, D.R. & Rowley, R.J. (1985). Severe storm disturbances and reversal of community structure in a southern California kelp forest. *Marine Biology*, **84**, 287–294.

Foster, M.S. (1982). Factors controlling the intertidal zonation of *Iridaea flaccida* (Rhodophyta). *Journal of Phycology*, **18**, 285–294.

Grigg, R.W. & Maragos, J.E. (1974). Recolonization of hermatypic corals on submerged lava flows in Hawaii. *Ecology*, **55**, 387–395.

Harris, L.G., Ebeling, A.W., Laur, D.R. & Rowley, R.J. (1984). Community recovery after storm damage: a case of facilitation in primary succession. *Science*, **224**, 1336–1338.
Hawkins, S.J. & Hartnoll, R.G. (1983). Grazing of intertidal algae by marine invertebrates. *Annual Review of Oceanography and Marine Biology*, **21**, 195–282.
Hawkins, S.J., Southward, A.J. & Barrett, R.L. (1983). Population structure of *Patella vulgata* L. during succession on rocky shores in south west England. *Fluctuation and Succession in Marine Ecosystems* (Ed. by L. Cabioch, M. Glemarec & J.-F. Samain), pp. 103–107. *Oceanologica Acta*, Vol. Special, Dec. 1983. Proceedings 17th European Symposium on Marine Biology, Brest, 1982.
Hughes, T.P. & Jackson, J.B.C. (1985). Population dynamics and life histories of foliaceous corals. *Ecological Monographs*, **55**, 141–166.
Hutchinson, G.E. (1951). Copepodology for the ornithologist. *Ecology*, **32**, 571–577.
Jackson, J.B.C. & Buss, L.W. (1975). Allelopathy and spatial competition among coral reef invertebrates. Proceedings of the National Academy of Sciences (U.S.A.) **72**, 5160–5163.
Kay, A.M. & Keough, M.J. (1981). Occupation of patches in the epifaunal communities on pier pilings and the bivalve *Pinna bicolor* at Edithburg, South Australia. *Oecologia (Berlin)*, **48**, 123–130.
Lawrence, J.M. (1975). On the relationships between marine plants and sea urchins. *Oceanography and Marine Biology Annual Review*, **13**, 213–286.
Littler, M.M. & Littler, D.S. (1980). The evolution of thallus form and survival strategies in benthic marine macroalgae: field and laboratory tests of a functional form model. *American Naturalist*, **116**, 25–44.
Loya, Y. (1976). The Red Sea coral *Stylophora pistillata* is an *r* strategist. *Nature (London)*, **259**, 478–480.
Lubchenco, J. (1983). *Littorina* and *Fucus:* effects of herbivores, substratum heterogeneity, and plant escapes during succession. *Ecology*, **64**, 1116–1123.
Lubchenco, J. & Gaines, S.D. (1981). A unified approach to marine plant-herbivore interactions. I. Populations and communities. *Annual Review of Ecology and Systematics*, **12**, 405–437.
Marks, P.L. (1983). On the origin of the field plants of the northeastern United States. *American Naturalist*, **122**, 210–228.
Menge, B.A. (1976). Organization of the New England rocky intertidal community: role of predation, competition and environmental heterogeneity. *Ecological Monographs*, **46**, 355–393.
North, W.J. (Ed.) **(1971).** The biology of giant kelp beds (*Macrocystis*) in California. *Beihefte zur Nova Hedwigia*, **32**, 1–600.
Paine, R.T. (1966). Food web complexity and species diversity. *American Naturalist*, **100**, 65–75.
Paine, R.T. (1976). Size-limited predation: an observational and experimental approach with the *Mytilus-Pisaster* interaction. *Ecology*, **57**, 858–873.
Paine, R.T. (1979). Disaster, catastrophe, and local persistence of the sea palm *Postelsia palmaeformis*. *Science*, **205**, 685–687.
Paine, R.T., Castillo, J.C. & Cancino, J. (1985). Perturbation and recovery patterns of starfish-dominated intertidal assemblages in Chile, New Zealand, and Washington state. *American Naturalist*, **125**, 679–691.
Paine, R.T. & Vadas, R.L. (1969). The effects of grazing by sea urchins, *Strongylocentrotus spp.*, on benthic algal populations. *Limnology and Oceanography*, **14**, 710–719.
Paine, R.T. & Levin, S.A. (1981). Intertidal landscapes: disturbance and the dynamics of pattern. *Ecological Monographs*, **51**, 145–178.
Pearson, R.G. (1981). Recovery and recolonization of coral reefs. *Marine Ecology: Progress Series*, **4**, 105–122.

Reed, D.C. & Foster, M.S. (1984). The effects of canopy shading on algal recruitment and growth in a giant kelp forest. *Ecology*, **65**, 937–948.

Reimer, A.A. (1976). Succession of invertebrates in vacant tests of *Tetraclita stalactifera panamensis*. *Marine Biology*, **35**, 239–251.

Russ, R.G. (1982). Overgrowth in a marine epifaunal community: competitive hierarchies and competitive networks. *Oecologia (Berlin)*, **53**, 12–19.

Rylaarsdam, K.W. (1983). Life histories and abundance patterns of colonial corals on Jamaican reefs. *Marine Ecology: Progress Series*, **13**, 249–260.

Sebens, K.P. (1985). Community ecology of vertical rock walls in the Gulf of Maine USA: small-scale processes and alternative community states. *The Ecology of Rocky Coasts* (Ed. by P.G. Moore & R. Seed), pp. 346–371. Hodder and Stoughton, Sevenoaks.

Shugart, H.H. (1984). *A Theory of Forest Dynamics: The Ecological Implications of Forest Succession Models*. Springer-Verlag, Berlin.

Sousa, W.P. (1979a). Experimental investigations of disturbance and ecological succession in a rocky intertidal algal community. *Ecological Monographs*, **49**, 227–254.

Sousa, W.P. (1979b). Disturbance in marine intertidal boulder fields: the nonequilibrium maintenance of species diversity. *Ecology*, **60**, 1225–1239.

Sousa, W.P. (1980). The responses of a community to disturbance: the importance of successional age and species' life histories. *Oecologia*, **45**, 72–81.

Sousa, W.P. (1984a). The role of disturbance in natural communities. *Annual Review of Ecology and Systematics*, **15**, 353–391.

Sousa, W.P. (1984b). Intertidal mosaics: patch size, propagule availability, and spatially variable patterns of succession. *Ecology*, **65**, 1918–1935.

Southward, A.J. (1976). On the taxonomic status and distribution of *Chthamalus stellatus* (Cirripedia) in the north-east Atlantic region: with a key to the common intertidal barnacles of Britain. *Journal of the Marine Biological Association, UK*, **56**, 1007–1028.

Southward, A.J. & Crisp, D.J. (1954). Recent changes in the distribution of the intertidal barnacles *Chthamalus stellatus* Poli and *Balanus balanoides* L. in the British isles. *Journal of Animal Ecology*, **23**, 163–177.

Stephenson, T.A. & Stephenson, A. (1933). Growth and asexual reproduction in corals. *Scientific Reports of the Great Barrier Reef Expedition*, **3**, 167–217.

Suchanek, T.H. (1985). Mussels and their role in structuring rocky shore communities. *The Ecology of Rocky Coasts* (Ed, by P.G. Moore & R. Seed), pp. 70–96. Hodder and Stoughton, Sevenoaks.

Svane, I. (1982). Ascidian reproductive patterns related to long-term population dynamics. *Sarsia*, **68**, 249–255.

Turner, T. (1983). Facilitation as a successional mechanism in a rocky intertidal community. *American Naturalist*, **121**, 729–738.

Vadas, R.L. (1977). Preferential feeding: an optimization strategy in sea urchins. *Ecological Monographs*, **47**, 337–371.

Woodley, J.D., Chornesky, E.A., Clifford, P.A., Jackson, J.B.C., Kaufman, L.S., Knowlton, N., Lang, J.C., Pearson, M.P., Porter, J.W., Rooney, M.C., Rylaarsdam, K.W., Tunnicliffe, V.T., Wahle, C.M., Wulff, J.L., Curtis, A.S.G., Dallmeyer, M.D., Jupp, B.P., Koehl, M.A.R., Neigel, J. & Sides, E.M. (1981). Hurrican Allen's impact on Jamaican coral reefs. *Science*, **214**, 749–755.

17. ARE COMMUNITIES EVER STABLE?

MARK WILLIAMSON
Department of Biology, University of York, York YO1 5DD

INTRODUCTION AND EXAMPLE

The question in the title does not have a simple yes or no answer, not only because the concepts of community and stability are each somewhat complex, but also because the answer depends on the time-scale. I shall try to illustrate points about communities, stability and time-scales with examples from bird and plankton communities, and bring in some points about the nature of environmental variability and its important consequences for the practical decision about whether a particular community can be said to be stable.

The example I start with is the breeding bird community of Eastern Wood, Bookham Common in Surrey, England (National Grid reference TQ1356). The data on the number of pairs breeding exists from 1950 to 1979 inclusive, with the exception of 1975 (Appendix 1 and Williamson 1981). The area surveyed is 16 ha of oak (*Quercus robur* L.), and forty-five species have been recorded as breeding. The management was such that the wood was becoming rather thicker during these three decades. The first point to make is that these are all bird species, and so only part of the community. All community studies are, perforce, limited taxonomically and May (1983) elaborates on the consequence of this for our understanding of communities. Even if the subsets studied can be shown to be stable, it does not follow that the community as a whole is stable. The mean abundance for each of those forty-five species is shown in Fig. 17.1, which shows the usual rough approximation to a cumulative log normal curve, even though the inflection point is around an average of one pair per year. The bottom end of the curve is not a natural end, and corresponds to two records over 29 years. As there are species–area curves, so there are species–time curves, and further observations could be expected to lead, very slowly, to an increase in the number of species recorded in this community. The data are shown in a less usual way in Fig. 17.2, where various features can be seen. In the second line, the decline of the wren (*Troglodytes troglodytes* (L.)) to one pair after the hard winter of 1962–63 can be seen, Further down, the decline of the willow warbler (*Phylloscopus trochilus* (L.)) and the increase of the starling (*Sturnus vulgaris* (L.)) from an initial period of 7 years' absence are,

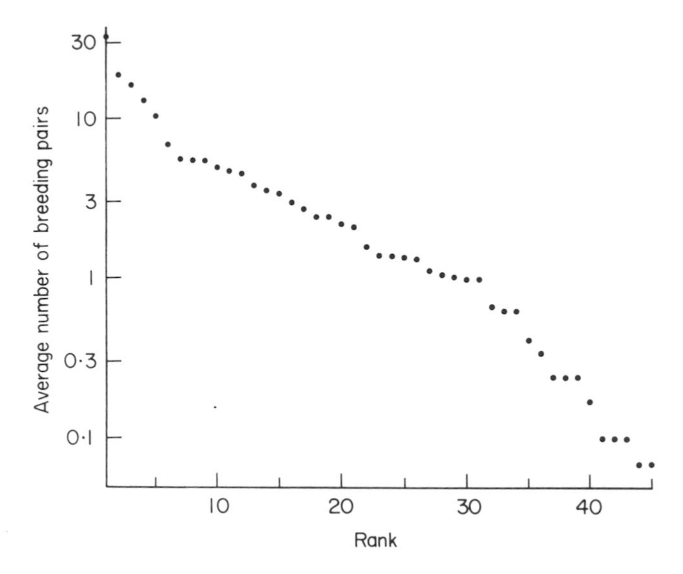

FIG 17.1. Mean number, over 29 years, of breeding pairs for forty-five species of birds censused at Eastern Wood, Bookham Common, on a logarithmic scale and in rank order.

between them, the most obvious indications of change in the community (lines 9 and 11.) Because the habitat is known to be changing, these changes show only that the community is not stationary, not that it is not stable, terms that will be defined more formally below. Towards the bottom of the figure it can be seen that breeding becomes more and more irregular, showing in another form that the community has no clearly defined limit.

Simberloff (1983) has examined this community, together with birds of Skokholm described below and at one other site (Diamond & May 1977), to see if the communities are in equilibrium, which is another concept. A community in equilibrium for any length of time is likely to be at a stable equilibrium, but a stable community need never be at equilibrium.

Following perhaps MacArthur & Wilson (1967), many authors have discussed the properties of communities in terms of their total species, and the time plot for that at Bookham Common is shown in Fig. 17.3. In the same figure are shown the total number of breeding pairs. The number of species is sensitive especially to variations in numbers of rare species—those with only one or two pairs at most—while the total number of breeding pairs is particularly sensitive to variations in the most abundant species. The sharp dip in the latter in 1963 reflects a crash in the wren population. Despite appearances, neither curve is stationary, i.e.

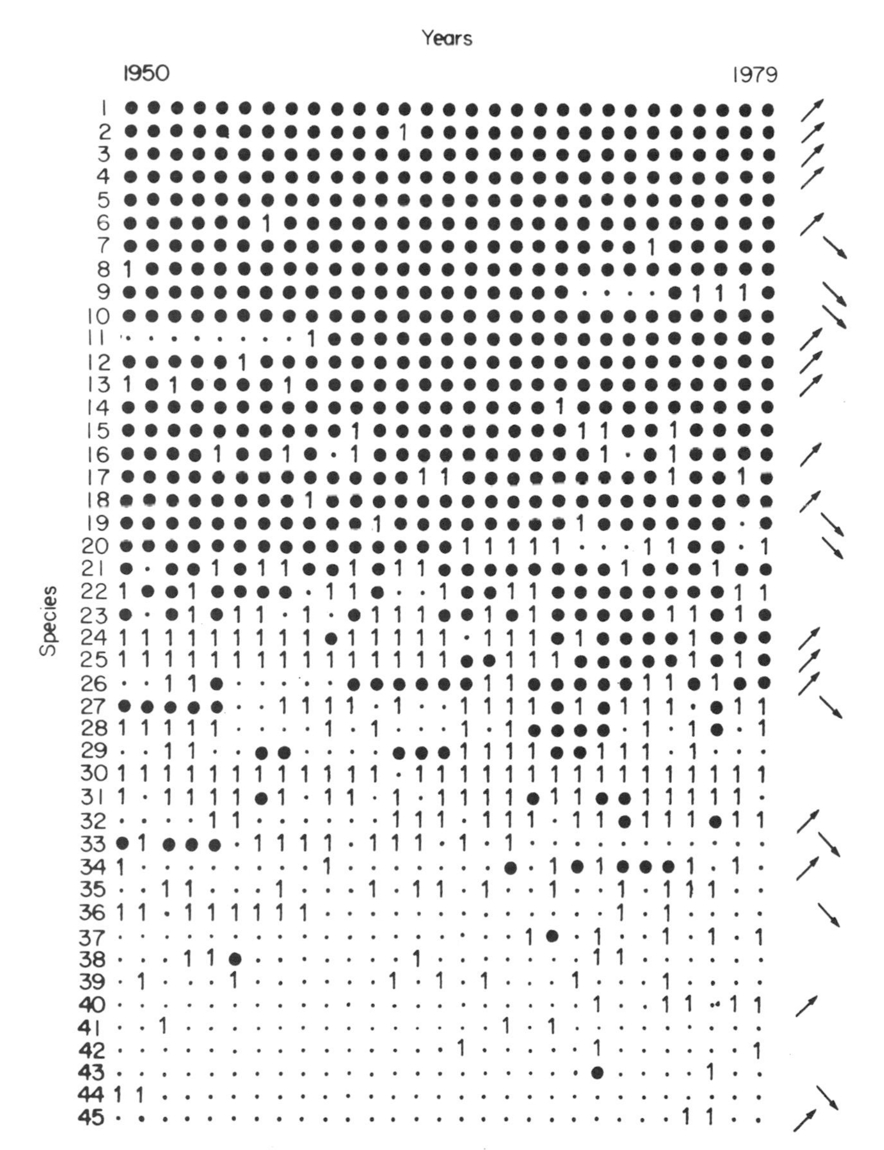

FIG 17.2. Diagram of population density for forty-five bird species over 29 years at Eastern Wood, Bookham Common: (●) two or more breeding pairs; (1) one breeding pair; (·) no breeding pair. The species are in the same order as in Fig. 17.1. The right hand column shows statistically significant increases or decreases over the period, from a runs test. Species numbers: 1, robin; 2, wren; 3, blue tit; 4, great tit; 5, blackbird; 6, wood pigeon; 7, chaffinch; 8, song thrush; 9, willow warbler; 10, jay; 11, starling; 12, coal tit; 13, nuthatch; 14, blackcap; 15, chiffchaff; 16, dunnock; 17, marsh tit; 18, great spotted woodpecker; 19, bullfinch; 20, garden warbler; 21, treecreeper; 22, long-tailed tit; 23, cuckoo; 24, magpie; 25, carrion crow; 26, mistle thrush; 27, green woodpecker; 28, pheasant; 29, turtle dove; 30, tawny owl; 31, willow tit; 32, lesser spotted woodpecker; 33, whitethroat; 34, goldcrest; 35, woodcock; 36, sparrow hawk; 37, mandarin; 38, nightingale; 39, spotted flycatcher; 40,

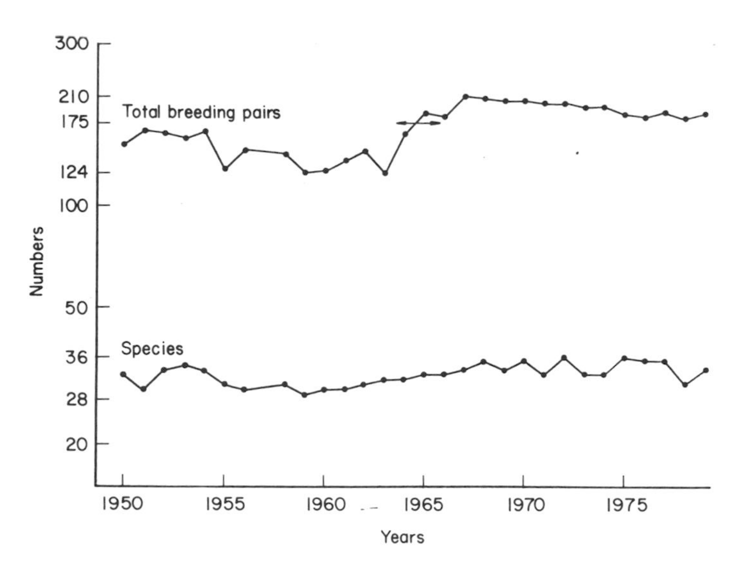

FIG. 17.3. Time plots of the toal number of breeding pairs, and the number of species breeding, on the same logarithmic scale, for the birds of Eastern Wood, Bookham Common.

neither has a constant mean. For the total number of breeding pairs this is obvious; there is no year after 1964 which has as few pairs as the maximum before that period. The total number of species is also greater in the later years, and the linear increase with time is significant at the 1% level, even though it only amounts to 1% per year. Various diversity measures are popular in the study of communities, but as these record variously the behaviour of the more abundant, or the less abundant species, the behaviour of these measures will lie between the two curves shown in Fig. 17.3. All these measures are of limited use in the study of the behaviour of communities because they ignore the information on species composition. This information is particularly important in the study of succession, which involves, by definition, changes in the relative abundance of particular species. (Nomenclature in this chapter follows British Ornithologists' Union (1971) for birds, Bruce, Colman & Jones (1963), Hendey (1964) and Parke & Dixon (1976) for plankton.)

DEFINITIONS

The question in the title can only be answered if the terms 'community' and 'stable' are well defined. By a community I mean no more than the populations that are found in a defined region (Williamson 1972a), as in the example of the birds of Bookham Common. Stability is an elusive concept for mathematicians and biologists, and many different but related definitions can be found. Hutson & Law (1985) suggest using instead 'permanent coexistence', while Nisbet & Gurney (1982) defined a persistent population as stable. Both these sets of authors were referring to one or a few species only. As the Bookham Common community shows, communities contain species that come and go. This is true whatever size area is taken because of the species–area effect, which seems insufficient reason for declaring communities unstable, and so I return to more standard mathematical definitions. Ecologists have been familiar with these problems at least since Lewontin (1969) and May (1973); the mathematical treatment of Barnett & Storey (1970) is used below.

If there is a stable equilibrium point, then there is a zone around it within which the trajectories describing the time course of the system converge on the equilibrium point. This is the zone of local stability; if it encompasses the whole of phase space the system is globally stable. However, a stable equilibrium point is not necessary in a wider meaning of stability. If a zone can be defined such that trajectories starting within it never (under constant conditions and with small perturbations) pass outside another definable zone, then the system is stable even if the critical point is an unstable equilibrium. For instance, a limit cycle defines a zone within which all points will move to the cycle locus. More complex trajectories, such as the irregularly regular ones called chaos (May 1986) can still be stable in this sense.

So a system in perpetual change under constant conditions can be stable. Change of itself does not indicate instability. Eventually, the system will return to where it has been before (or arbitrarily close to it). The annual cycle of plankton is, as will be seen, possibly close to stable in this sense, though there there is the complication of an annual cycle of environment. To my knowledge, no-one has demonstrated that any natural biological system shows classical limit cycles, let alone chaotic fluctuations. The only satisfactory demonstration of limit cycles in ecology is the analysis by Nisbet & Gurney (1982) of Nicholson's blow fly experiments. The well known 4- and 10-year cycles of some mammal populations (Williamson 1975) may be, variously, limit cycles, driven cycles or resonant quasi-cycles. Schaffer (1984) claims that the lynx (*Felis lynx*) cycle shows chaos, but the shape of that much-studied cycle is more consistent with a driven cycle with stochastic variation (Williamson 1972a; Galbraith 1977).

Stationarity is a rather different concept. A stationary time-series has constant mean, constant variance and constant time structure, i.e. a constant correlogram and spectrum (Vandaele 1983). Ignoring some mathematical niceties, a stationary series will be stable. A non-stationary series though may either represent the passage through a series of stable conditions, the passage being imposed by varying conditions, as possibly in community succession, or may represent instability.

Associated with stability there are a variety of other concepts (Pimm 1982, 1984; Williamson 1972a) such as resilience, return time and, particularly, equilibrium. Many ecologists have suggested that communities are often not in equilibrium (Powell & Richerson 1985), but that does no show that they are not stable. If the community is tending to approach its equilibrium, and especially if the stable state is a set of points, cyclic, chaotic or whatever, then non-equilibrium is consistent with stability. What the community will show under those conditions is variability. But as the environment is ever variable, all communities will be variable, whether they are stable or not.

The natural measure of the variability of a single population is the standard deviation, or variance, of logarithm population size (Williamson 1972a, 1984). The extension of this to the study of the variability of communities was discussed by Williamson (1972a); the variability of a community is most satisfactorily studied by a principal component analysis. The eigen values, when scaled, give the variance of the components, and are the basis of measures of community variability. It may be possible to divide the variability of the community into one or more non-stationary components defining trends and stationary components around those trends. If the community is in a moving stable state, it is then only necessary to show this for those components with trend.

ENVIRONMENTAL HETEROGENEITY HAS A REDDENED SPECTRUM

It is a commonplace that the further off in space the more different, on average, the environment is. It is perhaps not so immediately obvious that the same is true in time, though the history of life through the geological record shows that for organisms it is still true. The mathematical description of this is that environmental variables have a reddened spectrum. A spectrum is a measure of power at different frequencies or wave lengths. By analogy with the visible spectrum, long wave lengths are called red, short ones blue. The power at a spectral frequency is measured by a variance, and the difference between two objects is also a crude measure of variance. So the commonplace observation translates into longer wave lengths having greater variance, and so to the title of this section.

Many spectra of temperature in time were brought together by the US Committee for the Global Atmospheric Research Program (1975) who wrote 'the variance spectral density increases with increasing period over the entire frequency domain but is most pronounced for periods longer than about 30 years'. Fougere (1985) lists several physical phenomena well described by a straight line relationship between the logarithm of the variance and the logarithm of the frequency. The fractal geometry of the physical environment (Mandelbrot 1983) and of vegetation (Morse *et al.* 1985) gives rise naturally to a reddened spectrum, as is obvious from the mode of construction of the Koch snowflake curve.

The biological importance of this phenomenon for ecological systems has been recognized by Williamson (1981, 1983) and Steele (1985), even though the latter's contrast between marine and terrestrial systems seems only weakly supported by the present evidence. The reddened spectrum may be important for the species–area relationship. Williamson (1981) plotted the number of geological types against area for a small part of England. These types can be ordinated by the step-across method (Williamson 1978) and then examined by spectral analysis. However, this is an elaborate way of showing what is obvious: increase in the number of geological types with increase in area shows that the greater the distance the greater the probability of meeting a new geological type. In ecology, there is a classical way of showing the same effect. It has been shown repeatedly (Goodall 1952) 'that long narrow rectangles are more efficient sampling units than squares or circles of the same area, in that each is likely to include a wider range of local variation in the vegetation.' Botanical quadrats are usually measured in metres. Using geological quadrats of a unit 1 km^2 the same effect can be shown (Table 17.1); sets of long thin quadrats have a greater mean and a smaller variance in general. Increasing the within-sample variance by using a long narrow quadrat also decreases the between-sample variance. The phenomenon of environmental variation at different scales in space and time can be seen in all habitats, terrestrial and marine (Dayton & Tegner 1984).

This reddened spectrum has important implications for testing the stability of communities. Broadly, it is possible to distinguish three ways in which stability could be assessed in a constant environment. These are by observation, by experiment and by building mathematical models of the system. Observations can be made of the variability and the time structure of the community, using principal components. In a non-constant environment, all communities whether stable or not will be both variable and non-stationary (see also Usher, Chapter 2). Secondly, it is possible to envisage various experimental types of perturbation of the community.

TABLE 17.1. Geological types in samples of equal area but different shape in Cumbria, England. For further details of the area and data, see Williamson (1981)

Dimensions (km)	4 × 4	2 × 8	1 × 16
Mean number of types	10·29	11·50	15·04
Standard deviation	2·90	2·95	2·42

Bender, Case & Gilpin (1984) distinguished between a pulse and a press perturbation. Approximation to pulses may be found in natural conditions, such as the effect of the hard winter of 1962–63 at Bookham Common. Nevertheless, the reddened spectrum means that there will be a succession of pulses of various sizes and directions with the largest pulse being larger, in general, the longer the period of observation. Press perturbations are the permanent alteration in density of a species. Pimm (1982) and others discuss the effect of the removal of a species: in practice, the addition of a species, i.e. invasion of a foreign species, is much commoner. Simberloff (1981) showed that the commonest effect of introduced species is to produce no recorded change, and that in itself would indicate the communities involved are stable. An examination of one such invasion is given below.

For a linear or linearized system of differential or difference equations, the eigen values indicate stability or otherwise (May 1973). It should be noted that as the points in the left half of the complex plane (the stability zone for differential equations) can be mapped one to one into the unit circle (the stability zone for difference equations) and vice versa, both representations are in this sense equally likely to be stable, More generally, Liapunov (or Lyapunov) functions can be sought (Goh 1980).

TESTING FOR STABILITY

Despite the difficulties caused by the reddening of the environmental spectrum, data on natural communities do give some indications about stability, as I shall show by examining the birds of Skokholm and some plankton data, as well as looking again at the birds of Bookham Common.

Counts for the breeding pairs of land birds on the island of Skokholm (off the south-west tip of Wales, National Grid reference SM7304) are available from 1928 to 1979 (Williamson 1983). These birds are even less of a community than are those of Bookham Common, because the data set ignores the very much more abundant sea birds. For instance, the area covered by nesting lesser black-backed gulls (*Larus fuscus* L.) increased markedly between 1950 and 1980 (P.J. Conder, pers. comm.) so decreasing the nesting area available for some land birds. The counts of total land breeding species, total pairs, and the first two principal components derived by the step-along method are shown in Fig. 17.4. Step-along is a development of step-across (Williamson 1978) and allows a metric ordination of data with zeros and variations in abundance, but without the artificial distortion of some better known methods. As this particular ordination has already been discussed (Williamson 1983) perhaps it is sufficient to point out here that the first principal component is related to the total pairs, but is also sensitive to which particular species have increased. The ordinations suggest three periods of relative stationarity: (i) before the war, (ii) between 1949 and 1962, and (iii) from 1973 to the end of the observations. The marked transition from 1962 to 1973 saw the establishment of the jackdaw (*Covus monedula* L.) and the stock dove (*Columba oenas* (L.)). This is shown particularly in the first principal component; the second is sensitive to the change of composition, with a middle period characterized by abundance of birds with rather bleaker habitats, such as the wheatear (*Oenanthe oenanthe* (L.)) and lapwing (*Vanellus vanellus* (L.)) as opposed to the prominence of species of more mesic habitats such as blackbird (*Turdus merula* (L.)) and dunnock (*Prunella modularis* (L.)) The plots for total pairs and total species also seem to show three periods of relative stability, though the transition times, as shown in Fig. 17.4, do not quite agree with each other or with those of the more reliable ordination analysis.

The difficulty of showing stability in these time plots is similar to the difficulty of showing density dependence in time plots of population numbers (Bulmer 1975; Southwood 1978). A natural approach, which indeed works quite well under controlled conditions with large experimental perturbations (Williamson 1972a), is to plot the change of the numbers of a variable against the variable, i.e. Δs against s. This is shown for the species numbers both at Skokholm and at Bookham Common in Fig. 17.5. At Skokholm, it would appear that there were three periods of stability successively at 12, 8 and 14 species. At Bookham Common there appears to be just one at 32 species, though not closely defined, showing the weaknesses of the method. (i) As has already been shown, there is in fact a slight trend at Bookham Common which is quite lost in this plot. (ii) The plot of Δs against s appears to have a slope of -1, which means the expectation is always a return to equilibrium whatever the perturbation. The perturbations appear to have no influence on the succeeding

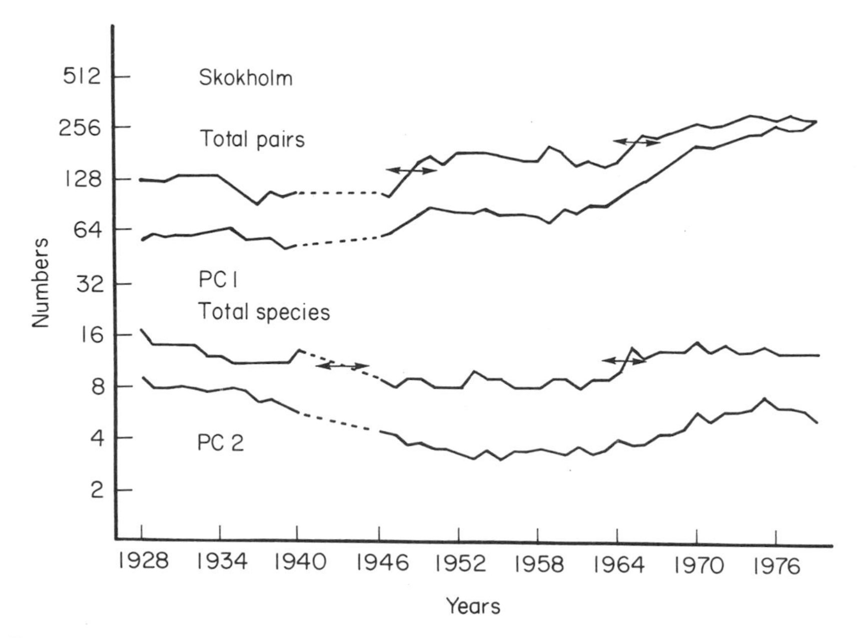

FIG. 17.4. Time plots for the total number of breeding pairs, the number of species breeding, and the first and second principal components from a step-along analysis (Williamson 1983) on the logarithms of the number of land birds on Skokholm Island, Wales.

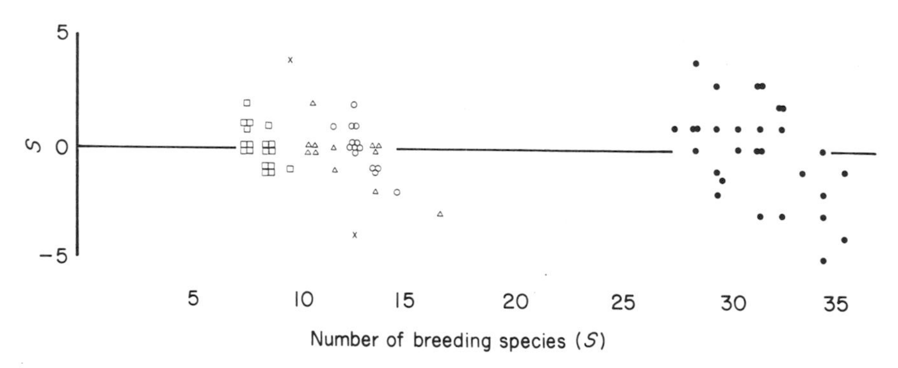

FIG. 17.5. The change in the number of breeding species (Δs) against the number of breeding species (s) at Skokholm and Eastern Wood: (Δ) Skokholm 1928–40; (□) Skokholm 1946–64; (○) Skokholm 1965–79; (●) Eastern Wood, Bookham Common 1950–79.

population. (iii) The lack of independence of the two axes, or of any variant on them, means that, as in the case of density dependence (Bulmer 1975), the sampling properties are intractable. Consequently, stability cannot be demonstrated by any statistical test applied to this plot.

As for modelling the community, these data are consistent with a random walk, even though non-random processes are undoubtedly involved. The data do not lend themselves to any sensible calculation of eigen values. Larger, and better understood, perturbations are needed, so I will consider next a community with such perturbations.

PLANKTON IN THE IRISH SEA

The annual cycle

Good and extensive data tend to be published only when it is not clear how best to analyse them. A case in point is the extensive collections of surface plankton in Port Erin Bay, in the Isle of Man (National Grid reference SC1969). Samples were collected on a standard line with standard nets, at frequencies varying from daily to weekly, systematically from January 1907 until December 1920. Monthly averages, and details of the work are given in Johnstone, Scott & Chadwick (1924). As with all community studies, the taxonomic extent of the data is limited. There are data on both phytoplankton and zooplankton, but the phytoplankton includes only the larger types, particularly diatoms, while zooplankton consists mainly of copepods and similar sized crustaceans. The smaller phytoplankton, which make an important contribution to total production, and the larger faster zooplankton are both missing. Nevertheless, it is still an excellent data set, and useful for considering the stability of communities.

The annual cycle in this plankton, as revealed by principal component analysis, is shown in Fig. 17.6. In order to use a standard method, I selected the twenty-two taxa that were present in almost every month and then did a principal component analysis with row means and column means standardized. This allows a direct comparison between the variables expressed as rows and columns (Williamson 1972b), although there are some complications since the data can be viewed as spanning a three-dimensional space of species, months and years.

In the principal component analysis of both plankton and Skokholm the arc hyperbolic sine of the data was used: this is a logarithmic transformation that allows for zeros more neatly than the customary method of adding 1 (Williamson 1981).

Fig. 17.6 gives the results simultaneously for taxa and for months. It takes two linear (one-dimensional) components to define a cycle (or any other two-dimensional figure), and the first two components here define the annual cycle neatly both in terms of months and taxa. By using a method which allows the direct comparison, the two can be superimposed

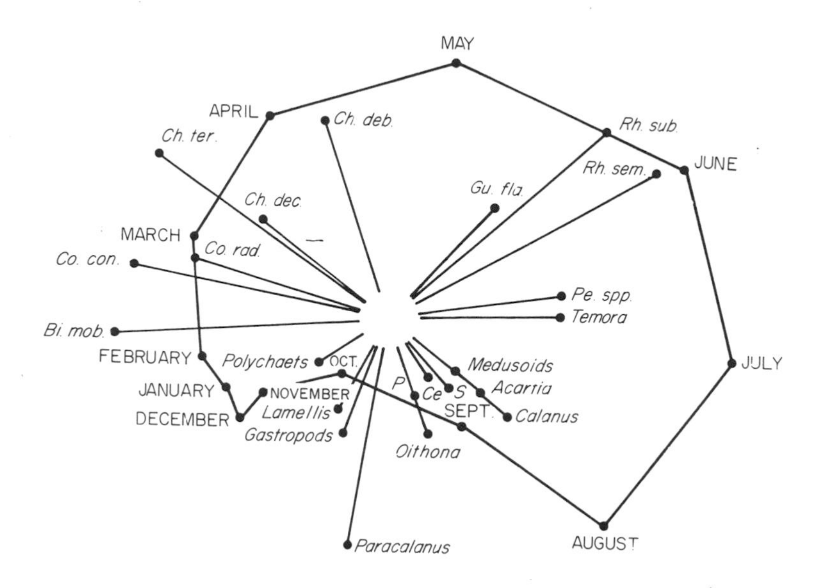

FIG. 17.6. Rows and columns results of a principal component analysis of the twenty-two commonest taxa taken in Port Erin Bay, Irish Sea from 1907 to 1920. The axes are the first and second components of the row and column standardized time × taxa logarithmic data matrix. Phytoplankton have two letter generic abbreviations, zooplankton otherwise. Bi mob *Biddulphia mobiliensis*; Ce *Ceratium tripos*; Ch *Chaetoceras*, deb *debile*, dec *decipiens*, ter *teres*; Co *Coscinodiscus*, con *concinnus*, rad *radiatus*; Gu fla *Guinardia flaccida*; Pe spp Peridinians, Rh *Rhizosolenia*, sem *hebetata* f *semispina*, sub *shrubsolei*; Acartia *A. clausi*; Calanus *C. finmarchicus* and *C. helgolandicus*; Gastropods, larvae; Lamellis, lamellibranch larvae; Medusoids, gonophores; Oithona *O. similis*; Paracalanus *P. parvus*; Polychaets, larvae; P *Pseudocalanus elongatus*; S *Sagitta elegans*; Temora *T. longicornis*.

showing, broadly, the time of the year when each taxon has its peak abundance. Some of these taxa are species, but others are groups. As I showed in an earlier principal component analysis of plankton (Williamson 1961) and as Felsenstein (1985) has emphasized in rather a different context, relatively little is lost by grouping closely related species in analyses of this sort. One reason why this grouping is satisfactory can be seen in Fig. 17.6. The various diatom species in the genera *Chaetoceras* and *Rhizosolenia* occur in two groups in the annual cycle in April and June respectively. Grouping these species together by genera would lose some information, but not distort it.

The annual cycle in plankton is an obvious reflection of the annual cycle of physical conditions in the sea. Although the causes of the spring bloom in phytoplankton, the summer peak in zooplankton, and the resurgence of phytoplankton in the autumn are all broadly explicable (Smayda 1980) it cannot be said that the cycle is understood in detail, and in general it has not even been well described. My major aim in this chapter is to consider what observable reactions to perturbation tell us about the stability of communities. Quantitative descriptions like Fig. 17.6 are essential for this; understanding the mechanisms can come later.

Although it is plausible to suppose that the community is always out of equilibrium, chasing, in a stable way, a moving equilibrium state, other interpretations are possible, e.g. the community might be near equilibrium at all times, keeping up with the changing physical conditions. In the next section, I show that the characteristic delay may be more than 1 month, which fits the first interpretation better. A third interpretation is that the community is not approaching a stable equilibrium at any time. The obvious contrary indication is the stability of the annual cycle. On a longer time-scale the simplest interpretation of the cycle shown in Fig. 17.6 is that the continued repetition of an annual cycle—albeit slightly different each year—has led to the selection of a set of species which are, over the year, in a stable state. That is, the community might well be in a stable equilibrium over the year as a whole, but not in any one particular month. The time-scale is important in considering stability. In evolutionary time not only will conditions change but new species will arrive or arise, the community will shift and, on that time-scale of more than tens of thousands of years, no ecological community is stable. In tens of years, the regular reaction of the plankton community suggests stability. On shorter time-scales—months for the plankton, years for the birds—the small and erratic physical perturbations make the interpretation of observational data difficult and uncertain. These are 'pulse' perturbations in the terminology of Bender, Case & Gilpin (1984). 'Press' perturbations may be more informative.

An invasion

The commonest 'press' perturbation of communities is the invasion of a new species rather than the elimination of one already present. Invasions are a special category of colonization. As part of the SCOPE Programme on the Ecology of Biological Invasions, I have been responsible for drawing together information on invasions in and around the British Isles, which was reported at a Royal Society Discussion Meeting in February 1986. There have been several successful invasions of British waters by foreign algae (Robinson *et al.* (1980). One invader well known to plankton biologists is the diatom *Biddulphia sinensis* Grev. This alga was known from tropical waters, from the Red Sea eastwards, in the nineteenth century, having first been found at Hong Kong. It is therefore surprising that it has invaded the much cooler waters round north-west Europe, appearing off the Danish coast in 1903 (Ostenfeld 1908). It was found in the Isle of Man plankton in November 1909. At first sight it is typical of Simberloff's (1981) commonest type of invader, a species that has no effect on the other members of the community. The population size comes to a peak in November and December, near to but ahead of the taxon called by Johnstone Scott & Chadwick. (1924) *Biddulphia mobiliensis*, (which appears to include *B. regia* as well). Because of the absence of *B. sinensis* in the first 34 months of the whole 168-month period, I did not include it in the principal component analysis shown in Fig. 17.6. It was therefore with some surprise that I realized interpretation of the third principal component reflected the presence and absence of *B. sinensis*.

In Fig. 17.7 the presence or absence of *B. sinensis* in the record and the positive or negative sign of the third component are shown simultaneously. A better fit of these two variables is found by shifting one in relation to the other, and the statistics of this are shown in Fig. 17.8. The maximum effect of *B. sinensis* as measured by the third component has a delay of somewhat more than a month. This delay is recorded in the population densities of the other plankton, and is an average of the delays in all effects, direct and indirect, in the interactions within the community. Looking at this effect in more detail, by examining the vector of taxon scores, and the details of the annual cycle for each taxon, it is evident that *B. sinensis* has an extensive though weak effect through the whole community. The effect is most noticeable around January, i.e. a month or two after the peak in *B. sinensis*. Some species are more abundant when *B. sinensis* is present, others less. The total effect is small, only a few per cent in the total variance of the community. To my mind this indicates the importance of weak and indirect effects in the community structure, the small but definite effects that invaders may produce, and is consistent with the view that the community was in a stable state and has been shifted to a new one by the arrival of *B. sinensis*. None of this can be proved without a detailed understanding of the population dynamics of the main species in the community, and that is many years away.

DISCUSSION AND CONCLUSIONS

Environmental heterogeneity and its reddened spectrum should colour all considerations of communities in time. Not only does it imply that perturbations will be continuous and on average larger the longer the time, it also means that stability can only be assessed over stated time-scales.

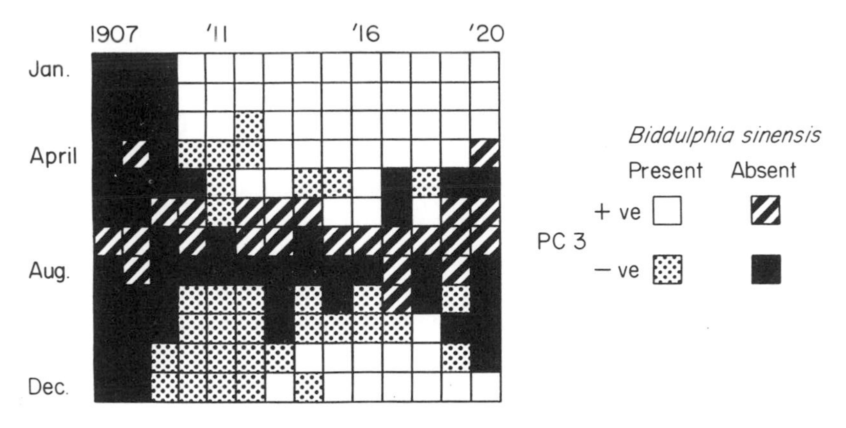

FIG. 17.7. A time diagram of the joint variation of the occurrence of the diatom *Biddulphia sinensis* in samples, and the sign of the third component of the analysis shown in Fig 17.6.

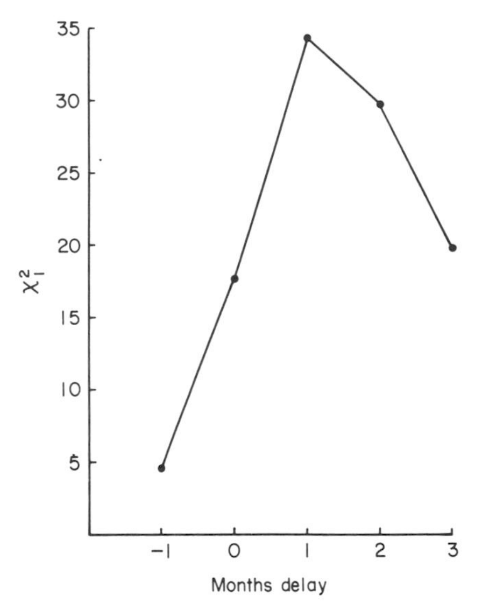

FIG 17.8. Chi-squared values for 2 × 2 tables from the joint occurrence of *B. sinensis* and the sign of the third component. The months delay 0 value refers to the four categories shown in Fig. 17.7, the other values to the component moved in time relative to the diatom.

The other main point I wish to stress is that communities are conveniently and sensibly studied by multivariate techniques, i.e. by principal component analysis and related methods. Other measures of the community, whether species number, total numbers of individuals or any diversity index, are insufficient because they ignore the particular species involved in community change, regarding all species as equivalent. Multivariate measures are also sensitive to the correlation structure of the community, and so to interactions direct and indirect, strong and weak, information which is lost by studying the individual species seriatim.

Over a longer time-scale than any of these three examples, Lamb (1984) has given a multivariate description of the change in pollen counts, and hence of vegetation, in Labrador. Over the same scale of thousands of years, but with less sophisticated techniques, Cole (1985) found vegetational inertia: a lag in the community change compared with the physical change. The larger changes over longer time-scales reported by Lamb and Cole would be expected from the reddened spectrum. These studies are important also for showing that the effect of physical change can be measured and assessed; prerequisites for using such continuously varying, natural, 'pulse' perturbations to decide whether or not a community is stable.

On the short time-scales available to most ecologists, some conclusions can be drawn: (i) communities can be stable even though not at equilibrium; (ii) as conditions are ever changing, so are communities never stationary; (iii) suitably regular 'pulse' perturbations, as in the annual cycle of the plankton, and the 'press' perturbations associated with invading species, throw light on the dynamics of the community and so, indirectly and obscurely, on the question of community stability; (iv) time-scales are important—in the long run no community has been stable to evolutionary change.

ACKNOWLEDGMENTS

For comment and discussion I am grateful to Prof. J.D. Currey, Dr M.M. Dodson, Dr R. Law, Mr D.R. Morse, Dr D. Orr, Dr M.B. Usher and Dr C. Williamson. For help with computing, figures and typing I thank Mr G. Atkinson, Ms M. Chapman, Miss K. Edwards, Mrs A. Harrison, Mr C. Haswell. Dr G. Beven kindly allowed the use of his unpublished data.

REFERENCES

Barnett, S. & Storey, C. (1970). *Matrix Methods in Stability Theory.* Nelson, London.

Bender, E.A., Case, T.J. & Gilpin, M.E. (1984). Perturbation experiments in community ecology: theory and practice. *Ecology*, **65** 1–13.

British Ornithologists' Union (1971). *The Status of Birds in Britain and Ireland.* Blackwell Scientific Publications, Oxford.

Bruce, J.R., Colman, J.S. & Jones, N.S. (1963). *Marine fauna of the Isle of Man.* Liverpool University Press, Liverpool.

Bulmer, M.G. (1975). The statistical analysis of density dependence. *Biometrics*, **31**, 901–911.

Cole, A. (1985). Past rates of change, species richness, and a model of vegetational inertia in the Grand Canyon, Arizona. *American Naturalist*, **125**, 289–303.

Dayton, P.K. & Tegner, M.J. (1984). The importance of scale in community ecology: a kelp forest example with terrestrial analogs. *A New Ecology: Novel Approaches to Interactive Systems* (Ed. by P.W. Price, C.N. Slobodchikoff & W.S. Gaud), pp. 457–481. Wiley, New York.

Diamond, J.M. & May, R.M. (1977). Species turnover rates on islands: dependence on census interval. *Science*, **197**, 266–270.

Felsenstein, J. (1985). Phylogenics and the comparative method. *American Naturalist*, **125**, 1–15.

Fougere, P.F. (1985). On the accuracy of spectrum analysis of red noise processes using maximum entropy and periodogram methods: simulation studies and application to geophysical data. *Journal of Geophysical Research*, **90**(A5), 4355–4366.

Galbraith, R.F. (1977). Discussion on the papers by Mr Campbell and Professor Walker, Dr Morris and Dr Tong. *Journal of the Royal Statistical Society* (A) **140**, 453.

Goh, B.-S. (1980). *Management and Analysis of Biological Populations*. Elsevier, Amsterdam.

Goodall, D.W. (1952). Quantitative aspects of plant distribution. *Biological Reviews*, **27**, 194–245.

Hendey, N.I. (1964). An introductory account of the smaller algae of the British coastal waters. V. Bacillariophyceae (Diatoms). *Fisheries Investigations* (4) **5**, 1–317.

Hutson, V. & Law, R. (1985). Permanent coexistence in general models of three interacting species. *Journal of Mathematical Biology*, **21**, 285–298.

Johnstone, J., Scott, A. & Chadwick, H.C. (1924). *The Marine Plankton*. Hodder and Stoughton, London.

Lamb, H.F. (1984). Modern pollen spectra from Labrador and their use in reconstructing Holocene vegetational history. *Journal of Ecology*, **72**, 37–59.

Lewontin, R.C. (1969). The meaning of stability. *Brookhaven Symposia in Biology*, **22**, 13–23.

MacArthur, R.H. & Wilson, E.O. (1967). *The Theory of Island Biogeography*. Princeton University Press, Princeton.

Mandelbrot, B.B. (1983). *The Fractal Geometry of Nature*. Freeman, New York.

May, R.M. (1973). *Stability and Complexity in Model Ecosystems*. Princeton University Press, Princeton.

May, R.M. (1983). The structure of food webs. *Nature*, **301**, 566–568.

May, R.M. (1986). When two and two do not make four: nonlinear phenomena in ecology. *Proceedings of the Royal Society Series B.*, **228**, 241–266.

Morse, D.R., Lawton, J.H., Dodson, M.M. & Williamson, M.H. (1985). Fractal dimension of vegetation and the distribution of arthropod body lengths. *Nature*, **314**, 731–733.

Nisbet, R.M. & Gurney, W.S.C. (1982). *Modelling Fluctuating Populations*. Wiley, New York.

Ostenfeld, C.H. (1908). On the immigration of *Biddulpha sinensis* Grev. and its occurrence in the North Sea during 1903–1907. *Meddeleser fra Kommissionen for Havundersøgelser, Serie: Plankton*, **1**(6), 1–51.

Parke, M. & Dixon, P.S. (1976). Check list of British marine algae—third revision. *Journal of the Marine Biological Association*, **56**, 527–594.

Pimm, S. (1982). *Food Webs*. Chapman & Hall, London.

Pimm, S. (1984). The complexity and stability of ecosystems. *Nature*, **307**, 321–326.

Powell, T. & Richerson, P.J. (1985). Temporal variation, spatial heterogeneity, and competition for resources in plankton systems : a theoretical model. *American Naturalist*, **125**, 431–464.

Robinson, G.A., Budd, T.D., John, A.W.G. & Reid, P.C. (1980). *Coscinodiscus nobilis* (Grunow) in continuous plankton records, 1977–78. *Journal of the Marine Biological Association* **60**, 675–680.

Schaffer, W.M. (1984). Stretching and folding in lynx fur returns: evidence for a strange attractor in nature? *American Naturalist*, **124**, 798–820.

Simberloff, D. (1981). Community effects of introduced species. *Biotic Crises in Ecological and Evoutionary Time* (Ed. by M.H. Nitecki), pp. 53–81. Academic Press, New York.

Simberloff, D. (1983). When is an island community in equilibrium? *Science*, **220**, 1275–1277.

Smayda, T.J. (1980). Phytoplankton species succession. *The Physiological Ecology of Phytoplankton* (Ed. by I. Morris), pp. 493–569. Blackwell Scientific Publications, Oxford.

Southwood, T.R.E. (1978). *Ecological Methods*, 2nd ed. Chapman & Hall, London.

Steele, J.H. (1985). A comparison of terrestrial and marine ecological systems. *Nature*, **313**, 355–358.

United States Committee for the Global Atmospheric Research Program (1975). *Understanding Climatic Change*. National Academy of Sciences, Washington.

Vandaele, W. (1983). *Applied Time Series and Box-Jenkins Models*. Academic Press, New York.

Williamson, M. (1961). A method for studying the relation of plankton variations to hydrography. *Bulletin of Marine Ecology*, **5**, 224–229.

Williamson, M. (1972a). *The Analysis of Biological Populations*. Edward Arnold, London.

Williamson, M. (1972b). The relation of Principal Component Analysis to the Analysis of Variance. *International Journal of Mathematical Education in Science and Technology*, **3**, 35–42.

Williamson, M. (1975). The biological interpretation of time series analysis. *Bulletin of the Institute of Mathematics and its Applications*, **11**, 67–69.

Williamson, M. (1978). The ordination of incidence data. *Journal of Ecology*, **66**, 911–920.

Williamson, M. (1981). *Island Populations*. Oxford University Press, Oxford.

Williamson, M. (1983). The land-bird community of Skokholm : ordination and turnover. *Oikos*, **41**, 378–384.

Williamson, M. (1984). The measurement of population variability. *Ecological Entomology*, **9**, 239–241.

APPENDIX

Census results for breeding birds at Eastern Wood, Bookham Common for 1976–79, courtesy of Dr G. Beven. For earlier results see Williamson (1981).

	1976	1977	1978	1979
Mandarin	0	1	0	1
Pheasant	1	2	0	1
Woodcock	1	1	0	0
Stock dove	1	0	1	1
Wood pigeon	11	14	6	12
Turtle dove	1	0	0	0
Cuckoo	1	2	1	2
Tawny owl	1	1	1	1
Green woodpecker	0	2	1	1
Great spotted woodpecker	3	3	3	3
Lesser spotted woodpecker	1	2	1	1
Carrion crow	1	2	1	2
Magpie	1	2	2	3
Jay	5	5	4	5
Great tit	15	14	13	16
Blue tit	18	15	14	19
Coal tit	11	7	7	5
Marsh tit	3	2	1	2
Willow tit	1	1	1	0
Longtailed tit	2	2	1	1
Nuthatch	4	5	3	5
Treecreeper	2	1	3	2
Wren	22	23	33	18
Mistle thrush	2	1	2	2
Song thrush	4	8	5	4
Blackbird	10	11	11	13
Robin	33	34	40	41
Blackcap	4	5	4	4
Garden warbler	2	2	0	1
Willow warbler	1	1	1	2
Chiffchaff	4	3	5	4
Goldcrest	1	0	1	0
Dunnock	2	2	4	4
Starling	6	7	7	7
Greenfinch	0	0	0	1
Redpoll	0	1	0	0
Bullfinch	2	2	0	2
Chaffinch	3	3	3	2
House sparrow	1	1	0	0

18. INVASIONS OF FOREST COMMUNITIES DURING THE HOLOCENE: BEECH AND HEMLOCK IN THE GREAT LAKES REGION

MARGARET BRYAN DAVIS
Department of Ecology and Behavioral Biology, University of Minnesota Minneapolis, Minnesota 55455, USA

INTRODUCTION

During the Holocene (the last 10 000 years) boreal and temperate tree species advanced into deglaciated regions of North America at rates averaging 100–400 m per year (Davis 1981a.) Similar rates are documented in Europe (Firbas 1949; Huntley & Birks 1983).

How were such rapid rates of range extension possible? Many ecologists feel that seed dispersal and establishment must have limited the speed of range extension (Iverson 1954; Davis 1961, 1976; Walker 1982), while others consider climatic boundaries and their changes in space and time the most important factors (T. Webb 1986). Van der Pijl (1969, p. 22) reviewed mechanisms of seed dispersal, and concluded that the estimated rates of spread seem appropriate only for pioneer species, 'exceeding regular possibilities' for late-successional trees. Walker (1982) placed emphasis on establishment and competition by suggesting that periodic disturbances could have facilitated rapid range extension by opening forest communities to invasion. The fossil record in North America shows that pioneer species did attain the most rapid rates of range extension. Nevertheless, late-successional trees diffused onto a forested landscape at rates as rapid as 200 m per year. It is my purpose here to discuss the mechanisms by which shade-tolerant, long-lived and generally slow-growing late-successional tree species were able to extend their ranges rapidly and to invade forest communities.

History of beech and hemlock in the Great Lakes region

For the past several years we have been studying fossil pollen at a network of sites, in order to learn how beech (*Fagus grandifolia*) and hemlock (*Tsuga canadensis*) moved into the Great Lakes region (Davis *et al.* 1986; K.D. Woods & M.B. Davis, unpubl.; Webb 1982; M.W. Schwartz, unpubl.). Beech and hemlock were chosen for study because (i) they entered the region at about the same time, (ii) they commonly co-occur in northern hardwoods forests, and (iii) they both reach their western limits in this region. The original focus of the research was on seed dispersal. The Great Lakes are geographical barriers 100 km wide: if trees cannot disperse across barriers, the Great Lakes will cause delays in range extensions. The two tree species provided a useful comparison, because beech seeds are dispersed by animals, whereas hemlock seeds are dispersed by wind.

Beech and hemlock produce pollen that is readily identified in sediments. We have compared pollen deposited in sediments 150 years ago, before logging and forest clearance, with nineteenth century forest records from the federal Land Office Survey. There are quantitative relationships between the pollen percentages and tree abundances (Schwartz 1985). Beyond the species limits pollen percentages drop rapidly to background levels below 1%, assuring us that species limits can be recognized within 30–50 km from the quantities of pollen in lake sediments (M.B. Davis, M.W. Schwartz & K.D. Woods, unpubl.). The latter relationship means that fossil pollen can be used to map the species' frontiers as they advanced into the region.

A recent paper summarizing our results presents maps of pollen percentages of beech and hemlock at 1000-year intervals (Davis *et al.* 1986). We included data from twenty-one sites we have investigated, plus thirty published studies by other authors that extend the area of study (Fig. 18.1). Figure 18.2 summarizes our interpretation of the results, showing the inferred locations of the species frontiers at 1000-year intervals. Hemlock pollen first appeared at scattered sites in Upper and Lower Michigan. These scattered colonies apparently originated from seeds transported long-distance from Ontario, where hemlock was established about 8000 years BP (McAndrews, 1970, 1981; Kapp, 1977; Liu, 1982). Hemlock populations expanded rapidly and spread over a large region (>50 000 km^2) by 5000 years BP . Its subsequent spread to the west and south-west occurred much more slowly; it reached its present limit in Wisconsin about 1500 years ago (unpubl. data).

Beech invaded lower Michigan from the south, extending its range northward at a relatively rapid rate. Between 6000 and 5000 years BP , an outlying population was established across the lake in Wisconsin. Beech dispersed to Wisconsin either directly across the lake, or around its southern end, across what is presently a disjunction in the beech range. At that time the predominant vegetation in the area of the disjunction was prairie, but beech may have dispersed distances of 25–50 km between scattered patches of forest growing in ravines and other protected sites (Webb 1983). A second dispersal event occurred 4000 years BP, when

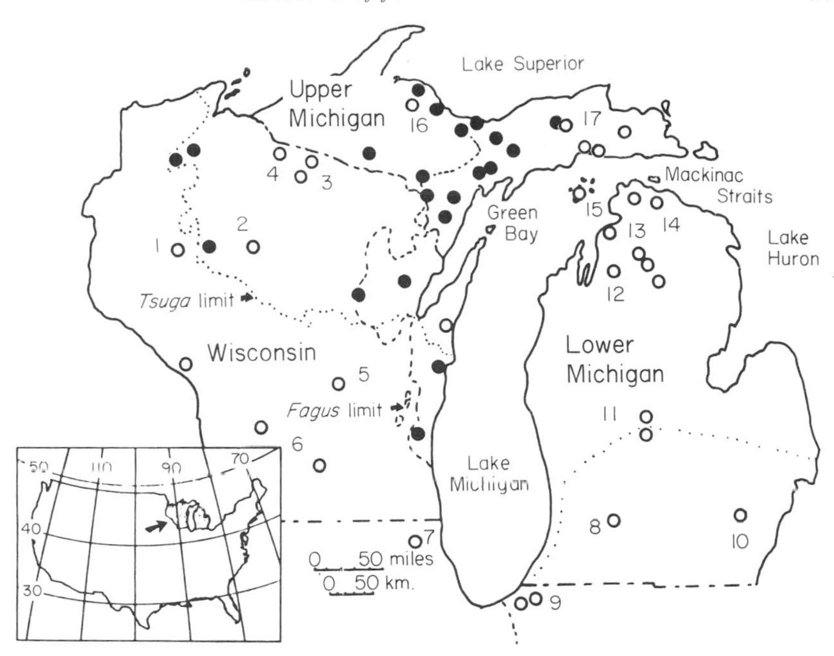

FIG. 18.1. Map of the northern Great Lakes region. Black dots indicate the locations of small lakes from which we have prepared pollen diagrams (Davis *et al.* 1986). Open circles indicate the locations of sites of pollen diagrams reported in the literature as follows: (1) Peters & Webb 1979; (2) Heide 1981; (3) Webb 1974; (4) Swain 1978; (5) Maher 1982; West 1961; (6) Davis 1977; (7) King 1981; (8) Manny, Wetzel & Bailey 1978; (9) Bailey 1972; (10) Kerfoot 1974; (11) Kapp 1977; Gilliam, Kapp & Bogue 1967; (12) Lawrenz 1975; (13) Bernabo 1981; (14) R.P. Futyma, unpublished; (15) Kapp *et al.* 1969; (16) Brubaker 1975; (17) Futyma 1982. (Figure modified from Davis *et al.* 1986.)

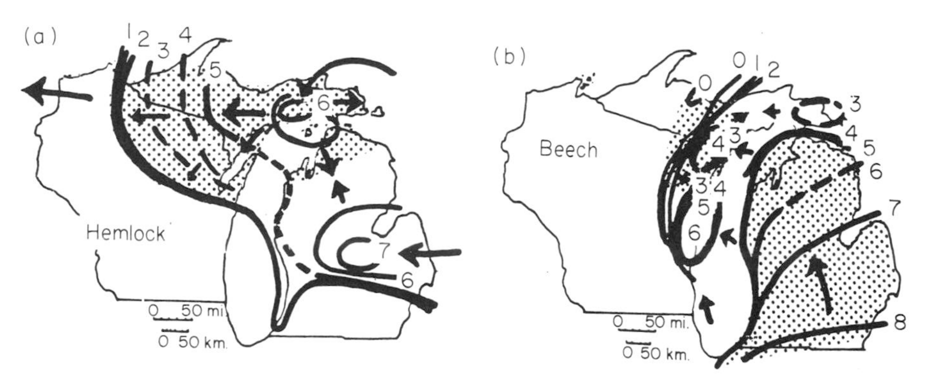

FIG. 18.2. Summary of the diffusion of beech and hemlock into the Great Lakes region. Black lines indicate the inferred positions of the species frontiers at 1000-year intervals. Shaded areas are the sectors shown in Fig. 18.4. (Figure modified from Davis *et al.* 1986.)

another outlying population became established on the west side of Green Bay in Upper Michigan. Rapid expansion of range occurred around 2500 years BP when beech colonized the eastern half of Upper Michigan. The range limit was fairly stable between 2000 and 500 years BP, but within the last several centuries an outlying population was established about 50 km to the west, and the limit of continuous populations advanced about 20 km (K.D. Woods & M.B. Davis, unpubl.).

The data are used in this chapter as the basis for a general discussion of factors that affect range extension and invasion of forest communities.

DIFFUSION MODELS

Parameters affecting the rate of advance of a species frontier are dispersal and population growth rate (Okubo 1980). These variables are the most important in attempting to explain the complex patterns of range expansion recorded by fossil pollen. The mode of seed dispersal is fixed for most species, but the probabilities of seed arriving at favourable sites vary, depending on seed production and on the availability and behaviour of biotic vectors and/or wind patterns. Population growth rates are also variable, depending on climate, soils, and the biotic environment.

To focus the discussion, I am considering two models of diffusion, a continuous front model (A), and a model where discontinuous populations are established far in advance of the front (B) (Fig. 18.3). In Model A, the rate of advance of the species frontier depends on the average distance of seed dispersal per generation and on the number of years per generation. The seed rain is assumed to decrease exponentially away from the source tree. Model B invokes occasional dispersal of seeds very great distances. No particular statistical distribution is assumed; the distribution of seeds might be patchy in space and time, for example. The mechanism for hyperdispersal (B) might be different from (A), for example, dispersal by a different species of bird that rarely carries seeds, but does fly long distances. In the case of wind-dispersed species, Model A would involve the winds and storms that occur in most years, whereas Model B would invoke unusual storms (e.g. tornados) that could carry seeds distances several orders of magnitude greater. The outlying populations established in Model B would serve as 'infection centres' for the surrounding landscape, and as a source for further long-distance-dispersal ahead of the species frontier.

In the real situation beech and hemlock were advancing into previously established forest, which is not shown in Fig. 18.3. Within a forest, competition with other species for light and nutrients usually lessens the chances for successful establishment, and lengthens the time before first

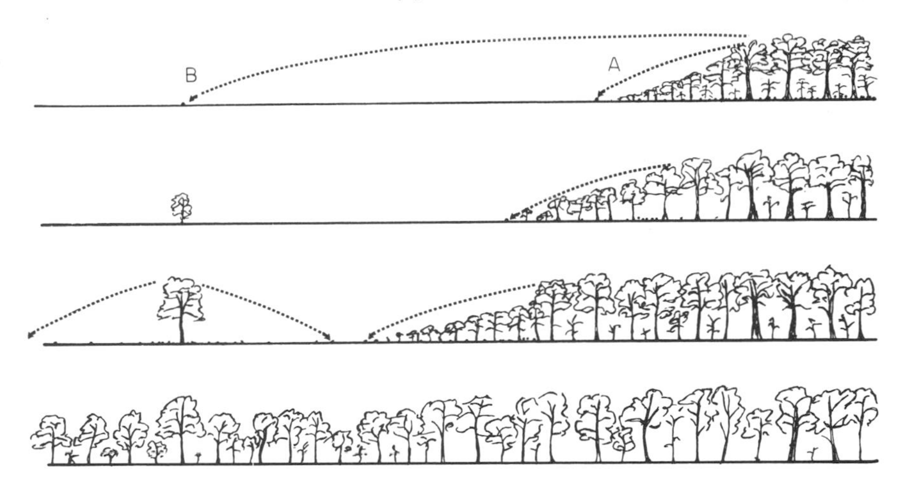

FIG. 18.3. Sketch of range extension for a hypothetical tree species: (A) advance as a continuous front; (B) advance as discontinuous populations, founded by long-distance dispersal of seeds. The vegetation that is being invaded is not shown in the sketch.

fruiting, thus slowing the rate of population growth. Slower population growth will affect the rate of advance of the species front and reduce the speed with which outliers act as 'centres of infection' expanding to form a continuous population over the entire landscape.

NATURE OF THE INVADING SPECIES FRONT

In order to consider beech and hemlock invasions using the diffusion models as a focus for discussion, we need to determine whether the species fronts were continuous or discontinuous. Pollen data from Davis *et al.* (1986) have been assembled from sites in sectors that include the longest radius of diffusion in the upper Great Lakes region. Different sectors were used for hemlock and beech (Fig. 18.4). Hemlock and beech pollen percentages at 1000-year intervals (from Davis *et al.* 1986) were plotted against the distances of the sites from the species front 150 years ago. Pollen percentages 150 years ago, from a somewhat larger data set including sites far outside the species limit (M.B. Davis, M.W. Schwartz & K.D. Woods, unpubl.), are also shown in Figs 18.5 and 18.7 for comparison.

Hemlock diffusion along an E–W transect across Upper Michigan and northern Wisconsin

Seven thousand years ago hemlock pollen appeared in trace quantities at a number of sites, and a significant percentage (>1·0%) appeared at one

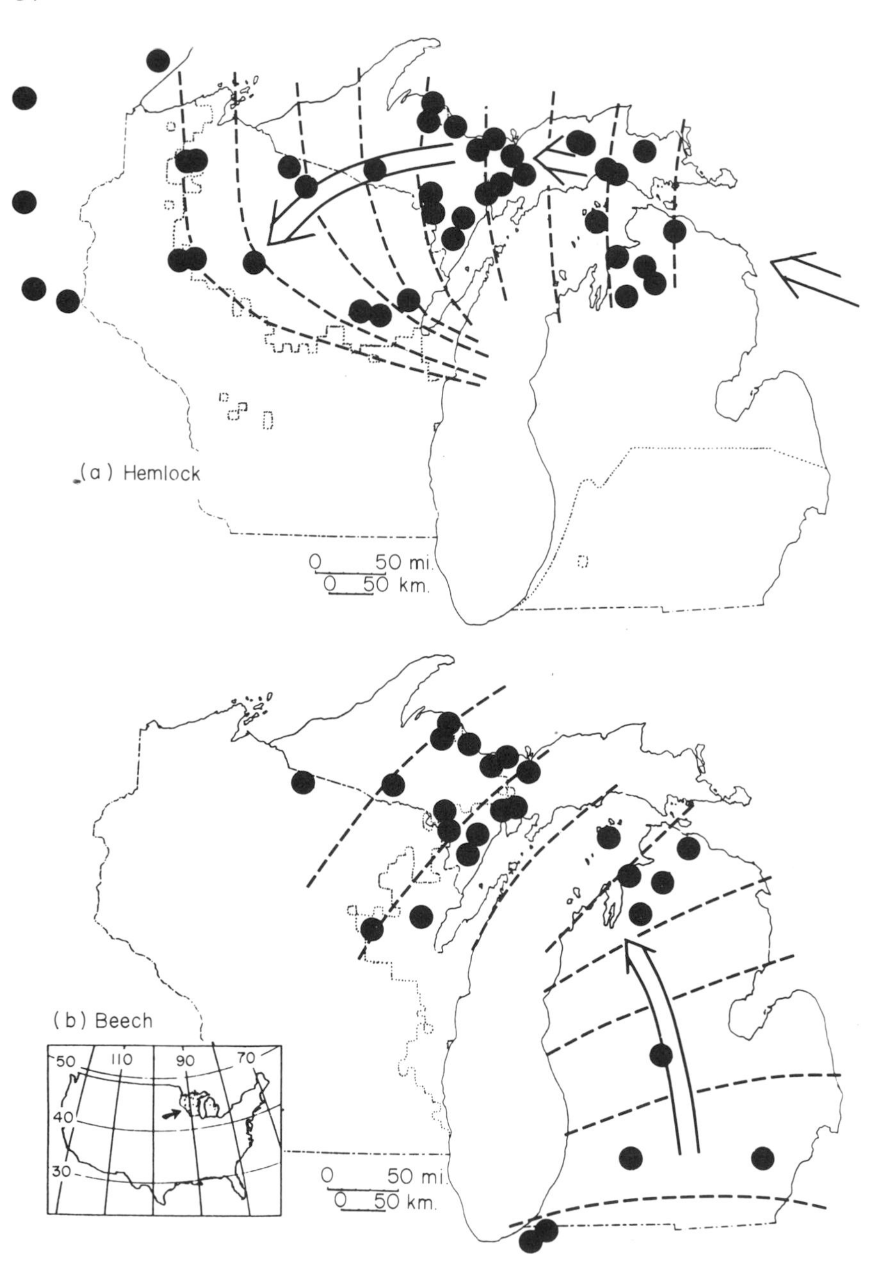

FIG. 18.4. (a) Outline map of Michigan and Wisconsin showing the sector along the E–W radius of diffusion of hemlock that was used for the tabulation of pollen percentages through time. Black dots indicate sites from which pollen data were summarized. (b) SE–NW radius of diffusion for beech. Black dots indicate the sites that were used for beech. Minnesota sites in (a) are from Birks (1976), Fries (1962), Waddington (1969), Wright & Watts (1969) and Wright, Winter & Patten (1963).

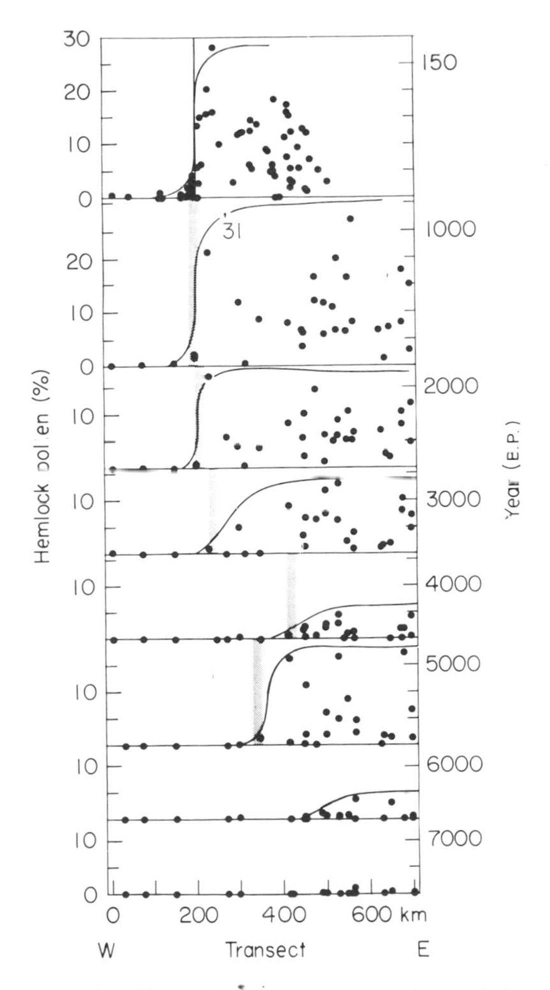

FIG. 18.5. Percentages of fossil hemlock pollen at 1000-year intervals from the series of small lakes across Upper Michigan and northern Wisconsin shown in Fig. 18.4a. Pollen percentages are plotted against distances from the species boundary 150 years ago. Data are from authors cited in legend for Fig. 18.1 and from Davis *et al.* (1986). Shaded lines indicate the probable positions of the species limit in the past.

site in Upper Michigan. Zero values characterized the western two-thirds of the transect (Fig. 18.5). Similar evidence from Lower Michigan suggests that isolated colonies may have become established before 6000 years BP in both Upper and Lower Michigan—a discontinuous front. During the next millennium, populations grew and coalesced, and a more or less continuous population was formed in the eastern part of Upper Michigan. By 5000 years ago a dense population had become established over an expanded arc with a steep species front resembling the hemlock front 150 years ago. Hemlock populations had spread rapidly, diffusing over an area larger than 50 000 km^2 in less than 2000 years.

During the next 2000 years hemlock populations declined steeply everywhere in eastern North America. A number of different lines of argument suggest that the decline resulted from the spread of a pathogen similar to the chestnut blight (Davis 1981b; 1983; Webb 1982; Allison, Moeller & Davis 1986). Along our transect pollen percentages declined, and there was an apparent retreat of the species front. By 3000 years BP populations had recovered and western advance resumed. The species front advanced about 50 km between 3000 and 2000 years BP, and about 25 km during the next millennium. We have not detected movement of the species front in Wisconsin during the last 1000 years.

Discussion of hemlock diffusion

The discontinuous model of diffusion describes hemlock during the first stages of invasion 6000 years ago. Apparently long-distance dispersal brought seeds 100–150 km from hemlock populations in Ontario. These early colonies acted as centres of infection for the remainder of the region, which was colonized quickly. Seed dispersal and the intrinsic rate of increase could have limited the spread of these rapidly expanding populations. Colonization occurred at about the same time in Upper and Lower Michigan, suggesting that Lake Michigan did not pose a significant barrier.

The initial colonization of Michigan through long-distance dispersal is not difficult to visualize for hemlock. Hemlock produces numerous small, winged seeds, which are shed throughout the autumn and winter months. They land on the surface of the snow and might easily be blown across the frozen surface of a lake, even a lake as large as Lake Michigan, the northern part of which freezes in winter.

Effective seed dispersal explains the large number of outlying populations to the west of the geographical range of hemlock (Fowells 1965). Prior to logging there were thirteen such populations in Minnesota, the farthest 150 km beyond the species limit (Fig. 18.6). The fossil record shows clearly that continuous populations of hemlock reached their present limit in Wisconsin 1800 years ago, with the subsequent establishment of a large outlier 20 km west of the limit 1300 years ago (Davis *et al.* 1986; unpublished data). The smaller outliers still farther west, in Minnesota, are not relict stands left from a previous continuous range, but new populations established by long-distance dispersal of seeds. Although prevailing winds blow from the west to the east, there are occasional

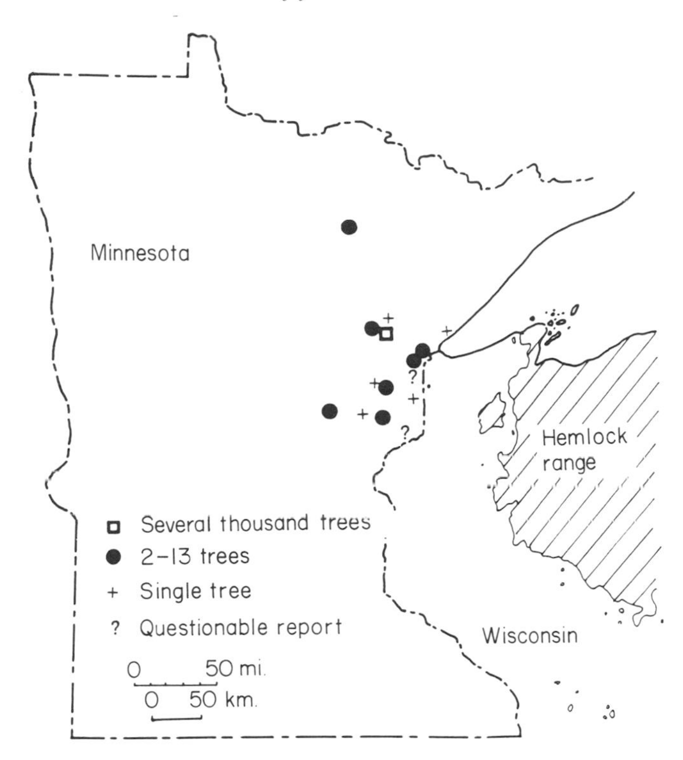

FIG. 18.6. Map of the distribution of hemlock in Minnesota and western Wisconsin in the early nineteenth century. Wisconsin occurrences are from the Land Office Survey Notes. The Minnesota occurrences were compiled from the records of the Heritage Program of the Minnesota Department of Natural Resources. Figure from Calcote (unpubl.).

storm winds in the opposite direction. These are apparently frequent enough to carry seed that becomes established in pockets of favourable habitat. The age of outliers to the south in Ohio and Indiana is not known, however. Some of these isolated populations could be relicts from a previously more extensive range.

Hemlock spread rapidly through eastern Michigan 6000–5000 years BP. Between 5000 and 1000 years BP, range extension occurred so slowly that seed dispersal can hardly have been a limiting factor. Problems in establishment, and slow population growth, apparently prevented hemlock from extending its range more rapidly. Once established, hemlock competes strongly with other plants because it casts a very dense shade. Young hemlock are more shade-tolerant than other species and can live for as long as 200 years before reproduction beneath a forest canopy.

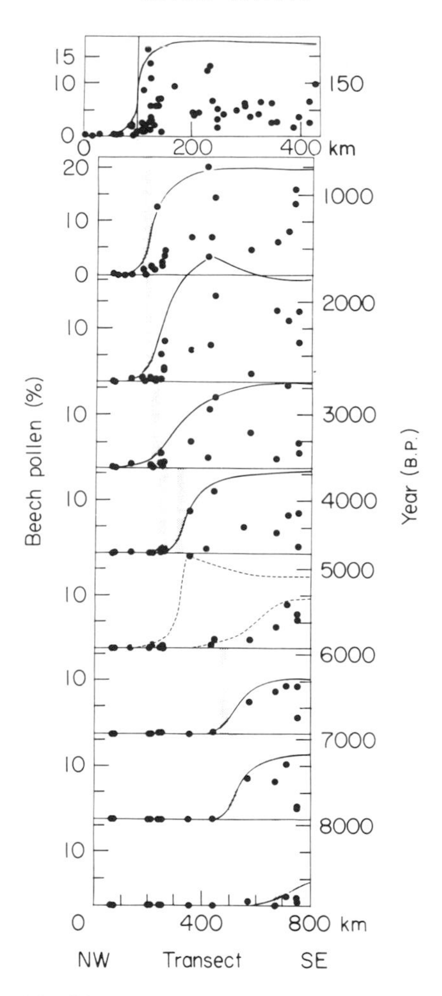

FIG. 18.7. Percentages of fossil beech pollen at 1000-year intervals from the series of small lakes shown in Fig. 18.4b. Pollen percentages are plotted against the distance of the site from the species boundary 150 years ago. Data are from authors cited in legend for Fig. 18.1 and from Davis *et al.* (1986). Shaded lines indicate probable locations of the species limit in the past.

Ring widths indicate that saplings can survive multiple episodes of suppression and release before reaching the canopy (Woods 1981). Once in the canopy, hemlock can live 600 or more years. Under these conditions age at first reproduction may be several centuries. As a result of the long generation length, the rate of population growth can be quite slow, and possibly limiting to the rate of spread.

Diffusion of beech along a SE–NW transect through Michigan and northern Wisconsin

Beech pollen percentages along the transect indicate that beech advanced northward through lower Michigan as a continuous population, reaching the straits that separate lower and upper Michigan by 5000 years BP. The total distance traversed was 400 km in 3000 years, an average rate of advance of 130 m per year. There was a general decline in percentages 5000 years ago; except for one tentatively dated site (Beaver Island: Kapp, Bushouse & Foster 1969) with very high percentages, the data from 5000 years BP fail to provide evidence for a steep front. The pollen percentages from 4000 years BP, however, indicate high abundances close to the species front. At this time an outlying population became established west of Green Bay on the far side of Lake Michigan. The newly established front on the west side of the lake was nearly stable for the next 1000 years. Between 3000 and 2000 years BP it advanced westward a few km, but there was extensive invasion of new territory in a direction perpendicular to the diffusion radius we are considering here (K.D Woods & M.B. Davis, unpubl.). The frontier moved westward slowly 2000–1000 years BP, advancing more rapidly between 1000 years BP and the present. During the last 4000 years, the rate of advance of the species front to the west has varied from 10 to 50 m per year, but there has always been some invasion of new territory.

Discussion of beech diffusion

Beech advanced along the diffusion radius as a continuous front. The only exception occurred as beech dispersed across Lake Michigan, establishing an outlying population on the western shore.

Beech nuts are dispersed by birds and mammals. Blue jays (*Cyanocitta cristata*) commonly feed on beech mast, which they cache near their nesting sites. The caches are placed just below the leaf litter layer on the forest floor, a site favourable for germination (Johnson & Adkisson 1985). Groups of about a dozen crowded beech seedlings germinating in the spring are not an uncommon sight, suggesting that birds often forget the locations of caches. The tendency of the birds to nest in woodland assures the placement of many cached nuts in habitats that are suitable for beech trees. Direct observations of jays in southern Wisconsin indicated that they transported beech nuts up to 4 km, making several 8 km round-trips each day (Johnson & Adkisson 1985). Dispersal by jays seems adequate for the advance of a continuous front at the rates we are reporting here, at least since 4000 years BP. Blue jays are probably adequate to explain the more rapid rates 8000–5000 years BP through Lower Michigan, although one is forced to assume a dispersal distance of 13 km and a generation length of 100 years, or dispersal 6–7 km and a generation length of 50 years, to explain the rate of advance. Although beech can fruit at 50 years when growing in an open situation (Fowells 1965), such short generations seem unlikely in a forest. A different means of dispersal might be necessary, however, to explain dispersal across open water. S.L. Webb (1986) suggests that the extinct passenger pigeon, a strong flyer that fed extensively on beech mast, could have been the agent for occasional long-distance dispersal events.

Beech is a strong competitor in northern hardwoods forests. It can persist in deep shade, growing slowly upwards toward the canopy. Saplings can often take advantage of small canopy gaps through rapid horizontal growth, outcompeting hemlock and sugar maple (*Acer saccharum*) (Thomas L. Poulson, University of Illinois-Chicago, pers. comm). Thus, it can persist and compete effectively for light gaps. In the absence of herbivores the survivorship of seedlings and saplings may be lower than hemlock, however, as beech cannot tolerate suppression by shade for as many years as hemlock (Woods 1981). It is difficult to assess seed production in beech near its western limit because it varies from year to year. 1984 was a good seed year in northern Michigan and beech populations were strongly represented by seedlings on the forest floor in the spring of 1985, indicating that reproduction by seed as well as by sprouting (Ward 1961) can occur near the western limit for the species.

INVASION OF FOREST COMMUNITIES

The pollen record from the northern Great Lakes region indicates that hemlock and beech were advancing into a forest mosaic where sugar maple and basswood (*Tilia americana*), and possibly yellow birch (*Betula lutea*) grew on wetter sites, and white pine (*Pinus strobus*), oak (*Quercus borealis*) and paper birch (*B. papyrifera*) grew on drier sites. During the last 3000 years, extension of beech and hemlock ranges was accompanied by a general expansion of mesic trees and a decrease of pine and oak. The changes suggest that a more moist climate permitted hardwood populations to expand, replacing pine, oak and paper birch on marginal sites (Davis *et al.* 1986; K.D. Woods & M.B. Davis, unpubl.). Although our data show regional invasion, they lack resolution to demonstrate changes within local communities. We do not know whether beech and/or hemlock moved first into newly established hardwood communities, participating in the displacement of pine, oak and paper birch, a successional sequence that can be observed today (T.L. Poulson, pers. comm.), or whether they immediately invaded old, established hardwood communities, where it was necessary to compete with longer-lived, shade-tolerant species such as sugar maple, basswood and yellow birch. Eventually they invaded these communities, as both beech and hemlock came to grow most abundantly on mesic sites. More detailed study utilizing small hollows that collect very local pollen could answer this question (M.W. Schwartz, unpubl.).

Because beech and hemlock seeds can tolerate dense shade, it seems reasonable that if their seeds were once dispersed into a forest, seedlings could become established and compete successfully with resident maple, basswood and yellow birch. The age-structure of beech in Dukes Experimental Forest, Michigan, suggests that this process may be occurring. The geographical limit of continuous beech populations occurs within an old-growth forest at Dukes that has been set aside as a scientific and natural area. Beech saplings can be found several tens of metres west of mature beech trees. Of course one has to make assumptions about the survival of the saplings to predict movement of the species frontier (Frederick Metzger, US Forest Service, pers. comm.). Similar observations are not available for hemlock, however. Hemlock seedlings are rare in most hemlock-hardwood forests, whether hemlock is dominant, or other hardwoods such as beech, sugar maple, basswood, or yellow birch are dominant. In these forests hemlock seedlings are established only in particular microsites, such as stumps or rotting logs (Curtis 1959). Because hemlock seedlings are preferentially browsed by deer in winter, mortality is high. Natural mortality rates are difficult to judge, however, because deer populations have been fluctuating due to human impact (Anderson & Loucks 1979).

Disturbance

Walker (1982) has suggested that periodic disturbance could have facilitated rapid range extension by opening communities to invasion. This may have been the case for beech in Europe, where the entry of beech into deciduous forests occurred at a time of intense prehistoric human disturbance (Iversen 1973). In North America, in contrast, recent human disturbance in the form of logging has caused a decrease in the abundance of both hemlock and beech. The change in hemlock density in Wisconsin is obvious in comparisons of maps prepared from Land Office Survey Notes (mid-nineteenth century) and maps of the present-day forest

(Davis, Schwartz & Woods, unpubl.). Pollen records in Wisconsin and New England also show decreased abundances of both species (Schwartz 1985; Davis 1985). Beech and hemlock appear to respond rather differently to natural disturbances, such as windthrow (Henry & Swan 1974). Furthermore, different kinds and frequencies of disturbances have different effects. The concept that disturbance promotes the entry of species into forest communities deserves discussion.

Frequency of disturbance

Frequent severe windstorms are a feature of the regional climate of north-central and north-eastern United States. The frequency of catastrophic windthrow of forest trees has been estimated from historical evidence such as maps of storm-damaged forests by nineteenth century Land Office surveyors, or from evidence within individual forest stands, such as fire scars on trees, tip-up mounds on the forest floor, or tree-ring sequences showing episodes of suppression and release (Lorimer 1985). Raup (1957), Stephens (1956), Lorimer (1977) and Schoonmaker & Foster (1985) all emphasize the importance of catastrophic windthrow damage to forests in New England. Lorimer (1977) used Land Office records to calculate the return time of large-scale catastrophic storms in Maine at 1150 years.

Lorimer recognized that vulnerable areas may experience severe storms frequently, while other sites may escape damage for longer periods than the regional average. In addition, frequency and intensity of storms varies from one geographical region to another. In central Massachusetts, tip-up mounds in a 0·25 ha forest plot were dated by ring-counts of trees growing on their surfaces. The age-classes of mounds suggested that strong winds had uprooted the larger trees in the forest approximately once per century for the last 450 years (Stephens 1956). Stephens was able to correlate three out of four episodes of windthrow with historical references to New England hurricanes in 1938, 1815, and 1635. An additional series of mounds were formed in 1850. The important point here is that hurricanes need not strike full force in a region, causing catastrophic damage, to have significant impact on the age structure of a forest community. Canham & Loucks (1984) make the same point.

In the northern Great Lakes region, tornados and violent downbursts of wind accompanying thunderstorms are major sources of disturbance. Tornados cause almost complete destruction to forest vegetation along a narrow path, twisting trees off at the base of the trunk. Downbursts affect larger areas, ranging from a few to 3800 ha (Stearns 1949; Canham &

Loucks 1984). Canham and Loucks calculate return times for catastrophic windstorms in northern Wisconsin at 1210 years. In southern Wisconsin downbursts are less common, but tornados, ice storms and fires occur at greater frequency (Canham & Louck 1984). The now extinct passenger pigeon was an additional source of disturbance, as forests (mainly oak or beech) used by nesting colonies were described as resembling sites of tornado damage (Bishop 1932). Roosts covered several km^2 and involved tens of millions of birds. The birds congregated so densely that branches were broken from trees, and droppings accumulated to a depth of several cm (Blockstein & Tordoff 1985; McKinley 1960). The return time is difficult to estimate, as the only records date from a period when the pigeon population was already in decline. Fire was the most important periodic disturbance along the prairie border (Grimm 1983), and to the north in the boreal forest, where the return time of fire averaged 100 years (Heinselman 1973).

Raup (1957) maintained that few if any American old-growth forests were all-aged, as might be expected if senescent trees were replaced one by one under equilibrium conditions. He argued that all stands that had been studied showed the effects of periodic disturbance: they were composed of cohorts, each of which originated following a disturbance that removed most of the canopy trees, permitting a new cohort of trees to grow to maturity. The age structure of a Wisconsin forest provides an example (Stearns 1949).

Disturbance and invasibility

Certainly the destruction of canopy trees helps invading species that are already present to grow rapidly to canopy size and thus to gain dominance more quickly (Davis & Botkin 1985). By increasing the light on the forest floor, or by increasing the frequency of gaps, disturbance might increase the growth rates of juvenile beech and hemlock, hastening maturation and fruiting. This mechanism would be effective in shortening the generation time for beech; maturation in just 50 years would have allowed diffusion of beech through lower Michigan at 130 m per year even if dispersal distance per generation were only 6–7 km. With a shortened generation time, the observed distances of seed dispersal by blue jays become sufficient to explain the rapid migration rates.

Entry to a forest community, however, is not necessarily facilitated by disturbance. Cut-over forests in the Great Lakes region often grow up into a dense stand of sugar maple saplings, or sugar maple and basswood sprouts. These successional stands cast dense shade and compete strongly for water and nutrients. They contain fewer young seedlings than mature forests (Bray 1956). The young, even-aged forest has fewer gaps than an all-aged stand, and thus fewer opportunities for rapid growth of invading seedlings. Invasion might succeed only if seeds arrived within the first year or two following disturbance, but even then beech and hemlock might be out-competed because their vertical growth rates in light gaps are slower than maple (Thomas L. Poulson, pers. comm.). In any case the probability seems small that a long-distance dispersal event would coincide in both space and time with a disturbance event.

Insight into the relationship between disturbance and invasion is provided by the Flambeau forest, an old-growth hemlock-hardwood stand west of the beach limit in Wisconsin that was flattened by a downburst storm in 1978 (Canham & Loucks 1984). Sprouts and seedlings of sugar maple and basswood are growing rapidly 7 years after the disturbance. Hemlock reproduction is poor, however (F.S. Stearns, University of Wisconsin-Milwaukee, pers. comm.). There were few hemlock seedlings in the stand prior to 1978, although hemlock was the dominant tree. Mature trees, which could have provided a seed source, were killed by the storm which occurred in July, before seed was mature. Of course there is no seed-bank in the soil for hemlock.

There is evidence, however, that hemlock establishment *can* be increased by certain kinds of disturbance. A small (0·4 ha) forest plot in southern New Hampshire was studied intensively by Henry & Swan (1974). The forest had originated following a windstorm and fire in the mid-seventeenth century. The ages of trees still living in 1967 show that beech recruitment is not correlated with disturbance: twenty-five beech trees have entered the stand since 1810, at the rate of about one per decade. The oldest hemlock trees entered the stand following a fire in the mid-seventeenth century. Several hemlocks were established in the centuries following, at apparently random intervals. The majority, however, became established during the past 150 years, just before or immediately following a series of windstorms which occurred in the early decades of the twentieth century. In this New Hampshire forest, the removal of canopy trees facilitated the growth and survival of previously suppressed hemlock in the understorey; frequent windstorms had the important effect of increasing the numbers of hemlock canopy trees (Henry & Swan 1974; Schoonmaker & Foster 1985; Lorimer 1985).

A similar response to windstorms has not been demonstrated in Michigan or Wisconsin. The disturbance regimes are different from New England, and frequently involve fire. It seems likely that forest succession following windstorm *and* fire is different from succession following windthrow alone. Intense fire sweeping through an area of blowdown would kill many of the hardwoods capable of sprouting, such as maple, basswood and beech. The seed bank in the soil would be destroyed, changing the competitive environment for invading species. In the Great Lakes region, reduction of the humus layer might provide a more favourable seedbed for hemlock. Experimental studies are needed to test these predictions.

A greater understanding of the processes of secondary succession in forests is essential to judge the chances of invasion through establishment following disturbance. Many studies of forest community structure emphasize gap-phase replacement in all-aged stands, compensatory recruitment, and the maintainance of diversity under equilibrium conditions. Succession following catastrophic disturbance deserves the greater emphasis that it is receiving in the most recent literature (Pickett & White 1985).

CONCLUSIONS

Fossil tree pollen in sediments records rapid range extensions, demonstrating that temperate forest communities of North America and Europe have been invaded repeatedly by forest species during the last 10 000 years. The rates at which species have been able to extend their ranges, and by inference to invade forest communities, have varied over space and time. In this chapter I have discussed some of the factors that could have limited range extensions of two species, beech and hemlock. At this point the most rapid rates recorded by fossil pollen can be explained by processes that have been observed in modern forests, such as dispersal of beech nuts by jays. The occurrence of outlying populations west of the hemlock range makes it clear that probabilities are relatively high for long-distance dispersal in this species, explaining hemlock's rapid invasion of Michigan between 6000 and 5000 years BP. Seed dispersal, even for trees that characterize old-growth forest, is surprisingly effective. Therefore, it seems likely that during most of the history of the invasion of the Great Lakes region, invasion occurred more slowly than if it were limited by seed dispersal *per se*.

Variables such as seed production, seedling establishment, seedling and sapling mortality, and generation length were far more important. The ways in which these parameters are influenced by climate must be understood before the fossil record of past changes in geographical distribution can be accurately interpreted as a record of climate. Much can be learned about the influence of community context on these parameters by studying secondary succession in forests, especially the succession that occurs in forest undisturbed by human activities following episodes of natural disturbance. Information about the establishment of seedlings, and the outcome of competition as forest stands mature and humus layers accumulate, is needed to understand the evidence contained in the fossil record regarding the rates at which forest communities were invaded by beech and hemlock.

ACKNOWLEDGMENTS

This work has been supported by the National Science Foundation, grants DEB79-09801, DEB82-00290, DEB84-07943. I wish to thank R.P. Futyma for giving me access to his unpublished data from Michigan, and to R.R. Calcote for preparing the map used as Fig. 18.5. K.D. Woods, A. Middeldorp, M.W. Schwartz, and R.R. Calcote provided useful discussion and criticism. I am grateful also for comments and suggestions made by Thomas Poulson, Craig Lorimer and Forest Stearns, who read the manuscript, and I acknowledge the participants on the ESA Section for Vegetation Study field trip in June, 1985, led by Orie L. Loucks, for their stimulating discussion of the problems of range extension and invasion of forest communities.

REFERENCES

Allison, T., Moeller, R.E. & Davis, M.B. (1986). Pollen in laminated sediment provides evidence for a mid-Holocene forest pathogen outbreak. *Ecology* **67**, 1101–1105.

Anderson, R. & Loucks, O.L. (1979). White-tailed deer (*Odocoileus virginianus*) influence on structure and composition of *Tsuga canadensis* forests. *Journal of Applied Ecology*, **16**, 855–861.

Bailey, R.E. (1972). *Vegetation history of northwest Indiana.* Ph.D. thesis, Indiana University, Bloomington, Indiana.

Bernabo, J.C. (1981). Quantitative estimates of temperature changes over the last 2700 years in Michigan based on pollen data. *Quaternary Research*, **15**, 143–159.

Birks, H.J.B. (1976). Late-Wisconsinan vegetational history at Wolf Creek in central Minnesota. *Ecological Monographs*, **40**, 395–429.

Bishop, H.O. (1932). The wild pigeon. *American Forests (Nov.)*, pp. 596–599.

Blockstein, D.E. & Tordoff, H.B. (1985). The passenger pigeon. *American Birds*, **39**, 845–851.

Bray, J.R. (1956). Gap phase replacement in a maple-basswood forest. *Ecology*, **37**, 598–600.

Brubaker, L.B. (1975). Post-glacial forest patterns associated with till and outwash in Northcentral Upper Michigan. *Quaternary Research*, **5**, 499–527.

Canham, C.D. & Loucks, O.L. (1984). Catastrophic windthrow in the presettlement forests of Wisconsin. *Ecology*, **65**, 803–809.

Curtis, J. T. (1959). *The Vegetation of Wisconsin.* University of Wisconsin Press, Madison, WI.

Davis, A.M. (1977). The prairie-deciduous forest ecotone in the upper Middle West. *Association of American Geographers Annals*, **67**, 204–213.

Davis, M.B. (1961). Pollen diagrams as evidence of late-glacial climatic change in southern New England. *New York Academy of Science Annals*, **95**, 623–631.

Davis, M.B. (1976). Pleistocene biogeography of temperate deciduous forests. *Geoscience and Man*, **13**, 13–26.

Davis, M.B. (1981a). Quaternary history and the stability of deciduous forests. *Forest Succession* (Ed. by D.C. West, H.H. Shugart & D.B. Botkin), pp. 132–177. Springer-Verlag, New York.

Davis, M.B. (1981b). Outbreaks of forest pathogens in Quaternary history. Vol. 3, pp. 216–227, *Proceedings of the IV International Palynological Conference* Lucknow, India.

Davis, M.B. (1983). Holocene vegetation history of the eastern United States. *Late-Quaternary Environments of the United States*, Vol. 2, *The Holocene* (Ed. by H.E. Wright, Jr), pp. 166–181. University of Minnesota Press, Minneapolis.

Davis, M.B. (1985). History of the vegetation on the Mirror Lake watershed. *An Ecosystem Approach to Aquatic Ecology* (Ed. by G.E. Likens), pp. 53–65. Springer-Verlag, New York.

Davis, M.B. & Botkin, D.B. (1985). Sensitivity of cool-temperate forests and their fossil pollen record to rapid temperature change. *Quaternary Research*, **23**, 327–340.

Davis, M.B. Woods, K.D., Webb, S.L. & Futyma, R.P. (1986). Dispersal versus climate: expansion of *Fagus* and *Tsuga* into the upper Great Lakes Region. *Vegetatio* (in press).

Firbas, F. (1949). *Spät und nacheiszeitliche Waldegeschichte Mitteleuropas nordlich der Alpen*, Vol. 1. Gustav Fisher, Jena.

Fowells, H.A. (1965). *Silvics of forest trees of the United States*. US Dept. of Agriculture, Agricultural Handbook 271, US Govt. Printing Office, Washington, D.C.

Fries, M. (1962). Pollen profiles of late-Pleistocene and recent sediments at Weber Lake, northeastern Minnesota. *Ecology*, **43**, 295–308.

Futyma, R.P. (1982). *Postglacial vegetation of eastern Upper Michigan*. Ph.D. thesis, University of Michigan, Ann Arbor, Michigan.

Gilliam, J.A., Kapp, R.O. & Bogue, R.D. (1967). A post-Wisconsin pollen sequence from Vestaburg Bog, Montcalm County, Michigan. *Michigan Academy of Sciences, Arts and Letters*, **52**, 3–17.

Grimm, E.C. (1983). Chronology and dynamics of vegetation change in the prairie and woodland region of southern Minnesota, USA. *New Phytologist*, **93**, 311–350.

Heide, K.M. (1981). *Late-Quaternary vegetational history of northcentral Wisconsin, USA: estimating forest composition from pollen data*. Ph.D. dissertation, Brown University.

Heinselman, M.L. (1973). Fire in the virgin forests of the Boundary Waters Canoe Area, Minnesota. *Quaternary Research*, **3**, 329–382.

Henry, J.D., & Swan, J.M.A. (1974). Reconstructing forest history from live and dead plant material: an approach to the study of forest succession in southwest New Hampshire. *Ecology* **55**, 772–783.

Huntley, B. & Birks, H.J.B. (1983). *An Atlas of Past and Present Pollen for Europe 0–13,000 Years Ago*. Cambridge University Press, Cambridge.

Iversen, J. (1954). The late-glacial flora of Denmark and its relation to climate and soil. *Geological Survey of Denmark Publication Series II*, No. 80, pp. 87–119.

Iversen, J. (1973). *The Development of Denmark's Nature Since the Last Glacial*. Geological Survey of Denmark, Series V, No. 7-C, Copenhagen.

Johnson, W.C. & Adkisson, C.S. (1985). Dispersal of beech nuts by blue jays in fragmented landscapes. *American Midland Naturalist*, **113**, 319–324.

Kapp, R.O. (1977). Late Pleistocene and post-glacial plant communities of the Great Lakes Region. *Geobotany* (Ed. by R.C. Romans), pp. 1–27. Plenum, New York.

Kapp, R.O., Bushouse, S. & Foster, B. (1969). A contribution to the geology and forest history of Beaver Island, Michigan. *Proceedings of the 12th Conference Great Lakes*

Research 1969. pp. 225–236, International Association for Great Lakes Research, Ann Arbor, Michigan.

Kerfoot, W.C. (1974). Net accumulation rates and the history of cladoceron communities. *Ecology*, **55**, 51–61.

King, J.E. (1981). Late Quaternary vegetational history of Illinois. *Ecological Monographs*, **51**, 43–62.

Lawrenz, R.W. (1975). *The development of Green Lake, Antrim County, Michigan*. M.S. thesis, Central Michigan University, Mount Pleasant, Michigan.

Liu, K-B. (1982). Palynology and paleoecology of the Boreal Forest/Great Lakes–St. Lawrence forest ecotone in northern Ontario. *AMQUA Abstracts* p. 122.

Lorimer, E.G. (1977). The presettlement forest and natural disturbance cycles of northeastern Maine. *Ecology*, **58**, 139–148.

Lorimer, E.G. (1985). Methodological considerations in the analysis of forest disturbance history. *Canadian Journal of Forest Research*, **15**, 200–213.

Maher, L.J., Jr (1982). The palynology of Devil's Lake, Sauk County, Wisconsin. *Quaternary History of the Driftless Area.* Fieldtrip Guide Book 5, Extension Geological and Natural History Survey, University of Wisconsin.

Manny, B.A., Wetzel, R.G. & Bailey, R.E. (1978). Paleolimnological sedimentation of organic carbon, nitrogen, phosphorous, fossil pigments, pollen, and diatoms in a hypereutrophic, hardwater lake: a case history of eutrophication. *Polskie Archiwum Hydrobiologii*, **25**, 243–267.

McAndrews, J.H. (1970). Fossil pollen and our changing landscape and climate. *Rotunda*, **3**, 30–37.

McAndrews, J.H. (1981). Late Quaternary climate of Ontario: temperature trends from the fossil record. *Quaternary Paleoclimate* (Ed. by W.C. Mahaney). Geoabstracts Ltd., Toronto.

McKinley, D. (1960). A history of the passenger pigeon in Missouri. *Auk*, **77**, 399–419.

Okubo, A. (1980). Diffusion and ecological problems: mathematical models. *Biomathematics*, **10**, 1–254.

Peters, A. & Webb, T., III. (1979). A radiocarbon-dated pollen diagram from west-central Wisconsin. *Ecological Society of America Bulletin*, **60**, 102.

Pickett, S.T.A. & White, P.S. (1985). *The Ecology of Natural Disturbance and Patch Dynamics*. Academic Press, Orlando, Florida.

Pijl, L. van der (1969). *Principles of Dispersal in Higher Plants*. Springer-Verlag, New York.

Raup, H.M. (1957). Vegetational adjustment to the instability of the site. *Proceedings of the 6th Technical Meeting*, International Union for Conservation of Nature and Natural Resources, Edinburgh 1956,. pp. 36–48. London.

Schoonmaker, P.K. & Foster, D.R. (1985). Disturbance history and vegetation dynamics in old-growth forests of central new England.*Ecological Society of America Bulletin*, **66**, 266.

Schwartz, M.W. (1985). *A critical investigation of regression techniques and data collection methods to improve estimates of the pollen/tree relationship*. M.S. thesis, University of Minnesota.

Stearns, F.S. (1949). Ninety years change in a northern hardwood forest in Wisconsin. *Ecology*, **30**, 350–358.

Stephens, E.P. (1956). The uprooting of trees: a forest process. *Soil Science Society of America Proceedings*, **20**, 113–116.

Swain, A.M. (1978). Environmental changes during the last 2000 years in north-central Wisconsin: analysis of pollen, charcoal and seeds from varved lake sediments. *Quaternary Research*, **10**, 55–68.

Waddington, J.C.B. (1969). A stratigraphic record of pollen influx to a lake in The Big Woods of Minnesota. *Geological Society of America Special Paper*, **12B**, pp. 263–282.
Walker, D. (1982). Vegetation's fourth dimension. *New Phytologist*, **90**, 419–429.
Ward, R.T. (1961). Some aspects of the regeneration habits of American beech. *Ecology*, **42**, 828–832.
Webb, S.L. (1983). *The Holocene extension of the range of American beech* (Fagus grandifolia) *into Wisconsin: paleoecological evidence for long-distance seed dispersal.* M.S. thesis, University of Minnesota.
Webb, S.L. (1986). Potential role of passenger pigeons and other vertebrates in the rapid Holocene migrations of nut trees. *Quaternary Research* (in press).
Webb, T., III. (1974). A vegetation history from northern Wisconsin: Evidence from modern and fossil pollen. *American Midland Naturalist*, **93**, 12–34.
Webb, T., III. (1982). Temporal resolution in Holocene pollen data. *Third North American Paleontological Convention Proceedings*, **2**, 569–572.
Webb, T. III. (1986). Is the vegetation in equilibrium with climate? An interpretative problem for late-Quaternary pollen data. *Vegetatio* (in press).
Woods, K.D. (1981). *Interstand and intrastand pattern in hemlock-northern hardwood forests.* Ph.D. dissertation, Cornell University.
Wright, H.E., Jr. & Watts, W.A. (1969). Glacial and vegetation history of northeastern Minnesota. *Minnesota Geological Survey Special Paper* **Sp-11**, pp. 1–59.
Wright, H.E., Jr., Winter, J.C. & Patten, H.L. (1963). Two pollen diagrams from southeastern Minnesota : problems in the late- and postglacial vegetational history. *Geological Society of America Bulletin*, **74**, 1371–1396.

19. THE SPATIAL CONTEXT OF REGENERATION IN A NEOTROPICAL FOREST

STEPHEN P. HUBBELL[1]
AND
ROBIN B. FOSTER[2]

[1] *Program in Evolutionary Ecology and Behavior, Department of Zoology, University of Iowa, Iowa City, Iowa 52242 USA, and Smithsonian Tropical Research Institute, Box 2072, Balboa, Republic of Panama; and*
[2] *Department of Botany, Field Museum of Natural History, Chicago, Illinois 60605 USA, and Smithsonian Tropical Research Institute, Box 2072, Balboa, Republic of Panama*

INTRODUCTION

There are few greater biological mysteries than the origin and maintenance of high tree species richness in tropical forests. The classical explanation for this phenomenon is that the great age of the tropics and its benign and stable climate have permitted a slow accumulation of tree species. Under these favourable climatic regimes, natural selection will be dominated by biotic interactions, leading to highly specialized species having unique means for exploiting limiting resources or modes of regeneration. Forests of these ecologically specialized trees are therefore in equilibrium, each species limiting its own growth to a greater extent than the growth of its competitors. The result is a stabilized assemblage of a particular taxonomic set of competitively coevolved species (Ashton 1969).

We have recently questioned whether such competitve niche differentiation is the principal means by which tropical tree species coexist in species-rich tropical forests (Hubbell & Foster 1986b). Many tree species in a Panamanian forest appear to be generalists, broadly overlapping in habitat and regeneration niche characteristics within a few major life-history guilds (Hubbell & Foster 1986c, d). We argued that such generalists are more likely to arise in species-rich communities in which the identity of one's competitors is spatially and temporally uncertain (Hubbell & Foster 1986b). When the selective environment in local biotic neighbourhoods is unpredictable, the evolutionary result of such diffuse competition may be the convergent coevolution of generalist tree species within a few regeneration guilds, rather than pairwise coevolution of specialists in competitive balance. Data on the Panamanian forest indicate that local tree neighbourhoods are indeed very diverse and differ substantially in species composition from one tree to the next (Hubbell & Foster 1986b).

We refer to the classical view as the 'equilibrium' hypothesis because tropical forests are assumed to be competitively stabilized assemblages of particular tree taxa. The view that tropical tree communities are composed mainly of generalist species which are not, or only weakly, stabilized by niche differentiation is one of several 'non-equilibrium' hypotheses for tropical forest organization and dynamics (Connell 1978; Hubbell 1979). It presupposes that tropical tree species within major life-history guilds coexist not in spite of, but because of, being functionally similar generalists, and that there are few forces besides drifting abundance which can eliminate such species once established (Hubbell & Foster 1986b).

Obviously real tropical forests do not conform to either extreme model (Hubbell 1984). Nevertheless, the central question remains: To what extent do biotic forces exist to stabilize particular assemblages of tree species in tropical forests? The spatial pattern of regeneration of tree species in tropical forests should provide important clues to the answer. If the equilibrium view is more nearly correct, then tropical tree species should exhibit strong local self-inhibition in regeneration. When a tree species becomes too common, density-dependent processes should reduce the local per capita reproductive performance of the species relative to less common competing trees in the area. If the non-equilibrium view is closer to the truth, then the correlation between local tree abundance and per capita reproductive performance should be weak or absent.

These ideas are implicit in Aubreville's (1938, 1971) 'mosaic' hypothesis of tropical forest regeneration. Aubreville's hypothesis states that the spatial variation of floristic composition in species-rich tropical forest is maintained by temporal variation in species composition at any given location. He argued that most tree species in the closed-canopy forests of the Ivory Coast appeared unable to regenerate beneath themselves or in their immediate vicinity. Thus, a succession of different species would have to occupy a given site in the forest before the same species could reoccupy the site. This suggests the possibility that predictable successional cycles of mature-forest tree species might occur, or that cyclic or non-transitive tree replacement processes might be involved (Acevedo 1981).

In this chapter we examine the evidence bearing on the mosaic hypothesis for a mapped 0·5 km^2 plot of tropical forest in Panama. The evidence is circumstantial at present rather than direct because it is derived from a single census and map of tree spatial patterns in the forest, not from following tree replacements in the forest through time. The analysis also assumes that replacements for individual mature trees must be drawn from existing saplings beneath the canopy of these trees. While this assumption is probably valid for many shade-tolerant species, it is less true for shade-intolerant pioneers and persistent late-secondary heliophiles. These species usually come into gaps only after they open, and are at much lower densities in the closed canopy forest understorey (Hubbell & Foster 1986a). Consequently, we have limited the present analysis to mature-forest species.

We ask the following general question: What changes in the present composition of the canopy layer of the forest will occur, as predicted from the current populations of juvenile trees in the understorey? If the forest is approximately in equilibrium, we might expect the predicted changes to be relatively slight. On the other hand, if tree populations are undergoing considerable flux, then signs of this change should be evident from the analysis. We can ask the above question in several different ways in order to evaluate the robustness of the result. (i) What is the probability that an individual of tree species *x* will be replaced by an individual of species *y*? (ii) Are these probabilities sensitive to the size-class of sapling considered? (iii) How invariant are these probabilities spatially from one part of the forest to another? (iv) Are some species more likely to replace themselves in the canopy than others?

In this chapter we focus on the replacement of only the most common mature-phase overstorey or midstorey tree species. It is particularly critical that self-limitation and stable replacement probabilities be present among the most common species for the community equilibrium argument to be tenable. If a majority of common species does not exhibit density dependence or constant species replacement probabilities, then successional patterns will not be predictable, and non-equilibrium hypotheses will be suggested.

STUDY SITE AND METHODS

The study was conducted in tropical moist forest on Barro Colorado Island (BCI), Panama. BCI is a 15 km^2 artificial island, created by damming the Rio Chagres to form the Panama Canal. BCI has a seasonal climate (Leigh, Rand & Windsor 1982), and the flora is very well known (Croat 1978). Half of the island is covered with old-growth mature forest which, from palaeoecological studies, is known to be at least 500 years old (Piperno, pers. comm.; Hubbell & Foster 1986a). In 1980 a permanent 50 ha plot was established in old-growth forest (Hubbell & Foster 1983); and by 1982 a complete census and map of all free-standing woody plants in the plot over 1 cm diameter at breast height (dbh) had been finished. Beginning in 1983, and continuing annually thereafter, a detailed map of canopy height, layering, and canopy gaps has also been made (Hubbell & Foster 1986c). The plot contains over 300 tree and shrub species in censusable size-classes, and over 235 000 individuals. A great range in relative species abundance was found, from twenty-five species each represented by a single plant, to one shrub species having nearly 40 000 individuals (Hubbell & Foster 1986d).

We conducted two types of analyses. In the first analysis, saplings of diameter classes 1–2 cm, 2–4 cm, and 4–8 cm dbh were tallied by species beneath individual canopy trees. The assignment of saplings was to the nearest tree over 30 cm dbh; in effect the plot was tesselated into polygons containing one tree over 30 cm dbh and its associated saplings. We then calculated probabilities of tree replacement according to two hypotheses of mechanism for the ten most common species. The remaining species were assigned to a 'black box' category for these analyses. From these replacement probabilities, the predicted equilibrium vector of species abundances was calculated and compared with existing species abundances.

The two hypotheses we constructed assume either a species effect but no site effect (Hypothesis 1), or a site effect but no species effect (Hypothesis 2). Horn's (1975) Markovian hypothesis is of the first type, namely that there is a species effect but no site effect. Thus, the proportion of saplings of species i beneath all adults of species j is taken as the estimate of the probability of replacement of species j by species i. For the second hypothesis (a site effect but no species effect), we divided the plot into eight subplots 250 m on a side, and tallied the number of saplings of each species. Here the proportion of each species among the saplings in the subplot was taken as the estimate of the replacement probability, regardless of which species in the canopy the saplings were beneath. After considering the two hypotheses separately, we combined them into one hypothesis asserting that both site and species effects exist. In this case, we considered not only the subplot but also the species beneath which the saplings were found.

The second type of analysis is to examine the pattern of dispersion of saplings around adults of conspecific and heterospecific trees. We can determine whether there is a relative scarcity of saplings in the vicinity of adults of the same or different species using a simple new dispersion statistic (Hamill & Wright 1986). The basis of the statistic is to calculate the fractional area of the plot which lies within radial distance r of any adult tree. This gives the expected proportion of saplings under the null hypothesis of no effect of adult location on sapling placement. As distance r is increased, the cumulative probability density function for the expected distribution can be plotted. This distribution can be compared with the observed sapling distribution and the differences evaluated with a Komolgorov-Smirnov test. We performed these tests for saplings 1–2 cm dbh within and among the five most common overstorey and midstorey tree species. The tests were performed between species to evaluate whether potential sapling avoidance of adults was truly a conspecific effect or simply the result of avoiding any large tree.

RESULTS

Tree replacement probabilities

The ten common species included in the analysis constitute 43·7% of the 4216 trees of mature-forest species over 30 cm dbh in the canopy layer of the BCI plot (Table 19.1). The replacement probabilities—under Horn's hypothesis of a species effect but no site effect (Hypothesis 1)—have been calculated separately for the sapling size-classes 1–2, 2–4, and 4–8 cm dbh, respectively (Tables 19.2–19.4). The rows are the original species of canopy tree, and the columns are the replacement species.

Trees in the 'other species' category, AAZZ, collectively have a high probability of replacing any of the ten common species, but much lower replacement probabilities when considered individually. Note also that the 'other species' collectively have a higher probability of replacing any of the ten common species than they have of replacing themselves, for all three sapling size classes. For the rest of the discussion we focus on the replacements only among the ten common species of Table 19.1. There are several immediate conclusions which can be drawn from these probabilities.

(a) In nearly all cases it is at least possible to replace any of the ten species with any of the other species in one step. There are a few zero entries in the table of probabilities derived from the 4–8 cm dbh saplings (Table 19.4), but these are probably due to low sample sizes.

(b) Some species are much more likely to replace a given species than others. For example, *Trichilia* is 3–10 times more likely to replace *Protium* than is *Guatteria* (Tables 19.2–19.4).

(c) Self-replacement is possible in all species. Moreover, a Mann-Whitney test reveals that the probabilities of self-replacement are not significantly lower than the probabilities of heterospecific replacement,

TABLE 19.1. Ten common BCI overstorey and midstorey tree species used in the analysis of tree replacement. Abundances are listed for all size-classes over 1 cm dbh and for trees over 30 cm dbh. Acronyms are used in later tables. The acronym AAZZ refers to all remaining mature-forest overstorey and midstorey tree species not list individually below

Species	Acronym	Family	Total abundance	Abundance Over 30 cm dbh
Alseis blackiana	ALSB	Rubiaceae	7563	279
Beilschmiedia pendula	BEIP	Lauraceae	2384	127
Guatteria dumetorum	GUAD	Annonaceae	1571	81
Poulsenia armata	POUA	Moraceae	3433	97
Prioria copaifera	PRIC	Leguminosae	1358	118
Protium tenuifolium	PROT	Burseraceae	2662	67
Quararibea asterolepis	QUAI	Bombacaceae	2401	388
Tetragastris panamensis	TET2	Burseraceae	3229	118
Trichilia tuberculata	TRI3	Meliaceae	12 932	538
Virola sebifera	VIRI	Myristicaceae	2404	30
Other species	AAZZ	------------	60 141	2373

TABLE 19.2. Matrix of replacement probabilities (expressed as percentages) for ten common overstorey and midstorey BCI tree species, based on saplings in the 1–2 cm dbh size-class. The acronym AAZZ is for all other species. Rows are the species being replaced; columns are the replacing species. Underlined entries are for probabilities of self-replacement

	ALSB	BEIP	GUAD	POUA	PRIC	PROT	QUAI	TET2	TRI3	VIRI	AAZZ
ALSB	4·80	2·13	0·64	0·89	1·44	3·47	1·44	4·95	15·45	1·98	62·80
BEIP	8·85	5·57	0·76	1·53	0·31	3·43	0·61	4·65	7·25	4·50	62·55
GUAD	7·48	4·38	1·64	2·37	0·55	4·02	1·64	3·83	13·32	2·56	58·21
POUA	6·35	5·10	1·77	2·50	0·83	5·20	1·25	3·75	9·68	2·71	60·87
PRIC	9·95	2·56	1·46	1·19	2·01	2·83	1·64	7·31	8·31	1·74	61·01
PROT	4·46	3·54	1·05	2·62	0·66	8·51	1·18	1·97	10·35	2·10	63·57
QUAI	7·44	2·04	0·98	1·32	1·17	4·08	1·77	3·65	17·60	1·29	58·66
TET2	8·89	3·39	1·36	1·78	1·36	3·40	0·85	3·05	10·33	2·88	64·52
TRI3	10·73	2·63	0·79	1·55	1·40	3·40	1·10	5·04	3·43	2·23	67·43
VIRI	8·05	4·60	1·53	1·92	1·15	1·15	0·38	4·60	9·58	0·77	66·28
AAZZ	12·00	3·47	1·07	1·40	0·82	3·03	1·41	4·28	12·36	2·04	58·13

for any sapling size-class. The critical value for the rank sum statistic for $P = 0{\cdot}05$ is 593 (ignoring the 'other species' replacements in the calculations). For each size-class, the observed rank sum statistic and its P value are: for 1–2 cm dbh saplings, $S = 480$ ($P < 0{\cdot}4$); for 2–4 cm dbh saplings, $S = 528$ ($P < 0{\cdot}2$); for 4–8 cm dbh saplings, $S = 543$ ($P < 0.15$).

(d) Although self-replacement is possible, the probabilities of self-replacement, or of any particular species replacement, are generally low.

TABLE 19.3. Matrix of replacement probabilities (expressed as percentages) for ten common overstorey and midstorey BCI tree species, based on saplings in the 2–4 cm dbh size-class. The acronym AAZZ is for all other species. Rows are the species being replaced; columns are the replacing species. Underlined entries are probabilities of self-replacement

	ALSB	BEIP	GUAD	POUA	PRIC	PROT	QUAI	TET2	TRI3	VIRI	AAZZ
ALSB	2·36	1·48	0·65	1·65	2·01	3·07	2·89	3·78	18·35	1·77	62·01
BEIP	5·65	1·60	2·46	3·84	0·85	3·31	1·17	4·27	10·89	2·88	63·07
GUAD	4·82	3·90	1·84	4·59	2·29	2·52	2·75	4·36	11·47	2·06	61·47
POUA	2·79	2·37	2·24	6·84	1·54	3·35	1·54	2·93	8·52	1·68	66·20
PRIC	6·93	1·61	1·11	2·23	1·61	1·36	2·10	5·20	13·00	1·73	63·12
PROT	3·72	0·85	2·70	6·59	1·52	3·55	2·20	2·87	8·95	3·04	64·02
QUAI	5·50	1·36	1·25	2·53	9·53	3·30	1·98	2·31	20·02	1·03	59·77
TET2	5·62	2·27	1·44	3·59	2·51	1·44	2·03	1·56	14·35	3·11	62·08
TRI3	7·28	2·20	1·24	2·10	1·97	2·04	1·97	4·14	5·34	1·54	70·19
VIRI	6·37	2·94	1·96	0·98	0·98	1·47	1·47	4·41	9·31	0·49	69·61
AAZZ	8·41	2·12	1·70	2·81	1·24	2·57	2·39	3·07	15·46	1·87	58·37

TABLE 19.4. Matrix of replacement probabilities (expressed as percentages) for ten common overstorey and midstorey BCI tree species, based on saplings in the 4–8 cm dbh size-class. The acronym AAZZ is for all other species. Rows are the species being replaced; columns are the replacing species. Underlined entries are probabilities of self-replacement

	ALSB	BEIP	GUAD	POUA	PRIC	PROT	QUAI	TET2	TRI3	VIRI	AAZZ
ALSB	3·12	1·61	1·13	3·12	1·70	1·42	3·50	1·61	15·71	1·80	65·28
BEIP	2·09	1·90	3·80	6·46	1·14	1·52	9·51	3·23	10·65	4·18	64·07
GUAD	5·19	1·85	0·37	5·56	1·85	2·22	4·82	2·22	6·30	4·82	64·82
POUA	2·17	2·89	1·69	11·08	0·96	1·69	2·41	2·17	8·43	2·65	63·86
PRIC	2·73	2·73	2·31	5·45	1·05	0·84	2·31	2·73	13·63	2·10	64·15
PROT	1·70	2·04	2·04	12·59	0·34	1·36	2·38	0·68	9·18	2·72	64·97
QUAI	3·77	1·82	1·95	5·18	1·68	2·15	2·22	2·02	17·77	1·48	59·96
TET2	2·99	1·59	2·59	5·78	2·79	7·97	2·19	0·00	13·55	3·79	63·94
TRI3	4·61	1·54	2·79	3·47	1·94	1·71	2·45	3·02	7·29	2·68	68·53
VIRI	1·65	0·00	4·13	3·31	2·48	0·00	0·00	2·48	7·44	2·48	76·03
AAZZ	5·39	1·43	2·14	5·03	1·51	1·68	2·95	1·86	13·14	2·34	62·54

In three species, *Trichilia*, *Alseis* and *Virola*, the probabilities of self-replacement are the lowest or second lowest of the ten replacement probabilities in at least two sapling size-classes.

(e) Because of these low probabilities of self-replacement, trees of several different species will usually occur between successive trees of the same species occupying a given site. This is true for even the most common species. From the matrix of replacement probabilities, we can calculate, for example, that the expected number of trees between successive *Trichilia* trees in the same spot is between 13·8 (estimate from 1–2 cm dbh saplings) and 29·2 (estimate from 4–8 cm dbh saplings). For a somewhat less common species such as *Prioria*, the number of intervening trees is greater (estimates of 49·8 trees for 1–2 cm dbh saplings to 95·2 trees for 4–8 cm dbh saplings).

(f) The generally low probabilities of pairwise species replacement (most are well below 0·1) means that the predictability of successional sequences of tree species in a given site will also be quite low. In spite of this, a few species pairs exhibit a weak tendency toward cyclical replacement. For example, *Alseis* is three times more likely to be replaced by a *Trichilia* than by itself, whereas *Trichilia* is about three times more likely to be replaced by *Alseis* than by itself (Table 19.2).

(g) Sapling size-class has a measurable effect on estimates of tree replacement probabilities, sometimes large. The estimate of the probability of self-replacement for *Poulsenia*, for example, increases from 2·50% for 1–2 cm dbh saplings (Table 19.2), to 6·84% for 2–4 cm dbh saplings (Table 19.3), and further to 11·08% for 4–8 cm dbh saplings (Table 19.4).

We can use these replacement probabilities to calculate an equilibrium relative abundance for the tree species in question, assuming that the replacement probabilities remain unchanged (Horn 1975). Because the matrix of replacement probabilities is different for the three sapling size-classes, the predicted equilibrium community is also different in each case (Fig. 19.1). The replacement probabilities derived from the 2–4 cm dbh size-class of saplings most closely predict the present community composition, as measured by mean percentage error per species. Whichever prediction is taken, however, the community is expected to change. For some species the indicated direction of change is the same regardless of sapling size-class used to make the projection. For example, *Quararibea* is expected to decline sharply in abundance, whereas *Virola* should increase. The direction of change is uncertain in several other species because of conflicting upward or downward predictions from the different sapling size-classes.

Thus far we have discussed Hypothesis 1, namely that there is a species effect but no site effect. Hypothesis 2, that there is a site effect but no species effect, requires a different analytical approach. Here we simply asked, for the eight 250 × 250 m subplots, what is the projected proportion of a species in the canopy of the subplot after all trees in the canopy are replaced once? The projection is based on the relative proportion of saplings in the subplot without regard to the species of tree beneath which the saplings are found. Site does make a difference to projected relative abundance, but more of a difference for some tree species than others (Table 19.5). Coefficients of variation (S.D./mean) over subplots range from a low of 0·296 for *Virola* (2–4 cm dbh saplings) to a high of 1·232 for *Prioria* (2–4 cm dbh saplings). The projections for changing relative

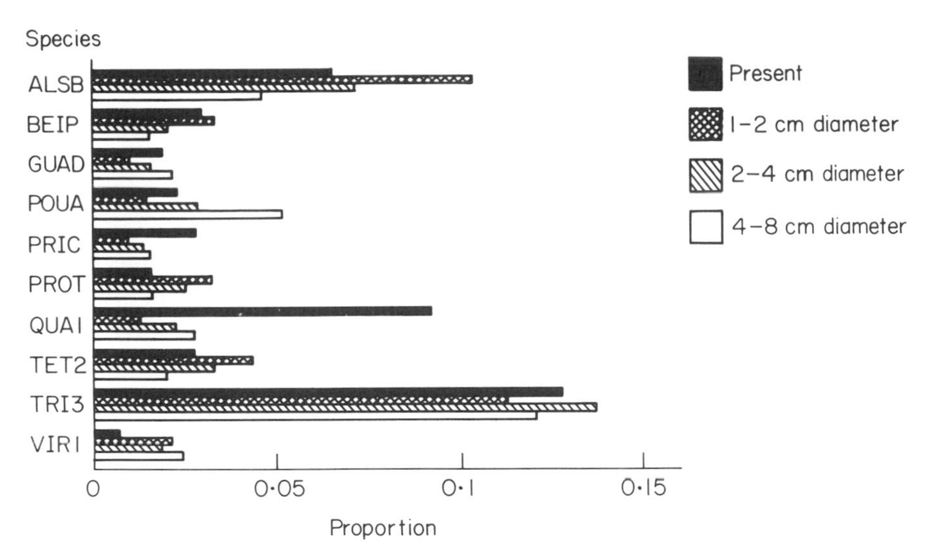

FIG. 19.1. Present and equilibrium relative species abundances of ten common overstorey and midstorey tree species. Present relative abundances are shown as solid bars. Equilibrium abundances are predicted from species replacement probabilities for three sapling size-classes: 1–2 cm dbh (cross-hatched bars), 2–4 cm dbh (single-hatched bars), and 4–8 cm dbh (open bars). *Quararibea* and *Prioria* are expected to decrease in abundance, whereas *Virola* is expected to increase, regardless of the sapling size-class used to make the projection. The most abundant overstorey tree species, *Trichilia*, is expected to remain about the same in abundance. Expected changes in the remaining species are less certain due to conflicting predictions from the different sapling size-classes.

TABLE 19.5. Spatial variation in present and projected relative species abundance for ten common overstorey and midstorey BCI tree species. Relative abundances are reported as percentages. Means and standard deviations are computed over the eight 250 × 250 m subplots

			Predicted					
	Present		1–2 cm dbh		2–4 cm dbh		4–8 cm dbh	
Species	Mean	S.D.	Mean	S.D.	Mean	S.D.	Mean	S.D.
ALSB	6·61	1·82	10·81	4·03	7·29	3·14	4·60	2·65
BEIP	2·99	2·34	3·35	2·27	2·03	1·18	1·54	0·87
GUAD	1·89	0·66	1·04	0·46	1·61	0·56	2·17	0·90
POUA	2·26	1·68	1·42	0·49	2·79	1·18	5·05	2·21
PRIC	3·02	4·59	1·07	1·55	1·43	1·76	1·57	1·35
PROT	1·56	0·97	3·03	2·22	2·46	1·73	1·65	0·87
QUA1	9·04	4·46	1·37	0·60	2·24	0·83	2·74	1·12
TET2	2·85	1·47	4·41	1·69	3·31	1·50	2·02	1·13
TRI3	12·87	4·89	11·41	5·68	14·11	8·94	12·64	7·78
VIR1	0·71	0·41	2·16	1·10	1·82	0·54	2·40	0·78

species abundance are similar to those made by Hypothesis 1. For example, *Quararibea* is expected to decline, and *Virola* to increase in abundance under both hypotheses.

These results suggest that the probabilities of species replacement are not homogeneous from one part of the plot to the next, namely, that both canopy species and site factors affect which tree species will next occupy a given place in the forest. We therefore constructed a composite hypothesis with both species and site effects. Sample size imposed limits on the accuracy with which we could estimate the species replacement probabilities, so we computed the complete matrices of probabilities only for the two 25 ha halves of the plot, rather than for each of the eight 250 × 250 m subplots separately. When the equilibrium relative tree species abundances are computed from these matrices, the results confirm the spatial inhomogeneity of tree replacement processes in the BCI forest (Table 19.6). For example, the commonest overstorey tree, *Trichilia,* fares much worse in the west half of the plot than in the east half, regardless of which sapling size-class is used to make the projection. Conversely, *Virola* is expected to increase in relative abundance in all parts of the plot, but to increase more markedly in the west than in the east (Table 19.6).

Relative dispersion patterns

The second type of analysis is to examine the dispersion pattern of juveniles around large conspecific or heterospecific trees. If BCI tree species have a strong inhibitory effect on their own recruitment, this may manifest itself in a depression of sapling numbers in the vicinity of adults. If such an effect is detected, then it must be determined whether the effect is intraspecific, or whether the saplings are depressed in the shade or root competition of any large tree. In the first case we have circumstantial evidence for the kind of density-dependence required for equilibrium; in the second case, we do not.

Of the five species examined, the two most common species show significant sapling avoidance of conspecific adults. *Trichilia* (Fig. 19.2) and *Alseis* (Fig. 19.3) both show a significant deficit of 1–2 cm dbh saplings in the vicinity of adults. In the figures, the *y*-axis is the cumulative probability of encountering saplings in the plot as a function of distance to the nearest adult in metres (*x*-axis). The solid line is the expected cumulative curve under the null model of no effect of adults on sapling dispersion. The dotted line is the observed cumulative curve. Stars at 5 m intervals along the observed curve indicate distances at which departures of the observed from the expected curve exceed the critical difference by the Komolgorov-Smirnov test at the $P = 0{\cdot}05$ level.

TABLE 19.6. The effects of species, site, and sapling size-class on the predicted equilibrium relative species abundance of ten common overstorey and midstorey BCI tree species. East and west halves of the 50 ha plot are compared for site effect. Equilibrium relative abundances (reported as percentages) are calculated from the replacement probability matrices for each half of the plot, and for each sapling size-class separately

	Predicted equilibrium relative abundance					
	1–2 cm dbh		2–4 cm dbh		4–8 cm dbh	
Species	East	West	East	West	East	West
ALSB	9·96	10·63	6·56	7·55	5·12	4·30
BEIP	2·45	4·44	1·30	2·93	1·06	2·02
GUAD	0·95	1·15	1·25	1·95	1·77	2·62
POUA	1·46	1·41	3·40	2·37	5·37	4·96
PRIC	0·23	1·76	0·57	2·34	0·57	1·00
PROT	4·52	1·75	3·12	1·75	2·02	1·16
QUAI	1·58	0·96	2·68	1·76	3·39	1·99
TET2	4·15	4·71	2·66	4·06	1·70	2·23
TRI3	14·13	7·88	17·57	9·07	15·85	7·88
VIRI	1·74	2·62	1·63	2·15	2·08	2·97

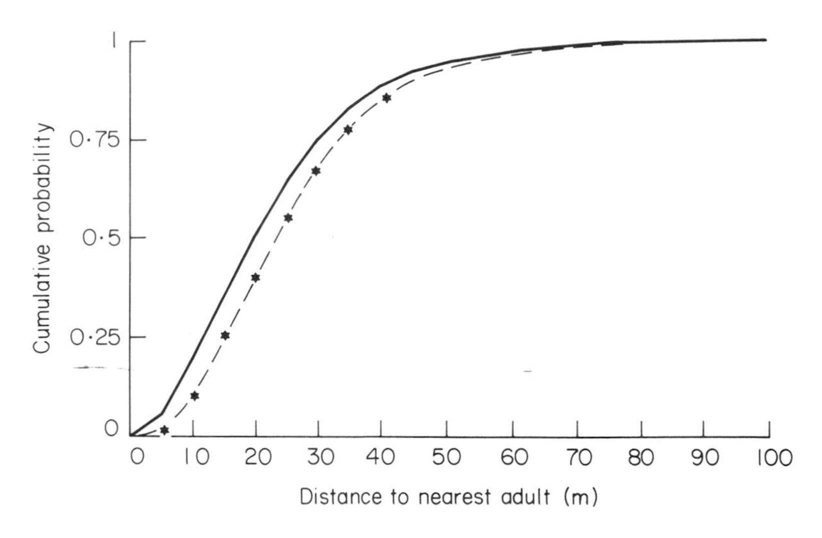

FIG. 19.2. Dispersion of 1–2 cm saplings of *Trichilia* in relation to *Trichilia* adult trees in the east half BCI plot. The solid curve is the expected cumulative probability of *Trichilia* saplings under the null hypothesis of no effect of adults on sapling dispersion, as a function of distance to the nearest adult in meters. The dotted line is the observed sapling curve of cumulative probability. The stars indicate test points (every 5 m) where the observed and expected curves differ significantly by a Komolgorov-Smirnov test at the $P = 0{\cdot}05$ level. The position of the observed curve indicates sapling avoidance of adults. The results in the west half were similar.

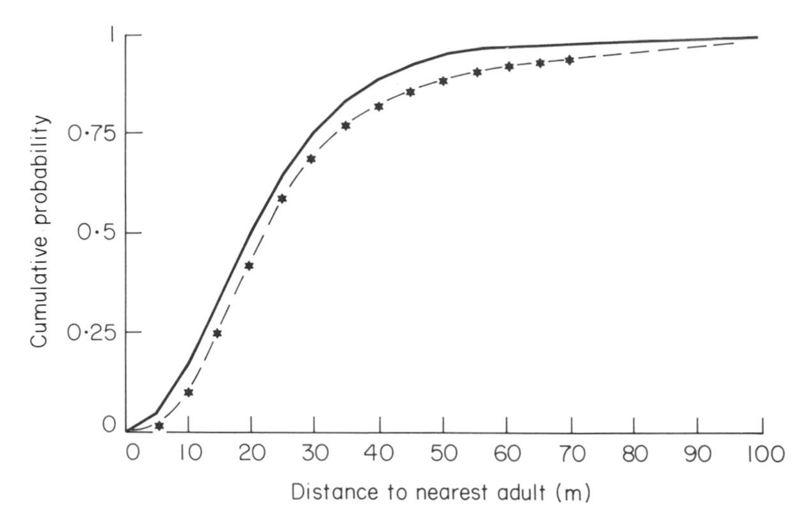

FIG. 19.3. Dispersion of 1–2 cm saplings of *Alseis* in relation to *Alseis* adult trees in the east half BCI plot. The results indicate significant avoidance by *Alseis* saplings of *Alseis* adults. The results in the west half of the plot were similar. See the legend to Fig. 19.2 for further explanation.

The fact that the observed curves lie significantly below and to the right of the expected curves for *Trichilia* and *Alseis* indicates that adults of these species have a negative effect on their own saplings. At first sight it appears that the negative effect persists to great distances from the adults, but this is not the case. The negative effect is a local phenomenon around adults, and the displacement in curves, due to this local effect, is passively propagated to greater distances from adults, where the saplings follow the null distribution. The curve represents cumulative probability, and it can be seen that the observed curve parallels the expected curve at distances greater than 10 m from adults.

The question now arises whether this effect is properly considered to be an intraspecific phenomenon. If we examine *Trichilia* juveniles in relation to *Alseis* adults, the negative effect disappears. In the east half of the plot, the observed distribution lies right on top of the expected distribution (Fig. 19.4). In the west half, *Trichilia* saplings are actually positively associated with *Alseis* adults. Thus, for *Trichilia* the effect is clearly an intraspecific one, at least when tested with one other species. However, when we examine *Alseis* juveniles in relation to *Trichilia* adults, we find a negative effect (Fig. 19.5)—not quite as strong as the intraspecific effect, but nevertheless significant.

Trichilia and *Alseis* are the two most common overstorey tree species in

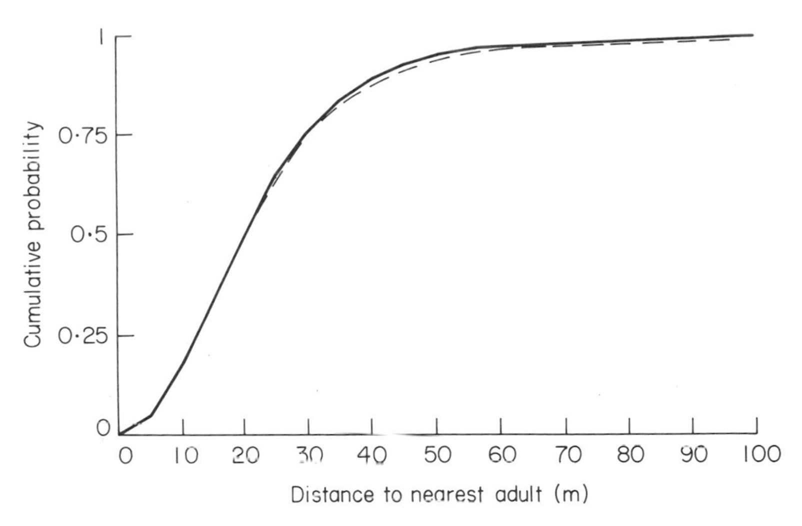

FIG. 19.4. Dispersion of 1–2 cm saplings of *Trichilia* in relation to *Alseis* adult trees in the east half BCI plot. *Trichilia* saplings show no avoidance of or aggregation with *Alseis* adults. The results differed in the west half of the plot, where *Trichilia* saplings were clumped around *Alseis* adults. See the legend to Fig. 19.2 for further explanation.

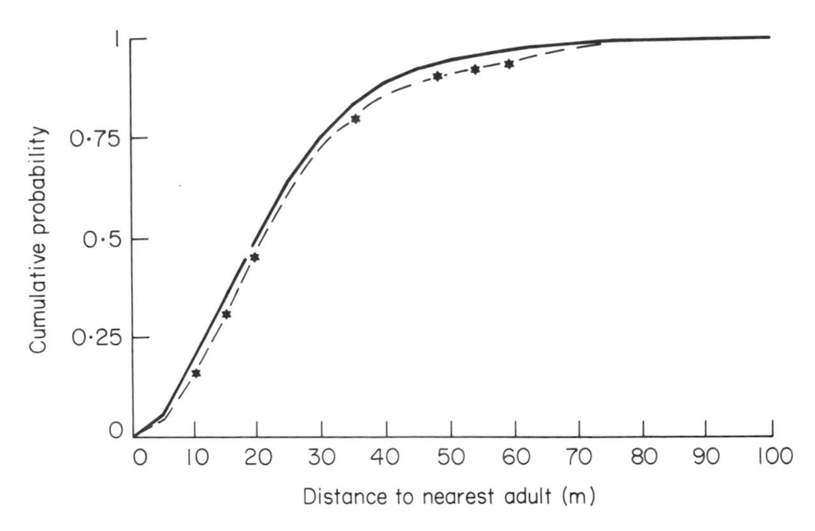

FIG. 19.5. Dispersion of 1–2 cm saplings of *Alseis* in relation to *Trichilia* adult trees in the east half BCI plot. The results show that *Alseis* saplings avoid *Trichilia* adults. The results differed in the west half of the plot, where *Alseis* saplings were neither spaced from nor clumped around *Trichilia* adults. See the legend to Fig. 19.2 for further explanation.

the BCI forest, and the next question is whether similar signs of self-inhibition are to be found in somewhat less common species. The complete results of all twenty-five reciprocal tests for the five most abundant species are presented in Table 19.7. Tests were done in each case on both the east and west halves of the plot to determine whether the results were spatially robust.

The answers for the next three most abundant tree species are different from those for *Trichilia* and *Alseis* (Table 19.7). *Tetragastris* showed no significant conspecific effects within 15m of adults. Both *Quararibea* and *Poulsenia* (Fig. 19.6) exhibited significant conspecific adult effects on juvenile dispersion, but the effects were positive, not negative. Several species exhibited strong positive aggregation of their saplings around the adults of other species, notably *Tetragastris* beneath *Trichilia* and *Quararibea*, and *Trichilia* beneath *Quararibea* and *Alseis*. Several species also showed negative interspecific associations in at least half of the plot. *Alseis* was particularly susceptible to negative interspecific effects, showing up in three out of the four heterospecific tests. In contrast, *Quararibea* saplings showed no significant positive or negative associations with adults of any other species.

These results support the conclusion from the first analysis that tree replacement processes are not homogeneous in the two halves of the plot. In eight out of the twenty-five tests, the results of the significance tests were different in the east and west halves, usually with one side showing significant avoidance or aggregation, and the other side showing non-significance. In at least one case, *Tetragastris* saplings vs. *Quararibea* adults, the results were significant in opposite directions, with avoidance in the east and aggregation in the west.

The two species showing the least sapling response to the adults of other species are *Quararibea* and *Poulsenia* (Table 19.7). Of the five species, these are the two most shade-tolerant species (Hubbell & Foster unpubl.). The species with consistently the highest sapling avoidance of adults of other species is *Alseis*. *Alseis* is the least shade-tolerant of the five species (Hubbell & Foster unpubl.), suggesting that saplings of this species are mainly avoiding the shade or root competition of any large tree, whether conspecific or not. On the other hand, *Trichilia* saplings are clearly avoiding large *Trichilia* trees; the mechanistic basis for this strong conspecific effect merits further investigation.

DISCUSSION

These results suggest that we can expect moderate change to occur in relative tree species abundance patterns in the BCI forest in the future.

TABLE 19.7. Patterns of sapling dispersion relative to large trees of the same and of different species, for the five most common overstorey and midstorey BCI tree species. Columns are the species of sapling; rows are the species of large tree. Entries in the table indicate whether significant departures from the null hypothesis exist at the $P = 0{\cdot}05$ level within 15 m of large trees. A negative (−) sign indicates sapling avoidance of large trees; a positive (+) sign indicates aggregation of saplings with adults. A zero (0) indicates no significant deviation from expected numbers within 15 m of large trees. Significance tests for dispersion pattern were performed separately for the east and west halves of the 50 ha plot, and the double entries in the table refer to the result of the test in the east and west halves, respectively

			Juvenile		
Large tree	*Trichilia*	*Alseis*	*Tetragastris*	*Quararibea*	*Poulsenia*
Trichilia	−/−	−/0	+/+	0/0	0/0
Alseis	0/+	−/−	−/0	0/0	−/0
Tetragastris	0/0	0/0	0/0	0/0	0/0
Quararibea	+/+	−/0	−/+	0/+	0/0
Poulsenia	0/0	−/0	0/0	0/0	+/+

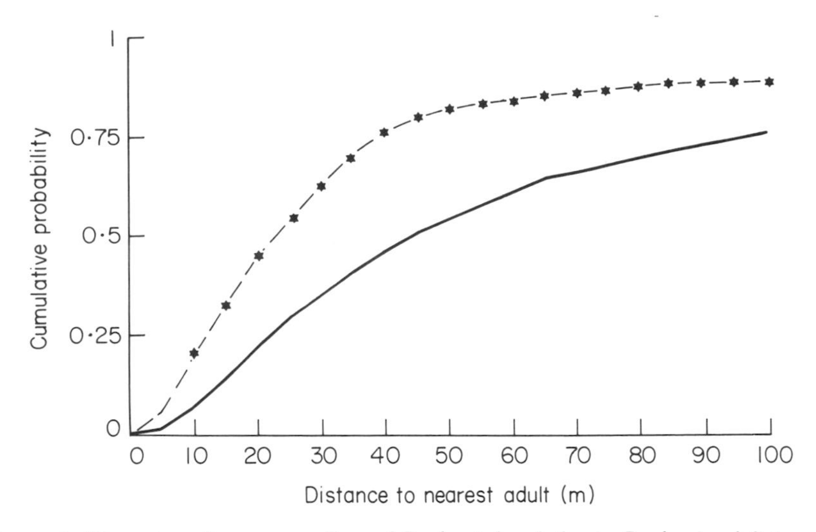

FIG. 19.6. Dispersion of 1–2 cm saplings of *Poulsenia* in relation to *Poulsenia* adult trees in the east half BCI plot. The results show that *Poulsenia* saplings are significantly clumped around conspecific adults. The results in the west half of the plot were similar. See the legend to Fig. 19.2 for further explanation.

Both analyses—probabilities of tree replacement and patterns of juvenile dispersion in relation to adult trees—come to the same general conclusion, that the BCI forest is not in equilibrium. They demonstrate that tree replacement processes are inhomogeneous spatially as well as temporally, as may be inferred from the varying results for the different subplots and sapling size-classes. Therefore, the present data taken as a whole are difficult to reconcile with the equilibrium view.

Nevertheless, certain results appear to support the equilibrium hypothesis. First, the 'other species' collectively were slightly more likely to replace any of the ten common species than they were to replace themselves. This suggests the possibility of cyclical replacement and self-inhibition. Second, the two most common overstorey tree species, *Trichilia tuberculata* and *Alseis blackiana*, gave strong evidence of self-inhibition of saplings in the vicinity of adults. Whether this phenomenon is due to differential mortality of saplings, to dispersal phenomena, or both cannot be ascertained from present data. The effect was more strongly an intra-specific phenomenon in *Trichilia* than in *Alseis*, which showed sapling avoidance of large trees in three of the remaining four common species tested. These results are supported by other data which show that *Trichilia* exhibits strong negative density dependence (Hubbell & Foster 1986b), and *Alseis* to a lesser degree (Hubbell & Foster, unpubl.).

However, in the remaining common species in the present study, there is no evidence of self-inhibition, and in two there is significant positive association of saplings and adults. Based on these and additional data, we have concluded that most of the BCI tree species, with the exception of *Trichilia* and *Alseis*, are at densities far below their respective carrying capacities, or at least below densities at which strong self-inhibition would be detectable (Hubbell & Foster 1986b). Our expectation therefore is that many BCI tree species followed through time will exhibit population fluctuations which, to all intents and purposes, are density independent. The conclusion that many species have non-stationary populations is supported by data showing radically different population size structures from one part of the plot to another in many species (Hubbell & Foster 1986c).

Aubreville's mosaic hypothesis receives only superficial support from these results. Because of the low probabilities of self-replacement, many different tree species will occupy a given site before it is reoccupied by the same species. However, the results show that as a class, probabilities of self-replacement are not statistically lower than the replacement probabilities for any pair of species. The universally low probabilities of pairwise species replacement guarantee that the local succession of tree species in any given spot in the forest will be unpredictable. The result that virtually any tree can replace any other in the BCI forest supports the view that many of the tree species in the forest are generalists, not specialists (Hubbell & Foster 1986b, d). Therefore, Aubreville's mosaic appears to arise primarily from a simple dilution effect due to the high diversity of generalist tree species in the BCI forest, and not from non-transitive tree replacement processes.

ACKNOWLEDGMENTS

We are grateful for the financial support of the National Science Foundation (US), the World Wildlife Fund, the Scholarly Studies Program of the Smithsonian, and grants from many other private donors. We also must acknowledge the help of more than seventy student assistants from the US, Panama, and several other Latin American countries who have participated in the research. We thank Brit Minor and David Hamill for computer assistance.

REFERENCES

Acevedo, M. (1981). On Horn's model of forest dynamics with particular reference to tropical forests. *Theoretical Population Biology*, **19**, 230–250.

Ashton, P.S. (1969). Speciation among tropical forest trees: some deductions in the light of recent evidence. *Biological Journal of the Linnean Society of London*, **1**, 155–196.

Aubreville, A. (1938). La foret coloniale: les forets de l'Afrique Occidentale Francais. *Ann. Acad. Sci. Colon.* (Paris), **9**, 1–245.

Aubreville, A. (1971). Regenerative patterns in the closed forest of the Ivory Coast. *World Vegetation Types* (Ed. by S.R. Eyre), pp. 41–55. Columbia University Press, New York.

Connell, J.H. (1978). Diversity in tropical rain forests and coral reefs. *Science*, **199**, 1302–1310.

Croat, T.R. (1978). *Flora of Barro Colorado Island*. Stanford University Press, Stanford, California.

Hamill, D.N. & Wright, S.J. (1986). Testing the dispersion of juveniles relative to adults: A new analytic method. *Ecology* (in press).

Horn, H.S. (1975). Markovian processes of forest succession. *Ecology and Evolution of communities* (Ed. by M.L. Cody & J.M. Diamond), pp. 196–211. Belknap Press, Cambridge, Massachusetts.

Hubbell, S.P. (1979). Tree dispersion, abundance, and diversity in a tropical dry forest. *Science*, **203**, 1299–1309.

Hubbell, S.P. (1984). Methodologies for the study of the origin and maintenance of tree diversity in tropical rainforest. *Biology International* (IUBS), Special Issue No. 6. (Ed. by G. Maury-Lechon, M. Hadley, & T. Younes), pp. 8–13.

Hubbell, S.P. & Foster, R.B. (1983). Diversity of canopy trees in a neotropical forest and implications for conservation. *Tropical Rain Forest: Ecology and Management*. (Ed. by S.L. Sutton, T.C. Whitmore & A.C Chadwick), pp. 25–41. Blackwell Scientific Publications, Oxford.

Hubbell, S.P. & Foster, R.B. (1986a). Canopy gaps and the dynamics of a neotropical forest. *Plant Ecology* (Ed. by M.J. Crawley), Ch.3. Blackwell Scientific Publications, Oxford.

Hubbell, S.P. & Foster, R.B. (1986b). Biology, chance, and history and the structure of tropical rain forest tree communities. *Community Ecology*. (Ed. by T.J. Case & J. Diamond), pp. 314–329. Harper and Row, New York.

Hubbell, S.P. & Foster, R.B. (1986c). La estructura en gran escala de un bosque neotropical. *Revista de Biología Tropical* (in press).

Hubbell, S.P. & Foster, R.B. (1986d). Commoness and rarity in a neotropical forest: implications for tropical tree conservation. *Conservation Biology* (Ed. by M. Soulé), pp. 205–231. Sinauer Associates, Sunderland, Massachussetts.

Leigh, E.G., Jr, Rand, A.S. & Windsor, D.M. (Eds) **(1982).** *The Ecology of a Tropical Forest: Seasonal Rhythms and Long-Term Changes*. Smithsonian Institution Press, Washington, D.C.

20. DOMINANT AND SUBORDINATE COMPONENTS OF PLANT COMMUNITIES: IMPLICATIONS FOR SUCCESSION, STABILITY AND DIVERSITY

J.P. GRIME
Unit of Comparative Plant Ecology (NERC), Department of Botany, The University, Sheffield S10 2TN

INTRODUCTION

An accepted research tactic in the analysis of plant and animal communities is to attempt to identify those differences between component species and populations which are likely to minimize competition and permit coexistence. Implicit in this approach is the assumption that resources and/or opportunities for regeneration exist in various forms and in a range of spatial and temporal patterns which are exploited in complementary ways by the constituent species.

In plant communities (at least) this conventional view of niche differentiation and exploitation must be enlarged to accommodate the fact that many communities harbour potential dominants with the capacity to monopolize opportunities for resource capture and regeneration. In all except the most severely disturbed or skeletal habitats it seems likely that the struggle between potential dominants provides a potent driving force for successional change and is a major determinant of the fate of subordinate species. In this chapter it will be argued therefore that studies of dominance should command high priority in our attempts to understand the structure and dynamics of vegetation. It appears necessary not only to analyse the very different mechanisms whereby plants attain dominance but also to identify the factors responsible for the decline and replacement of particular dominants (Watt 1947) and for their debilitation or exclusion from species-rich communities (Al-Mufti *et al.* 1977).

As a corollary we may also seek to understand why many species rarely if ever achieve a dominant status in vegetation. Is the lower relative abundance of such plants within communities a sign that they are perpetual 'also-rans' in the struggle for existence or are there distinct selective advantages associated with playing a subordinate role?

This chapter will attempt to identify some of the essential features of dominant and subordinate plants of terrestrial plant communities and will consider how knowledge relating to them can contribute to our understanding of vegetation characteristics.

DOMINANTS

Dominance implies not only a major contribution to the total biomass of the plant community but the tendency (as an individual or a population) to ramify throughout the edaphic and aerial environments and to influence the identity, quantity and local distribution of the other organisms present. This argument leads to the proposition that dominance mechanisms have two major components.

(a) The mechanism whereby the dominant plant achieves a biomass greater than that of its associates; this mechanism will vary according to the species and habitat conditions.

(b) The effects which dominant plants exert upon the fitness of neighbours; these include deleterious effects (e.g. resource depletion and release of phytotoxins) and promotory effects (e.g. provision of surfaces for epiphytes, release of substrates to parasites, symbionts and decomposing organisms and production of food exploited by herbivores).

These definitions provide for considerable variety in the types of vegetation dominants. The C–S–R model of primary strategies (Grime 1974) permits potential dominants to be distinguished from other plants (Fig. 20.1) and then subdivided into three functional classes which for simplicity are described respectively as ruderal, competitive and stress-tolerant dominants. The differences between these three types of plants are described in Grime (1979) and will not be reviewed in detail here. However, reference to certain of the distinguishing characteristics of the three classes of dominants is essential to an analysis of their role in vegetation dynamics.

Ruderal dominants

Where the vegetation developing in a productive terrestrial habitat suffers frequent and severe destruction it is likely that all of the plant populations present will be too sparse and ephemeral to be capable of exerting dominance as defined under the previous heading. However, where disturbance appears in the form of a single predictable annual event (e.g. winter flooding of river banks, autumn tillage of arable fields), the seasonal cycle may contain an uninterrupted growing season long enough to allow certain annual plants of tall stature and originating by synchronous germination of a large seed population to achieve a biomass sufficient to suppress the performance of neighbouring species. Examples here include *Impatiens glandulifera* on river terraces and crops and weeds of arable land such as *Hordeum vulgare* and *Bromus sterilis*.

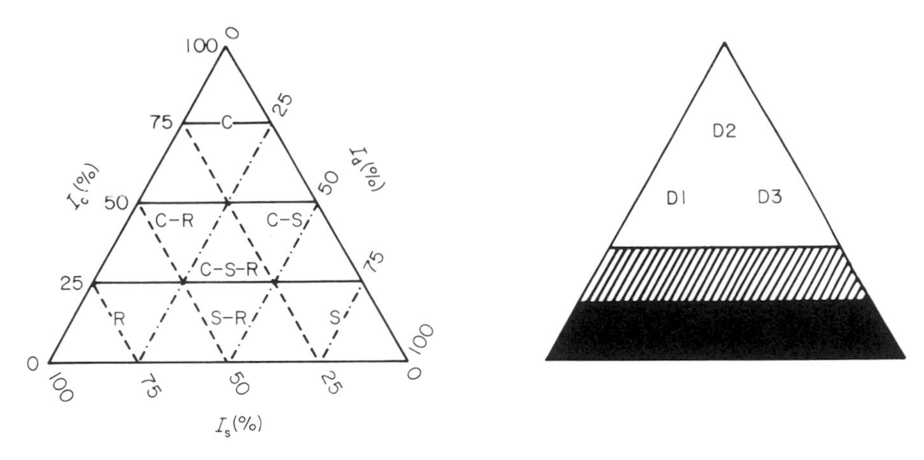

FIG. 20.1 (a) Model describing the various equilibria between competition, stress and disturbance in vegetation and the location of primary and secondary strategies. C, competitor; S, stress-tolerant R, ruderal; C–R, competitive–ruderal; S–R, stress-tolerant–ruderal; C–S, stress-tolerant–competitor; C–S–R, 'C–S–R strategist'. I_c, relative importance of competition (——); I_s, relative importance of stress (·—·—); I_d, relative importance of disturbance (– – –). (b) The distribution of three floristic elements within the triangular model: (□) potential dominants; (■) plants highly adapted to extremely disturbed and/or unproductive conditions; (▨) subordinates. D1, ruderal dominants; D2, competitive dominants; D3, stress-tolerant dominants.

Competitive dominants

In circumstances where productive habitats are subjected to occasional catastrophic disturbance (e.g. sites experiencing floods or fires every 5–30 years) the intervals during which vegetation recovery occurs are long enough to allow development of a rapidly-expanding biomass monopolized by herbs (e.g. *Urtica dioica*, *Reynoutria japonica*), shrubs (e.g. *Sambucus nigra*, *Piper hispidum*) and trees (e.g. *Populus grandidentata*, *Cecropia obtusifolius*) with high relative growth rates. Three features of competitive dominants ('active foraging', covariance in competitive abilities and life-form diversity) deserve special comment since they have profound implications for ecological theory.

'Active foraging'

Competitive dominants achieve high rates of resource capture in productive environments by means of exceedingly dynamic root systems and leaf canopies. From year to year, and in most cases within each growing season, there is continuous replacement of the leaves and roots. This feature, coupled with the rapid morphogenetic responses of expanding leaves and roots to local patchiness in resource concentration (Grime 1979; Crick 1985) results in a continuous adjustment of the spatial distribution of the absorptive surfaces above and below ground. This ability to maintain leaves and roots in the resource-rich zones of a changing environment is of key significance in the success of early successional perennials of productive habitats since it explains the ability of these dominants to sustain high rates of resource capture, growth and reproduction despite the development of local zones of resource depletion originating from the activity of the dominant itself and from encroachments by neighbours.

It is clear that the costs involved in such active foraging for light energy, mineral nutrients and water are considerable in terms of the high rates of reinvestment of captured resources in the construction of new leaves and roots and in their rapid senescence. To these costs we must add those associated with the high rates of herbivory experienced by the weakly-defended tissues of many competitive dominants (for evidence of this phenomenon and an evolutionary explanation for it see Grime 1979, Coley 1983). We may suspect, therefore, that there are severe penalties attached to active foraging by early successional dominants; the significance of these will be explored later in this paper.

Covariance in competitive abilities

The resources for which autotrophic plants compete are numerous (light energy, water and various mineral elements) and they are often located in different parts of the plant's environment. This has prompted some ecologists, most notably Newman (1973, 1983), to suggest that a distinction should be drawn between plants which compete effectively for light in the dense vegetation of fertile relatively undisturbed sites and those which occupy unproductive habitats and might be supposed to have superior competitive ability for below-ground resources. This view has been contested (Grime 1973b; Mahmoud & Grime 1976; Chapin 1980), with the argument that successful competition in productive vegetation depends upon simultaneously high rates of capture of light, water and mineral nutrients under circumstances in which zones of resource depletion are developing both above and (less conspicuously) below ground.

Evidence of the strong interdependence of competitive abilities above and below ground is available from a recent experiment in which turf microcosms providing localized access to subsoil moisture and growing under high and low soil fertility were subjected to 7 days of drought. The results (Fig. 20.2) show consistent differential mortality among the twenty species of herbaceous plants established in the microcosms. Fatalities were considerably higher in the more productive turf, presumably as a result of the larger transpiring shoot biomass. Of more critical interest, however, is the fact that at both high and low fertility the pattern of drought mortalities among the species was directly correlated with an index of competitive dominance (Grime 1973a) based upon above-ground attributes (height, lateral spread and litter accumulation) of the plant. This strongly suggests that species such as *Dactylis glomerata* and *Festuca rubra* owe their success in productive habitats not only to monopoly of the aerial environment but also to the capacity for extensive exploration and effective resource capture below ground.

Life-form diversity

Active foraging and high rates of resource capture and growth do not necessarily confer the potential for vegetation dominance. Ruderal dominants such as those described on p. 414, and even many small ephemeral weeds, exhibit active foraging and high rates of resource capture but they rapidly succumb when exposed to competition by large perennial plants. Failure here can be attributed to the diversion of captured resources to seeds rather than into the construction of the matrix of robust vegetative

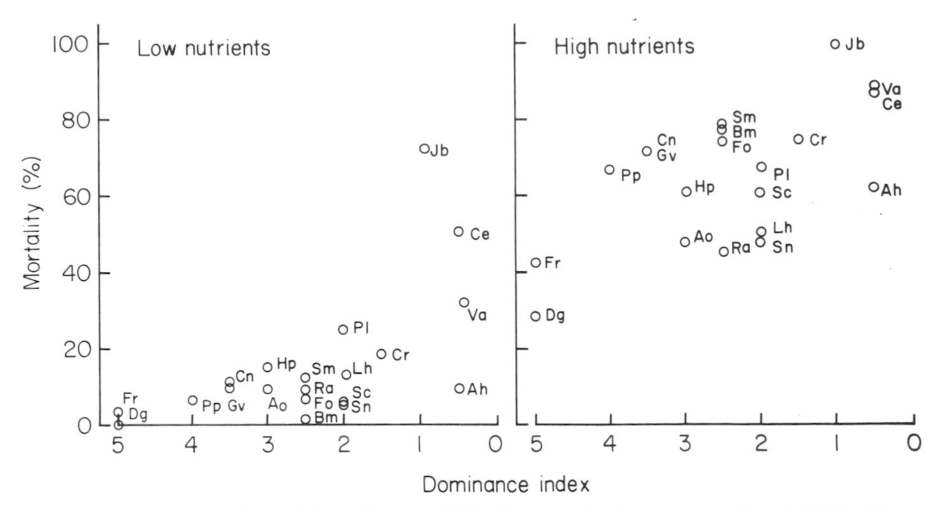

FIG. 20.2 A comparison of drought mortalities in twenty herbaceous species established in nutrient-poor (left) and nutrient-rich (right) microcosms. Ah, *Arabis hirsuta*; Ao, *Anthoxanthum odoratum*; Bm, *Briza media*; Ce, *Centaurium erythrea*; Cn, *Centaurea nigra*; Cr, *Campanula rotundifolia*; Dg, *Dactylis glomerata*; Fo, *Festuca ovina*; Fr, *Festuca rubra*; Gv, *Galium verum*; Hp, *Hieracium pilosella*; Jb, *Juncus bufonius*; Lh, *Leontodon hispidus*; Pl, *Plantago lanceolata*; Pp, *Poa pratensis*; Ra, *Rumex acetosa*; Sc, *Scabiosa columbaria*; Sm, *Sanguisorba minor*; Sn, *Silene nutans*; Va, *Veronica arvensis*.

tissues necessary for exclusive occupation of an undisturbed site over an extended period. This exemplifies the principle that dominance depends not merely upon resource acquisition but upon the way in which captured resources are utilized by the plant. Hence, from the experimental studies of Mahmoud & Grime (1976) and Crick (1985) we may deduce that the ability of the common grass *Arrhenatherum elatius* to dominate extensive areas of productive derelict grassland in Europe is not attributable solely to rapid rates of resource capture and growth, since many other common grasses are superior in these respects (Grime & Hunt 1975; Crick 1985). Rather it would appear that dominance by *A. elatius* is related to additional features such as the capacity to develop massive persistent tussocks which are impenetrable by other species. This same concept is highly relevant to our understanding of the life-form diversity in the sequence of competitive dominants (perennial herbs–shrubs–trees) often observed during the early stages of secondary succession in productive habitats in temperate regions. Accepting the argument of Egler (1954), many of the woody plants which eventually displace the herbaceous perennials originate from seedlings present at the early stages of recolonization. This suggests that the most likely explanation for the delayed dominance by woody species is the result of three factors. The first is the longer period required for the development of the elevated leaf canopies of trees and shrubs. The second is the slower growth rates due to the production of wood at the expense of leaves and roots. The third factor is the suppression of woody seedlings by the herbaceous dominants which, through extensive lateral spread above and below ground, are better equipped to dominate in the short-term. It is interesting to note that in tropical forests herbaceous dominants are often less conspicuous in the early stages of secondary succession; this observation may be related to the fact that here many of the pioneer trees of disturbed sites delay the development of wood until the late sapling stage and produce seedlings with multiple stems which allow lateral spread of the leaf canopy at a very early stage of development.

Stress-tolerant dominants

Earlier in this paper it was suggested that ecological penalties are associated with the active foraging mechanisms of competitive dominants. These may be expected to arise from the fact that heavy expenditure of captured resources in new leaves and roots will be of selective advantage only in circumstances where active foraging gains access to large reserves of light energy, water and mineral nutrients. From this argument we may suspect that competitive dominants will be excluded from habitats in which productivity is low and resource availability is brief and unpredictable (e.g. light as sunflecks, mineral nutrients as short pulses from decomposition processes or water as occasional rainshowers). In these circumstances conservation of captured resources is of primary significance and dominant plants are likely to be those which owe their status to their capacity to harvest *and retain* scarce resources in a continuously hostile physical environment. In keeping with this prediction, we find that the leaves of plants of unproductive environments tend to be comparatively long-lived, morphologically-implastic structures which are strongly defended against herbivory (Grime 1979; Bryant & Kuropat 1980; Coley 1983; Cooper-Driver 1985).

Stress-tolerant dominants, despite their many common features of life-history and physiology (Grime 1977), are associated with a wide range of life-forms and ecologies. This diversity may be related to the fact that in addition to the numerous lichens, bryophytes, herbs and small shrubs of unproductive biomes (arctic, alpine, arid) and habitats (cliffs, rock outcrops, bogs, heathlands), there are also stress-tolerant trees with slow growth rates and long life-spans which occupy relatively unproductive sites (e.g. Currey 1965).

Implications for succession, stability and diversity

Succession

Theories of plant succession in terrestrial environments have been dominated by concepts relating to life-histories, plant architecture and mechanisms of regeneration (Clements 1916; Egler 1954; Drury & Nisbet 1973; Connell & Slatyer 1977; Whitmore 1982; Finegan 1984). One objective of this chapter is to emphasize that some of the structural changes which we observe in vegetation are confounded with physiological transitions, knowledge of which is exceedingly helpful in understanding the mechanisms controlling species replacements and reactions to particular types of perturbation.

In Fig. 20.3 various familiar successional phenomena are portrayed in an attempt to illustrate the potential of strategy concepts to admit physiological criteria into the study of vegetation dynamics. In each succession diagram the strategies of the dominant plants at particular points in time are indicated by the position of arrowed lines within the triangular model. The passage of time in years during succession is represented by numbers on each line and shoot biomass at particular points is reflected in the size of the circles.

Fig. 20.3f depicts the course of secondary succession in a forest clearing

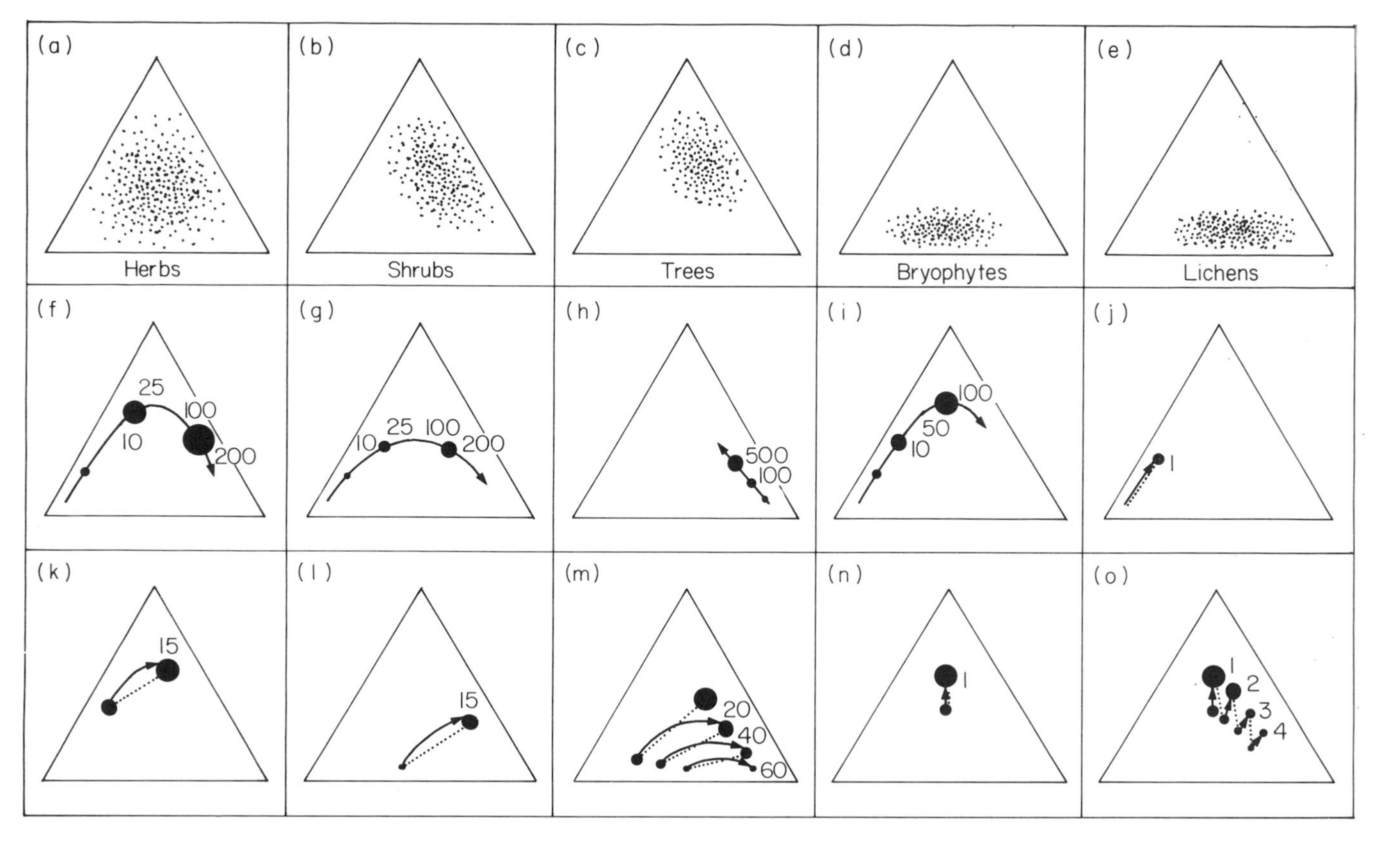

FIG. 20.3 Models describing the strategic range of five life forms (a–e) and representing various successional phenomena (f–o). (a) herbs, (b) shrubs, (c) trees (d) bryophytes, (e) lichens. For a description of the succession diagrams see text.

situated on a moderately fertile soil in a temperate climate. Biomass development is initially rapid and there is a fairly swift replacement of species as the community experiences successive phases of competitive dominance by rapidly growing herbs, shrubs and trees. At a later stage, however, the course of succession begins to deflect downwards towards the stress-tolerant corner of the triangular model. This process begins even during stages where the plant biomass is expanding appreciably and reflects a change from vegetation exhibiting high rates of resource capture and loss, to one in which resources, particularly mineral nutrients, are efficiently retained in the biomass.

In Fig. 20.3g secondary succession is described for a site of lower soil fertility. Here the course of events is essentially the same as that described in Fig. 20.3f except that the successional parabola is shallower and the plant biomass smaller as a consequence of the earlier onset of mineral nutrient limitation of the initial phase of competitive dominance.

Primary succession in a skeletal habitat such as a rock outcrop is represented in Fig. 20.3h. In this case the initial colonists, probably lichens and bryophytes, are stress-tolerators of low biomass and they occupy the site for a considerable period, giving way eventually to small slow-growing herbs and shrubs. This sequence coincides with the process of soil formation and provides an example of the facilitation model of vegetation succession (Connell & Slatyer 1977).

In the examples so far considered plant succession has been interpreted mainly as an interaction between resource availability and the characteristics of established plants. Greater sophistication can be introduced into the models by including circumstances where ineffective seed dispersal or failure in seedling establishment limits access of potential dominants into successional pathways. Fig. 20.3i, for example, describes the common situation (Niering & Goodwin 1962; Kochummen & Ng 1977) in which the development of a dense herbaceous cover prior to the arrival of the propagules of woody species prevents their establishment and strongly delays succession.

The range of models can be extended further to include loops representing vegetation responses to major perturbations. These vary from simple truncation of succession by annual harvesting and fertilizer input in an arable field (Fig. 20.3j) to the cycles of vegetation change associated with coppicing *Fraxinus excelsior* woodland (Fig. 20.3k) or rotational burning of *Calluna vulgaris* moorland (Fig. 20.3l). Where the vegetation of moderately unproductive habitats is subjected to repeated cycles of destruction by burning, browsing or cropping (Fig. 20.3m), the declining mineral nutrient capital of the soil may be expected to bring about a series of arcs of progressively lower trajectory in successive cycles of vegetation recovery.

Finally, models can be drawn to represent circumstances in which proclimax communities are maintained by orderly sublethal damage to the vegetation. Fig. 20.3h describes a productive fertilized meadow in which the expansion of competitive dominants is restricted by annual mowing. In Fig. 20.30 we see the sequence of events where the meadow is not fertilized and there is a drift towards lower productivity and incursion by stress-tolerant species.

Stability

Reference to some of the essential differences between ruderal, competitive and stress-tolerant dominants, reviewed earlier in this chapter (pp. 414–419) provides the basis for predictions of vegetation stability. These predictions have been discussed (Grime 1979) and tested (Leps, Osbornova-Kosinova & Rejmanek 1982) elsewhere. Here it will suffice to make the essential distinction between resistance to perturbation (low in ruderal and competitive dominants, higher in stress-tolerant dominants) and resilience (highest in the first two, lower in the latter) and to add that the regenerative strategies of plants, particularly those involving the maintenance of persistent banks of seeds of buds within the soil, are a major additional determinant of resilience.

Diversity

Where fast-growing herbs or shrubs with the capacity for clonal expansion are able to exercise competitive dominance it is not unusual to find exceedingly low floristic diversity. Mechanisms which have been proposed to explain the low invasibility of these 'near-monocultures' include intense shading and deposition of litter and it has been suggested already (pp. 416–418) that local resource depletion below ground may be important also. In addition we may suspect that the very dynamic 'foraging' responses of the leaves and roots of competitive dominants creates a highly unpredictable and hazardous environment for smaller plants of narrower niche width. In marked contrast, we may suspect that where vegetation is dominated by stress-tolerant plants, the more predictable matrix of living and dead materials which they provide will allow greater opportunity for other plants to exploit the habitat. This prediction appears to be confirmed by observations of high diversity in mature communities of relatively unproductive forests and grasslands (Holdridge *et al.* 1971; Bratton 1976; Furness 1980) in both of which co-existence with stress-tolerant dominants often involves well defined spatial relationships with or attachment to living or non-living surfaces of the dominant plants.

SUBORDINATES

In Fig. 20.1 it is suggested that plants which are incapable of dominance fall into two broad classes. First, there are the plants characteristic of extreme habitats where the intensities of stress or disturbance, or of various combinations between the two, are sufficient to exclude potential dominants. Second, there are subordinate plants which through a variety of mechanisms coexist with dominant plants. As stated in Grime (1979), some 'by virtue of their morphology or phenology, escape the main impact of the stresses generated by the dominant (e.g. vernal herbs beneath deciduous trees) whilst others (e.g. evergreen herbs in woodland) are adapted to tolerate these stresses. Coexistence in certain other plants is achieved by exploitation of areas within the habitat where the environment is locally unfavourable to the dominant. In addition there are subordinate species which owe their presence to microhabitats created by the dominant (e.g. epiphytic angiosperms, ferns, mosses and lichens).'

Complementing the definition of dominance on p. 414, subordination may be recognized therefore as the consequence of those specializations, of many different kinds, which prevent the majority of plants from monopolizing the edaphic and aerial environments. These specializations have morphological, phenological and life-history concomitants and restrict the activity of the plant to part of the micro-environmental mosaic and/or to part of the available growing season.

Fitness of subordinates

The Introduction to this paper questioned whether there were distinct selective advantages associated with playing a subordinate role in communities. Reference to Table 20.1, which identifies the twenty-four most commonly occurring herbaceous plants of the Sheffield region, provides strong evidence that some subordinates (*Cerastium fontanum*, *Heracleum spondylium*, *Plantago lanceolata*, *Poa trivialis* and *Rumex acetosa*) are as widely successful as the most familiar dominants. Clearly many factors determine the fitness of species in an area of varied landscape and vegetation cover. These include genetic variation within and between populations, flexibility in mechanisms of regeneration and capacity to

TABLE 20.1. The commonest herbaceous plants of the Sheffield region. The values tabulated refer to the percentage of 2748 m^2 samples found to contain the species. Subordinate species are indicated by heavy type

Species	%	Species	%
Poa trivialis	**22·0**	*Rubus fruticosus* agg.	11·5
Deschampsia flexuosa	20·3	*Festuca ovina*	11·2
Festuca rubra	19·4	***Plantago lanceolata***	**9·6**
Agrostis capillaris	18·3	***Cerastium fontanum***	**9·2**
Holcus lanatus	18·1	*Elymus repens*	9·2
Dactylis glomerata	17·9	*Hyacinthoides non-scripta*	9·1
Poa pratensis	15·0	***Heracleum shondylium***	**8·5**
Arrhenatherum elatius	14·3	*Ranunculus repens*	8·2
Holcus mollis	12·8	***Rumex acetosa***	**7·9**
Chamerion angustifolium	12·6	*Lolium perenne*	7·8
Poa annua	12·1	*Pteridium aquilinum*	7·8
Taraxacum officinale agg.	12·0	*Mercurialis perennis*	7·0

exploit the most commonly-occurring habitats. In the case of common subordinates, however, two additional factors deserve consideration.

(a) Many subordinate species are capable of exploiting a similar niche in a variety of habitats and in association with a wide range of dominants. *Poa trivialis*, for example, appears as a vernal component in eutrophic woodlands, grasslands and marshes with an extraordinarily wide range of dominants including taller grasses, broad-leaved herbs and woody plants.

(b) The reduced morphology and restricted phenology of many of the most successful subordinate species reduces the risk and severity of damage due to factors such as drought, flooding, grazing, mowing and burning. In consequence, populations of subordinates may be expected to show a degree of homeostasis greater than that of the dominants when exposed to fluctuating conditions of climate and management.

Implications for succession and diversity

Succession

Earlier in this paper it has been suggested that it is the interactions of the various kinds of dominant plants with their environments and with each other which provides the main impetus for succession. Against this background it is tempting to regard subordinates as mere dependents or opportunists. This view would be mistaken, however, since there is abundant evidence that subordinate plants often influence the fate of dominant plants. It is generally accepted (p. 421, Fig. 20.3h) that in primary succession relatively inconspicuous plants such as lichens and mosses have a critical role in soil formation and thus facilitate succession to vegetation dominated by vascular plants. Similarly nitrogen enrichment of the soil through the presence of subordinate leguminous plants could eventually favour one dominant at the expense of another. It is also relevant to point out that in order to invade potentially favourable habitats many of the most effective competitive dominants which appear early in secondary successions (*Chamerion angustifolium*, *Phragmites australis*, *Typha* spp., *Epilobium hirsutum*, *Salix* spp.) depend upon establishment by tiny and extremely vulnerable seedlings. Recent field studies (Hillier 1984; Keizer, van Tooren and During 1985) have shown that selective effects on invading seedling populations can be exercised by bryophytes. In view of this it seems reasonable to suggest that subordinate plants could often play a critical role in succession through their influence upon the colonizing success of various dominants. Selective effects on seedling establishment may arise from the presence of even less conspicuous members of the plant community. This may be illustrated by reference to a recent experiment in which the mortality of seedlings of various herbaceous plants was measured in microcosms in which the soil surface was colonized to various extents by a film of blue-green algae. The results (Fig. 20.4) reveal that there were strong selective effects. Small-seeded species such as *Centaurium erythraea* and *Juncus bufonius* failed completely in circumstances where grasses such as *Festuca ovina* and *Dactylis glomerata* established successfully.

Diversity

Already it has been predicted (p. 422) that in successional communities diversity will be greater where the dominant plants are relatively stress-tolerant and the less dynamic behaviour of the shoots and roots will create microhabitats of sufficient duration for exploitation by subordinate plants. This phenomenon achieves its greatest expression in nutrient-stressed tropical rainforest (Holdridge *et al.* 1971) where even the living leaves of the dominant trees may provide niches for epiphytic subordinates.

It thus appears that in some forests high diversity arises, at least in part, from the presence of numerous relatively stable niches for subordinate plants. Diversity in tropical and temperate forests also, of course, depends upon many additional factors such as the presence of local successional sequences initiated where canopy gaps arise from tree falls.

In grasslands, high densities of subordinate plants are characteristic of species-rich communities, a relationship formalized in the humped-back model (Grime 1973a) which, as shown in Fig. 20.5, can be related directly

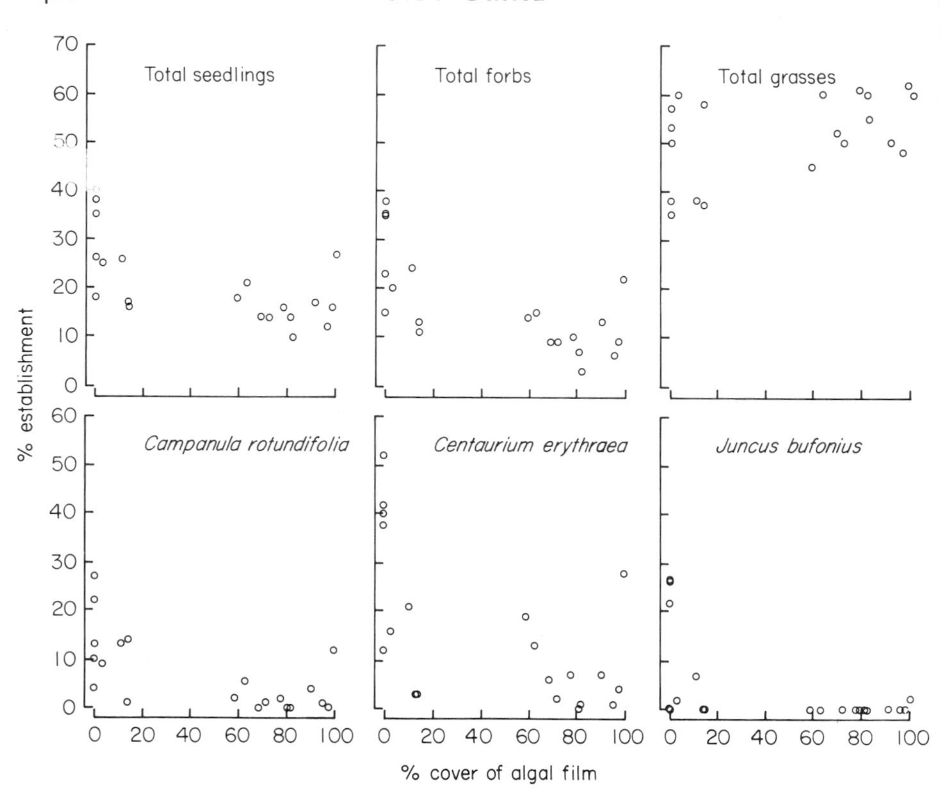

FIG. 20.4 Establishment of seedlings in microcosms varying in percentage cover of the soil surface by an algal film: (a) all species; (b) forbs; (c) grasses; (d) *Campanula rotundifolia*; (e) *Centaurium erythraea*; (f) *Juncus bufonius*.

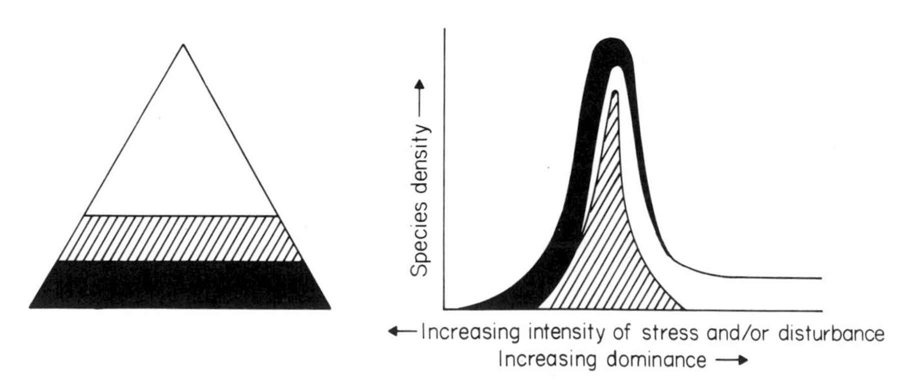

FIG. 20.5 Scheme describing the distribution of three floristic elements in the C–S–R model (left) and in the humped-back model (right). The axes and positions of strategies in the C–S–R model are as in Fig. 20.1a: (□) potential dominants; (■) plants highly adapted to extremely disturbed and/or unproductive conditions; (▨) subordinates.

to the C–S–R model of primary plant strategies. Although the humped-back model has been tested by quantitative studies (Al-Mufti *et al.* 1977), uncertainties remain with regard to the processes which often allow large numbers of subordinate plants to coexist within 'the hump'. As the reviews of Grubb (1977), Grime (1979), Huston (1979), Pickett (1980), and Tilman (1982) make clear, there is considerable diversity of opinion with regard to the relative importance of spatial heterogeneity, vegetation disturbance, regeneration opportunities and specific environmental stresses in the maintenance of species-richness. It seems possible that further penetration into this problem may not be possible by conventional studies in the field. Accordingly, current work on grassland diversity at Sheffield involves experimental manipulation of species-rich microcosms synthesized under controlled conditions.

ACKNOWLEDGMENTS

I wish to thank Mrs J.M.L. Mackey who participated in some of the experimental work. This paper is based upon research supported by the Natural Environment Research Council.

REFERENCES

Al-Mufti, M.M., Sydes, C.L., Furness, S.B., Grime, J.P. & Band, S.R. (1977). A quantitative analysis of shoot phenology and dominance in herbaceous vegetation. *Journal of Ecology*, **65**, 759–791.

Bratton, S.P. (1976). Resource division in an understory herb community: reponses to temporal and microtopographic gradients. *American Naturalist*, **110**, 679–689.

Bryant, J.P. & Kuropat, P.J. (1980). Selection of winter forage by subarctic browsing vertebrates: the role of plant chemistry. *Annual Review of Ecology and Systematics*, **11**, 261–285.

Chapin, F.S. (1980). The mineral nutrition of wild plants. *Annual Review of Ecology and Systematics*, **11**, 233–260.

Clements, F.E. (1916). *Plant Succession: an Analysis of the Development of Vegetation*. Carnegie Institute, Washington.

Coley, P.D. (1983). Herbivory and defensive characteristics of tree species in a lowland tropical forest. *Ecological Monographs*, **53**, 209–232.

Connell, J.H. & Slatyer, R.O. (1977). Mechanisms of succession in natural communities and their role in community stability and organisation. *American Naturalist*, **111**, 1119–1145.

Cooper-Driver, G. (1985). Anti-predation strategies in pteridophytes: a biochemical approach. *Proceedings of the Royal Society of Edinburgh*, **86B**, 397–402.

Crick, J.C. (1985). *The role of plasticity in resource acquisition by higher plants*. Ph.D. thesis, University of Sheffield.

Currey, D.R. (1965). An ancient bristlecone pine stand in Eastern Nevada. *Ecology*, **46**, 564–566.

Drury, W.H. & Nisbet, I.C.T. (1973). Succession. *Journal of the Arnold Arboretum, Harvard University*, **54**, 331–368.

Egler, F.E. (1954). Vegetation science concepts. I. Initial floristic composition, a factor in old-field vegetation development. *Vegetatio*, **4**, 412–417.

Finegan, B. (1984). Forest succession. *Nature*, **312**, 109–114.

Furness, S.B. (1980). *Ecological effects of temperature in bryophytes and herbaceous plants*. Ph.D. thesis, University of Sheffield.

Grime, J.P. (1973a). Competitive exclusion in herbaceous vegetation. *Nature*, **242**, 344–347.

Grime, J.P. (1973b). Competition and diversity in herbaceous vegetation: a reply. *Nature*, **244**, 310–311.

Grime, J.P. (1974). Vegetation classification by reference to strategies. *Nature*, **250**, 25–31.

Grime, J.P. (1977). Evidence for the existence of three primary strategies in plants and its relevance to ecological and evolutionary theory. *American Naturalist*, **111**, 1169–1194.

Grime, J.P. (1979). *Plant Strategies and Vegetation Processes*. John Wiley & Sons, Chichester.

Grime, J.P. & Hunt, R. (1975). Relative growth-rate: its range and adaptive significance in a local flora. *Journal of Ecology*, **63**, 393–422.

Grubb, P.J. (1977). The maintenance of species-richness in plant communities: the importance of the regeneration niche. *Biological Reviews*, **52**, 107–145.

Hillier, S.H. (1984). *A quantitative study of gap recolonization in two contrasted limestone grasslands*. Ph.D. thesis, University of Sheffield.

Holdridge, L.R., Grenke, W.C., Hatheway, W.H., Liang, T. & Tosik, Jr., J.A. (1971). *Forest Environments in Tropical Life Zones: a Pilot Study*. Pergamon Press, Oxford.

Huston, M. (1979). A general hypothesis of species diversity. *American Naturalist*, **113**, 81–101.

Keizer, P.J., van Tooren, B.F. & During, H.J. (1985). Effects of bryophytes on seedling emergence and establishment of short-lived forbs in chalk grassland. *Journal of Ecology*, **73**, 493–504.

Kochummen, K.M. & Ng, F.S.P. (1977). *Malaysian Forester*, **40**, 61–78.

Leps, J.J., Osbornova-Kosinova & Rejmanek, K. (1982). Community stability, complexity and species life-history strategies. *Vegetatio*, **50**, 53–63.

Mahmoud, A. & Grime, J.P. (1976). An analysis of competitive ability in three perennial grasses. *New Phytologist*, **77**, 431–435.

Newman, E.I. (1973). Competition and diversity in herbaceous vegetation. *Nature*, **244**, 310.

Newman, E.I. (1983). Interactions between plants. *Physiological Plant Ecology III: Responses to the Chemical and Biological Environment. Encyclopedia of Plant Physiology: New Series*, Vol. 12C (Ed by O.L. Lange, P.S. Nobel, C.B. Osmond & H. Zeigler), pp. 679–710. Springer-Verlag, Berlin.

Niering, W.A. & Goodwin, R.H. (1962). Ecological studies in the Connecticut Arboretum Natural Area. I. Introduction and survey of vegetation types. *Ecology*, **43**, 41–54.

Pickett, S.T.A. (1980). Non-equilibrium co-existence of plants. *Bulletin of the Torrey Botanical Club*, **107**, 238–248.

Tilman, D. (1982). *Resource Competition and Community Structure*. Princeton University Press, Princeton, New Jersey.

Watt, A.S. (1947). Pattern and process in the plant community. *Journal of Ecology*, **35**, 1–22.

Whitmore, T.C. (1982). On pattern and process in forests. *The Plant Community as a Working Mechanism* (Ed. by E.I. Newman), pp. 45–60. Blackwell Scientific Publications, Oxford.

21. WHAT MAKES A COMMUNITY INVASIBLE?

MICHAEL J. CRAWLEY
Department of Pure and Applied Biology, Imperial College at Silwood Park, Ascot, Berkshire

INTRODUCTION

In this paper I shall address two related questions and, in doing so, attempt to draw together some of the strands from other topics covered in the Symposium. I shall ask: (i) What attributes of communities make them more likely to be invaded? and (ii) What attributes of individual species make them more likely to be successful invaders? I shall draw upon data from two vast, but until now little-used sources, namely, the large-scale, but unintentional introduction of alien plant species into British plant communities, and the recorded attempts to establish introduced insects for the biological control of weeds.

THEORY

Put simply, a community is invasible when an introduced species is able to increase when rare. Most models of community dynamics assume that the principal barriers to invasion are: (i) competition with established native species; (ii) losses caused by generalist natural enemies (including diseases); (iii) lack of necessary mutualists to pollinate, disperse or otherwise facilitate the invader; or (iv) deleterious low density effects (Allee effects) operating on the invader itself (e.g. difficulty in finding mates). The development of this body of theory can be traced from MacArthur (1970), through May (1973), to Turelli (1981). While these studies have contributed to our understanding of what factors allow equilibrium persistence of invaders, they throw little light on the question of what makes a community invasible in the first instance. For example, we know the threshold population size for the persistence of measles in human populations (Anderson & May 1984), but many communities which are too small to support an endemic measles population are readily invasible by the disease. It simply runs its course, then goes extinct once the whole population is immune.

In terms of niche theory, the glib (and tautological) answer to the question of what makes a community invasible, is that there exists a 'vacant niche' for the invading species. There are three obvious problems with this concept. First, botanists and zoologists tend to mean two different things by 'vacant niche'. To a zoologist it tends to mean 'unexploited resources' whereas to a botanist it tends to mean 'unexploited space'. Second, a community may be readily invasible even though neither unexploited resources nor open space are available (e.g. when the invader *displaces* the current residents). Third, what appear, superficially, to be unexploited resources may be unexploitable for a host of reasons (low resource quality, abundant natural enemies, missing mutualists, missing habitat requisites, and so on). In short, it is impossible to recognize the existence of a vacant niche without the empirical attempt to establish a given species in a given environment. The parameters which define niche in the mathematical models (above) are themselves functions of the *structure* of the system (i.e. they depend upon precisely which species are present), and cannot be estimated in ways which allow us to predict whether or not a particular invading species will be successful in practice.

It seems clear that two different theoretical approaches may allow fresh insights into the question of invasibility. G. Sugihara (unpublished) provides an elegant example of the 'top down' approach, which enables him to make powerful predictions about the topology of food webs subject to invasion. The alternative, 'bottom up', approach holds out the possibility, at least in principle, of predicting whether or not a particular community is invasible by a specified species, based on a detailed, population-dynamics approach. Here the invader's rates of birth, death, immigration and emigration are specified as functions of the current configuration of the community. Examples of this approach are provided in the following two sections.

ALIEN PLANT SPECIES IN THE BRITISH ISLES

The European Garden Flora (Walters *et al.* 1984) contains over 12 000 different plant species. Most of these have been cultured in the British Isles without any attempt at containment or quarantine. In addition, countless thousands of species have been introduced unintentionally by a variety of means (Table 21.1). At a conservative guess, an order of magnitude more plant species have been introduced and grown in this country than are represented in the entire native flora (about 1500 species). A vast number (probably the great majority) of these plants are quite incapable of growth and reproduction when exposed to the full rigours of the British weather and to competition with native vegetation. While we can never hope to know the number of introductions, we can

TABLE 21.1 Means of introduction of alien plant species into the British flora, with examples of plants thought to have arrived by each route

Means of introduction	Example
Escapes from gardens	*Buddleja davidii*
Outcast from gardens	*Reynoutria japonica*
Contaminated foreign seed	*Datura stramonium*
Bird seed	*Amaranthus retroflexus*
Grain processing; milling & distilleries	*Scandix pecten-veneris*
Machinery	*Chamomilla suaveolens*
Outcast aquarium plants	*Myriophyllum brasiliense*
Tan-bark outcast	*Herniaria hirsuta*
Deliberate plantings	*Mahonia aquifolium*
Skins & wool	*Xanthium spinosum*
Relic of cultivation	*Linum usitatissimum*
Discarded ships' ballast	*Trifolium resupinatum*
Railway gravel & cinder	*Senecio squalidus*
Animal fodder and bedding	*Cardaria draba*
Trouser turn-ups	????

TABLE 21.2. Comparison of the numbers of wool alien species which grow to recognizable size in the south of France and on Tweedside. This provides us with a crude estimate of the failure rate of seeds introduced to the two environments that failed to germinate and grow to attain recognizable size. Data from Hayward & Druce (1919)

Source region	Montpellier	Tweedside
Mediterranean	416	113
East Europe & West Asia	17	48
Asia	2	14
South Africa	6	43
Australasia	5	21
North America	4	23
Central America	11	8
South America	49	43
Unknown	16	5
Total	526	348

obtain a crude estimate of the rate of failure of introduced seeds to germinate and grow to a recognizable size, by comparing the numbers of species from similar samples of wool aliens which survive in Britain and in their native habitats (Table 21.2). For almost every source region, the number of species introduced as propagules was at least three times as great as the number of species recognized at flowering time.

Do any patterns emerge from the successes and failures of these species which allow us to rank native plant communities in terms of their invasibility? Table 21.3 shows all the plant communities used in the *Flora of the British Isles* (Clapham, Tutin & Warburg 1962) and the percentage of their entire floras made up by alien species. These data were arrived at by drawing up comprehensive species lists for all plant communities (using a wide range of county floras supplemented by field experience). The clear (and not unexpected) pattern to emerge from this table is that the most invasible are: (i) communities with low average levels of plant cover; and (ii) communities subject to frequent disturbance. Similarly, in comparing communities close to potential sources of alien plants (cities, especially centres of communication with docks and railway yards) with isolated communities (as in the comparison between lowlands and uplands), clear biogeographic patterns emerge. Communities close to large sources of potential immigrants have many aliens, while isolated areas have few. Furthermore, the pool of potential species capable of growth and reproduction in the different plant communities is variable in size. For instance, the pool of immigrants capable of growing in arable fields is likely to be substantially larger than the pool of species capable of growing in high altitude peatlands. It is interesting that, as a group, seaside communities are particularly poor in aliens. This, presumably, reflects the fact that most maritime plants are very widespread simply because dispersal by sea is so effective, and has already introduced most potential colonists to most suitable habitats.

The role of disturbance is clearly seen in comparing the two kinds of woodlands. In managed woodlands, most of the dominant tree species are themselves aliens, intentionally planted by man; some of these regenerate by seed. In managed deciduous woodlands, many ground-layer plants have also been introduced for amenity or to provide cover for game, and have subsequently spread. Tracks, rides and clearings in these managed woodlands provide extra habitats for alien plants. In contrast, native woodlands contain very few alien plant species. This may be partly a function of their geographic isolation (e.g. all the native pine woodlands are in the heart of the Scottish Highlands), but also reflects the lower rates of disturbance and higher average levels of plant cover in native woodlands (e.g. absence of roadways, cut-over areas or timber-extraction trails).

Biogeographic patterns of alien plants can be seen from an analysis of local county floras. These works differ widely in their treatment of introduced plants, some covering only naturalized species maintaining their numbers by natural seed or vegetative reproduction, while others include many casual species which, although they flower, do not recruit from

TABLE 21.3. Percentage of the total flora of different British habitats made up by introduced plant species. The habitat classification employed follows Clapham, Tutin & Warburg (1962) but excludes ambiguous or little-used habitat names. L = low; M = moderate; H = high; V = variable

Class	Habitat	% alien	Cover	Disturbance
Man-made	Waste	78	L	H
	Walls	46	L	L
	Fields	37	V	H
	Hedgerow	22	H	H
Woodland	Conifers	56	H	H
	Parkland	19	H	V
	Deciduous	5	H	L
	Pine	0	H	L
By running water	River shingle	39	L	H
	Shady banks	30	H	M
	Open banks	24	H	M
	Ditches	13	M	M
Wetland	Bog	5	M	L
	Peat	5	L	V
	Marsh	4	H	L
	Fen	2	H	L
Lowland	Damp grass	13	H	L
	Dunes	13	V	V
	Dry grass	5	H	L
	Heath & moor	2	H	V
Upland	Fast streams	12	L	H
	Springs	4	H	M
	Grassland	2	H	L
	Summits	0	L	H
	Rock ledges	0	L	M
Rocks	Sea cliffs	18	M	M
	Limestone pavement	4	L	M
	Shady rocks	4	L	L
	Upland screes	0	L	H
Seaside	Shingle	5	L	H
	Salt marsh	4	V	V
	Brackish	0		
	Sea water	0		
Freshwater	Olig. margins	12		
	Eutrophic	9		
	Eut. margins	5		
	Oligotrophic	0		

seed. These impermanent species rely on continual immigration of propagules for their continued representation in the flora. Notwithstanding these shortcomings, analysis shows a species–area curve for the aliens of British and Irish counties with a slope of 0·13 (Table 21.4), as we go from

Rutland up to Yorkshire. Significant outliers for which there are modern, critical Floras, include Kent (where the docks and industries of the Thames Estuary provide both high rates of immigration and ample waste ground on which the aliens can grow), and Connemara and the Burren (a large, isolated area of western Ireland with rather uniform vegetation cover, which appears to be rather difficult to invade). It is interesting to note that the famous limestone pavements of the Burren, for all that they are rather open and rather disturbed, have extremely few aliens compared to the man-made 'equivalent' habitat of walls and mortar (Table 21.4). Further, although the peat bogs of Connemara support few vascular plant aliens, they are certainly not 'uninvasible', and the alien moss *Campylopus introflexus* is extremely common on peat throughout the region.

When the vast areas of north-eastern US and Western Australia are

TABLE 21.4. Species–area relationships for county Floras in Britain show slopes of 0·13 on log–log axes for aliens, and this slope is not altered by the inclusion of two vast continental areas (Western Australia and north-east US). For natives, the county floras show *no* species–area relationship, but adding the continental areas gives a slope of 0·26 (see text for details)

Place	Area (km^2)	Native species	Alien species	Reference
Berkshire	1876	961	632	Bowen 1968
Connemara and the Burren	3100	515	170	Webb & Scannell 1983
Cornwall	3800	1024	629	Margetts & David 1981
Essex	3988	890	490	Jermyn 1974
Gloucester	3151	976	784	Riddelsdell, Hedley & Price 1948
Kent	4176	927	916	Philp 1982
Monmouth	1500	860	314	Wade 1970
Moray, Nairn & E. Inverness	6777	944	568	Webster 1978
Northumberland & Durham	7858	846	301	Tate & Baker 1868
Rutland	394	744	276	Messenger 1971
Scarborough	1300	867	170	Walsh & Rimington 1953
Shropshire	4400	955	250	Sinker *et al.* 1985
Surrey	1964	902	439	Lousley 1976
West Yorkshire	7148	892	354	Lees, 1888
North-east USA	2428000	4425	1098	Fernald 1970
Western Australia	2527633	6819	750	Beard 1969

included in the species–area relationship, there is no significant change in the slope of the curve for alien plants. Note that, for its area, the north-eastern US is richer in aliens than Australia, consistent with its closer climatic and edaphic similarity to the main source of aliens (central and Mediterranean Europe), and the (presumably) higher rate of species introduction (lower isolation).

For native plants, the British county Floras give the curious result of a species–area curve with zero slope! On reflection this is not so peculiar, since many large administrative areas are large simply *because* they are barren! Including the huge areas of the US and Western Australia gives a species–area curve of slope 0·26. In other words, at this Continental scale, native plants increase in number with area more steeply than do aliens. This is presumably because the beta diversity of native plants is greater than the beta diversity of aliens (e.g. native plants may be replaced by different species in different parts of a large region, whereas the aliens are likely to be more or less ubiquitous; see Crawley 1986a). Note that the relative positions of the two large areas are reversed; Western Australia is richer in native species than the north-eastern US.

The non-equilibrium nature of alien plant assemblages can be seen clearly in the study of alien arable weeds in the north-western US (Forcella & Harvey 1983). Both total species richness and species equitability have increased continuously since the turn of the century, despite the fact that the railways and other alleged means of seed transportation had been established long before this (Fig. 21.1). In the British Isles, too, alien plant species continue to change dramatically in abundance. One of the classic case histories concerns the Canadian pondweed, *Elodea canadensis*. This was first recorded in about 1840, increased throughout the canals and slow-flowing rivers of the country to the point where it was a hindrance to navigation, and then, quite inexplicably, declined in abundance to its current status (widespread but rarely abundant; Simpson 1984).

Other species are still increasing rapidly (Fig. 21.2). The success of American willowherb *Epilobium ciliatum* is relatively easy to understand, since it produces copious quantities of plumed seeds capable of wide dispersal by wind and, in addition, produces detachable vegetative rosettes each autumn at the base of the dead flowering shoot, which are spread during cultivation and in attempts at weeding. In contrast, the spread of *Veronica filiformis* is more difficult to understand, since the plant has never been observed to set seed in this country. Originally introduced as a rockery plant, it has become widespread as a weed of lawns, alleged to be spread about by vegetative fragments on gardeners' wellingtons, and on

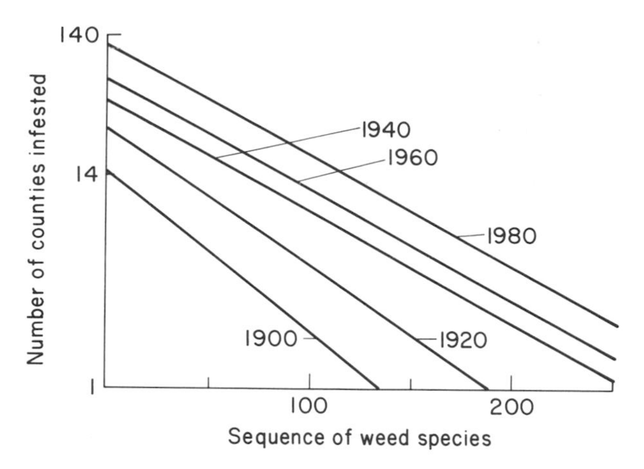

FIG. 21.1. The changing relative abundance of weed species in arable communities in the north-western US. New weed species are still becoming established, and the relative abundance of long-established aliens still changing (Forcella & Harvey 1983).

the blades of County Council mowing machines! How *Reynoutria japonica*, another non-seeding plant allegedly spread only through direct movement of its rootstocks by man, came to be in so many remote (and apparently undisturbed) roadside habitats in the Highlands of Scotland is a complete mystery.

The impermanence of many apparently successful establishments is vividly demonstrated by the fact that of the 348 species of wool aliens identified from the gravel banks of the River Tweed below the mills of Galashiels (Hayward & Druce 1919), none now remains. The vast majority of these plants flowered but (for reasons which are not understood) failed to produce any viable seed.

The taxonomic composition of alien floras is a distinctly non-random sample from the pool of available immigrants. Some families are strongly over-represented (e.g. Cruciferae, Labiatae and Caryophyllaceae) while others are vastly under-represented (e.g. ferns, Orchidaceae and Cyperaceae; see Table 21.5). It is a vivid testimony to the power and applicability of Murphy's Law that the 'world's worst weed' belongs to the Cyperaceae; the nut sedge *Cyperus rotundus* is a native of India, yet is a weed of fifty-two crops in ninety-two countries (Holm *et al.* 1977)! Most of the over-represented families are closely associated with man and his crops; they are weeds of arable or waste ground. It is more difficult to generalize about the under-represented species. Some are primitive (the ferns) but others are highly advanced (the orchids). Some orchids may require highly specialized mutualistic associations with soil fungi in order

TABLE 21.5. Representation of different plant families as aliens in the north-eastern US. Certain families are significantly over-represented as aliens (H) while others are significantly under-represented (L). Families were only included in the analysis if they contained fifty or more native species (from data in Fernald 1970)

Family	Natives	Aliens	% alien	Difference
Compositae	564	139	20	—
Rosaceae	489	62	11	—
Cyperaceae	451	22	5	L
Gramineae	389	98	20	—
Leguminosae	168	69	29	—
Scrophulariaceae	112	40	26	—
Ranunculaceae	88	15	15	—
Liliaceae	85	25	23	—
Labiatae	85	62	42	H
Orchidaceae	74	1	1	L
Cruciferae	71	76	52	H
Ericaceae	70	6	8	L
Umbelliferae	66	19	22	—
Juncaceae	65	2	3	L
Polypodiaceae	64	1	2	L
Onagraceae	60	6	9	L
Saxifragaceae	59	9	13	—
Caryophyllaceae	56	44	44	H
Salicaceae	52	13	20	—

to become established, but clearly not all species do (*Anacamptis pyramidalis* and *Dacylorhiza incarnata* quickly established on alkaline spoil heaps in Lancashire over 40 km from the nearest source of seed). Primitive groups like ferns and horsetails are conspicuous by their absence from the kinds of man-made habitats where aliens are most abundant (though *Pteridium aquilinum* and *Equisetum arvense* are important exceptions), perhaps because there are so few annual species among their number. Families like Compositae and Onagraceae, on the other hand, provide many species of annuals which are currently increasing rapidly in abundance in Britain.

The 'top twenty' British aliens are shown alphabetically in Table 21.6, based on a subjective assessment of their distribution and abundance. The plants come from a wide range of plant families and life forms (trees and herbs, perennials and annuals). The list contains several breeding systems (inbreeders, outbreeders and not-breeders-at-all; see Brown & Burdon, Chapter 6) and many modes of dispersal (though wind dispersal

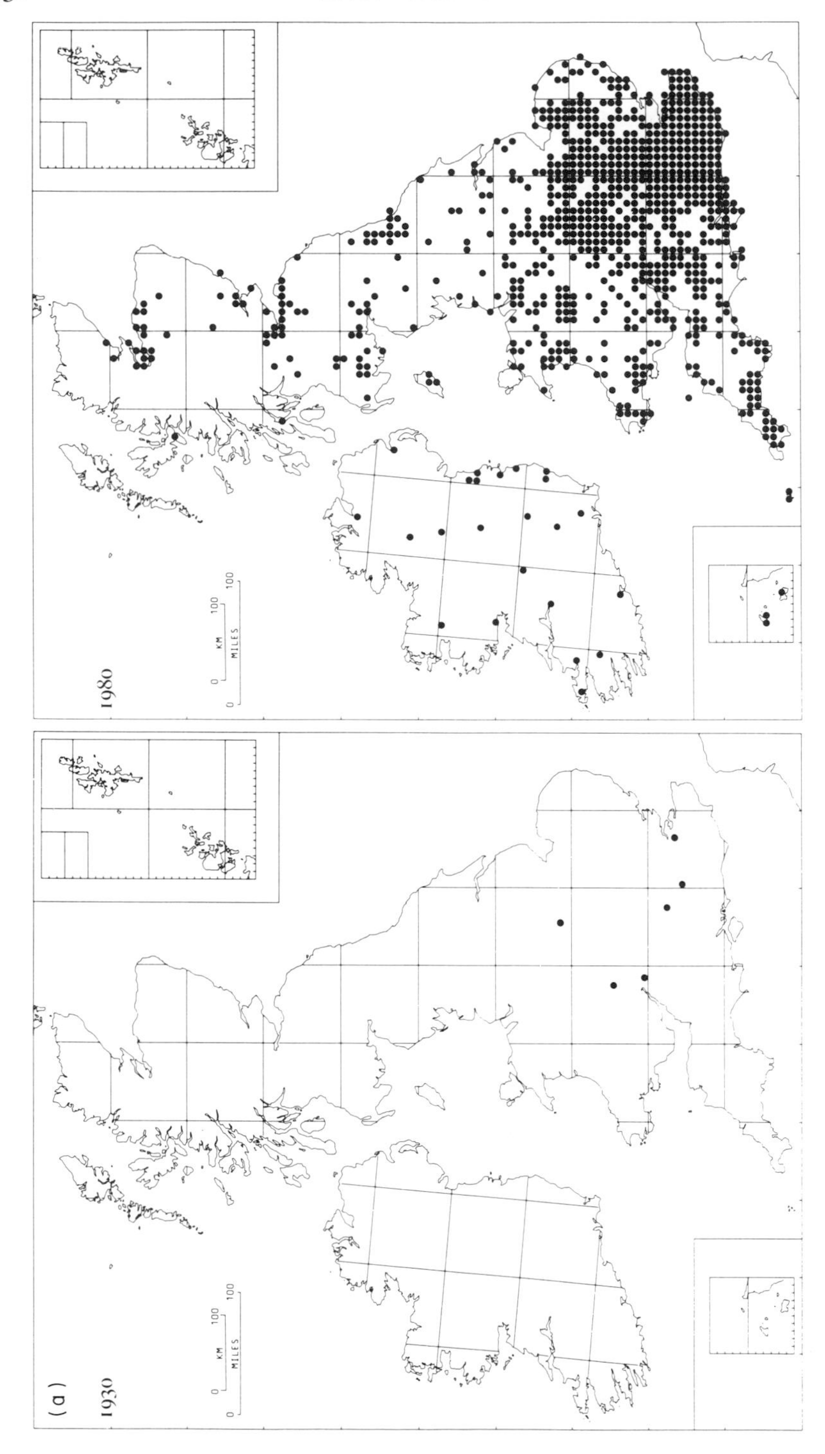
1980
0 100
KM
MILES
0 100
(a)
1930
0 100
KM
MILES
0 100

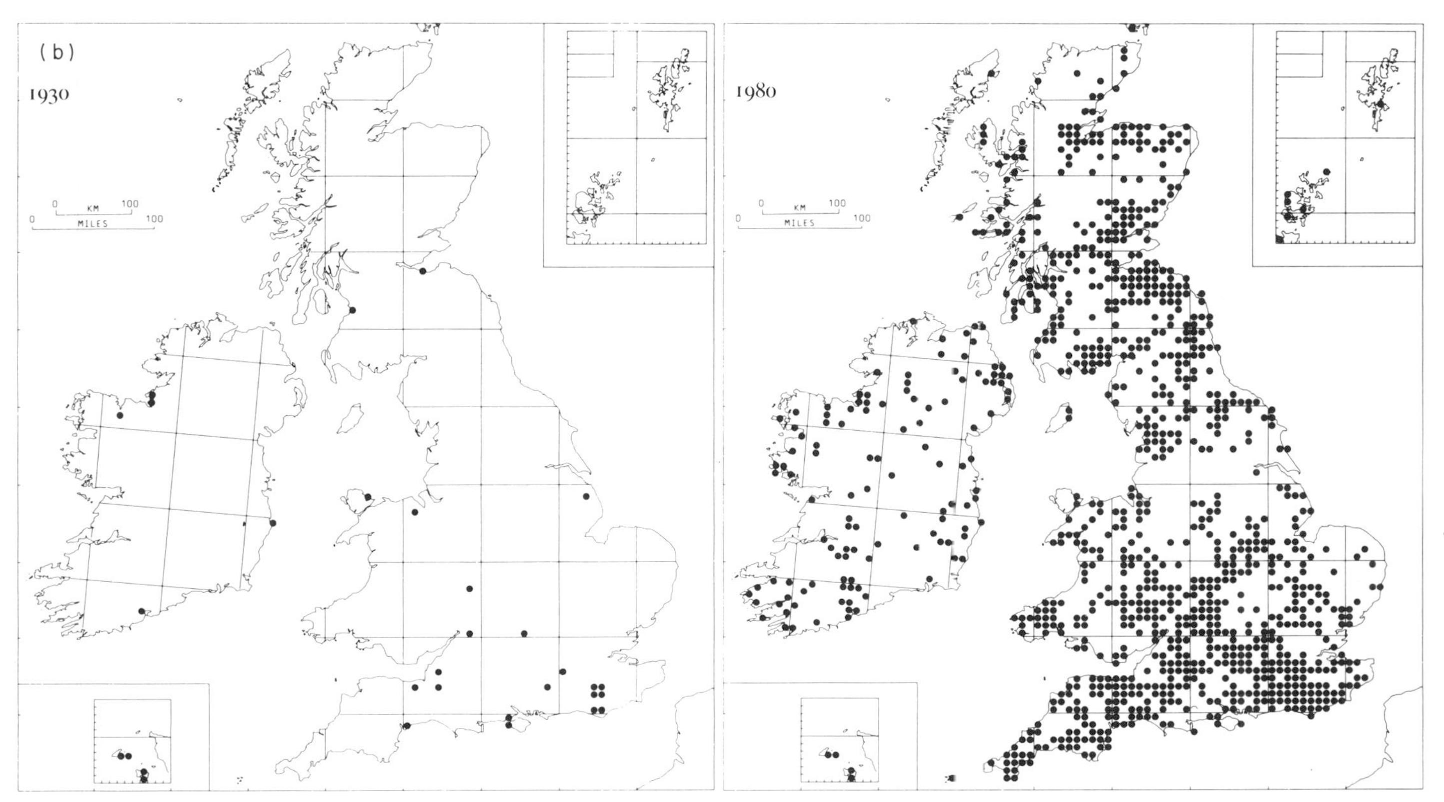

Fig. 21.2. The rapid spread of two aliens in Great Britain. (a) *Epilobium ciliatum* which spreads via copius winged seeds and prolific production of vegetative rosettes. (b) *Veronica filiformis* which sets no seed at all and is transported as vegetative fragments attached to footwear or mowing machines. Maps produced by the Biological Records Centre of the Institute of Terrestrial Ecology, Monks Wood.

TABLE 21.6. The 'top twenty' British alien plant species. Those forming dense thickets are marked Y (yes) under the heading 'T'. Those reproducing by seed are marked Y under the heading 'S'. The species come from a wide variety of plant families, possess a wide range of life-forms and breeding systems, and are found in all types of habitats

Species	Family	Life-form	T	S	Habitat
Acer pseudoplatanus	Acer	Tree	Y	Y	Woodland
Aegopodium podagraria	Umbe	Rhizo. perennial	Y	Y	Gardens, waste, etc
Avena fatua	Gram	Annual	N	Y	Cultivated
Buddleja davidii	Budd	Shrub	?	Y	Walls, railways, etc
Centranthus ruber	Vale	Herb. perennial	Y	Y	Walls, cliffs
Crepis vesicaria	Comp	Monocarp	N	Y	Waysides, waste, etc
Elodea canadensis	Hydr	Hydrophyte	Y	N	Slow water
Epilobium brunnescens	Onag	Herb. perennial	?	Y	Gravel stream banks
E. ciliatum	Onag	Herb. perennial	Y	Y	Cultivated, waste, etc
Erigeron canadensis	Comp	Annual	N	Y	Cultivated, etc
Impatiens glandulifera	Bals	Annual	Y	Y	Muddy streams
Matricaria suaveolens	Comp	Annual	?	Y	Tracks, etc
Mimulus guttatus	Scro	Herb. perennial	Y	Y	Rivers and streams
Reynoutria japonica	Poly	Rhizo. perennial	Y	N	Wayside and waste
Rhododendron ponticum	Eric	Shrub	Y	Y	Woodland
Senecio squalidus	Comp	Monocarp	?	Y	Waste, walls
Smyrnium olusatrum	Umbe	Herb. perennial	Y	Y	Waysides near sea
Symphoricarpos albus	Capr	Shrub	Y	Y	Shady waste
Veronica filiformis	Scro	Herb. perennial	Y	N	Lawns, etc
V. persica	Scro	Annual	N	Y	Cultivated

and direct movement by man appear to predominate). The only noteworthy attribute possessed by the majority of the species is their tendency to form dense thickets where regeneration by other plants is precluded. Whether allelopathic effects are involved in the maintenance of site tenure (particularly by the more short-lived species) has yet to be seen (Rice 1984).

From comparative data such as these it is virtually impossible to assess the relative importance of such processes as natural-enemy-release and competitor-release in the successful spread of alien plants. Certain facts do provide tantalizing clues, however. For example, European botanists travelling in California gain the distinct impression that many European aliens grow larger there than in their native lands! Australians visiting California are struck by how well the *Eucalyptus* trees grow there in the absence of herbivorous insects. In Table 21.7 I have listed the numbers of European native species for which maximum size data are available in the native habitats and as aliens in California. Significantly more species are

TABLE 21.7. Plant size as a native and an alien. By a comparison of plant sizes in the *Flora Europea* and in Munz (1968) it appears that more species attain larger size as aliens than grow to lesser stature. Assuming an expected 50:50 split between species appearing larger and smaller as aliens gives a chi-squared value of 6·6 ($P < 0{\cdot}01$)

	Number	Percentage
Larger as aliens	99	43·4
Smaller as aliens	66	28·9
No difference	63	27·7

larger as aliens in California than are smaller! Notwithstanding the statistical difficulties of comparing maximum plant heights (and the practical problem that plant taxonomists are likely to select large, herbivore-free plants as their herbarium specimens), it does seem that something interesting is going on. It could be that the Californian environment is more conducive to growth than the European, or that the preadapted phenology of the aliens allows them to grow for a larger part of the year than the natives (e.g. many Californian aliens appear to stay green later into the summer than do the natives). Finally, release from herbivorous invertebrates could account for the superior growth (and the rapid spread) of plants in alien environments. We await the results of controlled field experiments to test these assertions.

In terms of ecological theory, these successful plant invaders throw rather little light on such matters as niche shifts and competitive exclusion. There is *no* evidence, to date, of a niche shift in a native plant species brought about as a result of competitive displacement by an alien invader. Where there are suggestive correlations (e.g. the rapid spread of the alien *Veronica persica* as an arable weed in Britain and the decline of the native, weedy *Veronica* spp.), these can be attributed to other, much simpler factors (e.g. habitat alteration, herbicide susceptibilities, etc). It is clear, of course, that population densities of native plants are reduced in habitats where alien plants establish. In native woodlands invaded by dense thickets of *Rhododendron ponticum* it is obvious that the populations of all shrub and ground-layer plant species are dramatically reduced.

There is no evidence, however, that any plant species has been excluded from any plant community by competition with aliens, or that competitive exclusion has led to any plant extinctions. This is hardly surprising, perhaps, given the short time (only 300 years or so) that even the longest-established aliens have been around.

INSECTS INTRODUCED FOR THE BIOLOGICAL CONTROL OF WEEDS

The data base for this analysis was drawn up by a team of biocontrol practitioners (see Acknowledgments) as part of the Silwood Project on the Biological Control of Weeds. The data were gathered to supplement the initial synthesis of published weed control studies draw up by Julien (1982). All known insect releases, worldwide, were catalogued, using a questionnaire completed, as far as possible, by the people who had been responsible for the release attempts. The questionnaire contained 118 entries covering the population biology of the weed and the agent, the status of the weed, and the effort expended in collection, screening, breeding and release. The data base and a detailed analysis (including a full consideration of its statistical shortcomings) is to be described in a forthcoming book (Crawley & Moran, in press).

The present analysis is limited to a consideration of the demographic attributes of the insects in relation to the likelihood of successful establishment. The question of what makes a *successful* weed control agent (i.e. one that brings about a substantial reduction in equilibrium weed abundance once established) will be dealt with elsewhere (Crawley 1986b). The releases represent intentional (and often repeated) attempts to establish a species in the presence of almost limitless food resources, and we shall ask what factors other than resource-availability are responsible for the failure of these introductions.

For the purposes of analysis the data are divided into three classes, to deal with prickly pears (cacti of the genus *Opuntia*), lantana (*Lantana camara*) and 'all other weeds'. This is done because *Opuntia* and *Lantana* represent the bulk of all weed biocontrol attempts, and the demographic attributes of their commonly employed control agents would otherwise swamp the analysis (Table 21.8). Also, the *kind* of control achieved differs dramatically between these two cases. The majority of successful *Opuntia* control programmes tend to be quite dramatic, and to involve only one species (either the pyralid moth *Cactoblastis cactorum* or a species of cochineal insect in the genus *Dactylopius*), whereas control of *Lantana* is less spectacular and tends to be achieved in a gradual, stepwise manner, using a large group of insects attacking different tissues, often using different species to attack the same tissue in different microhabitats. The animal species released against *Opuntia* and *Lantana* are shown in Table 21.9. They come from a wide range of taxa, with a considerable variety of sizes and life-histories. The group of 'other weeds' is a heterogeneous mixture including some notable successes (e.g. the thistle *Carduus nutans* controlled by the seedhead-feeding weevil *Rhynocillus conicus* (Harris 1983), and the klamath weed *Hypericum perforatum* controlled by the leaf-feeding beetle *Chrysolina quadrigemina* (Huffaker & Kennet 1959), as well as some persistent (and oft-repeated) failures, like nut-sedge *Cyperus rotundus*, the 'world's worst weed' (Cock 1985).

The numbers of species of insects released against these groups of weeds are shown in Table 21.8, along with the percentage of release attempts which led to successful establishment. The data base probably contains all of the successful establishments, but inevitably some of the early failures will not have been documented, so the establishment percentages are biased on the optimistic side.

The commonest causes of failure to establish are bad weather and generalist predators. Programmes have been devastated by severe cold spells which have killed the entire release, or by sudden, violent downpours which have washed away all the released insects. The greatest threat to establishment, however, appears to be from generalist predators, especially ants (see also Goeden & Louda 1976). For releases involving insect eggs, ants may consume the entire consignment overnight. The ants appear to regard this as beneficence on the part of philanthropic humans determined to ply them with insect caviar! The data base also shows that insects which lay their eggs in batches are significantly *less* likely to establish successfully than species which space-out their eggs, again reflecting the importance of generalist predators which could easily eliminate whole batches of eggs, but might have more difficulty disposing of an entire clutch of widely dispersed eggs.

In the case of *Opuntia* species, it is common practice to move a successfully established control agent from one species of prickly pear to another species growing nearby (Moran & Zimmermann 1984). In countries like South Africa, where there are many weedy species of *Opuntia*, this has led to large numbers of failures of establishment being put down

TABLE 21.8. Numbers of herbivore species introduced as potential control agents against *Opuntia* species, *Lantana camara*, and the group of 'all other weeds', showing the percentages of release attempts that led to successful establishment

	Species of agents employed	Establishment (%)
All other weeds	173	66
Lantana camara	30	59
Opuntia species	22	65

TABLE 21.9. The principal insect species released to control (a) *Lantana camara* and (b) *Opuntia* species, showing the kind of insect, the date and place of first release, and whether or not the release led to a substantial (Y) level of control

Insects released	Taxon	Date	Place	Control
(a) *Lantana camara*				
Aerenicopsis championi	Cerambycid	1902	Hawaii	—
Apion spp.	Weevils	1902	Hawaii	—
Autoplusia illustrata	Noctuid	1976	Australia	—
Calcomyza lantanae	Agromyzid	1974	Hawaii	—
Cremastobomycia lantanella	Gracillarid	1902	Hawaii	—
Diastema tigris	Noctuid	1954	Fiji	—
Epinotia lantana	Tortricid	1914	Australia	?
Eutreta xanthochaeta	Tephritid	1902	Hawaii	partial
Evander xanthomeles	Cerambycid	1902	Hawaii	—
Hypena strigata	Noctuid	1965	Australia	?
Lantanophagha pusillidactyla	Pterophorid	1902	Hawaii	?
Leptobyrsa decora	Tingid	1969	Australia	—
Neogalea esula	Noctuid	1957	Australia	—
Octotoma championi	Chrysomelid	1978	South Africa	—
O. scabripennis	"	1966	Australia	Y
Ophiomyia lantanae	Agromyzid	1914	Australia	—
Orthezia insignis	Bug	1902	Hawaii	—
Plagiohammus spinipennis	Cerambycid	1966	Australia	—
Pseudopyrausta acutangulalis	Pyralid	1954	Fiji	—
Salbia haemorrhoidalis	Pyralid	1958	Australia	Y
Strymon bazochii	Lycaenid	1914	Australia	—
Teleonemia elata	Tingid	1969	Australia	—
T. harleyi	"	1969	Australia	—
T. prolixia	"	1974	Australia	—
T. scrupulosa	"	1936	Australia	Y
Thmolus echion	Lycaenid	1921	Fiji	—
Uroplata girardi	Chrysomelid	1966	Australia	Y
(b) *Opuntia* spp.				
Archlagocheirus funestus	Cerambycid	1943	South Africa	—
Cactoblastis cactorum	Pyralid	1926	Australia	Y
Chelinidea tabulata	Coreid	1922	Australia	?
C. vittiger	"	1925	Australia	—
Dactylopius ceylonicus	Cochineal	1795	India	Y
D. austrinus	"	1933	Australia	Y
D. confusus	"	1832	South Africa	—
D. opuntia	"	1926	India	Y
D. tomentosus	"	1925	Australia	Y
Melitara doddalis	Pyralid	1949	Hawaii	—
M. prodenialis	"	1928	Australia	—
Metamasius spinolae	Weevil	1948	South Africa	?
Mimorista pulchellalis	Pyralid	1979	South Africa	—
M. flavidissimalis	"	1925	Australia	—
Moneilema armatum	Cerambycid	1950	Hawaii	—
M. crassum	"	1950	Hawaii	—
M. ulkei	"	1926	Australia	—
M. variolare	"	1932	Australia	—
Olycella junctolineella	Pyralid	1924	Australia	—
Tucumania tapiacola	Pyralid	1935	Australia	—

to 'host-plant incompatibility'. These attempts to shift insects to related, but not closely matched host-plants are understandable (and inexpensive), but reflect the general attitude in biocontrol of 'suck it and see'. Despite various classification schemes for potential release agents (Harris 1973; Wapshere 1982, 1985; Goeden 1983), ecological theory has contributed little or nothing to the day-to-day practice of choosing potential weed control agents.

The clear picture to emerge from analysis of the data base is that the most important single parameter influencing the probability of successful establishment is the insect's intrinsic rate of increase. In all three classes, agents with higher rates of increase were more likely to establish successfully (Tables 21.10, 21.11). Correlated with this, small insects were significantly more likely to establish on *Lantana* and on 'other weeds'. In the case of *Opuntia*, the two most commonly used insects (*Cactoblastis* and *Dactylopius*) are so different in size, and fail to establish with roughly equal probability, that the size correlation does not emerge. Other components of the intrinsic rate of increase include voltinism and obligatory diapause. Multivoltine insects are more likely to establish on *Lantana* but not on the other weeds. Obligatory diapause may slightly reduce the probability of establishment on *Lantana* but, again, not in the other two classes.

Part of the conventional wisdom of weed biocontrol has been the notion that individual insects with large appetites make the best control agents (Wapshere 1982). The data base does not support this notion, and species with high per capita feeding rates are significantly *less* likely to establish on *Lantana* and on 'other weeds' (there are insufficient data to test this for *Opuntia*). The reason for this presumably lies in the correlations between feeding rate, body size, and intrinsic rate of increase, rather than on the plausible (but unproven) increase in risky plant-to-plant movement consequent upon individual insects causing substantial damage to whole plants.

Attributes of the native community from which the insect was collected also influence the likelihood of successful establishment. It has been assumed by some biocontrol practitioners that close 'climatic matching' between the home and source ranges is likely to improve success (Wapshere 1985), and also that insects kept rare by natural enemies in their native habitats are likely to be successful when freed from predators, parasites and diseases in their new homes (Harris 1973). The data base does not support either of these contentions. Insect species which are widespread in their native lands are significantly more likely to become established than species with local or patchy distributions. Also, animals

TABLE 21.10. Are smaller agent species more likely to establish than larger species? Data show the mean length of species failing and succeeding in establishment for the three categories of weeds

	Failed vs. Established
All other weeds	Yes (7·4 vs. 5·5 mm; SE 0·3)
Lantana camara	Yes (9·0 vs. 5·3 mm; SE 0·8)
Opuntia species	No

TABLE 21.11 Insect demography and the probability of establishment. Intrinsic rate of increase was calculated *indirectly* by a consideration of the insects' size, coupled with measures of fecundity, voltinism and development rate. We have no reliable data on comparative mortality schedules for the different insect species. Note that widespread, common species are most likely to become established

Question	*Opuntia*	*Lantana*	Other weeds
Are insects with high intrinsic rates of increase more likely to establish?	Yes	Yes	Yes
Are small insects more likely to establish than larger species?	No difference	Yes	Yes
Are multivoltine species more likely to establish than univoltine species?	No	Yes	No
Are species with obligatory diapause less likely to establish?	No	Perhaps	No
Are species with voracious individuals less likely to establish?	No data	Yes	Yes
Is there a threshold release size necessary for establishment?	No	No	No
Are species which lay their eggs in batches less likely to establish?	Yes	No difference	Yes
Are species which are widespread in their native land more likely to establish?	Yes	Yes	Yes
Are species which are abundant in their native land more likely to establish?	Yes	No difference	Yes
Are species regulated at low density by natural enemies in their native land more likely to establish?	No	No	No

which are numerically abundant in their native environment make better invaders than species which occur at low densities as natives. Insects which are thought to be maintained at low densities by natural enemies in their native habitats appear, in fact, to be significantly *less* likely to become established.

This does not mean that climatic matching is irrelevant, and local races within widespread populations may indeed have a higher rate of success. Similarly, it would be wrong to suggest that release from natural enemies is irrelevent to the rapid spread of biocontrol agents. The important message, however, is that common, widespread insects are substantially more likely to make successful invaders than are scarce, patchily distributed species. The demographic attributes which allow a species to become widespread and abundant in one country appear to be transferrable to another land, suggesting that there *is* an element of predictability to be achieved from an understanding of the insects' population biology.

In none of the reported cases was interspecific competition with native or introduced herbivore species found to be a significant influence on establishment. Given that these (usually carefully chosen) insects are released into habitats containing vast quantities of unexploited food resources, this is far from surprising. The debate about single versus multiple releases in biocontrol concerns communities of herbivores at, or close to, equilibrium. As we shall discuss elsewhere (Crawley & Moran, in press) there is rather little to suggest that interspecific competition has *ever* reduced the effectiveness of a biological control programme as a result of multiple releases. During a successful control attempt, interspecific competition for food may become important once weed abundance has been substantially reduced, and agent species may even be driven to extinction. There is no evidence, however, that release of multiple species has ever led to the replacement of an economically effective species by one which is economically less effective (Pemberton & Willard 1918; Levins 1969; Keller 1984).

The data base provides no evidence for a threshold population size for successful introduction of any of the insect species released for weed control. We see no 'Allee effects', for example, where impaired reproductive performance at low densities stems from difficulties in mate-finding (such species would probably have been weeded out before introduction, at the screening or breeding stages). Successful establishment (and even spectacular control) can be achieved following tiny releases, as when introduction of a single gravid female of *Dactylopius* has led to the control of vast tracts of *Opuntia* cactus in Mauritius (Moutia & Mamet 1945).

It is abundantly plain, however, that there is a large stochastic element to these failures of establishment. This is reflected in the fact that the probability of establishment increases both with the size of individual releases, and with the number of releases. Clearly, in a heterogeneous world, the more microsites in which releases are made, the more likely it is that one colony will become successfully established. Similarly, given a genetically heterogeneous agent species, the larger the release, the bigger the chance of the 'right genotype' being represented. A number of potentially successful invaders will have failed to establish simply for want of better fortune. On the other hand, repeated releases in themselves are no guarantee of establishment, and some control agents perversely refuse to establish, no matter how successful the same insects may have been in similar habitats elsewhere.

CONCLUSIONS

Survey of the data relating to the spread of alien plants reveals several clear patterns. We can rank plant communities in terms of their invasibility on the average proportion of bare ground (a rough and ready measure of the intensity of competition for establishment microsites), and on the frequency and intensity of soil surface disturbance. Thus, aliens make up more than 50% of the flora on urban waste ground, but less than 5% in unmanaged, native woodland. This much is straight-forward, and will come as no surprise to field naturalists. It is clear from the rapid rate of spread of newly introduced alien plants through natural communities (e.g. *Epilobium brunnescens* on the banks of gravelly, upland streams) that plant species could be 'reshuffled' into novel combinations quite quickly. This observation prompts the question as to whether plant communities which today we might regard as stable, equilibrium assemblages have originated either: (i) by prolonged, intimate coevolution; or (ii) by the invasion of preadapted species which evolved elsewhere. The rapid range expansions of these successful aliens brings home the important point that plant species which grow together did not necessarily evolve together.

The rate at which particular plant communities are invaded (given their characteristic levels of ground cover and soil disturbance) clearly depends upon biogeographic factors such as the size of the available pool of alien species, and the rate of propagule immigration. This, in turn, depends upon the isolation of the site, and the area of the target plant community. The time which has elapsed since the last major disturbance (the successional age of the site) will also influence the richness of the alien flora, but in ways which, as yet, are not at all clear. Whether alien plants are lost from mature communities as a result of interspecific competition remains to be seen. The percentage of aliens is certainly higher in the first few years of secondary succession (Brown & Southwood, Chapter 15), but since *all* early-successional species go extinct eventually, the absence of aliens from many late-successional communities may reflect reduced pool sizes or reduced immigration rates rather than competitive exclusion.

What is clear, however, is that all communities *are* invasible. The so-called 'invasion resistant' model communities can be invaded either by substituting a resident species with a superior invader, or by adding to the community species with novel demographic attributes, superior to any of those in the original pool. Our problem lies in the fact that we are totally unable to *predict* whether a particular introduction will succeed or fail. Our broad patterns of generalization let us down time and again in specific cases, and our theory of niche structure fails conspicuously to provide us with the parameters which might define a truly 'vacant niche'. Reluctantly, it has to be admitted that at the moment we have virtually no predictive ability in relation to the species characteristics of plants associated with invasibility in a particular kind of community.

A glimmer of hope is seen from the weed biocontrol data, in the fact that common and widespread insects make good invaders; in other words, 'commonness' appears to be transferrable (and thus, in principle, predictable) from one community to another. Given several potential control agents with similar native distributions and abundances, we would bet on the species with the highest rate of increase (on *Lantana*, this would probably be a small, multivoltine insect without obligatory diapause) as being the most likely to establish. Beyond this, we would struggle.

As a final example, it is instructive to consider one of the most successful recent invasions of the British insect fauna. This involves the gall-forming cynipid wasp *Andricus quercuscalicis* which attacks the acorns of *Quercus robur* during its agamic generation (Collins, Crawley & McGavin 1983). The insect has spread since its accidental introduction into southern England in the early 1960s, so that it now covers virtually all of England and Wales. It is so abundant that in the autumn of 1983 it destroyed over 90% of what was a reasonably large crop of acorns on English oak (Crawley 1984, 1985). Conventional wisdom about competitor-free space and enemy-free space would have led us to predict that this species stood virtually no chance of successful establishment in Britain! The community of cynipid wasps on oak was already large (Askew 1961). Further, the large guild of abundant, generalist parasitoids which are the natural enemies of *A. quercuscalicis* in its native, European habitats, were already well established in Britain where they attacked other cynipid species. In addition to the apparent lack of both enemy-free space and competitor-free space, the insect had an extraordinarily complex life-cycle, in which the alternating sexual and asexual generations are passed on different species of oak trees! Between each generation the females must fly between *Quercus robur* and the alien (but now extensively naturalized) *Q. cerris*. There is no way that we would have been able to predict the establishment, let alone the spectacular spread of this species.

In hindsight we can rationalize that the wasp did, in fact, have competitor-free space, since no other gall-wasps attack either the acorns of *Q. robur* or the male flowers of *Q. cerris* (where the sexuals make their galls). Also, it turns out to have enemy-free space in the agamic generation because, for reasons we do not yet understand, the resident guild of parasite species (allegedly the same as those attacking the wasp on the Continent) do not attack it here, so there is virtually no parasitism or predation of the knopper gall generation in England (R.S. Hails, unpublished). But this is post-hoc-ism of the most blatant kind!

In summary, our two sets of data do show broad patterns which conform with ecological intuition. For biological weed control agents they show the fundamental importance of the intrinsic rate of increase in determining the likelihood of successful invasion, and highlight the importance of generalist natural enemies, and the unimportance of interspecific competition in determining failures in establishment. For perennial alien plants, on the other hand, interspecific competition appears often to be of vital importance, and successful alien species tend to be thicket-forming plants capable of prolonged site pre-emption, and of preventing the establishment of other species beneath them. Evidence of herbivore-release for plants in alien environments is suggestive but by no means conclusive. Patterns consistent with niche shifts or competitive exclusion in native species following the spread of introduced plants have *not* been found.

What little we *do* know tends to suggest that we shall never be able to predict which of a set of invaders is likely to establish, and which, having become established, is likely to become the most abundant. At least part of the reason for this lies in the fact that the supposed constants of our theoretical models are in fact variables dependent upon the composition of the community, and are *not* fundamental species' attributes. Thus, for example, competition coefficients relating species A and C in Lotka-Volterra models cannot be computed in any straightforward manner from the coefficients relating species A and B, and species B and C. Similarly, addition of species C to a community is likely to alter the competition coefficients relating the resident species A and B. Until we can develop models in which the competition coefficients are context-specific, it is unlikely that community ecology will become genuinely predictive.

ACKNOWLEDGMENTS

The Silwood Project on the Biological Control of Weeds was the brainchild of Cliff Moran and Jeff Waage. It involves close cooperation between CIBC, CSIRO and USDA and aims to obtain comprehensive documentation on every attempted weed control project, whether or not it led to successful establishment of the agent. Information on 627 cases up to 1980 is held in a computer data base at Imperial College, and the data base is updated continuously as new attempts are evaluated. Those most closely associated with the planning and execution of the project are Matthew Cock, Mick Crawley, David Greathead, Mike Hassell, Peter Harris, John Lawton, Cliff Moran, Dieter Schroeder, Vicky Taylor and Jeff Waage, with help from Jim Cullen, Dick Goeden, Peter Room and Tony Wapshere.

REFERENCES

Anderson, R.M. & May, R.M. (1984). Spatial, temporal and genetic heterogeneity in host populations and the design of immunization programmes. *IMA Journal of Mathematics Applied to Medicine and Biology*, **1**, 233–266.

Askew, R.R. (1961). On the biology of the inhabitants of oak galls of Cynipidae (Hymenoptera) in Britain. *Transactions of the Society of British Entomologists*, **14**, 237–268.

Beard, J.S. (1969). *A Descriptive Catalogue of West Australian Plants*, 2nd edn. SGAP. Perth.

Bowen, H.J.M. (1968). *The Flora of Berkshire*. H.J.M. Bowen. Oxford.

Clapham, A.R., Tutin, T.G. & Warburg, E.F. (1962). *Flora of the British Isles*, 2nd edn. Cambridge University Press, Cambridge.

Cock, M.J.W. (1985). *A Review of Biological Control of Pests in the Commonwealth Caribbean and Bermuda up to 1982*. Commonwealth Institute of Biological Control Technical Communication No. 9. Commonwealth Agricultural Bureau, Farnham Royal.

Collins, M., Crawley, M.J. & McGavin, G.C. (1983). Survivorship of the sexual and agamic generations of *Andricus quercuscalicis* on *Quercus cerris* and *Q. robur*. *Ecological Entomology*, **8**, 133–138.

Crawley, M.J. (1984). The gall from Gaul. *International Dendrological Yearbook 1983*, 135–137.

Crawley, M.J. (1985). Reduction of oak fecundity by low density herbivore populations. *Nature*, **314**, 163–164.

Crawley, M.J. (1986a). The structure of plant communities. *Plant Ecology* (Ed. by M.J. Crawley). Blackwell Scientific Publications, Oxford.

Crawley, M.J. (1986b). The population biology of invaders. *Philosophical Transactions of the Royal Society of London, Series B*, in press.

Crawley, M.J. & Moran, V.C. (in press) *The Biological Control of Weeds. A Theoretical and Practical Appraisal.*

Fernald, M.L. (1970). *Gray's Manual of Botany: A Handbook of the Flowering Plants and Ferns of the Central and Northeastern United States and Adjacent Canada*. Van Nostrand, New York.

Forcella, F. & Harvey, S.J. (1983). Relative abundance in an alien weed flora. *Oecologia*, **59**, 292–295.

Goeden, R.D. (1983). Critique and revision of Harris' Scoring System for selection of insect agents in biological control of weeds. *Protection Ecology*, **5**, 287–301.

Goeden, R.D. & Louda, S.M. (1976). Biotic interference with insects imported for weed control. *Annual Review of Entomology*, **21**, 325–342.

Harris, P. (1973). The selection of effective agents for the biological control of weeds. *Canadian Entomologist*, **105**, 1495–1503.

Harris, P. (1983). *Carduus nutans*, nodding thistle, and *C. acanthoides*, plumeless thistle (Compositae). *Biological Control Programmes Against Insects and Weeds in Canada, 1961–1980* (Ed. by J.S. Kelleher & M.A. Hulme), pp. 115–126. Technical Communication of the Commonwealth Institute for Biological Control, Farnham Royal.

Hayward, I.M. & Druce, G.C. (1919). *The Adventive Flora of Tweedside*. Buncle, Arbroath.

Holm, L.G., Plucknett, D.L., Pancho, J.V. & Herberger, J.P. (1977). *The World's Worst Weeds: Distribution and Biology*. University of Hawaii Press, Honolulu.

Huffaker, C.B. & Kennet, C.E. (1959). A ten year study of vegetational changes associated with biological control of klamath weed. *Journal of Range Management*, **12**, 69–82.

Jermyn, S.T. (1974). *Flora of Essex*. Essex Naturalists Trust, Colchester.

Julien, M.H. (1982). *Biological Control of Weeds. A Catalogue of Agents and their Target Weeds*. Commonwealth Agricultural Bureau, Slough.

Keller, M.A. (1984). Reassessing evidence for competitive exclusion of introduced natural enemies. *Environmental Entomology*, **13**, 129–195.

Lees, F.A. (1888). *The Flora of West Yorkshire*. Lovell Reeve, London.

Levins, R. (1969). Some demographic and genetic consequences of environmental heterogeneity for biological control. *Bulletin of the Entomological Society of America*, **15**, 237–240.

Lousley, J.E. (1976). *Flora of Surrey*. David & Charles, Newton Abbot.

MacArthur, R.H. (1970). Species packing and competitive equilibrium for many species. *Theoretical Population Biology*, **1**, 1–11.

Margetts, J.L. & David, R.W. (1981). *A Review of the Cornish Flora 1980*. Institute of Cornish Studies, Redruth.

May, R.M. (1973). *Stability and Complexity in Model Ecosystems*. Princeton University Press, New Jersey.

Messenger, G. (1971). *Flora of Rutland*. Leicester Museums, Leicester.

Moran, V.C. & Zimmermann, H.G. (1984). The biological control of cactus weeds: achievements and prospects. *Biocontrol News and Information*, **5**, 297–320.

Moutia, L.A. & Mamet, R. (1945). A review of twenty-five years of economic entomology in the Island of Mauritius. *Bulletin of Entomological Research*, **36**, 439–472.

Munz, P.A. (1968). *A Californian Flora and Supplement*. University of California Press, Berkeley.

Pemberton, C.E. & Willard, H.F. (1918). A contribution to the biology of fruit-fly parasites in Hawaii. *Journal of Agricultural Research*, **15**, 419–465.

Philp, E.G. (1982). *Atlas of the Kent Flora*. Kent Field Club, Maidstone.

Rice, E.L. (1984). *Allelopathy*, 2nd edn. Academic Press, New York.

Riddelsdell, H.J., Hedley, G.W. & Price, W.R. (1948). *Flora of Gloucestershire*. Chatford House Press, Bristol.

Simpson, D.A. (1984). A short history of the introduction and spread of *Elodea* Michx. in the British Isles. *Watsonia*, **15**, 1–9.

Sinker, C.A., Packham, J.R., Trueman, I.C., Oswald, P.H., Perring, F.H. & Prestwood, W.V. (1985). *Ecological Flora of the Shropshire Region*. Shropshire Trust for Nature Conservation, Shrewsbury.

Tate, G. & Baker, J.G. (1868). A New Flora of Northumberland and Durham. *Natural History Transactions of Northumberland and Durham 1867*, Vol. 2, pp. 1–316.

Turelli, M. (1981). Niche overlap and invasion of competitors in random environments. I. Models without demographic stochasticity. *Theoretical Population Biology*, **20**, 1–56.

Wade, A.E. (1970). *The Flora of Monmouthshire*. National Museum of Wales, Cardiff.

Walsh, G.B. & Rimmington, F.C. (1953). *The Natural History of the Scarborough District*. Vol. 1 Scarborough Field Naturalists Society, Scarborough.

Walters, S.M., Brady, A., Brickell, C.D., Cullen, J., Green, P.S., Lewis, J., Matthews, V.A., Webb, D.A., Yeo, P.F. & Alexander, J.C.M. (1984). *The European Garden Flora: A Manual for the Identification of Plants Cultivated in Europe, Both Out-of-doors and Under Glass*, Vol. II. *Monocotyledons* (Part II). Cambridge University Press, Cambridge.

Wapshere, A.J. (1982). Priorities in the selection of agents for the biological control of weeds. *Proceedings of the British Crop Protection Conference: Weeds. Brighton 1982*, pp. 779–785.

Wapshere, A.J. (1985). Effectiveness of biological control agents for weeds: present quandaries. *Agriculture, Ecosystems and Environment*, **13**, 261–280.

Webb, D.A. & Scannell, M.J.P. (1983). *Flora of Connemara and The Burren*. Royal Dublin Society and Cambridge University Press, Cambridge.

Webster, M. McC. (1978). *Flora of Moray, Nairn and East Inverness*. Aberdeen University Press, Aberdeen.

AUTHOR INDEX

Figures in italics refer to pages where full references appear.

SUBJECT INDEX